AF833713

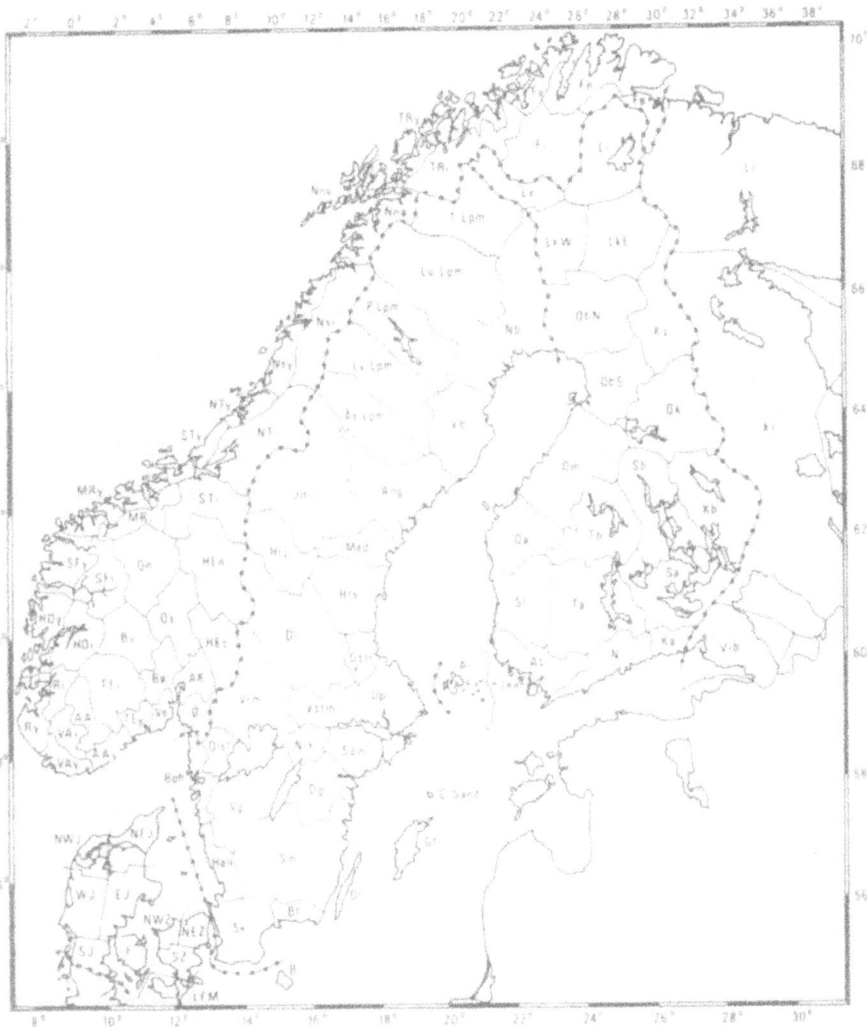

List of abbreviations for the provinces used throughout the text, on the map and in the following tables.

DENMARK

SJ	South Jutland	LFM	Lolland, Falster, Møn
EJ	East Jutland	SZ	South Zealand
WJ	West Jutland	NWZ	North West Zealand
NWJ	North West Jutland	NEZ	North East Zealand
NEJ	North East Jutland	B	Bornholm
F	Funen		

SWEDEN

Sk.	Skåne	Vrm.	Värmland
Bl.	Blekinge	Dlr.	Dalarna
Hall.	Halland	Gstr.	Gästrikland
Sm.	Småland	Hls.	Hälsingland
Öl.	Öland	Med.	Medelpad
Gtl.	Gotland	Hrj.	Härjedalen
G. Sand.	Gotska Sandön	Jmt.	Jämtland
Ög.	Östergötland	Ång.	Ångermanland
Vg.	Västergötland	Vb.	Västerbotten
Boh.	Bohuslän	Nb.	Norrbotten
Dlsl.	Dalsland	Ås. Lpm.	Åsele Lappmark
Nrk.	Närke	Ly. Lpm.	Lycksele Lappmark
Sdm.	Södermanland	P. Lpm.	Pite Lappmark
Upl.	Uppland	Lu. Lpm.	Lule Lappmark
Vstm.	Västmanland	T. Lpm.	Torne Lappmark

NORWAY

Ø	Østfold	HO	Hordaland
AK	Akershus	SF	Sogn og Fjordane
HE	Hedmark	MR	Møre og Romsdal
O	Oppland	ST	Sør-Trøndelag
B	Buskerud	NT	Nord-Trøndelag
VE	Vestfold	Ns	southern Nordland
TE	Telemark	Nn	northern Nordland
AA	Aust-Agder	TR	Troms
VA	Vest-Agder	F	Finnmark
R	Rogaland		

n northern s southern ø eastern v western y outer i inner

FINLAND

Al	Alandia	Kb	Karelia borealis
Ab	Regio aboensis	Om	Ostrobottnia media
N	Nylandia	Ok	Ostrobottnia kajanensis
Ka	Karelia australis	ObS	Ostrobottnia borealis, S part
St	Satakunta	ObN	Ostrobottnia borealis, N part
Ta	Tavastia australis	Ks	Kuusamo
Sa	Savonia australis	LkW	Lapponia kemensis, W part
Oa	Ostrobottnia australis	LkE	Lapponia kemensis, E part
Tb	Tavastia borealis	Li	Lapponia inarensis
Sb	Savonia borealis	Le	Lapponia enontekiensis

USSR

Vib Regio Viburgensis Kr Karelia rossica Lr Lapponia rossica

FAUNA ENTOMOLOGICA SCANDINAVICA
Volume 24 1991

Blowflies (Diptera, Calliphoridae) of Fennoscandia and Denmark

by

Knut Rognes

E. J. Brill/Scandinavian Science Press Ltd.
Leiden · New York · København · Köln

The paper in this book meets the guidelines for permanence and durability of the Committee on Production Guidelines for Book Longevity of the Council on Library Resources.

© *Copyright 1991 by E. J. Brill, Leiden, The Netherlands*

All rights reserved. No part of this book may be reproduced or translated in any form, by print, photoprint, microfilm, microfiche or any others means without written permission from the publisher.
Authorization to photocopy items for internal or personal use is granted by E.J. Brill provided that the appropriate fees are paid directly to Copyright Clearance Center, 27 Congress Street, SALEM MA 01970, USA. Fees are subject to change.
PRINTED IN THE NETHERLANDS

Editor-in-chief: N.P. Kristensen
Desk editor: V. Michelsen

ISBN 90 04 09304 4
ISSN 0106-8377

Library of Congress Cataloging-in-Publication Data

Rognes, Knut.
Blowflies (Diptera, Calliphoridae) of Fennoscandia and Denmark by Knut Rognes.
p. — cm. — (Fauna entomologica Scandinavica. ISSN 0106-8377: v. 24)
Includes bibliographical references and index.
ISBN 9004093044 (cloth)
1. Blowflies—Scandinavia—Classification. I. Title.
II. Series.
QL537.C24R64 1990
595.77'4—dc20 90-40345
CIP

Cover illustration: female *Calliphora terraenovae* Macquart. (From "Manual of Nearctic Diptera. Volume 2", Agriculture Canada 1987)

Author's address:

Knut Rognes
Stavanger Lærerhøgskole
Postboks 2521 Ullandhaug
N-4004 Stavanger
Norway
or:
Havørnbrautene 7a
N-4048 Hafrsfjord
Norway

Contents

Abstract 7
Introduction 7
Acknowledgements 8
Material and methods 9
Relationships, monophyly and diagnosis of the Calliphoridae 11
Subfamilies of the Calliphoridae 12
Morphology of the adults 17
Morphology of the immature stages 23
Biology 27
Collection, preservation and identification 29
Abbreviations used in text 30
Key to North European subfamilies of Calliphoridae 30
Subfamily Calliphorinae 32
  Key to genera of Calliphorinae 33
  Genus *Bellardia* Robineau-Desvoidy 34
  Genus *Calliphora* Robineau-Desvoidy 59
  Genus *Cynomya* Robineau-Desvoidy 92
  Genus *Onesia* Robineau-Desvoidy 97
Subfamily Chrysomyinae 103
  Key to genera of Chrysomyinae 105
  Genus *Phormia* Robineau-Desvoidy 105
  Genus *Protocalliphora* Hough 106
  Genus *Protophormia* Townsend 128
  Genus *Trypocalliphora* Peus 137
Subfamily Helicoboscinae 140
  Genus *Eurychaeta* Brauer & Bergenstamm 141
Subfamily Luciliinae 147
  Genus *Lucilia* Robineau-Desvoidy 147
Subfamily Melanomyinae 184
  Key to genera of Melanomyinae 185
  Genus *Angioneura* Brauer & Bergenstamm 188
  Genus *Eggisops* Rondani 193
  Genus *Melanomya* Rondani 197
  Genus *Melinda* Robineau-Desvoidy 199
Subfamily Polleniinae 206
  Key to genera of Polleniinae 206
  Genus *Morinia* Robineau-Desvoidy 209
  Genus *Pollenia* Robineau-Desvoidy 212
Subfamily Rhiniinae 241
  Genus *Stomorhina* Rondani 241
Catalogue 247
Literature 252
Taxonomic index 267
Biological index 272

A colour plate is given on p. 246.

Abstract

The taxonomy and nomenclature of the species of Calliphoridae of Fennoscandia, Denmark and neighbouring parts of the Karelian U.S.S.R. are revised and their distribution catalogued. Male and female genitalia are illustrated in detail, many of them for the first time, and are exploited at all taxonomic levels. A new subfamily classification is proposed in which 8 subfamilies are recognised: Calliphorinae, Chrysomyinae, Helicoboscinae, Luciliinae, Melanomyinae, Polleniinae, Rhiniinae and Rhinophorinae. The species of Rhinophorinae are not dealt with. *Morinia* Robineau-Desvoidy is transferred to the Polleniinae. *Angioneura* Brauer & Bergenstamm, *Eggisops* Rondani, *Melanomya* Rondani and *Melinda* Robineau-Desvoidy are united as the Melanomyinae. The monophyly of the subfamilies has been largely established on the basis of the structure of the female genitalia. 17 genera and 52 species are recognised and keys to all taxa are provided. 75 primary types (holotypes, lectotypes, neotypes) have been examined and 31 lectotypes have been designated. Neotypes have been designated for 6 nominal taxa: *Musca amentaria* Scopoli, *Onesia viarum* Robineau-Desvoidy, *Onesia vulgaris* Robineau-Desvoidy, *Onesia viridicyanea* Robineau-Desvoidy, *Pollenia labialis* Robineau-Desvoidy and *Polleniella griseotomentosa* Jacentkovský. 50 specific and 4 generic synonyms and 7 new combinations have been established. A few previously unrecognised old names have been brought into use. The name *Bellardia vespillo* (Fabricius) is used for a recently re-discovered species. *Bellardia pubicornis* (Zetterstedt) is recorded as new for the fauna of the U.S.S.R. *Pollenia labialis* Robineau-Desvoidy (= *intermedia*: authors) is reported for the first time from the Nearctic Region.

Introduction

Members of the Calliphoridae are among the most abundant, familiar and oldest known insects. According to ancient Egyptian Papyrus their development from larvae was known already 1550 B.C. In Homer's "Iliad" Achilles expressed his fear of blowflies because of their habit of laying eggs and producing "worms" in dead bodies and thus making them unrecognisable (Hennig, 1973).

The following workers from Fennoscandia and Denmark have described calliphorid species from this area: Linnaeus (1758, 1761), De Geer (1776), Fabricius (1794, 1805), Fallén (1817), Zetterstedt (1838, 1844, 1845, 1859), Siebke (1863), Ringdahl (1931).

A comprehensive treatment of the calliphorid fauna of Denmark was given by Lundbeck (1927) who, to a certain extent, took the male genitalia into account when characterising the species. This work is now out of date. A faunistic contribution was made by Lyneborg (1965), who first recorded *Calliphora loewi* from Denmark. A few faunistic notes were given by Ardö (1957). Studies on the attraction of flies (including blowflies) to the stinkhorn fungus (*Phallus impudicus* Pers.) were made by Nielsen (1963, 1967, 1968), who first recorded *Calliphora subalpina* from Denmark. Biological studies on myiasis and related phenomena in Denmark were provided by Cragg (1950) and Nielsen, Nielsen & Walhovd (1978). Accounts of the veterinary importance of blowflies in Denmark, and notes on their medical significance, have been given by Nielsen (1984a, 1984b) and Tikjøb & Haarløv (1985).

The blowfly fauna of Norway was investigated by several authors during the 19th century and the early part of the present century, usually as a minor part of their study of the dipterous fauna in general or of its ecology (Zetterstedt, 1838, 1845; Walker, 1849; Siebke, 1853, 1863, 1866a, 1866b, 1870, 1873, 1877; Schøyen, 1889; Storm, 1891, 1896, 1907; Bidenkap, 1892, 1898, 1901; Strand, 1900, 1903, 1906; 1914). In more recent times, Duffield (1937), Ringdahl (1944a, 1944b, 1954), who based his determinations on Lundbeck's work and his knowledge of the Zetterstedt types, Natvig (1950, 1959), Davies (1954) and Ardö (1957) have contributed to the knowledge of the Norwegian fauna. In a series of papers, Rognes (1980, 1981, 1982, 1983a, 1983b, 1985b, 1986d) provided new records of known species and recorded species new to the Norwegian fauna, as a result of several collecting trips; some type studies were also included. A checklist of Norwegian calliphorids based on a revision of Norwegian collections was published by Rognes (1985b). Soot-Ryen (1925) described the effect of certain blowflies on the dried fish industry in Northern Norway. Brinkmann (1976) provided a study of blowfly myiasis in sheep in Western Norway.

In modern times the Swedish fauna has been studied by Ringdahl (1921, 1931, 1934, 1937, 1939, 1945a, 1945b, 1947, 1951, 1952, 1958, 1960). He revised the Zetterstedt types and other material in Stockholm and Lund, and collected in many parts of Sweden himself. A few faunistic notes were given by Ardö (1957) and Persson (1983).

Work on the Finnish dipterous fauna was pioneered by Bonsdorff (1866), who included records of blowflies. Frey (1909), Tiensuu (1938), Brander (1964), Hedström (1964), and Tiensuu (1964) published faunistic data on higher Diptera, including blowflies. Checklists of the Calliphoridae have been provided by Tiensuu (1941) and Hackman (1980). In numerous papers Nuorteva and his coworkers have studied the sarcosaprophagous blowflies mostly from faunistic, ecological, epidemiological and forensic points-of-view (see bibliography for references). Contributions to their ecology in Finland have also been given by Hanski (1976), Hanski & Kuusela (1977, 1980), Kuusela & Hanski (1982) and Kuusela (1983, 1984). Frey (1925), Grunin & Nuorteva (1969), Grunin, Nuorteva & Rajala (1969), Grunin (1972), Hakanen, Grunin & Nuorteva (1974), Itämies & Ojanen (1975) have contributed to the knowledge of the bird parasitic blowfly fauna of Finland.

The most recent taxonomic treatment of the entire North European fauna is that of Zumpt (1956a), who covered the whole Palaearctic area and who based his treatment to a large extent on the work of Séguy (1928a, 1941). Individual calliphorid genera that still presented taxonomic problems have subsequently been revised by Schumann (1964, 1973, 1974a, 1974b) (*Onesia*, *Melinda*, *Bellardia*), and Rognes (1985a, 1987a, 1987b, 1988) (*Trypocalliphora*, *Pollenia*). Mihályi (1977) made it possible to identify certain females of European *Lucilia* and made important contributions to the European *Pollenia* (Mihályi, 1976). The identity of a few nominal species based on types from Fennoscandia was clarified by Rognes (1983c, 1986b), and the genus *Eurychaeta* Brauer & Bergenstamm (*Helicobosca* Bezzi), previously included in the Sarcophagidae, was transferred to the Calliphoridae by Rognes (1985b, 1986a, 1986c). The Palaearctic Calliphoridae were catalogued by Schumann (1986). However, problems still remain in the fields of classification, taxonomy and nomenclature. Furthermore, the detailed structure of the female genitalia has remained insufficiently known or has not been investigated at all.

The purpose of the present work is to present a revision of the North European calliphorid fauna and its nomenclature, propose a new subfamily classification based on cladistic principles, provide detailed illustrations of male and female genitalia and other important features of adult morphology for all the species, provide keys to taxa at all levels, and provide information on local and world distribution and biology of each species.

Acknowledgements

It is great pleasure for me to be able to express my long and very deeply felt gratitude to Adrian C. Pont for all his help (material, nomenclatural and spiritual), trust, encouragement and never failing interest throughout the years I have worked on the taxonomic, nomenclatural and other problems met with during the preparation of this book and the papers preparing the ground for it. My sincere thanks are also due to Stig Andersen for his generosity and unselfishness in discussing all kinds of problems and for putting vital information at my disposal, including help with xerox copies of rare literature. He generously left it to me to present his own discoveries regarding the status of the nominal species *Musca vespillo* Fabricius and to redescribe the recently re-discovered *Bellardia* species to which the Fabrician name is applied and of which he captured a large series of specimens. My deep gratitude is also due to Thomas Pape, who similarly shared and discussed his views on numerous matters arising during this study and also showed me important material of a new Japanese species of *Eggisops*. Verner Michelsen kindly explained the technique and potential significance of the study of female ovipositors. I am also deeply indebted to Leif Lyneborg who entrusted me with the task of preparing this volume for the Fauna Entomologica Scandinavica series and for encouragement and patience.

I would also like to thank the following friends and colleagues for their invaluable help and their willingness to discuss various problems: H. Andersson, P. Ferrar, F. Gregor, B. A. Holloway, L. Greve Jensen, T. Jonassen, H. Kurahashi, P. Lauterer, H. de Souza Lopes, F. Midtgaard, T. R. Nielsen, P. I. Persson, D. Povolný, C. W. Sabrosky, G. E. Shewell and J. R. Vockeroth.

I also thank the following colleagues for placing material in their care or in their private collections at my disposal (depository acronyms are explained below), or for the exchange or gift of material: A. C. Pont (BMNH), B. A. Sagvolden (BS), C. Bergström (CB), B. E. Cooper (CNC), R. Rozkošný (DESCS), J. O. Solem (DKNVS), A. Freidberg (DZTAU), C. Milkint (now Salvino) (FMC), H. Franz (HF), F. Mihályi (HNHM), J. H. Simonsen (JHS), M. Baez (MB), P. Lauterer & J. Stehlík (MMBCS), L. Matile (MNHN), P. I. Persson (NHRM), J. Ježek (NMP), P.Straumfors (RM), R.Mehl (RML), S. Andersen (SA), A. Fjellberg (TM), N. Woodley (USNM), F. Sick (ZIKIEL), A. G. Zinov'ev (ZILEN), L. Greve Jensen (ZMB), H. Schumann (ZMBERL), S. Andersen, L. Lyneborg, V. Michelsen and T. Pape (ZMC), A. Albrecht (ZMH), H. Andersson & R. Danielsson (ZML), A. L. Ozerov (ZMMSU), J. E. Raastad (ZMO) and H. Kurahashi.

My grateful thanks also go to my wife Agnete Rognes and my sons, Torbjørn, Øystein and Stig, for their interest and for their assistance with summer fieldwork over several years. Their high collecting morale on several occasions proved to be indispensable. Torbjørn also helped me with computer work, and designed important programs in dBASE III PLUS or improved on my own.

Blackwell Scientific Publications permitted reuse of some of my figures of *Pollenia*.

All the work on this project has taken place in the "entomological department" in the basement of my private home. The project has been financed mainly privately, but has also received important contributions from other sources. A proportion of a one-year grant ("vikarstipend"), kindly contributed by Stavanger Lærerhøgskole and Det Regionale Høgskolestyret i Rogaland in 1982-1983 in support of an earlier project on Norwegian Calypterate Diptera, was devoted to a revision of the Calliphoridae in Norwegian museum collections. Stavanger Lærerhøgskole also kindly lent me a LEITZ microscope (40-630x), paid part of running expenses in connection with correspondence, and for the last 2 years has allowed me to spend one day a week on the project undisturbed by other duties. Jæren Entomologklubb, Sandnes, financed some fieldwork in 1984. Travel grants provided by Norges Almenvitenskapelige Forskingsråd (D.68.40-023) in 1984 and by Stavanger Lærerhøgskole and Det Regionale Høgskolestyret i Rogaland in 1987 and 1990 enabled me to visit Copenhagen, Lille, Lund, London and Paris for type studies. Many thanks to L. Lyneborg, S. Andersen and V. Michelsen (ZMC), B. Radigois (MHNL), H. Andersson and R. Danielsson (ZIL), A. C. Pont (BMNH) and L. Matile (MNHN) for helping me in all possible ways during my visits.

Thanks also to my Commodore PC 10 computer, boosted with a 20MB Plus Hardcard, my Brother M-1509 printer, and the programs WordPerfect, ver. 4.2 and dBASE III PLUS ver. 1.1 for perfect functioning and service.

Material and methods

The study has been based mainly upon blowfly material in the following collections (those marked with an asterisk have formed the basis for the Catalogue):

BMNH	- British Museum (Natural History), Department of Entomology, London
*BS	- private collection of Bjørn Sagvolden, Rollag
*CB	- private collection of Christer Bergström, Uppsala
CNC	- Canadian National Collection, Biosystematic Research Centre, Agriculture Canada, Ottawa
DESCS	- Department of Environmental Studies, Faculty of Science, J.E. Purkyně University, Brno
*DKNVS	- DKNVS-Museet, Universitetet i Trondheim, Zoologisk Avdeling, Trondheim
DZTAU	- Department of Zoology, Tel Aviv University, Tel Aviv
FMC	- Field Museum of Natural History, Chicago
HF	- private collection of Dr. H. Franz, Mödling
HNHM	- Hungarian Natural History Museum, Zoological Department, Budapest

*JHS - Private collection of Jan Henrik Simonsen, Oslo
*KR - private collection of Knut Rognes, Stavanger
MB - Universidade de la Laguna, Departamento de Zoologia, Facultad de Biologia, Tenerife - Islas Canarias, La Laguna
MHNL - Musée d'Histoire Naturelle, Lille
MMBCS - Moravské Museum, Oddělení entomologické, Brno
MNHN - Muséum Nationale d'Histoire Naturelle, Laboratoire d'Entomologie, Paris
NHRM - Naturhistoriska Riksmuseet, Sektionen for Entomologi, Stockholm
NMP - National Museum in Prague, Museum of Natural History, Department of Entomology, Prague
*RM - Rana Musem, Naturhistorisk avdeling, Mo i Rana
*RML - private collection of Reidar Mehl, Statens institutt for folkehelse, Laboratorium for medisinsk entomologi, Oslo
*SA - private collection of Stig Andersen, Copenhagen
*TM - Universitetet i Tromsø, Tromsø Museum, Zoologisk Avdeling, Tromsø
USNM - United States National Museum, Smithsonian Institution, Washington
ZIKIEL - Christian-Albrechts-Universität zu Kiel, Zoologisches Institut, Lehrstuhl für Ökologie, Kiel
ZILEN - Zoological Institute, Academy of Sciences, Leningrad
*ZMB - Universitetet i Bergen, Zoologisk Museum, Entomologisk avdeling, Bergen
ZMBERL - Museum für Naturkunde an der Humboldt-Universität zu Berlin, Berlin
*ZMC - Universitetets Zoologiske Museum, afd. 3, Copenhagen
*ZMH - Zoological Museum, Division of Entomology, Helsinki
*ZML - Lund University, Museum of Zoology and Entomology, Lund
ZMMSU - Moscow Lomonosov State University, Zoological Museum, Moscow
*ZMO - Universitetet i Oslo, Zoologisk Museum, Entomologisk avdeling, Oslo

The catalogue is based on specimens examined personally (about 15000 specimens). No literary records have been accepted, with these exceptions: (1) a few records from Sweden published by Ringdahl (1952) and apparently based on specimens in ZML, NHRM or elsewhere, but not present in coll. Ringdahl (in ZML); (2) records by Nuorteva, Grunin and coworkers (see bibliography) from within the present borders of Finland (a few records also from Sweden) and dealing with species of *Calliphora*, *Cynomya*, *Protocalliphora*, *Protophormia*, *Lucilia*; (3) the record by Koskela, Itämies & Pasanen (1974) for *Lucilia bufonivora* from Ostrobottnia media; (4) the records by Tiensuu (1964) of *Eurychaeta palpalis* from various Finnish provinces; (5) a few published records from Norway in the ornithological literature of a bird parasitic blowfly which without any doubt belongs to *Trypocalliphora braueri*; (6) the record from EJ of *Calliphora loewi* by Nielsen (1967, 1968). Data from Germany and Great Britain have been taken from Pont (1976) and Schumann (1986) or from specimens collected or examined personally. Most of the data for Russian Karelia are also based on my own determinations, but a few records have been taken from reliable sources (Stackelberg, 1962; Grunin, 1970a; Kovalev & Verves, 1987).

The abbreviations for faunistic regions in Norway are to be understood in the sense of Økland (1981) and not in the sense of Strand (1943) or Løken (1973). So even though the catalogue uses small letters in the province abbreviations, they are to be understood as consisting only of capital letters.

Numerous calliphorid types for nominal taxa named by Fabricius, Fallén, Meigen, Macquart, Robineau-Desvoidy, Zetterstedt, Walker, Pandellé, Villeneuve, Séguy, Rodendorf, Shannon, Ringdahl, Wainwright, Baranov, Jacentkovský, Hall and Zumpt have been revised. Neotypes have been designated for 6 nominal taxa for which no types are known to exist, in order to create a stable nomenclature.

As a rule the keys have as a rule been made rather long, stimulating the user to consult as many diagnostic characters as possible. In my experience, misidentifications invariably result from too much reliance on single characters in this family.

Features described in the keys or at higher taxonomic levels are as a rule not repeated in the descriptions at lower levels. To obtain a full description of a particular species, both keys and higher level descriptions should therefore be consulted.

The International Code of Zoological Nomenclature (Third edition, dated February 1985, adopted by the XX General Assembly of the International Union of Biological Sciences) has been abbreviated as "ICZN" or "Code" in the following.

For bibliographical consistency the names of Russian authors (e.g. Khitsova, Rodendorf, Zinov'ev) have been transcribed from their Russian original spellings. The foreign-language spellings used by these authors themselves in their own publications (e.g Khitzova, Rohdendorf, Zinovjev) are not adopted.

Relationships, monophyly and diagnosis of the Calliphoridae

The Calliphoridae contains more than 1000 species in about 150 genera, and the family is worldwide in distribution (Hennig, 1973; Pont, 1980; Shewell, 1987). More than 250 species occur in the Palaearctic Region (Schumann, 1986). In Fennoscandia and Denmark 52 species in 17 genera are recognised in the present work, but there are 8 further species and 6 further genera in the subfamily Rhinophorinae (cf. Rognes, 1986d) which is not dealt with even though I include it in the Calliphoridae (see discussion below).

They are typical members of the larger systematic unit named Calliphoroidea by Hennig (1958, 1973, 1976) (= Tachinidae s.lat. of Griffiths, 1972) (= Oestroidea of McAlpine et al., 1981) which according to Griffiths (1972) is characterised by the following ground-plan apomorphies:

(1) Row of meral setae present, secondarily absent or transformed into a broad zone of densely set long thin hairs in the Oestridae (s.lat.) of Fennoscandia and Denmark.
(2) ST8 in ovipositor entire.
(3) Vein M sharply bent forwards.
(4) Vein A_1+CuA_2 (anal vein) not reaching margin of wing.
(5) T6 in ♂ shortened.

Andersen (1982, 1983) suggested that a complete anal vein is part of the ground-plan of the Tachinidae. Pape (1986) even thought that it might be a ground-plan feature of all the Calliphoroidea. I find Griffiths' view to be the most parsimonious and accept his interpretation of item (4).

Within this assemblage the families Tachinidae, Sarcophagidae and Calliphoridae (including the Rhinophorinae) probably form a monophyletic subgroup (at present unnamed) on the basis of at least the following autapomorphy:

(1) Presence of a "Gelenkfortsatz" i.e. a separate sclerotisation at the base of the postgonite (Tschorsnig, 1985a, 1985b) (cf. Fig. 20, "basal piece").

The Oestridae (s.lat), including the bot flies and warble flies, the well-known mammalian parasites, is possibly the sister-group of this three-family group (Hennig, 1976). Note, however, that Hackman & Väisänen (1985) have found a costal chaetotaxy in Oestridae (s.lat.) that is remarkably primitive, and that the systematic position of this group still is enigmatic (see also Zumpt, 1957). I can confirm the absence of a "Gelenkfortsatz" in *Oedemagena tarandi* (Linnaeus) (Hypodermatinae) and in *Cephenemyia trompe* (Modeer) (Oestrinae) (cf. Tschorsnig, 1985a: 10, 1985b: 48).

The monophyly of the Calliphoridae has been discussed by Rognes (1986c) and is possibly founded on three ground-plan apomorphies:

(1) Adults with thoracic surfaces prosternum, proepisternal depression, postalar wall and metasternal area hairy.
(2) Adult ♂ with distal end of each paraphallic process well sclerotised, free from wall of distiphallus for a considerable distance and in the shape of a long narrow process, a hook or a denticulate or bare plate.
(3) Metallic ground colour of thorax and abdomen.

The position of the outer *ph* seta, outside a line through the *prs* seta in many blowflies, was first used by Girschner (1893) and Villeneuve (1924) to define the group. This position is probably apomorphous but will not define the whole of the Calliphoridae in the currently accepted sense (see discussion by Rognes, 1986c).

In practical work in Fennoscandia and Denmark members of the Calliphoridae can be separated from the remaining Calliphoroidea by the following combination of characters:

Mouthparts well-developed. Subscutellum weakly or not developed. Body metallic green, blue or other shades, or non-metallic. If non-metallic, then hairs are usually present on at least one of the surfaces prosternum, proepisternal depression and postalar wall, and the hind coxa is bare behind. If all these surfaces are bare, then the lower calypters are small, tongue-like or subcircular with inner edge diverging backwards from the longitudinal axis of the fly. Outer *ph* seta often outside a line through *prs*.

The sister-group of the Calliphoridae is not known. Zumpt (1956b), following earlier authors (e.g. Séguy, 1941), proposed placing the Calliphoridae and Sarcophagidae together in a large group Calliphoridae s.lat., thus implicitly regarding the Calliphoridae and Sarcophagidae as sister-groups, i.e. more closely related to each other than to the Tachinidae. Herting (1957) also adopted this system. However, the Calliphoridae (s.lat.) when defined in this way appears to be based on plesiomorphies (lack of postscutellum, primarily saprophagous larvae) and therefore has no place in a phylogenetic system.

Griffiths (1982) suggested that the Calliphoridae might be the sister-group of a monophyletic group formed by the Sarcophagidae, Tachinidae and Oestridae on the basis of an alleged common synapomorphous modification of the spermathecal ducts, each spermatheca having a separate duct into the uterus. In the Calliphoridae and other groups, two of the ducts are joined so that "only two ducts branch from the vagina" (assumed ground-plan for the Schizophora) (Griffiths, 1982: 76). The data on which Griffiths based his suggestions appear to have been provided in the form of drawings prepared by Dr. B. A. Holloway (New Zealand) from her own dissections (Holloway, in litt. November 7, 1986). In the Sarcophagidae, only *Parasarcophaga crassipalpis* was studied, a member of the subfamily Sarcophaginae which is apomorphic as regards the aedeagus; consequently, any findings regarding the receiving female organs are not likely to be part of the ground-plan of this family. Dr. Holloway also tells me that the three spermathecal ducts had separate origins on a common stalk. According to figures published by Lopes (1941) for members of the subfamily Sarcophaginae, the spermathecal ducts may enter the uterus (vagina) separately or be joined at the same level to a common long stalk before reaching the uterus. In *Sarcotachina subcylindrica* Portschinsky (subfamily Paramacronychiinae according to Verves, 1986) the spermathecal ducts are joined at the same level to a common long stalk before reaching the uterus (Lopes, 1981). As far as I know the course of the ducts in Miltogrammatinae is not known, but to judge from the aedeagus one would expect a primitive calliphorid-like system. In *Oestrus ovis* (Oestrinae) three spermathecal ducts arise at the end of a common stalk, whereas in *Gasterophilus intestinalis* (Gasterophilinae) only two spermathecae are present, their ducts uniting before entering the uterus (Holloway, unpublished drawings in litt. November 7, 1986). The Hypodermatinae have three spermathecae, but the course of their ducts near the uterus is not known (the figure in Grunin, 1964-1969 is not sufficiently detailed). In most Calliphoridae that I have seen, a single duct branches off from the uterus and successively sends off two branches (cf. Figs. 523, 553). This system is in agreement with figures given by Graham-Smith (1938), although another system is also found (Fig. 500). Although I agree with Griffiths' interpretation that the calliphorid duct system is primitive, I do not accept the sister-group arrangement he proposes as it is obviously based on far too meagre evidence. Furthermore, it conflicts with the above interpretation of the "Gelenkfortsatz" character of Tschorsnig which I find very convincing.

Subfamilies of the Calliphoridae

In the fauna of Fennoscandia and Denmark 8 subfamilies of the Calliphoridae are recognised: Calliphorinae, Chrysomyinae, Helicoboscinae, Luciliinae, Melanomyinae, Polleniinae, Rhiniinae and Rhinophorinae. Their monophyly and characteristics are discussed separately in the text below. The groups are also believed to be valid when the world fauna of Calliphoridae is considered. When more is known about the interrelationship of these groups, which I find quite unclear at the moment, the rank of several of them will probably be reduced.

The characters found useful for establishing these groups and their assumed polarity, are as follows (p - plesiomorphies, a - apomorphies; a', a": same transformation series; a^1, a^2: different transformation series).

Colour

(1) Metallic body colour (p), black non-metallic body colour (a^1), yellow non-metallic ground-colour (a^2).

Head

(2) Parafacialia with small hairs only (p), with very strong setae (a).
(3) Parafacialia with small hairs (p), bare (a).
(4) Facial carina undeveloped (p), more or less developed (a).
(5) Lower part of face not or weakly protruding (p), protruding strongly in front of vibrissae (a).
(6) Upper half of occiput evenly dusted (p), with a bare glossy submarginal band across almost its entire width (a).

Thorax

(7) Outer *ph* seta in line with *prs* (p), outside a line through *prs* (a).
(8) Proepisternal depression, prosternum, metasternal area and postalar wall all hairy (p), one or more of these surfaces bare (a).
(9) Suprasquamal ridge posteriorly bare (p), posteriorly with a green metallic sclerite with numerous black setae (Fig. 397) (a).
(10) Lower half of anterior part of anepimeron bare (p), with numerous hairs (Fig. 196) (a).
(11) 2+1 *kepst* setae (p), 2+1+1 *kepst* setae (a).
(12) Coxopleural streak present (p), absent (a).
(13) Anteroventral edge of posterior spiracle quite bare (p), numerous fine setae in a row along this edge (a).
(14) Metakatepisternum (area immediately above hind coxa) bare (p), hairy in extreme lower part (a).

Wing

(15) Lower calypter posteriorly broad, inner edge converging with longitudinal axis of fly (p), narrow, inner edge diverging from the longitudinal axis of the fly or more or less subcircular (a).
(16) Lower calypter bare above (p), hairy above (a).
(17) Costa hairy below, also far beyond junction with R_1 (p), costa bare below beyond junction with R_1 (a'), costa bare below beyond junction with Sc (a").
(18) Stem-vein bare on dorsal surface of wing (p), with a row of setae (a).
(19) Subcostal sclerite only with microtrichiae, no setulae (p), with black setulae (Fig. 9) (a).
(20) Vein M sharply bent forwards in apical part of wing, reaching costa in front of wing apex (p), loss of fixed calliphoroid ground-plan of wing veins, especially as regards the course of vein M in apical part of wing (a), e.g. vein M in apical part of wing weakly curved and reaching costa at apex of wing (a^1), or vein M very sharply bent forwards, joining vein R_{4+5} before apex so that cell r_{4+5} is stalked (a^2).

Legs

(21) Hind coxa bare on posterior surface (p), with hairs on posterior surface (a).

Male terminalia

(22) Base of surstylus internally and ventrally without special vestiture (p), with a bundle of long setae internally and ventrally (a).
(23) Postgonite unmodified where the basal seta is situated (p), with a more or less distinct process bearing the basal seta (a).
(24) Acrophallus armed with minute denticles (p), without denticles, unarmed (a).
(25) Paraphallic processes not radically modified, free from wall of aedeagus and shaped as a long slender process, a hook or a denticulate or bare plate (p), paraphalli fused, distally broadened into a kind of shield on dorsal side (a).

Ovipositor

(26) Epiproct without microtrichiae (p), microtrichiose all over (a).
(27) Cerci without microtrichiae (p), more or less microtrichiose (a).
(28) Cerci oval, moderately broad (p), very short and broad, wart-like (a^1), long and narrow (a^2).
(29) Epiproct separated from T8 (p), epiproct fused with T8, the combined sclerite often appearing shovel-shaped posteriorly (a).
(30) T6 microtrichiose along hind edge, ST6 microtrichiose in posterior half (p), these sclerites all without microtrichiae (a).
(31) T7, T8, ST7 and ST8 without microtrichiae (p), more or less invaded by microtrichiae (a).
(32) ST8 apically complete (p), apex bilobed or bifid (a).
(33) ST8 sclerotised for its full length (p), anterior half of ST8 unsclerotised or very weakly sclerotised (a).
(34) ST8 present (p), ST8 completely absent or very strongly reduced (a).
(35) T6, T7, T8, ST6, ST7 and ST8 of moderate

length (p), strongly elongated (a).

(36) T7, T8, ST7 and ST8 with setae (p), devoid of setae or setulae, these replaced by numerous small sensillae (a).

(37) Pleural membranes of ovipositor microtrichiose except for a narrow zone anteriorly on segment 7 and a broad zone anteriorly on segment 8 (p), entirely without microtrichiae (a^1), wholly microtrichiose, no zones without microtrichiae anteriorly (a^2).

(38) Intersegmental membranes in ovipositor with ordinary microtrichiae (p), without microtrichiae of usual form, these entirely absent or replaced with armature of peculiar form (a).

Larvae

(39) First instar larva with labrum dorsally convex and connected to tentoropharyngeal sclerites through long and slender parastomal sclerites (p), separated from tentoropharyngeal sclerite (a^1), dorsally concave (i.e. upturned) and firmly fused with tentoropharyngeal sclerites (a^2).

(40) Accessory oral sclerite absent in third instar larvae (p), present (a).

(41) Posterior spiracle of third instar larva with complete peritreme, ecdysial scar distinct (p), peritreme incomplete medially, ecdysial scar tending to become indistinct or absent (a).

(42) Posterior spiracular plates of third instar with strongly sclerotised peritreme and slits, intermediate areas unsclerotised or weakly sclerotised (p), plates completely sclerotised (a).

The ground-plan features of the hypothetical ancestral calliphorid can be gathered from the list of plesiomorphies.

The subfamily classification differs in several respects from others in current use (Shewell, 1987; Schumann, 1986; Crosskey, 1977), and the most important points are outlined below.

In most modern classifications of the Diptera, the woodlouse-flies (Rhinophoridae) are regarded as a separate family (Crosskey, 1977; Tschorsnig, 1985a; Pape, 1986). I believe that on the basis of the aedeagal structure (presence of a mesohypophallus, ring-shaped sclerotisation at the base of the distiphallus, long narrow paraphallic processes, sometimes reduced) it is better regarded as a subgroup within the Calliphoridae characterised by numerous secondary changes or reductions. In accordance with a previous suggestion (Rognes, 1986c) it is listed as a subfamily in the present work (although the species are not treated). Pape (1986) regards the fusion in the midventral line of the ventral plates of the aedeagus as an autapomorphy of the Rhinophoridae. This feature occurs in many Calliphoridae (e.g. in Ameniinae, Rhiniinae, Melanomyinae and Calliphorinae) (cf. Rognes, 1986c), and it may be one of the ground plan features of this group; Pape's dismissal of these instances as "easily rejected as convergencies" cannot be accepted. The presence of a mesohypophallus is regarded by Pape (1986) as possibly "an important argument for a close affinity to the Calliphoridae". I agree with this.

In the present work the genera *Angioneura* and *Melanomya* have been removed from the Rhinophorinae, where they were placed by Herting (1961), Crosskey (1977) and Tschorsnig (1985a), and together with *Melinda* and *Eggisops* have been grouped into a revived subfamily Melanomyinae on the basis of their ovipositor structure (cf. also discussion of these genera by Rognes, 1986c).

Morinia has usually also been regarded as a rhinophorid (Stackelberg, 1970; Crosskey, 1977; Tschorsnig, 1985a), but Pape (1986) and Rognes (1986a, 1986b) argued that it is a calliphorid genus, because of the setose postalar wall and operculate posterior thoracic spiracle. I have transferred it to the Polleniinae.

The groups Rhiniinae and Chrysomyinae are well-defined and treated in this work as subfamilies. For the time being I have not adopted Boyes & Shewell's (1975) and Shewell's (1987) group "Chrysomyinae" (= Chrysomyini + Phormiini + Rhiniini), although based on an apparent synapomorphy (row of setae on upper surface of the stem-vein). For reasons discussed below I believe that the groups "Phormiini" and "Chrysomyini" as conceived at present cannot be maintained, as *Phormia* may be more closely related to "Chrysomyini" than to "Phormiini" (Rognes, 1985a). I also hesitate to accept that the "Rhiniini" and Chrysomyinae are more closely related to each other than to any other Calliphoridae, especially on the basis of the aedeagal and ovipositor structure and the structure of the posterior spiracles. I therefore find it best to leave them as separate groups until the phylogeny of the Chrysomyinae and Rhiniinae has been more fully analysed. Note that a hairy stem-vein is also present in the Polleniinae (*Pollenia atramentaria*), so it is conceivable that it was independently acquired by Chrysomyinae and Rhiniinae.

The Helicoboscinae was established as a calliphorid subfamily by Rognes (1985b, 1986a, 1986c). Its relationship with other groups is quite obscure.

Most students of the Calliphoridae (e.g. Zumpt, 1956a, 1956b; Shewell, 1987) include a group in their system (named "Calliphorinae" or "Calliphorini") defined by the absence of a row of setae on the dorsal surface of the stem-vein. This enormous assemblage is based on what is very clearly a plesiomorphy and is therefore unacceptable in a strictly cladistic system. I have found the time ripe for dismemberment of this artificial group and propose its replacement by smaller, probably monophyletic groups.

In the fauna of Fennoscandia and Denmark, 4 such groups, each given the rank of subfamily, are recognised: Calliphorinae, Luciliinae, Melanomyinae, and Polleniinae. This system in many ways resembles that of Shewell (1987). He gives Polleniini, Angioneurini, Luciliini and Calliphorini equal tribal rank within the "Calliphorinae", and the first three groups appear to be equivalent to my subfamilies Polleniinae, Melanomyinae, Luciliinae. My Calliphorinae appears in many ways to be similar to the group named Calliphorini in Lehrer's (1970) system, although he defines it differently.

The abandonment of the Calliphorinae/Calliphorini in the traditional sense creates problems as far as the fauna outside Fennoscandia and Denmark is concerned. The groups affected by this action must at least provisionally receive some formal rank and I propose that they be treated as subfamilies. This involves:

(1) The Afrotropical and Oriental group Auchmeromyiinae of Patton (1935, 1936), which on the basis of the aedeagal structure (Patton, 1935, 1936; Zumpt, 1953) appears to include the genera *Auchmeromyia* Brauer & Bergenstamm, *Pachychoeromyia* Villeneuve, *Booponus* Aldrich (= *Elephantolaemus* Austen) (= *Pavlovskiomyia* Grunin), *Cordylobia* Grünberg, *Neocordylobia* Villeneuve, *Hemigymnochaeta* Corti and *Tricyclea* Wulp. This assemblage appears equivalent to the Tricycleinae of Lehrer, 1970, but Patton's name has priority. They are flies with a yellow or brown ground-colour. According to Villeneuve (1924) several of these genera have a complete vein A_1+CuA_2 in the wing (i.e. anal vein reaching wing-margin), and this may constitute an autapomorphy for the group. Unfortunately I have not been able to study this feature. Most of the genera contain species which are obligatory parasites of mammals and man, either myiasis-producers or blood-suckers in the larval stages (Zumpt, 1965). The two latter genera include species which develop in various kinds of organic matter such as mushrooms, dung and faeces, and some appear biologically associated with ants and termites (Zumpt, 1953).

(2) The Afrotropical, Oriental and Australian group Phumosiinae (*Phumosia* Robineau-Desvoidy and *Euphumosia* Malloch), given tribal rank by e.g. Lehrer (1970) and Tumrasvin, Kurahashi & Kano (1979) and subfamily rank by Ferrar (1978). It includes metallic green or blue species as well as others with brownish colouring (Torgerson & James, 1967). The ovipositor has shortened sclerites and very elongate spermathecae (King & Kurahashi, 1977; Singh, Kurahashi & Kano, 1979; Tumrasvin, Kurahashi & Kano, 1979). Some species are macrolarviparous (unilarviparous) (Ferrar, 1978).

(3) The Afrotropical and Oriental group Bengaliinae, given subfamily rank by James (1966), Lehrer (1970) and tribal rank by e.g. Tumrasvin, Kurahashi & Kano (1979). It contains one genus of species with yellow or brown ground-colour and very enlarged mouthparts (*Bengalia* Robineau-Desvoidy). *Catapicephala* Macquart, containing species with brightly metallic colouration, was included in this group by Tumrasvin, Kurahashi & Kano (1979). *Bengalia* also has a complete vein A_1+CuA_2 in the wing (i.e. reaching margin) (Villeneuve, 1924; Pape, 1986). *Bengalia* adults are aggressive predators, feeding on termites and ant pupae or snatching prey from the ants themselves (Senior-White, Aubertin & Smart, 1940; Pont, 1980).

Additional calliphorid subfamilies are the Neotropical Mesembrinellinae (revised by Guimarães, 1977) and Toxotarsinae (revised by Dear, 1979; cf. also the critique by Lopes & Albuquerque, 1982), the Oriental and Australasian Ameniinae (revised by Crosskey, 1965), the remarkable New Zealand group Mystacinobiinae of Holloway (1976) (assigned to the Calliphoridae by Griffiths, 1982), and the enigmatic Neotropical and Afrotropical group Prosthetosomatinae, known only from the larval stages which infest nest-mounds of termites (Smith, 1975; Pont, 1980). According to Herting (in litt. 22.ii.1985), *Mimodexia* Rodendorf is a senior synonym of *Callidesia* Kugler and thus a rhinophorine (cf. also Tschorsnig, 1985a). Mimodexiinae Rodendorf thus falls as a synonym of Rhinophorinae and does not deserve status as a separate tribe or subfamily as maintained by Zumpt (1952), Lehrer (1972) and Lopes & Albuquerque (1982).

A thorough discussion of the monophyly and status of these groups is beyond the scope of the present work and must be left to future workers.

Morphology of the adults

The adult morphology and the terminology used for describing it are fully explained in Figs. 1-21. I believe that these illustrations are selfexplanatory. The sclerite named prosternum is situated midventrally between the fore coxae. Intrapostocular ciliae are minute hairs often present on the area between posterior eye margin and the postocular setae (Fig. 3). The ejaculatory sclerite is joined to the sperm duct proximad of the basiphallus (cf. Fig. 550). The phallapodeme, being very uniform, is not illustrated for every species. The terminology of McAlpine (1981) is adopted, with a few exceptions (particularly for the setae at the ante-

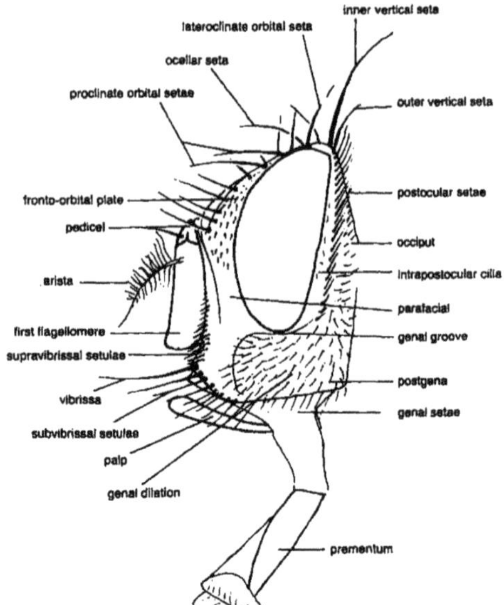

Fig. 3. Head of *Calliphora subalpina* (Ringdahl), left lateral view.

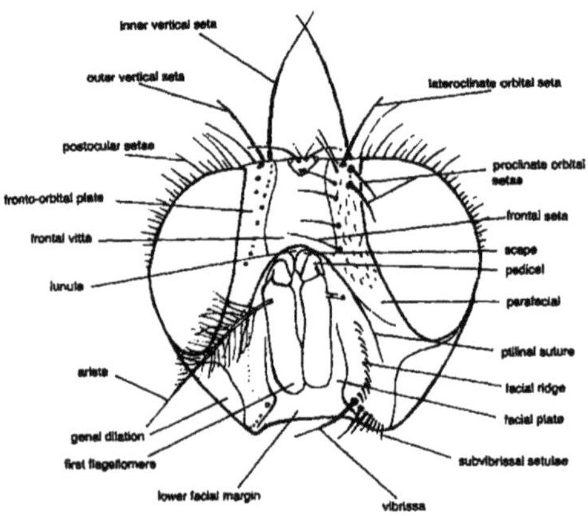

Fig. 1. Head of *Calliphora subalpina* (Ringdahl), anterior view.

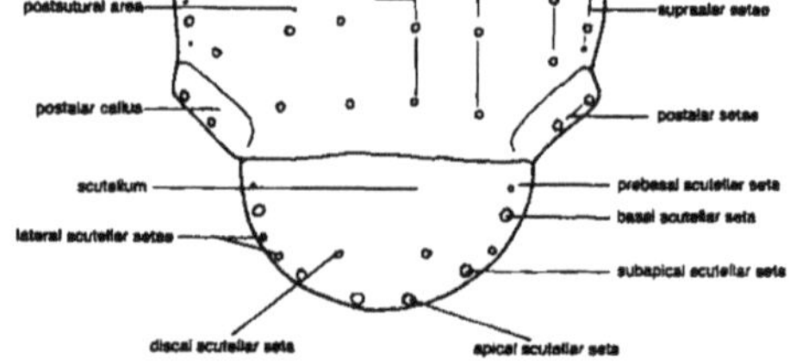

Fig. 4. Thorax of *Calliphora subalpina* (Ringdahl), dorsal view.

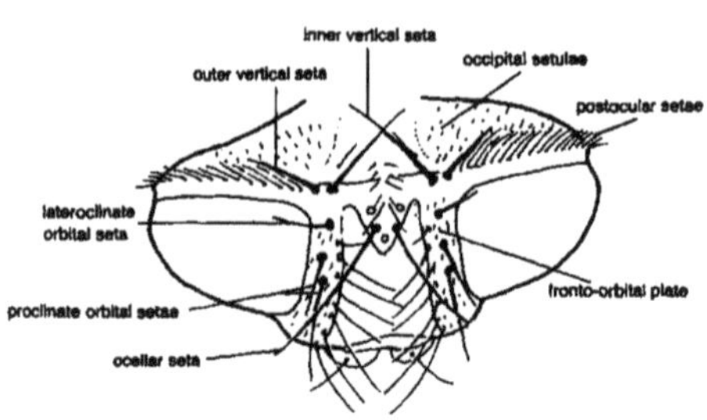

Fig. 2. Head of *Calliphora subalpina* (Ringdahl), dorsal view.

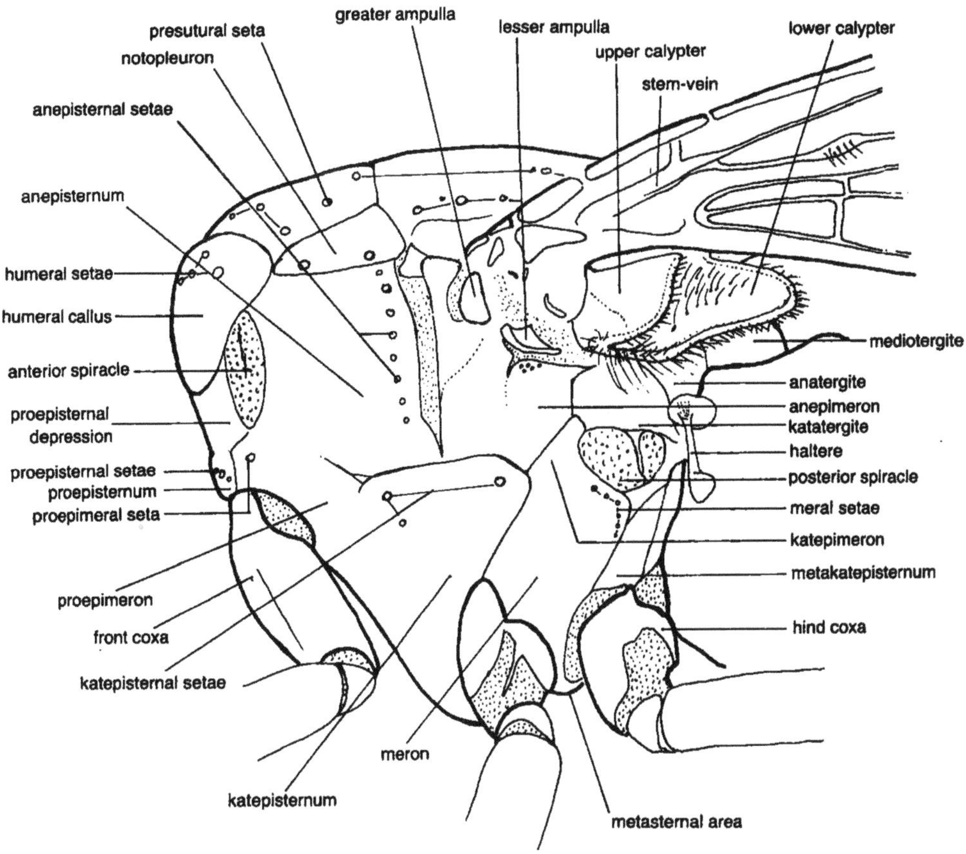

Fig. 5. Thorax of *Calliphora subalpina* (Ringdahl), left lateral view.

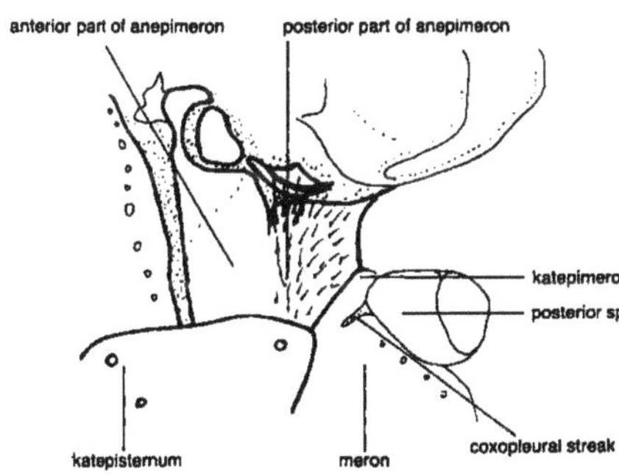

Fig. 6. Thorax of *Lucilia caesar* (Linnaeus), left lateral view of anepimeral area.

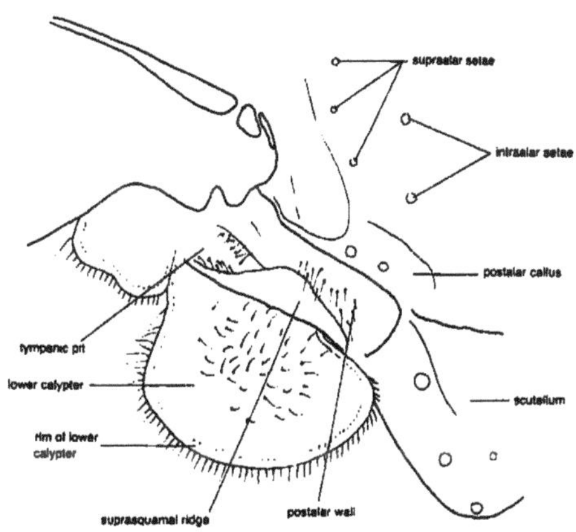

Fig. 7. Thorax of *Calliphora genarum* (Zetterstedt), left posterodorsolateral view of area behind wing base.

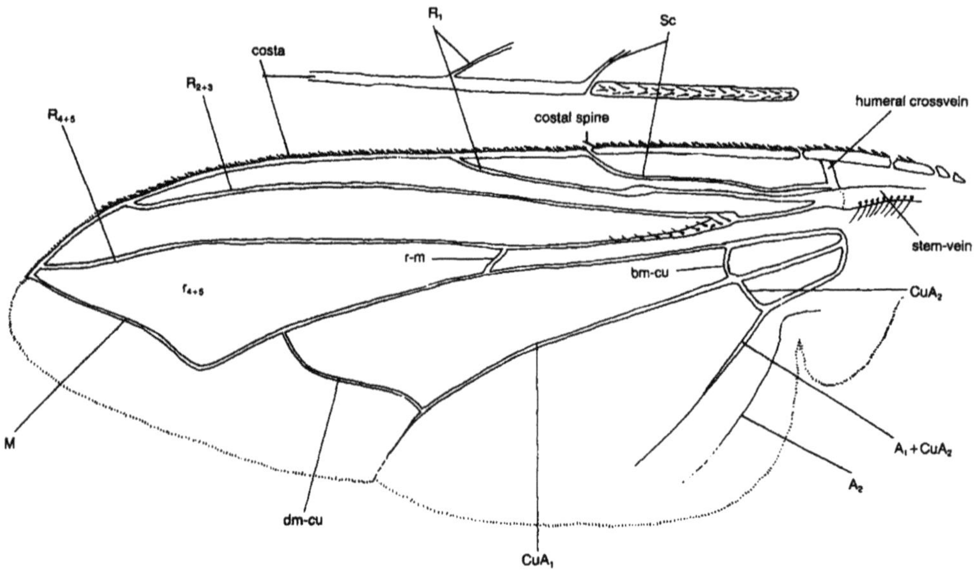

Fig. 8. Wing of *Trypocalliphora braueri* (Hendel), from above. Inset: parts of costa from below showing distribution of small hairs on its underside.

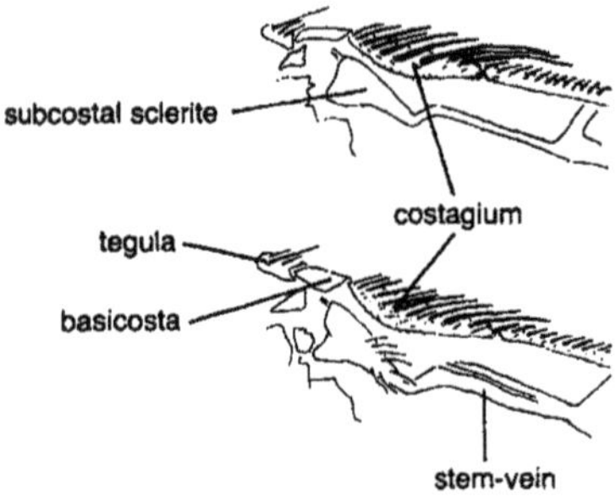

Fig. 9. Base of left wing from below of *Lucilia sericata* (Meigen) (upper figure) and *Lucilia caesar* (Linnaeus).

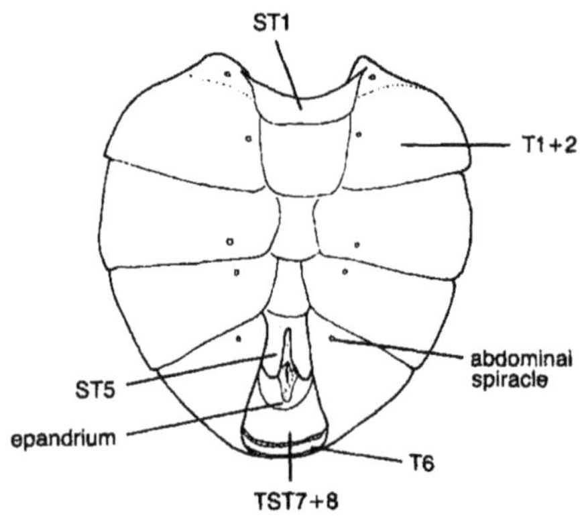

Fig. 11. Abdomen of *Pollenia rudis* (Fabricius), ventral view.

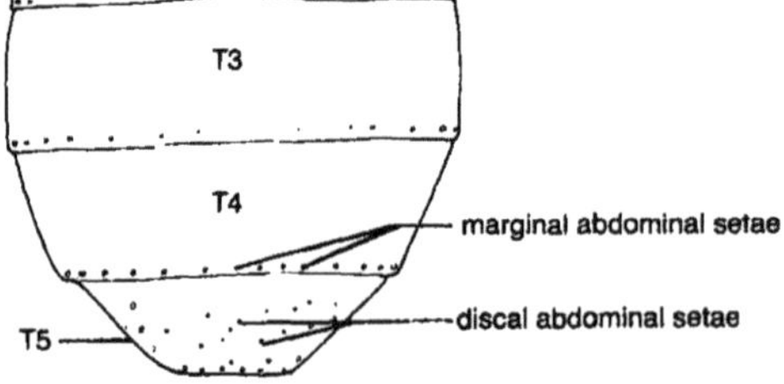

Fig. 10. Abdomen of *Pollenia rudis* (Fabricius), dorsal view.

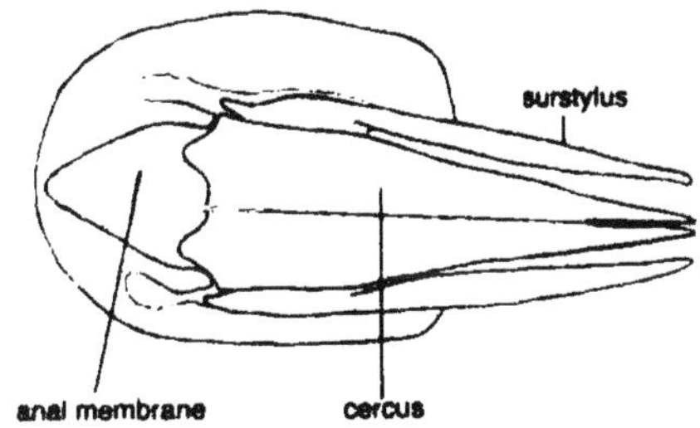

Fig. 12. Male postabdomen of *Lucilia magnicornis* (Siebke), posterior view.

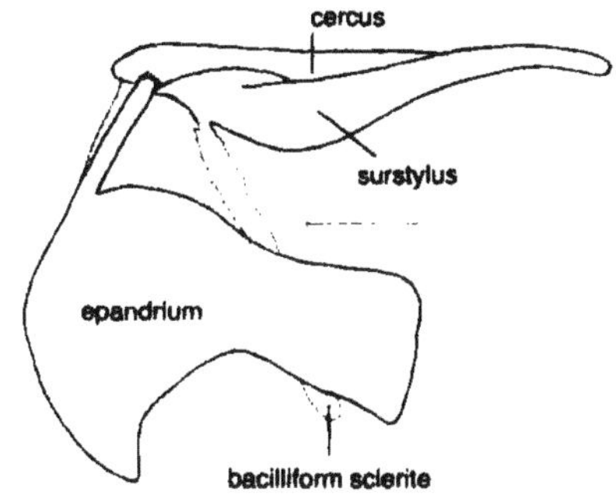

Fig. 13. Male postabdomen of *Lucilia magnicornis* (Siebke), left lateral view. Scale = 0.2 mm.

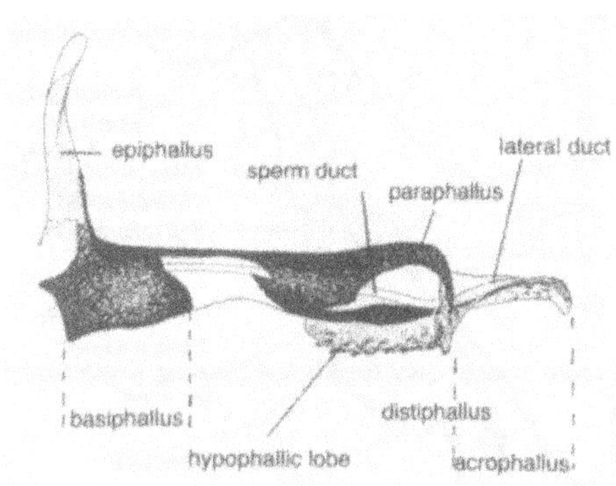

Fig. 14. Aedeagus of *Lucilia magnicornis* (Siebke), left lateral view.

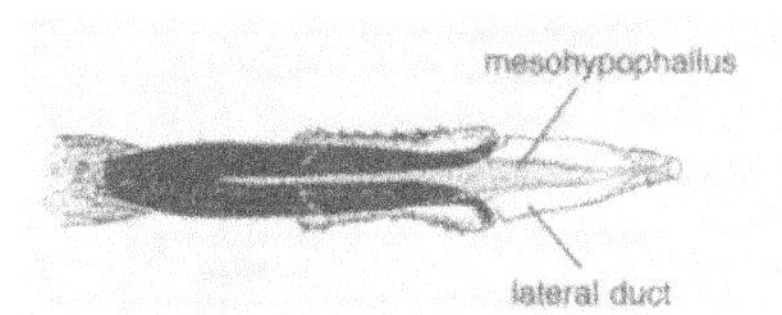

Fig. 15. Distiphallus of *Lucilia magnicornis* (Siebke), dorsal view.

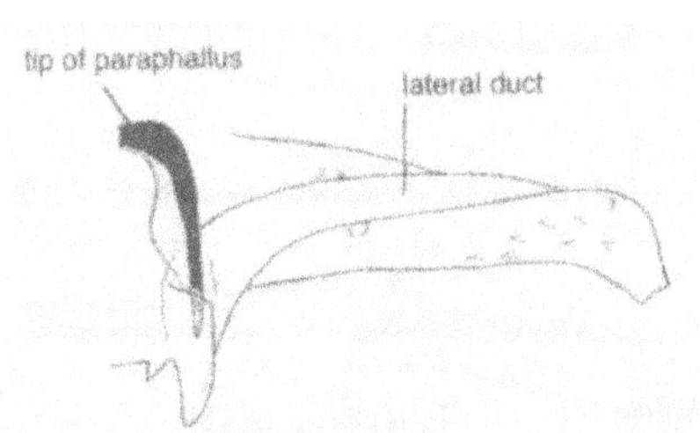

Fig. 16. Acrophallus of *Lucilia magnicornis* (Siebke), left lateral view.

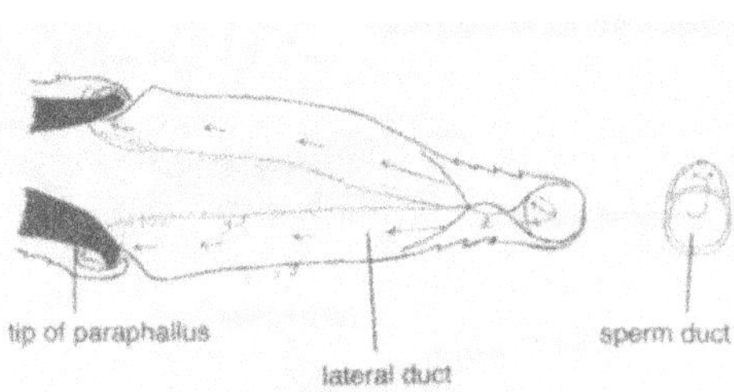

Fig. 17. Acrophallus of *Lucilia magnicornis* (Siebke), dorsal view. Inset: opening of sperm duct, with opening of lateral ducts above it.

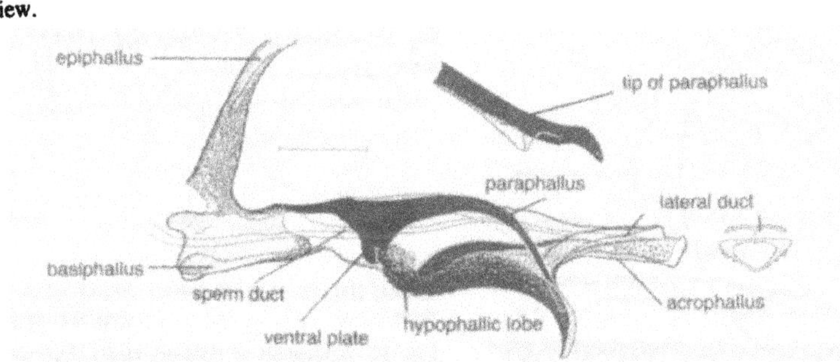

Fig. 18. Aedeagus of *Calliphora vomitoria* (Linnaeus), left lateral view. Insets: tip of paraphallus (enlarged) and opening of sperm duct with opening of lateral ducts above it. Scale = 0.2 mm.

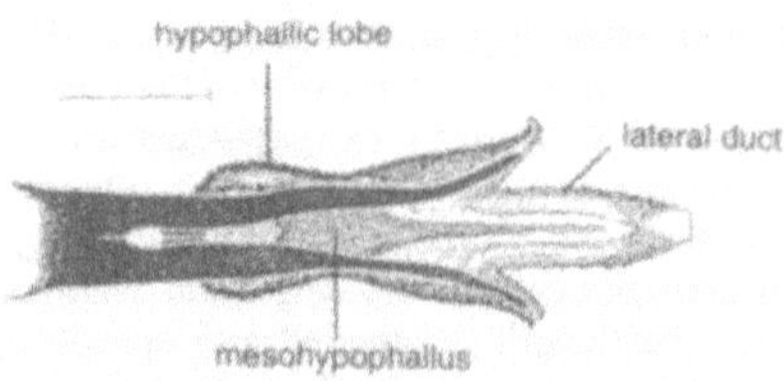

Fig. 19. Distiphallus of *Calliphora vomitoria* (Linnaeus), dorsal view. Scale = 0.2 mm.

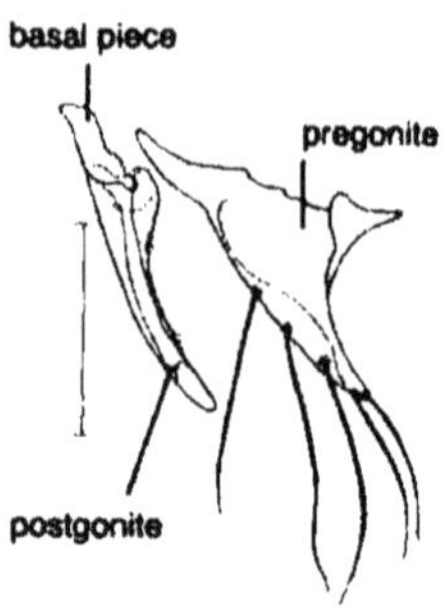

Fig. 20. Post- and pregonites of *Lucilia magnicornis* (Siebke). Scale = 0.2 mm.

rolateral corner of the thorax, where I follow O'Hara (1982)). Some features of the male and female postabdomen will be described in more detail below as they seem to be important from a comparative point-of-view.

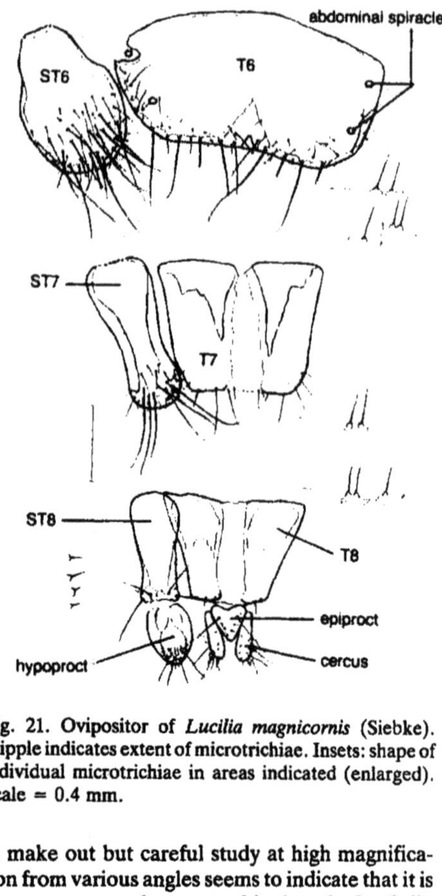

Fig. 21. Ovipositor of *Lucilia magnicornis* (Siebke). Stipple indicates extent of microtrichiae. Insets: shape of individual microtrichiae in areas indicated (enlarged). Scale = 0.4 mm.

Aedeagus

The presence of small denticles directed posteriorly on the ventral and ventrolateral surface of the distiphallus and acrophallus appears to be part of the ground-plan of the Calliphoridae (trend 24, above). A tendency to reduce the distribution of these denticles is exhibited by most subfamilies. In the Melanomyinae and Polleniinae they are wholly absent from the acrophallus.

In some species of *Calliphora* (*C. genarum*, *C. loewi*, *C. uralensis*, *C. vicina*) and *Lucilia* (*L. ampullacea*, *L. bufonivora*, *L. magnicornis*, *L. regalis*, *L. richardsi*, *L. sericata*, *L. silvarum*) the acrophallus has a structure here termed lateral duct (Figs. 14-19, 143, 144, 153, 154, 163, 164, 201, 202, 211, 212, 381, 382, 391, 392, 416, 417, 418, 437, 438, 439, 448, 449, 450, 459, 460, 461, 471, 472). Lewis & Pollock (1975) first directed attention to this structure. They also found it in *Protophormia terraenovae* (Chrysomyinae), but I have been unable to observe it there. It is difficult to make out but careful study at high magnification from various angles seems to indicate that it is an open groove for most of its length. In *Calliphora* the ducts are bordered apically by a common median and separate dorsal walls. In *Lucilia* the duct is bordered laterally and dorsally by a large lateral flange on each side. In *Pollenia* there is apparently a similar duct along the distal edge of the hypophallus in the *rudis*, *semicinerea*, *tenuiforceps* and *intermedia* species groups, and also in *Pollenia atramentaria*, but it is extremely difficult to make out. In *Lucilia* the ducts have their distal opening just above the exit of the ejaculatory duct (cf. also scanning electron micrographs by Lewis & Pollock, 1975), whereas in *Calliphora* they start a short distance behind the apex (e.g. Figs. 211, 391). The proximal opening appears to be into the strongly sclerotised closed canal (*L. bufonivora* and *L. silvarum*) or partly roofed-over passage (L. *sericata*, *L. regalis*, *L. richardsi* and *L. magnicornis*) or shelf-like support (*C. genarum*, *C. loewi*, *C. uralensis*, *C. vicina*, *C. vomitoria*, *L. am-*

pullacea), guiding and supporting the longitudinal movement of the distal part of the paraphallus. Lewis & Pollock (1975) apparently have overlooked the existence of this "paraphallus guide". Their "accessory secretion exit" at the apex of the cornu of the hypophallus in *L. sericata* (Lewis & Pollock, 1975: fig. 3) is the lower end of the "paraphallus guide" and in fact is not a completely encircled opening in this particular species. I do not agree with the view that this should be the external opening of a duct leading from somewhere within the aedeagus near the apex (i.e. from the sperm duct) to the outside, and that the aedeagus has three openings. The groove is morphologically external along its entire course. Accessory duct secretion appears first to leave the ejaculatory duct opening at the apex of the aedeagus (this is denied by Lewis & Pollock, p. 142), then to enter the apical opening of the lateral ducts above it (cf. Figs. 211, 391). From there it seems to follow the ducts backwards and downwards and finally to empty into the "paraphallus guide". During copulation the paraphallic tips are inserted into the lateral sacs on the dorsal side of the female uterovaginal tube. The secretion is deposited there possibly with their aid. The presence of lateral ducts on the outer surface of the acrophallus may be correlated with the presence of lateral sacs in the dorsal wall of the uterus, as there seems to be no conflict between the above data concerning the distribution of acrophallic lateral ducts within Calliphoridae and Hori's (1961) data concerning the distribution of lateral sacs in this family. However, I have seen conspicuous unsclerotised lateral sacs in a female *Protocalliphora proxima* (G.pr.173), which appears to lack lateral ducts in the acrophallus. The matter obviously needs further study. Tschorsnig (1985a) found a tripartite ejaculatory duct opening in some Rhinophorinae. How this is to be interpreted, must be left to future studies. An examination of the aedeagal structures with a scanning electron microscope is urgently needed for all groups.

Ovipositor

The structure of the ovipositor has been used for identification purposes by Lobanov (1976), but the distribution of the microtrichiae, which has proved of great significance in the taxonomy of the Muscidae (Zinov'ev, 1981, and references cited by him), has not received attention by workers on Calliphoridae.

Medium length sclerites with complete rows of marginal setae appear to belong to the ground-plan condition for the ovipositor of the Calliphoridae (trend 35, above). In viviparous forms, the sclerites are shortened in *Bellardia*, *Onesia* and *Eurychaeta* but elongate in *Eggisops*. They are also very elongate in the oviparous Melanomyinae. A full-length, undivided and apically complete ST8 is assumed to be the ground-plan condition of the Calliphoridae (trends 32, 33, 34, above). Such a ST8 occurs in all subfamilies except the Calliphorinae, where it is apically bifid, the Melanomyinae, where it is absent or very strongly reduced and the Chrysomyinae where only the posterior half is sclerotised.

I interpret the following distribution of microtrichiae as the ground-plan for the ovipositor of the Calliphoridae: All intersegmental membranes with microtrichiae, which are directed forwards (trend 38, above). All pleural membranes with microtrichiae, except for a narrow zone anteriorly on segments 7 and a broad zone anteriorly on segment 8; these also directed forwards (trend 37, above). The microtrichiae situated on the ovipositor sclerites are directed backwards. T6 with microtrichiae on posterior edge, T7 and T8 without microtrichiae, ST6 with microtrichiae on posterior half, ST7 and ST8 without microtrichiae (trends 30, 31, above). ST8 with small sensillae on posterior margin. Cerci oval in shape, moderately long, without microtrichiae, but with numerous sensillae of different types (as described by Wallis, 1962, for *Phormia regina*) (trends 27, 28, above). Epiproct without microtrichiae (trend 26, above). Hypoproct with microtrichiae, which are directed backwards.

Especially interesting are the apomorphous modifications of the cerci (trends 27, 28, above). The plesiomorphous oval-elongate type without microtrichiae occurs not only in the primarily saprophagous vertebrate carrion breeding genera *Calliphora*, *Cynomya*, *Phormia*, *Protophormia*, *Lucilia*, but also in the bird-parasitic obligatory myiasis producing genera *Protocalliphora*, *Trypocalliphora* and the amphibian parasite *Lucilia bufonivora*. This provides evidence for regarding blowfly myiasis and parasitism as phenomena derived from saprophagy, as suggested by Zumpt (1956a, 1965). In groups with different larval habits, the cerci may be shortened and wart-like (trend 28, a^1) (Melanomyinae - snail parasitism) or covered with microtrichiae (trend 27) (*Bellardia*, *Onesia* in the Calliphorinae, Helicoboscinae, Polleniinae, Rhiniinae - associated with invertebrates such as earthworms, snails, locusts) and narrowed (Polleniinae) (trend 28, a^2).

Internal reproductive organs

The uterus is simple and cylindrical in most species and this appears to be part of the ground-plan of the Calliphoridae. Apomorphous modifications appear in viviparous forms, where it may be either very elongate (Helicoboscinae, some Melanomyinae) or shortened with large lateral incubatory pouches (*Bellardia*, *Onesia*) (cf. Thompson, 1921a, 1921b; Rognes, 1986c). In some forms a pair of variously-shaped lateral sacs is present, and these receive the paraphallic tips during copulation (Graham-Smith, 1938; Hori, 1961; Rognes, 1988) (cf. discussion above). In some species of *Pollenia* the lateral sacs are wholly sclerotised and their shape is important for the identification of females (Rognes, 1987b; 1988). In some species of *Lucilia* only the apertures are sclerotised (Hori, 1961). Otherwise the lateral sacs are unpigmented.

Morphology of the immature stages

Eggs shining white, 0.9-1.5 mm long and 0.3-0.4 mm wide. In lateral view flat or slightly concave dorsally and convex ventrally. Chorion with a reticulate pattern. A narrow median area mid-dorsally, bounded by parallel carinae which unite posteriorly. The median area functions as a plastron, enabling the egg to respire when occasionally submerged in water.

The larvae of most Calliphoridae are typical pale maggots with a tapering anterior and a truncate posterior end. The morphology is illustrated in Figs. 22-41. A general account of the larval morphology of the Diptera and a key to families are given by Teskey (1981a, 1981b). A detailed survey of the breeding habits and immature stages of cyclorrhaphan families of Diptera, including the Calliphoridae, has recently been published by Ferrar (1987a, 1987b). Descriptions of larvae of the species of Calliphoridae that occur in Fennoscandia and Denmark are given by Nielsen (1917), Keilin (1919), Hall (1948), Kano & Sato (1952), Schumann (1954), Rodendorf (1957), Zumpt (1965), Ishijima (1967), Čepelák & Rozkošný (1968), Yahnke & George (1972), Erzinçlioğlu (1985), Rognes (1985a, 1986c), Smith (1986), Shewell (1987). The functional morphology of the mouthparts of mature larvae of some blowflies has been described by Miller (1932) and Roberts (1971).

First instar less than 2 mm long. Mouthparts with well-developed labrum and mandibles (median and lateral hooks) (*Calliphora*, *Cynomya*, *Phormia*, *Protophormia*, *Lucilia*, *Angioneura*, *Eggisops*, *Melinda*) (Figs. 23-24), with very strong labrum and smaller, toothed mandibles (*Eurychaeta*) (Fig. 22), with well-developed labrum and anteriorly reduced mandibles (some *Pollenia*) (Fig. 25), or with strong mandibles and the labrum vestigial or absent (*Bellardia*, *Onesia*, *Protocalliphora*) (Figs. 26-28). Labrum dorsally convex and connected to tentoropharyngeal sclerites through long and slender parastomal sclerites (*Calliphora*, *Cynomya*, *Protophormia*, *Lucilia*) or separated from tentoropharyngeal sclerites (*Eurychaeta*, *Pollenia*), sometimes dorsally concave (i.e. upturned) and firmly fused to tentoropha-

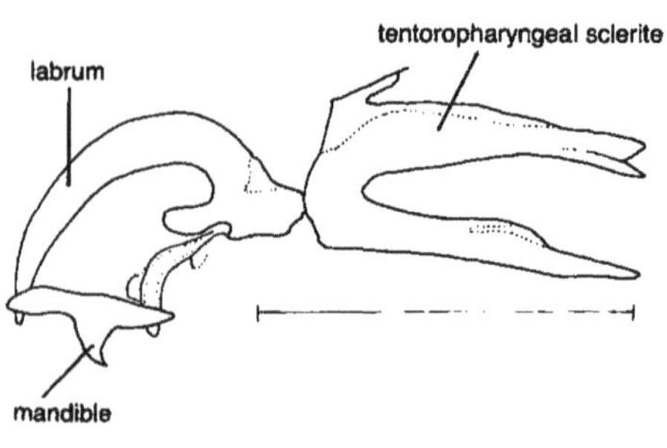

Fig. 22. First instar larva. Mouthparts of *Eurychaeta palpalis* (Robineau-Desvoidy), left lateral view. Scale = 1.0 mm.

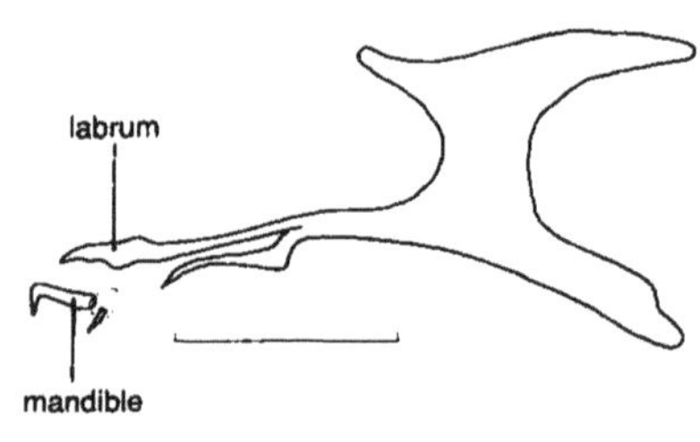

Fig. 23. First instar larva. Mouthparts of *Lucilia sericata* (Meigen), left lateral view. (Redrawn from Schumann, 1954: fig. 5). Scale = 0.1 mm.

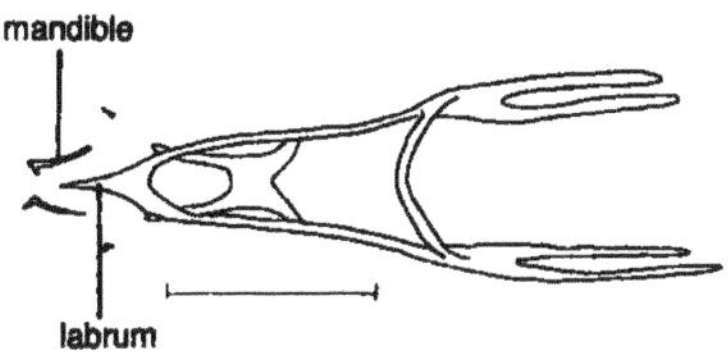

Fig. 24. First instar larva. Mouthparts of *Lucilia sericata* (Meigen), dorsal view. (Redrawn from Schumann, 1954: fig. 19). Scale = 0.1 mm.

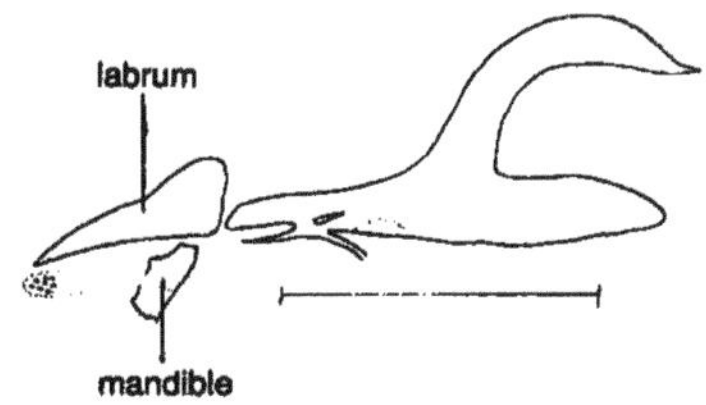

Fig. 25. First instar larva. Mouthparts of *Pollenia* sp., left lateral view. (Redrawn from Yahnke & George, 1972: fig. 3). Scale = 0.1 mm.

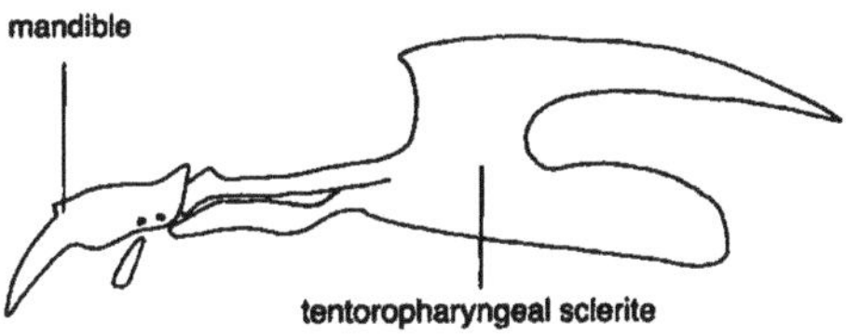

Fig. 26. First instar larva. Mouthparts of *Onesia floralis* (Robineau-Desvoidy), left lateral view. (Redrawn from Schumann, 1964: fig. 4a).

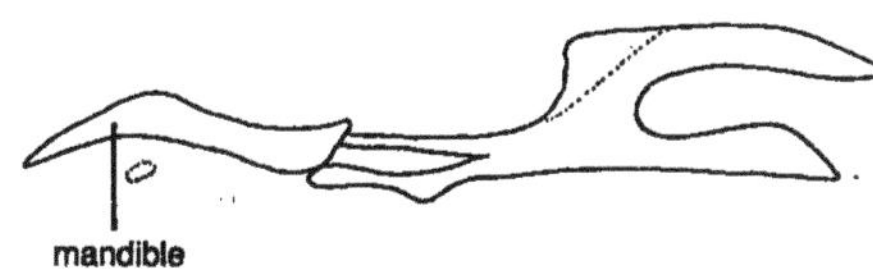

Fig. 27. First instar larva. Mouthparts of *Bellardia vulgaris* (Robineau-Desvoidy) left lateral view. (Redrawn from Schumann, 1974b: fig. 4).

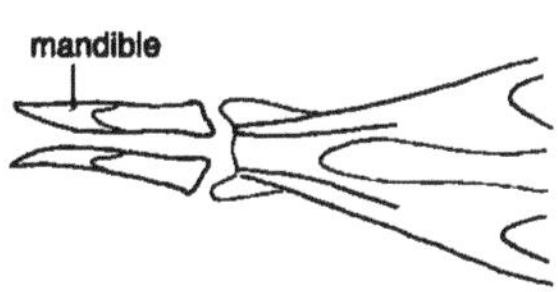

Fig. 28. First instar larva. Mouthparts of *Bellardia vulgaris* (Robineau-Desvoidy), dorsal view. (Redrawn from Shewell, 1987: fig. 46).

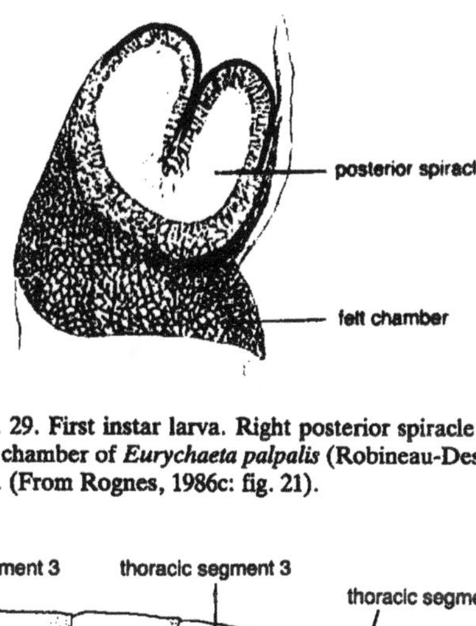

Fig. 29. First instar larva. Right posterior spiracle and felt chamber of *Eurychaeta palpalis* (Robineau-Desvoidy). (From Rognes, 1986c: fig. 21).

Fig. 30. Third instar larva of Calliphoridae. Diagram, right lateral view. Stipple indicates segmental spine bands.

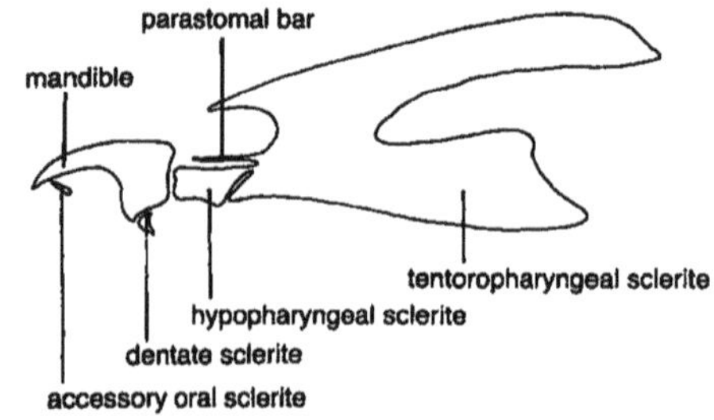

Fig. 31. Third instar larva. Mouthparts of *Calliphora vomitoria* (L.), left lateral view. (Redrawn from Hall, 1948: Plate 44, fig. H.)

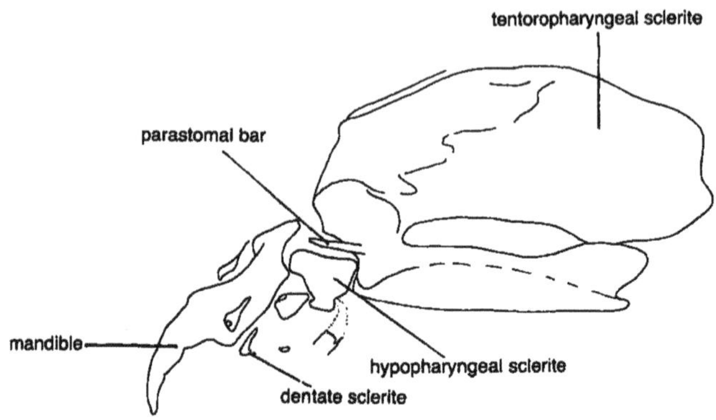

Fig. 32. Third instar larva. Mouthparts of *Eurychaeta palpalis* (Robineau-Desvoidy), left lateral view. (From Rognes, 1986c: fig. 27).

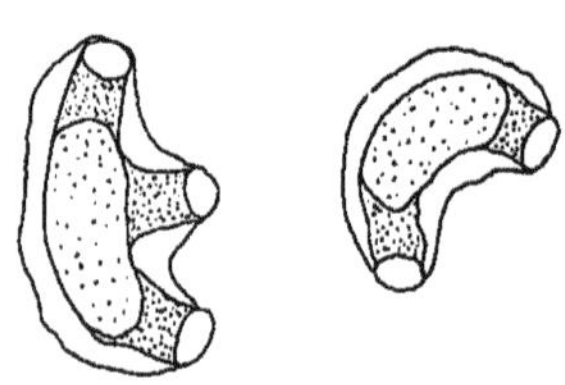

Fig. 33. Third instar larva. Anterior spiracles of *Trypocalliphora braueri* (Hendel). (From Rognes, 1985a: fig. 22.)

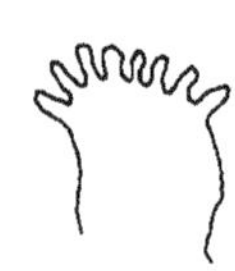

Fig. 35. Third instar larva. Anterior spiracle of *Lucilia ampullacea* (Villeneuve). (Redrawn from Kano & Sato, 1952: fig. 3).

Fig. 34. Third instar larva. Anterior spiracle of *Eurychaeta palpalis* (Robineau-Desvoidy). (From Rognes, 1986c: fig. 26).

Fig. 36. Puparium. Anterior spiracle of *Pollenia pediculata* Macquart. (From Rognes, 1987b: fig. 97).

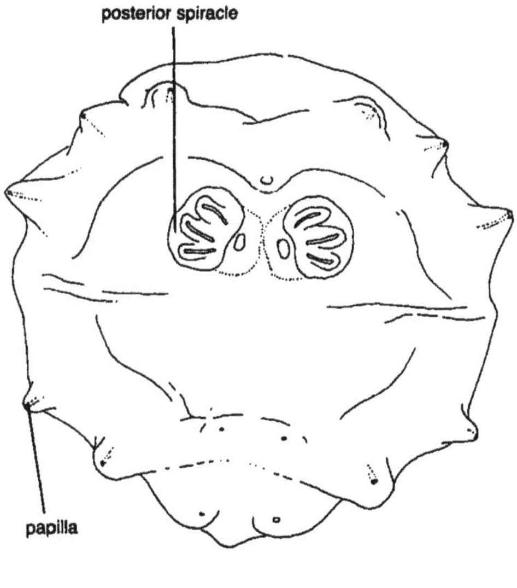

Fig. 37. Third instar larva. Posterior spiracular field of *Eurychaeta palpalis* (Robineau-Desvoidy). (From Rognes, 1986c: fig. 24).

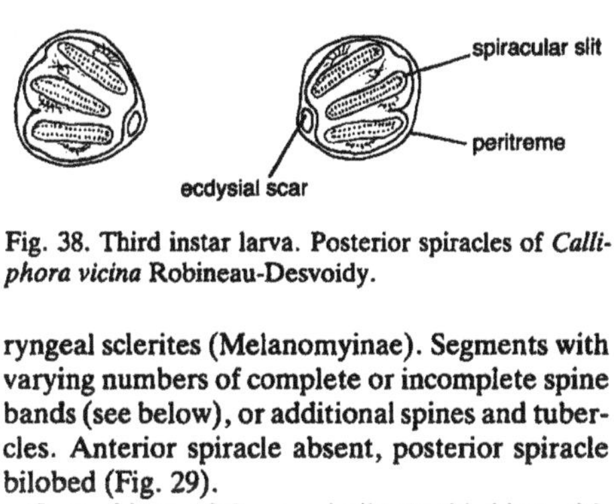

Fig. 38. Third instar larva. Posterior spiracles of *Calliphora vicina* Robineau-Desvoidy.

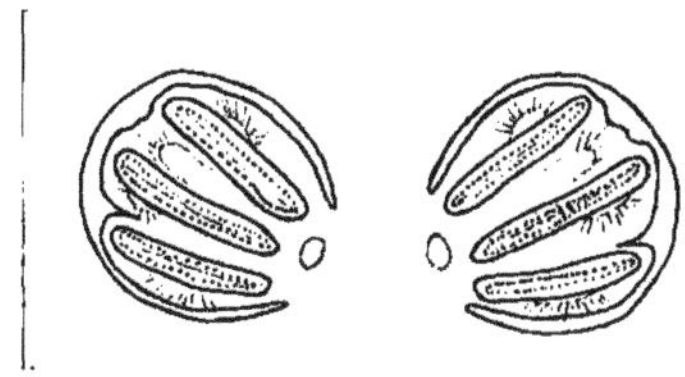

Fig. 39. Third instar larva. Posterior spiracles of *Protophormia terraenovae* (Robineau-Desvoidy). Scale = 0.5 mm.

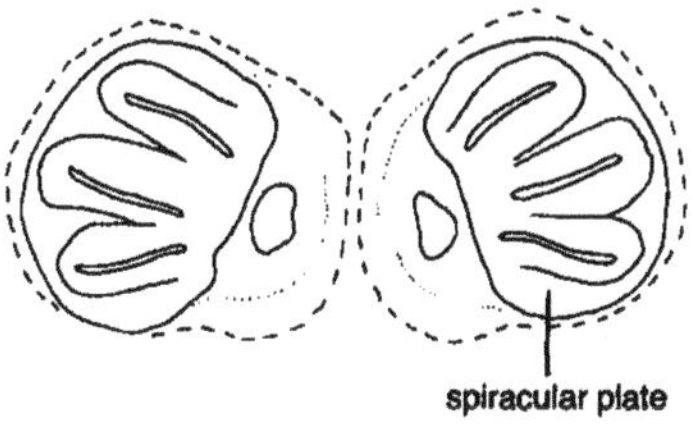

Fig. 40. Third instar larva. Posterior spiracles of *Eurychaeta palpalis* (Robineau-Desvoidy). (From Rognes, 1986c: fig. 25).

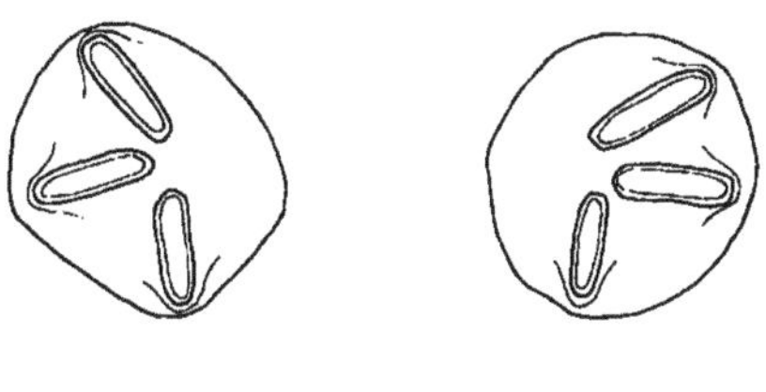

Fig. 41. Puparium. Posterior spiracles of *Pollenia angustigena* (Wainwright). (From Rognes, 1987b: fig. 89).

ryngeal sclerites (Melanomyinae). Segments with varying numbers of complete or incomplete spine bands (see below), or additional spines and tubercles. Anterior spiracle absent, posterior spiracle bilobed (Fig. 29).

Second instar 2-9 mm, similar to third but with weaker mouthparts and two slits in posterior spiracles.

Third instar 9-22 mm long. Most segments with more or less complete anterior bands of posteriorly-directed spines and posterior bands of anteriorly-directed spines (Fig. 30). Rarely, segments completely covered with spines (*Protocalliphora*), or most segments without spines (*Angioneura*, *Pollenia*). Anterior end in one genus with a crown of setae on the cephalic segment forming a kind of "sucker" through the centre of which the mouthparts protrude (*Protocalliphora*). Mouthparts well-developed. Mandibles often large, sometimes with a single accessory oral sclerite between them apically (Fig. 31). Tentoropharyngeal sclerite varying in shape (Figs. 31, 32), usually without an "incision" of weak sclerotisation in dorsal arm (except *Angioneura* and *Stomorhina*). Anterior spiracle variously shaped, with varying numbers of openings (Figs. 33-36). Posterior truncate end of maggot with a spiracular field surrounded by 7 pairs of papillae (Fig. 37), an additional pair of papillae present next to the anus. The spiracular field is usually flat, but sometimes it can be closed by a transverse folding and bringing together of dorsal and ventral halves. Posterior spiracles usually flush with surrounding area. Ecdysial scar median and more or less distinct. Peritreme complete (Fig. 38), incomplete (Fig. 39), or entire spiracular plate completely sclerotised (Fig. 40). 3 straight, more or less horizontal slits, angle between upper and lower slit much less than a right angle, occasionally lower slit vertical and angle more than a right angle (*Angioneura*, *Pollenia*) (Fig. 41).

Puparium reddish-brown to black, retaining most characters of the third instar larva. Puparial horns protruding to various degrees.

No taxonomic treatment of the larval stages or any key enabling larvae to be identified is given in this work, but references to descriptions are given at the appropriate place below.

Biology

The family Calliphoridae exhibits a rich diversity of life-cycles. A survey is given below of the different kinds encountered among Fennoscandian and Danish species. The biology of some genera (*Melanomya* and *Morinia*) is not known.

Necrophagy, coprophagy, saprophagy, myiasis

Primitively blowflies are oviparous and lay their eggs on the exposed dead bodies of various animals, especially vertebrates, irrespective of size. Being very numerous, their larvae play an extremely important part in natural communities as decomposers and scavengers. It has been estimated that the offspring of 2000 blowflies need 300 kg of meat in 3 days to mature (Schumann, 1965). In Fennoscandia and Denmark the genera *Calliphora*, *Cynomya*, *Phormia*, *Protophormia* and *Lucilia* belong to this group. *Calliphora vicina* may breed in exclusively faecal material (Zumpt, 1965). *Lucilia sericata* may breed in garbage of exclusively vegetable origin (Norris, 1965). The time spent in the different stages of development varies with species, temperature and other factors. At 27°C and 50% RH *Calliphora vicina* spends 24 hrs in the egg stage, 24 hrs as first instar larva, 20 hrs in the second instar, a little more than 7 days in the third instar (including prepupa) and about 11 days as pupa. This makes a total of about three weeks for the immature stages. When fully grown, the third instar larvae stop eating, migrate up to 6.5 m away from the breeding-site (usually a carcass), and settle under the ground surface. After a rest period, the larvae gradually evolve into white puparia (prepupae). Thereafter the skin becomes tanned and then changes into a brownish-black puparium, characteristic of most higher flies, within which true pupation occurs. After emergence the adults feed on carbohydrates at first and then later on protein, which is required for the maturation of the ova. After mating the females begin to oviposit. A single female *vicina* may lay up to 800 eggs during her life-time. Oviposition is completely inhibited if a layer of soil thicker than 2.5 cm covers the carcass (data from Kamal, 1958; Schumann, 1965; Nuorteva, 1977, 1987).

Blowflies of this group (especially *Calliphora*, *Lucilia*, *Protophormia*) are among the first insects to arrive on dead bodies, including humans, and are therefore of great significance in forensic medicine, an aspect of blowfly biology recently reviewed by Smith (1986). In Fennoscandia and Denmark the works by Nuorteva and coworkers are especially noteworthy (Nuorteva, Isokoski & Laiho, 1967; Nuorteva, Schumann, Isokoski & Laiho, 1974; Nuorteva, 1977, 1987).

Many species also feed on faeces, and indoors they may visit and oviposit on fresh or cooked meat, fish or dairy products. They are therefore of the greatest hygienic importance as potential vectors of bacteria, viruses, protozoans and helminths causing various enteric and other diseases (Schumann, 1963, 1965, 1971; Greenberg, 1971, 1973; Hall & Townsend, 1977). Blowflies have even been suspected of transmitting poliomyelitis (Nuorteva, 1959b, 1959c, 1959d, 1959e, 1960c, 1960d, 1961). Nuorteva's (1963) "synanthropic index" and Mihályi's (1967) "danger index" are methods of evaluating the hygienic importance of flies, including blowflies.

Blowflies have caused important losses for the dried-fish industry in Northern Norway (Soot-Ryen, 1925) and pose a permanent threat to all kinds of stored meat-products. Nielsen & Nielsen (1976) report cases of blowfly maggots having been found in vacuum-packed ham and have performed experiments on the viability, hatching and development of blowflies on such media under various conditions. It should be noted that direct access to the meat is not a condition for successful infestation as the female is able to force the ovipositor through the meshes of a fly net or through minute holes in plastic wrapping covering dishes containing meat and to drop eggs on to the substrate.

In cities, garbage may be an important source of

blowfly species. All offal must be stored in a way that ensures that it does not become infected with blowfly eggs or larvae, and garbage collection at regular intervals is necessary to maintain the blowfly population at minimal levels. Dumping of infested garbage on refuse depots results in the emergence of large numbers of blowflies which then invade gardens and houses in search of carbohydrates (flower nectar) (cf. Nuorteva, 1971a) and pose a hygienic threat to the neighbourhood, even if garbage is properly covered with soil and sand on the refuse depot to prevent oviposition on uninfested garbage by local blowflies.

Dipterous larvae may infest the bodies of living human or vertebrate animals and feed, at least for a certain period, on the host's dead or living tissue, liquid body-substances, or ingested food. This phenomenon is called myiasis and is an important aspect of the biology of the Calliphoridae. It has been reviewed by Zumpt (1965) for the Old World.

In Fennoscandia and Denmark a few blowflies are obligatory myiasis producers. *Lucilia bufonivora* oviposits on living amphibians and its larvae cause their eventual death (Lundbeck, 1927; Brumpt, 1934; Koskela, Itämies & Pasanen, 1974). The external blood-sucking larvae of *Protocalliphora* feed intermittently on the blood of nestling birds. The subcutaneous larvae of *Trypocalliphora* make permanent burrows in the bodies of nestling and juvenile birds. Hicks (1959; 1962; 1971) has given a bibliography on the occurrence of insects in birds' nests.

Facultative myiasis agents consist of normally free-living maggots (e.g. *Calliphora*, *Cynomya*, *Protophormia*, *Lucilia*) developing in carcasses or other decaying organic matter, which occasionally gain access to the living bodies of wild and domesticated animals through oviposition in wounds, lesions, natural body openings or wool, and there cause myiasis. Hence the importance of blowflies in veterinary medicine (Eichler, 1980; Soulsby, 1982).

As for birds, a case of wound-myiasis in the night jar (*Caprimulgus europaeus* L.) involving *Lucilia richardsi* and a case of myiasis in a living crane (*Grus grus* (L.)) involving a *Lucilia* species (probably *sericata*) have both been recorded from Finland (Nuorteva, 1959a; Itämies & Merilä, 1984). Regarding mammals, a case of myiasis in the hedgehog (*Erinaceus europaeus* L.) involving *Calliphora vicina*, *Lucilia illustris*, *L. ampullacea*, *L. caesar* and *Helicophagella melanura* (Sarcophagidae) has been reported in Denmark (Nielsen, Nielsen & Walhovd, 1978). In Finland a case of dermal myiasis in the hare (*Lepus timidus* L.) involving *Cynomya mortuorum* has been reported by Itämies & Koskela (1980). In Western Norway and Denmark sheep are not uncommonly attacked by *Lucilia caesar* and *L. illustris* (reviews by Cragg, 1950; Brinkmann, 1976; Nielsen, 1984a, 1984b). A single outbreak of sheep-strike was reported from southernmost Sweden during the period 1943-1949 (Cragg, 1950). No cases are known from Finland (Brinkmann, 1976). In sheep myiasis, other blowfly species (in Denmark *Calliphora vicina* and *C. vomitoria*) may oviposit in the lesions caused by the primary attack (secondary and tertiary strike) and if unattended the attacks are fatal within a short time. Sheep-strike is a serious problem in New Zealand, Australia, South Africa, England, Scotland and other sheep-rearing countries, and has recently been discussed by Zumpt (1965), Soulsby (1982) and Heath (1986).

A recent review of cases of human wound myiasis involving *Lucilia sericata* in Denmark has been given by Tikjøb & Haarløv (1985), and some references were also given by Nielsen, Nielsen & Walhovd (1978). A case has been reported from Denmark of intestinal myiasis in a two-year-old child involving *Lucilia sericata* (Haarløv, 1961). Cases have been reported from Finland involving intestinal myiasis in a baby caused by *Calliphora vicina* (Nuorteva & Auvinen, 1968) and wound myiasis in an old woman involving *Lucilia illustris* (Laitinen, Nuorteva & Renkonen, 1970). In Norway *Lucilia caesar* has been bred from a larva removed from a cancerous wound in the external auditory meatus (Reidar Mehl, pers.comm., identification by K.R.).

Norris (1965) has reviewed the bionomics of the necrophagous, coprophagous and myiasis-producing blowfly species. Their ecology in Finland has been studied over several years by Nuorteva and coworkers mainly through trapping experiments. The results have been described in numerous papers listed in the bibliography. No comparable studies exist for Denmark, Norway or Sweden.

Association with gastropods

So far as their biology is known, members of the Melanomyinae appear to be parasitic upon living snails. They are oviparous or viviparous. *Melinda* species place their eggs in the pallial cavity of snails. The larva lives as an internal parasite, ultimately causing the death of the snail and devouring it. Members of Helicoboscinae are viviparous and place a single very large first-instar larva on

3(2) Costa hairy below as far as junction with Sc, bare below beyond this point (Fig. 248); subcostal sclerite setulose; arista always with hairs on lower side, usually rather long (Figs. 242-246); at most a very narrow keel between basal antennal segments; lower part of face not protruding strongly below vibrissae; palpi yellow or brownish; posterior spiracle with fine setae along lower margin (sometimes inconspicuous in *Trypocalliphora*) and coxopleural streak absent (Figs. 247, 259-263, 312-313, 339); abdomen unicolorous metallic blue, green, brassy-green, blackish-green or blackish-blue

Chrysomyinae (p. 103)

– Costa hairy below as far as junction with humeral crossvein, bare below beyond this point (Fig. 673); subcostal sclerite with microscopic pubescence only; arista bare on lower side (Fig. 670); a broad double warty protuberance between basal antennal segments; lower part of face shining black and protruding strongly below vibrissae (Fig. 670); palpi black; posterior spiracle without fine setae along lower margin and coxopleural streak present (Fig. 672); abdomen with a yellow or white pattern on a black background (Fig. 674)

Rhiniinae (p. 241)

4(2) Lower calypter with hairs above (occasionally bare in *Bellardia pubicornis*) (Fig. 7); costa always bare below beyond junction with R_1 (Figs. 66, 170, 218, 230); lower parts of anterior part of anepimeron with bundle of hairs (often bare in *Bellardia pubicornis*) (Fig. 196); coxopleural streak absent (Figs. 48-54, 132-138, 217, 229); parafacialia with hairs, at least in upper part; palpi yellow; abdomen blue, green or olive-green metallic in colour, rarely (*Bellardia pubicornis*) non-metallic black with a weak bronze sheen through the dusting

Calliphorinae (p. 32)

– Lower calypter bare above (Fig. 397); lower parts of anterior part of anepimeron almost always quite bare (Fig. 6)

5

5(4) Coxopleural streak absent (Figs. 481-486); body non-metallic or dark blue metallic; palpi dark; costa hairy below beyond junction with R_1 (Figs. 491, 503, 514, 525, 536); cell r_{4+5} opening very close to apex of wing in non-metallic species (Figs. 491, 503, 514, 525); prementum short (if prementum long and slender and palpi yellow, see *Bellardia pubicornis* in Calliphorinae which may key out here)

Melanomyinae (p. 184)

– Coxopleural streak present (absent in *Lucilia ampullacea*, which is green metallic); palpi yellow or dark

6

6(5) Bright green metallic; surface immediately above hind coxae with small hairs (Figs. 370, 372); parafacialia bare; prosternum, proepisternal depression and postalar wall hairy; suprasquamal ridge with a distinct posterior sclerite bearing a tuft of setae (Fig. 397)

Luciliinae (p. 147)

– Non-metallic; surface immediately above hind coxae bare; parafacialia hairy or bare; prosternum and proepisternal depression bare; suprasquamal ridge without a posterior sclerite bearing a tuft of setae

7

7(6) Postalar wall hairy (in *Morinia* with rather few stiffish hairs); posterior thoracic spiracle usually rather large, with posterior lappet somewhat larger than anterior one (Figs. 560-569); lower calypter usually large, with inner edge converging backwards with longitudinal axis of fly, except in *Morinia* which uniquely lacks hairs on the node at base of R_{4+5} on both sides of the wing (Fig. 558); often long, yellow crinkly hairs present in addition to the normal and black ground vestiture (*Pollenia*)

Polleniinae (p. 206)

– Postalar wall bare; posterior thoracic spiracle small and subcircular, anterior and posterior lappets subequal; lower calypter small, tongue-like or subcircular, inner edge diverging backwards from the longitudinal axis of fly; node at base of R_{4+5} with hairs on both sides of wing

Rhinophorinae (not treated in this volume)

Subfamily Calliphorinae

The subfamily Calliphorinae in the sense adopted here is probably a monophyletic group supported by the following ground-plan apomorphies:

(1) Outer *ph* seta external to a line through *prs* seta.
(2) Lower half of anterior part of anepimeron usually with numerous hairs (Fig. 196).
(3) Coxopleural streak absent.
(4) Costa bare below beyond junction with R_1.
(5) Lower calypter hairy above.
(6) ST8 in ovipositor with bifid or bilobed apex.
(7) Accessory oral sclerite present in third instar larvae (except *Bellardia* and *Onesia*).

The following genera probably belong to this group: *Abago* Grunin, *Bellardia* Robineau-Desvoidy, *Calliphora* Robineau-Desvoidy, *Cyanus* Hall, *Cynomya* Robineau-Desvoidy, *Cynomyomima* Rodendorf, *Onesia* Robineau-Desvoidy, *Onesiomima* Rodendorf, *Polleniopsis* Townsend, *Ptilonesia* Bezzi, *Tainanina* Villeneuve, *Xenocalliphora* Malloch and *Pericallimyia* Villeneuve (cf. Zumpt, 1956a; Kurahashi, 1971, 1972b, 1978). The inclusion of several genera is only based on the hairiness of the lower calypter, and confirmation from study of the ovipositor and other body-parts is needed. *Tricycleopsis* Villeneuve has a few, inconspicuous, easily abraded hairs on the lower calypter (Senior-White, Aubertin & Smart, 1940; Kurahashi, 1970, 1972a), and might be a member of the Calliphorinae. However, Kurahashi (1970) treated it as closely related to *Melinda* Robineau-Desvoidy and included it within his "*Melinda*-group". I follow Kurahashi and regard it as a member of the Melanomyinae, even though its ovipositor is not known.

A probably monophyletic subgroup, the "*Onesia*-group" of Kurahashi, defined by the presence of a short ovipositor and a uterus with a pair of lateral incubatory pouches, is formed by the genera *Bellardia*, *Onesia*, *Polleniopsis* and *Tainanina* (Kurahashi, 1970, 1972b, 1978). The two former genera have microtrichiose cerci and epiproct in the ovipositor, fully microtrichiose intersegmental and pleural membranes and T6 strongly invaded by microtrichiae. The distribution of microtrichiae is not known in the others. All are viviparous so far as their biology is known.

Herting (1957) first pointed out the importance of the bifid apex of ST8.

In Fennoscandia and Denmark large to medium sized flies with black ground-colour and metallic blue, green, olive-green or bronze sheen. Head profiles as in Figs. 42-47. Eyes bare. Male frons narrow, without proclinate orbitals, with or without a single lateroclinate orbital seta. Female frons with 2-4 proclinate orbitals and 1 lateroclinate orbital seta. Arista long plumose to short pubescent. Facial carina absent or only slightly developed. Parafacialia hairy at least in upper half; no strong setae in lower half. Palpi yellow. Proepisternal depression, prosternum, metasternal area and postalar wall hairy. Hindmost *prst acr* seta far in front of suture. 3+1 *h*. Outer *ph* seta, when present, outside a line through *prs* seta parallel with longitudinal axis of fly. 2 *npl*. 2+1 *kepst* setae. Lower half of anterior part of anepimeron usually with numerous hairs (Fig. 196). Katatergite (supraspiracular convexity) pubescent only, without long hairs. Anatergite with short hairs below lower calypter. Coxopleural streak absent. No row of fine setae along the anteroventral edge of the metathoracic (posterior) spiracle. Area above hind coxa (metakatepisternum) bare. Tympanic pit hairy. Posterior part of suprasquamal ridge bare. Stem-vein bare above. Costa bare below beyond apex of R_1 (not in the East Palaearctic subgenus *Onesioides* Schumann of *Onesia* Robineau-Desvoidy). Subcostal sclerite pubescent, without short setae. R_{4+5} bare beyond junction with R_{2+3}. Cell r_{4+5} most often open, cell opening well in front of apex of wing (Figs. 66, 170, 218, 230). Lower calypter white, infuscated brown or very dark, posteriorly broad, inner edge converging backwards with longitudinal axis of fly, almost always hairy above. Mid tibia with 1 *v* seta, with a few exceptions. Hind coxa bare on posterior surface. Lateral margins of ST2 overlapping ventral margins of T2. Aedeagus with long, posteriorly serrate hypophallus, and well-developed or reduced paraphallus. Acrophallus with denticles. ST8 of ovipositor apically bilobed. Cerci of ovipositor elongate-oval (*Calliphora*, *Cynomya*) or rather broad and truncated apically (*Bellardia*, *Onesia*). Epiproct and cerci shining black without microtrichiae (*Calliphora*, *Cynomya*), or microtrichiose (*Bellardia*, *Onesia*). Oviparous or viviparous. Larvae saprophagous, or parasites or predators of earthworms.

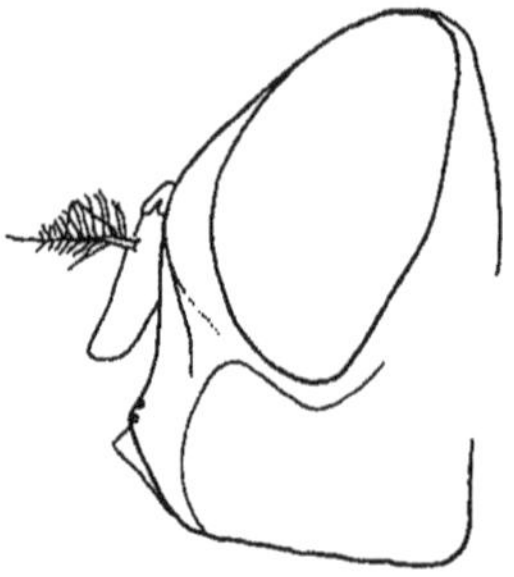

Fig. 42. *Calliphora uralensis* (Villeneuve), head profile, male.

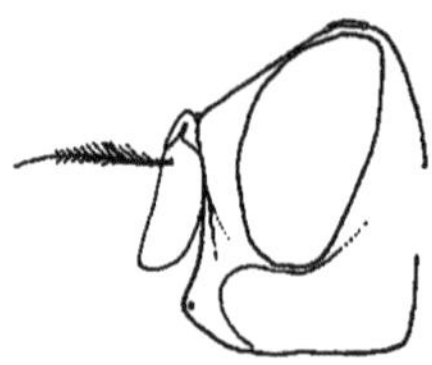

Fig. 43. *Calliphora genarum* (Zetterstedt), head profile, male.

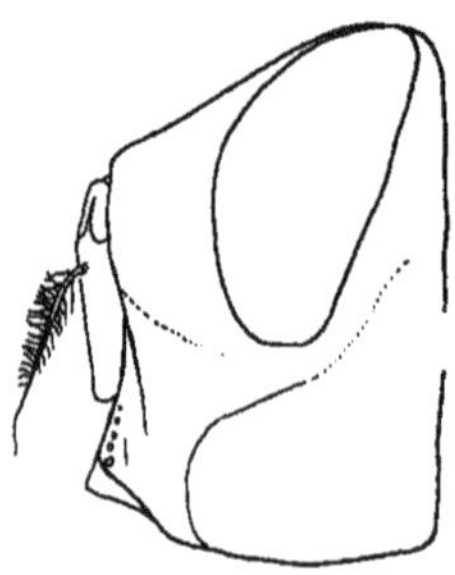

Fig. 44. *Onesia floralis* (Robineau-Desvoidy), head profile, male.

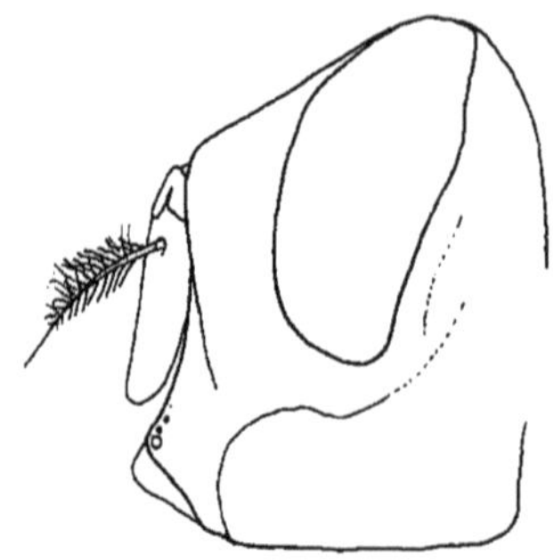

Fig. 45. *Cynomya mortuorum* (Linnaeus), head profile, male.

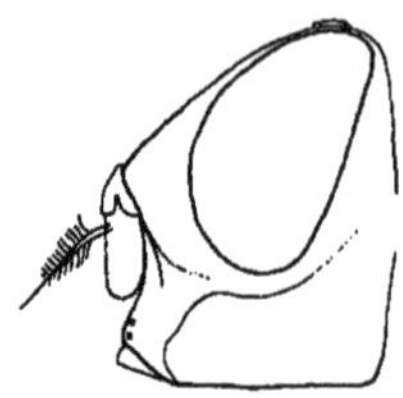

Fig. 46. *Bellardia vulgaris* (Robineau-Desvoidy), head profile, male.

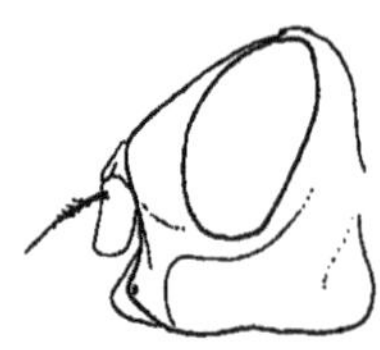

Fig. 47. *Bellardia pubicornis* (Zetterstedt), head profile, male.

Widespread, with representatives in all faunal regions. Four genera in Fennoscandia and Denmark.

Key to genera of Calliphorinae

1 1 *post acr*; abdomen shining metallic bluish green, practically without dusting; anterior half of fronto-orbital plate and frontal vitta, whole of facial plate and parafacial, and anterior two thirds of gena with bright yellow-orange ground-colour and golden dusting; lower calypter white with black hairs on most of disc

Cynomya Robineau-Desvoidy (p. 92)

– 3 *post acr*; abdomen most often conspicuously dusted

2

2(1) 0+3 *ia* (anterior *post ia* sometimes weak); costa setulose below to junction with R_1, bare below beyond this point (Fig. 230); first flagellomere much more than 2x as long as pedicel (Fig. 44); abdomen olive-green, with a shifting pattern; lower calypter pale with dark hairs except on outer and posterior third or half of disc

Onesia Robineau-Desvoidy (p. 97)

– 1+2 *ia*

3

3(2) First flagellomere much more than 2 times as long as pedicel, often very large (Figs. 42, 43); costa setulose below to junction with R_1, bare below beyond this point (Fig. 170); abdomen blue, with or without a shifting pattern; lower calypter pale or dark, with hairs on most of disc

Calliphora Robineau-Desvoidy (p. 59)

– First flagellomere at most 2 times as long as pedicel (Figs. 46, 47); costa usually setulose below only to junction with Sc, bare below beyond this point (Fig. 66) (except often in *Bellardia pandia* and rarely in *Bellardia vulgaris*); abdomen olive-green, with a more or less shifting pattern, bright green almost without dust, or almost black with faint bronze sheen; lower calypter pale with hairs usually confined to median and anterior half (occcasionally bare or with only 1-2 odd hairs in *Bellardia pubicornis*)

Bellardia Robineau-Desvoidy (p. 34)

Genus ***Bellardia*** Robineau-Desvoidy, 1863

Bellardia Robineau-Desvoidy, 1863b: 548.
Type-species: *Bellardia vernalis* Robineau-Desvoidy, 1863b: 548, by monotypy; = *Tachina obsoleta* Meigen, 1824.
Maravigna Lioy, 1864: 891.
Type-species: *Onesia clausa* Macquart, 1835: 234, by monotypy; = *Tachina obsoleta* Meigen, 1824.
Pseudonesia Villeneuve, 1920a: 296.
Type-species: *Musca puberula* Zetterstedt, 1838: 654, by monotypy and original designation; = *Tachina pubicornis* Zetterstedt, 1838. SYN.N.
Ambodicria Enderlein, 1933: 124.
Type-species: *Onesia polita* Mik, 1884, by original designation.
Miaspia Enderlein, 1933: 125.
Type-species: *Miaspia latigena* Enderlein, 1933: 126, by original designation; = *Onesia brevistylata* Villeneuve, 1926.
Melindopsis Kurahashi, 1964: 484.
Type-species: *Polleniopsis menechma* Séguy, 1934: 12, by original designation. SYN.N.
Feideria Lehrer, 1970: 15.
Type-species: *Polleniopsis menechma* Séguy, 1934: 12, by original designation. SYN.N

In Fennoscandia and Denmark a distinct genus of medium-sized to large flies with black groundcolour, silvery pollinosity and olive-green or weakly bronze metallic abdominal colour. Palpi yellow or (more rarely) darkened. Eyes bare. First flagellomere at most 2 times as long as pedicel. Arista bare in distal third, hairing of variable length, never very long, sometimes very short, especially on underside. Parafacialia with numerous black hairs. Occiput with mostly black hairs, but some pale hairs always present in lower part. Basicosta dark brown to black. Costa usually bare below between tips of Sc and R_1 (as in Fig. 66). Subcostal sclerite bare, without short black setulae. Cell r_{4+5} usually open, sometimes closed. 1-2+3 *acr*, 2-3+3 *dc*, 0-1+2 *ia*, 3 *sa*, 2+0-1 *ph*, 3 *h*, 3-4 *sc*. Hairing on lower calypter usually absent from at least outer and posterior half. Posterior spiracle with lappets of about equal size (Figs. 48-54). Mid tibiae with 2-3 *ad* and 1 *v* setae. Mid femora without *a* preapical seta in both sexes. Hind tibiae with 2-3 *av* setae and *ad*, *d*, and *pd* preapical setae of about same size. Abdomen most often olive-green with conspicuous dusting, with a few exceptions. Aedeagus with massive hook-shaped paraphalli and well-developed mesohypophallic rod. Acrophallus short. Ventral surface of hypophallus and acrophallus covered with spines. Spermathecae elongate-oval, inner wall papillate (as in Fig. 63). Ovipositor with shortened sclerites. All intersegmental and pleural membranes fully microtrichiose. T6 strongly invaded by microtrichiae. ST8 short, completely invaded by microtrichiae and consisting almost solely of the apical lobes. Female epiproct and cerci usually completely covered with microtrichiae. Uterus short with large lateral incubatory pouches.

All species are viviparous and numerous first instar larvae can be recovered from the double-sac heart-shaped uterus of the females. Larvae are

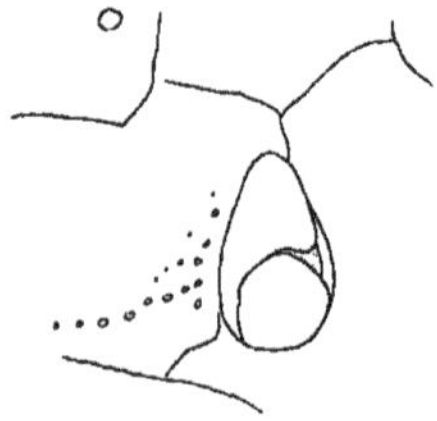

Fig. 48. *Bellardia viarum* (Robineau-Desvoidy), posterior spiracle.

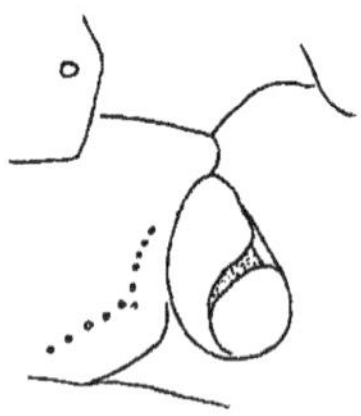

Fig. 49. *Bellardia vulgaris* (Robineau-Desvoidy), posterior spiracle.

Fig. 50. *Bellardia bayeri* (Jacentkovský), posterior spiracle.

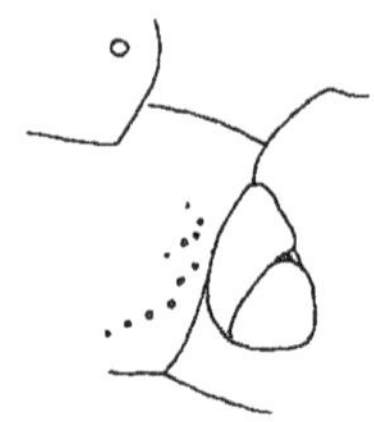

Fig. 51. *Bellardia pandia* (Walker), posterior spiracle.

Fig. 52. *Bellardia pubicornis* (Zetterstedt), posterior spiracle.

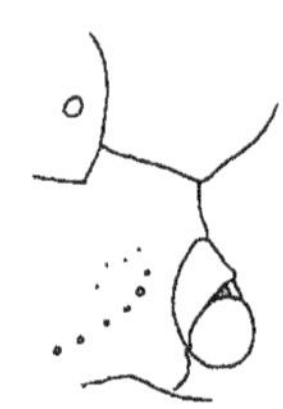

Fig. 53. *Bellardia stricta* (Villeneuve), posterior spiracle.

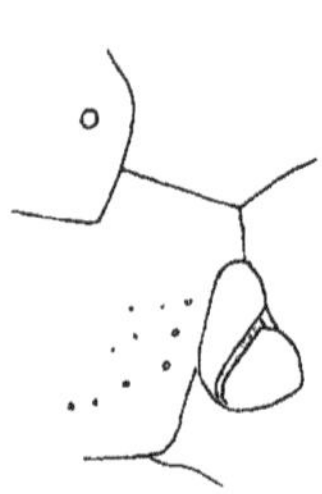

Fig. 54. *Bellardia vespillo* (Fabricius), posterior spiracle.

apparently parasites or predators of earth worms. Direct evidence for this was first given by Keilin (1915, 1919, as "*Onesia sepulchralis*", larva "A" - first instar, larva "B" - third instar). He found first instar larvae within the vesicula seminalis of "*Allolobophora foetida* Eisen, *A. caliginosa* Savigny and *A. chlorotica* Savigny". These he identified with first instar larvae which he had removed from the abdomen of adult females of "*Onesia sepulchralis*" from the Cambridge University collections. As these adults had 2 *post ia* setae and as "*Onesia sepulchralis*" (= *O. floralis*) does not occur in Great Britain (Emden, 1954), the larvae must belong to a species of *Bellardia*, of which there are three British species. According to Keilin (1915, 1919) the first instar larva subsequently migrates to the prostomium of the host, where its posterior end is able to gain access to the air. In this position the larva passes through two moults, feeding upon and ultimately destroying the host more or less completely. Thompson (1921a) and Schumann (1974b) also give information on the first instar larvae of *Bellardia*. According to reared material available, *Eisenia foetida* (Savigny) serves as the host for *B. bayeri* (Jacentkovský). Overwintering occurs in the third instar.

The genus is widely distributed in the Palaearctic Region. Schumann (1986) catalogued 29 species; 2 species described by Chen, Fan & Cui (1982) from China were not included. More species probably await description. A few of the Palaearctic species also occur in the eastern U.S.A. Shewell (1987) regards them as introduced and parasitic upon an introduced European earthworm. The problem of regarding *Bellardia* and *Pollenia* species as being introduced into North America together with their earthworm hosts is discussed below under *Pollenia* (p. 214 ff.).

The generic limits are poorly understood. The Australian *Onesia accepta* (Malloch) appears to belong to *Bellardia*, to judge from the ♂ genitalia figured by Malloch (1927). The larval stages and the earthworm parasitic habit described by Fuller (1933) are fully compatible with this. The "*Onesia*" species recorded by James & Kurahashi (1976) from the Bismarck Islands also appear to belong to *Bellardia*.

Species of *Bellardia* are very similar in external appearance and examination of the genitalia is essential for reliable identification of both sexes.

7 species in Fennoscandia and Denmark.

Note on synonymy. Of *Polleniopsis menechma* Séguy, type-species of *Melindopsis* and *Feideria*, I have examined the holotype ♂, including the accompanying genital preparation (Séguy slide 304), and two paratype ♀, including a preparation (Séguy slide 303) of numerous first instar larvae, all in MNHN. It is definitely a species of *Bellardia*. The figures of the ovipositor published by Kano & Shinonaga (1968) support this assignment.

Key to species of *Bellardia*

1 Abdomen black or bronze, not green, hardly metallic, with conspicuous dusting and shifting pattern; lower facial margin and lower anterior corner of head strongly protruding (Fig. 47); lower calypter bare or with a few hairs; arista (including hairing) less than half as wide as first flagellomere (Fig. 47), with short hairs above but hairs almost absent below; ♂: terminalia as in Figs. 100-106; ♀: ovipositor (Figs. 108, 109) with complete row of about 20 marginal setae on T8; T8 large, longer than T6; ST7 shorter than broad

5. *pubicornis* (Zetterstedt)

– Abdomen green or olive-green metallic, more or less dusted; lower facial margin less strongly projecting; ♀: no or only very few marginal setae on T8, which is much smaller

2

2(1) Abdomen shining green, with very weak dusting, tessellations absent or very weak; abdominal ground vestiture rather sparse, especially on T4 and T5; parafacials evenly and densely dusted white (♂) or yellow (♀), with rather few hairs which are almost absent along eye-margin; ♂: numerous discal setae on T4; terminalia as in Figs. 111-117; ♀: fewer discal setae on T4; ovipositor (Figs. 119-120) with T6 sclerotised in two widely separate halves, each half short and irregular with an inwardly-pointing process which bears 2-3 strong and a few shorter marginal setae; epiproct, which is often divided, shorter than broad, with 2 strong and at most a few short additional setae; ST6 shorter than broad; ST7 shorter than broad

6. *stricta* (Villeneuve)

– Abdomen most often conspicuously dusted with a more or less pronounced shifting pattern; parafacials less densely dusted, with a weak shifting pattern according to view; ♀: ST7 longer than broad

3

3(2) Upper part of parafacials with a dark brown spot which does not disappear when parafacial is viewed tangentially from above; lower calypter heavily and evenly infuscated, especially in ♂; ♂: terminalia very small, as in Figs. 79-84; ♀: ovipositor (Figs. 86-87) with weak sclerotisation in middle of T6, epiproct divided, no sclerotisation mid-ventrally in front of ST8 lobes.

3. *bayeri* (Jacentkovský)

– No dark spots present on upper parts of parafacialia, where any darkenings disappear when parafacial is viewed tangentially from above; lower calypter pale

4

4(3) Arista (including hairing) about half as wide as first flagellomere, the hairs short above and almost absent below; palpi dark; ♂: terminalia as in Figs. 122-128; surstyli shorter than cerci, rather strongly curved, with an external groove present all the way to tip; ♀: ovipositor (Figs. 130-131) with numerous discal setae on T6, and also on its anterior half; T6 with no line of weak sclerotisation mid-dorsally; epiproct shorter than broad, with 2 strong and at most 2-3 additional small setae; cerci devoid of microtrichiae in basal half, thus appearing glossy black basally and dusted apically in dried specimens; ST6 as long as broad

7. *vespillo* (Fabricius)

– Arista plumose, with long hairs above and below; palpi usually yellow; ♀: cerci dusted all over

5

5(4) Front tibia almost always with 2 *pv* setae; costa often with hairs below between the junction with Sc and R_1; ♂: terminalia as in Figs. 89-95, cerci in posterior view heart-shaped basally, surstyli strongly curved, depression on external side not reaching apex; ♀: ovipositor (Figs. 97-98) with numerous discal setae on T6 but these are absent from the anterior half, no line of weak sclerotisation mid-dorsally; epiproct shorter than broad, with two strong and numerous additional medium-sized setae; ST6 1.5-2.0 times as long as broad

4. *pandia* (Walker)

– Front tibia usually with only 1 *pv* seta; costa almost always bare below between the junction with Sc and R_1; ♀: ST6 as long as or slightly longer than broad

6

6(5) T4 usually without discal setae; wing-veins pale; ♂: terminalia as in Figs. 68-74; surstyli longer than cerci, weakly curved in profile, external groove not reaching apex; ventral epandrial corner slightly less than a right angle (arrow in Fig. 68); ♀: ovipositor (Figs. 76-77) with T6 shaped as a narrow band, most often with a narrow zone of weak sclerotisation mid-dorsally, giving it an angularly bent appearance in dried specimens; T6 with marginal setae along whole of hind edge, setae absent from anterior parts of disc; epiproct triangular, divided along mid-line and as long as broad; ST8 with a strong mid-ventral sclerotisation in front of lobes; cerci and hypoproct rather large; in dried specimens ovipositor conspicuously protruding behind T5

2. *vulgaris* (Robineau-Desvoidy)

– T4 usually with discal setae; wing-veins very dark, almost black; ♂: terminalia as in Figs. 56-62; ventral epandrial corner much more than a right angle (arrow in Fig. 56); ♀: ovipositor (Figs. 64-65) with T6 sclerotised as two widely separated halves, each with numerous marginal setae and each broadest medially; epiproct shorter than broad, undivided; ST8 without conspicuous mid-ventral sclerotisation in front of lobes; cerci and hypoproct smaller; in dried specimens ovipositor almost concealed between adjoining ventral edges of T5

1. *viarum* (Robineau-Desvoidy)

1. ***Bellardia viarum*** (Robineau-Desvoidy, 1830), COMB.N.

Figs 48, 55-65.

Musca pusilla Meigen, 1826: 71. Junior primary homonym of *Musca pusilla* Gmelin, 1790: 2879 (Syrphidae). SYN.N.
Onesia viarum Robineau-Desvoidy, 1830: 367.
Calliphora aculeata Pandellé, 1896: 210. SYN.N.
Onesia krameri Mueller, 1922: 60. SYN.N.
Xerophilophaga zetterstedti Enderlein, 1933: 123. SYN.N.
Onesia longelobata Jacentkovský, 1941b: 6 (Czech), 57 (German). SYN.N.

♂♀. A rather dark species, similar to *vulgaris* but more hirsute, and abdomen less pollinose. Lower calypter infuscated. Wings very dark. Costa always bare below between tips of Sc and R_1. Front tibia with 1 *pv* seta. T4 with numerous long discal setae.

♂. Frons 0.075-0.091 times head-width (n=5). Frontal vitta slightly narrower than each fronto-orbital plate. Ventral epandrial corner blunt, much more than a right angle (arrow in Fig. 56). Cerci straight, tapering evenly; surstyli long and rather narrow, weakly bent, usually diverging from cerci, especially in dried specimens (Figs. 56, 57).

♀. Frons 0.291-0.331 times head-width (n=5). Ovipositor usually concealed between adjoining edges of T5 in dried specimens. T6 sclerotised in two widely separated halves, each with numerous

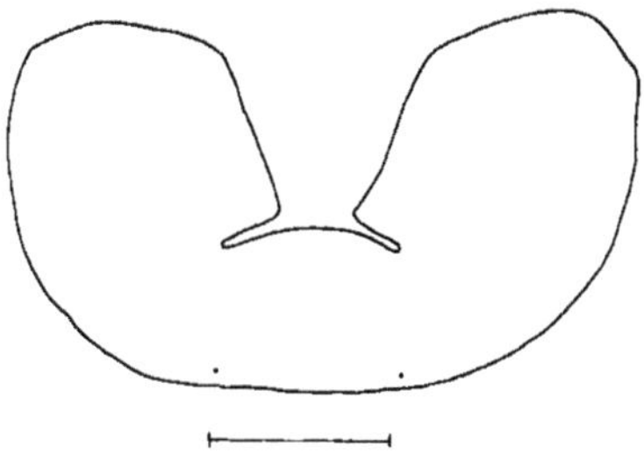

Fig. 55. *Bellardia viarum* (Robineau-Desvoidy), male. ST5. Scale = 0.4 mm.

Fig. 56. *Bellardia viarum* (Robineau-Desvoidy), male. Epandrium, cerci, surstylus and bacilliform sclerite, left lateral view. Scale = 0.2 mm.

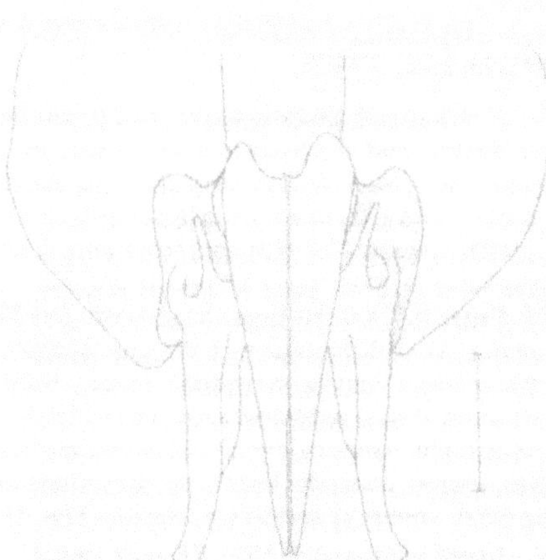

Fig. 57. *Bellardia viarum* (Robineau-Desvoidy), male. Cerci and surstyli, posterior view. Scale = 0.2 mm.

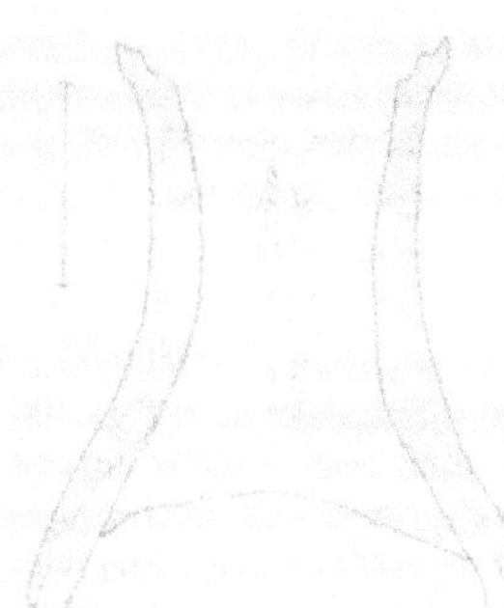

Fig. 58. *Bellardia viarum* (Robineau-Desvoidy), male. Bacilliform sclerites and membrane between them. Scale = 0.2 mm.

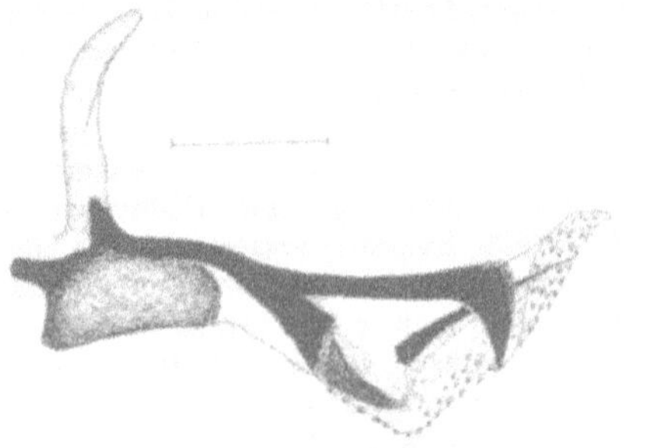

Fig. 59. ***Bellardia viarum*** **(Robineau-Desvoidy), male. Aedeagus, left lateral view. Scale = 0.2 mm.**

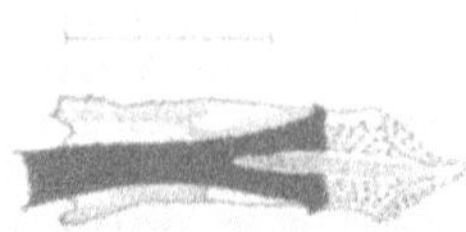

Fig. 60. ***Bellardia viarum*** **(Robineau-Desvoidy), male. Distiphallus, dorsal view. Scale = 0.2 mm.**

Fig. 61. ***Bellardia viarum*** **(Robineau-Desvoidy), male. Ejaculatory sclerite. Scale = 0.1 mm.**

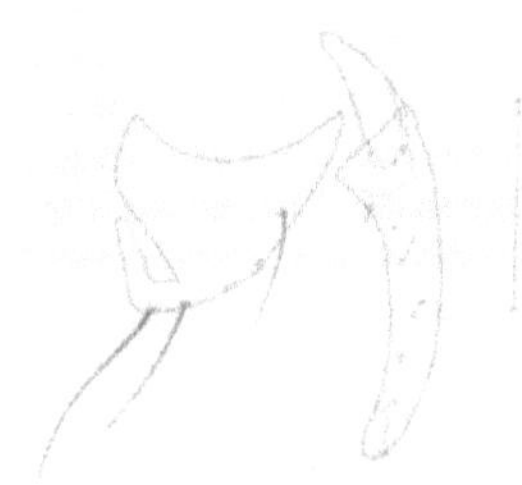

Fig. 62. ***Bellardia viarum*** **(Robineau-Desvoidy), male. Pre- and postgonites. Scale = 0.2 mm.**

Fig. 63. ***Bellardia viarum*** **(Robineau-Desvoidy), female. Spermathecae. Inset: papillate inside wall. Scale = 0.1 mm. G.pr. 217.**

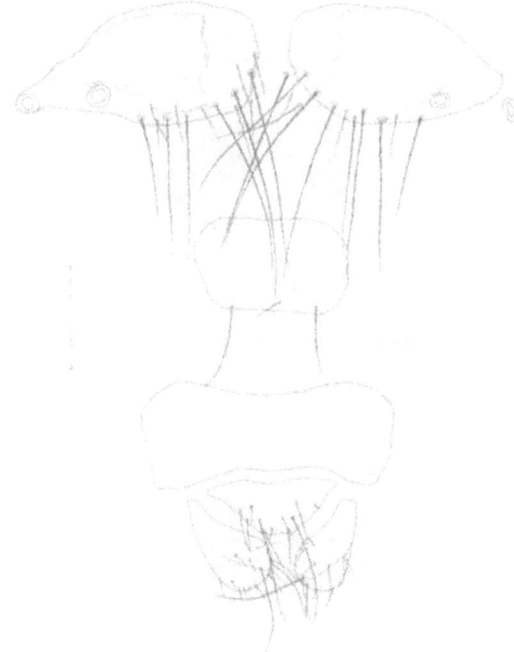

Fig. 64. *Bellardia viarum* (Robineau-Desvoidy), female. Ovipositor, dorsal sclerites. Stipple indicates extent of microtrichiae. Scale = 0.2 mm. G.pr. 217.

marginal setae, and each broadest medially. Epiproct shorter than broad, undivided, with numerous setae. Cerci wholly covered with microtrichiae. ST6 and ST7 slightly longer than broad. [9 ovipositor slides examined.]

Length. 5-9 mm.

Distribution. Common and widely distributed in Fennoscandia and Denmark. North to TRY and TRI (Norway). Ascends to the subalpine birch-forest (750 m a.s.l.) in South Norway. - All over the Palaearctic Region (Schumann, 1986).

Biology. Larval hosts unknown. Mouthparts of the first instar larva have been figured by Schumann (1974b). However, his figure of the third instar larval mouthparts (Schumann, 1974b: fig. 21), which was based on a specimen in coll. Lundbeck (Schumann, 1974b: 275), is actually of *bayeri* (see below under that species). The third instar larva of *viarum* is thus still unknown. Adults have been collected from April to October, with a maximum in June, July and August.

Nomenclature. *Musca pusilla* was described from ♂♀ syntypes. A ♂ was figured on Meigen Plate 119, fig. 9 (Morge, 1976b). In Meigen's col-

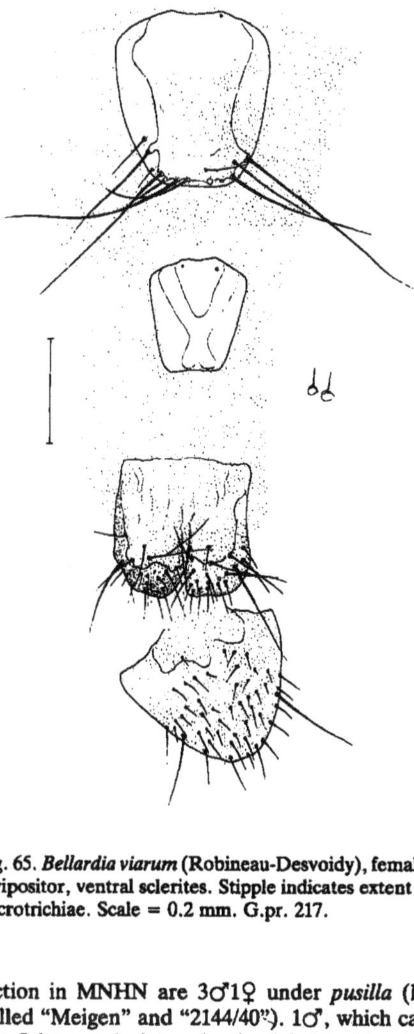

Fig. 65. *Bellardia viarum* (Robineau-Desvoidy), female. Ovipositor, ventral sclerites. Stipple indicates extent of microtrichiae. Scale = 0.2 mm. G.pr. 217.

lection in MNHN are 3♂1♀ under *pusilla* (labelled "Meigen" and "2144/40"). 1♂, which carries Schumann's determination label and has the genitalia visible, is the present species; 1♂ = *Melinda viridicyanea* Robineau-Desvoidy (*cognata* sensu Schumann); 1♂1♀ are *Bellardia* with concealed genitalia. To fix the identity of the name *pusilla* in the current sense (Schumann, 1974b), the ♂ with Schumann's determination label should be designated as lectotype. *Onesia viarum* was described by Robineau-Desvoidy (1830) as the fourth species under *Onesia*, and reported as "très-commune au printemps". It is likely to be a species of *Bellardia*. No specimens are left in Robineau-Desvoidy's collection in MNHN, and the name was listed as a nomen dubium by Schumann (1986: 38). In the General collection in MNHN under *Onesia* are two males under *viarum*. One is labelled "190" and "viarum" in Macquart's hand. It has lost both front legs and is rather mouldy but the genitalia are partly visible. It is the present species. A second male without labels = *Bellardia vespillo* (Fabricius). Séguy (1928a: 129, footnote) suggests that the labelled specimen was given by Robineau-Desvoidy to Macquart. I believe that neither of these specimens is a syntype. In the interests of nomenclatural stability and in accordance with Recommendation 75E of the ICZN I have brought forward the unused name *viarum* of Robineau-Desvoidy and here designate a recent specimen in good condition as neotype to fix its identity. It is labelled "MUSEUM PARIS / Fontenay-aux- / Roses (Seine) / Coll. E. SÉGUY 1919" and the genitalia are visible. I have given it a neotype label and placed it under *Onesia* in box 12 of Robineau-Desvoidy's collection in MNHN. In Pandellé's collection in MNHN under No. 2668 are 2♂3♀ under *Calliphora aculeata*. 2♂ = the present species, but the females cannot be identified with certainty unless dissected. I have labelled and here designate a male labelled "♂ / 2134" in Pandellé's hand and carrying Schumann's determination label as lectotype, the remaining specimens have been labelled as paralectotypes. *X. zetterstedti* Enderlein was based on 1♂1♀ syntypes (Enderlein, 1933) (not examined by me). The ♂ = *viarum* (*aculeata*), the ♀ = *pandia* Walker according to Schumann (1974b: 253, 277). Schumann (1986) established the synonymy *zetterstedti* = *aculeata*. However, he seems not have designated a lectotype for *zetterstedti*. The names listed in the synonymy of *viarum* are already established synonyms of the present species, even if formally new synonyms of the name *viarum*.

2. ***Bellardia vulgaris*** (Robineau-Desvoidy, 1830), COMB.N.

Figs 27, 28, 46, 49, 66-77.

Musca agilis Meigen, 1826: 70. Junior primary homonym of *Musca agilis* Harris, 1780 (Tachinidae). SYN.N.
Onesia vulgaris Robineau-Desvoidy, 1830: 367.
Onesia bisetosa Hall, 1948: 322. SYN.N.

♂♀. Paler than *viarum* and less hirsute. Abdomen more strongly dusted and with more conspicuous tessellations, especially in female. Costa very rarely with some hairs on underside between tips of Sc and R_1. Wing-veins yellow or brown. Front tibia usually with 1 *pv* seta, but occasionally with 2. T4 without discal setae.

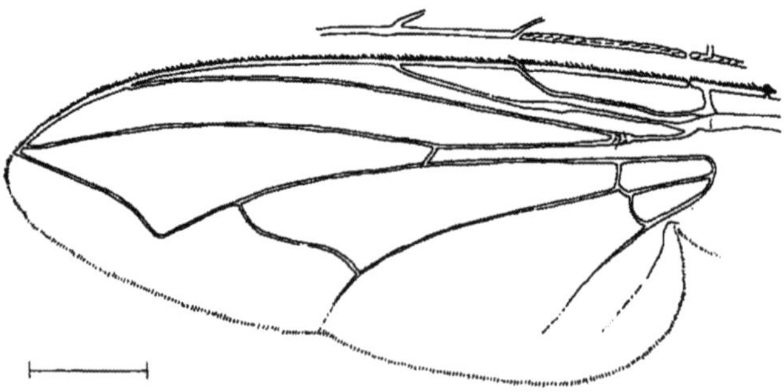

Fig. 66. *Bellardia vulgaris* (Robineau-Desvoidy). Left wing from above. Inset: costa from below showing distribution of small hairs on its underside. Scale = 1.0 mm.

Fig. 67. *Bellardia vulgaris* (Robineau-Desvoidy), male.
ST5. Scale = 0.4 mm.

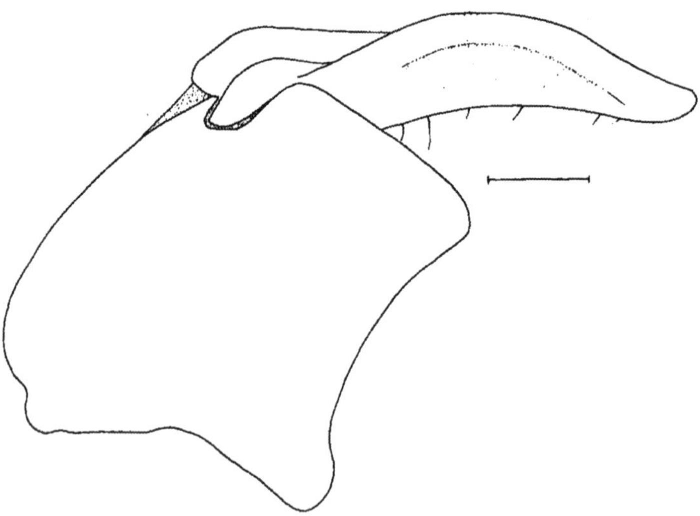

Fig. 68. *Bellardia vulgaris* (Robineau-Desvoidy), male. Epandrium, cerci and surstylus, left lateral view. Scale = 0.2 mm.

Fig. 69. *Bellardia vulgaris* (Robineau-Desvoidy), male. Cerci and surstyli, dorsal view. Scale = 0.2 mm.

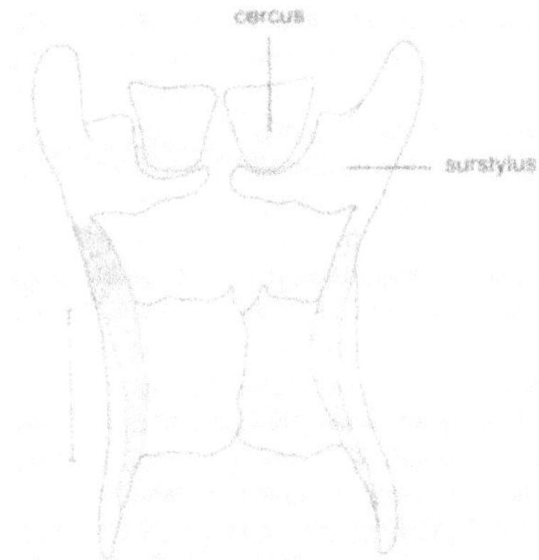

Fig. 70. *Bellardia vulgaris* (Robineau-Desvoidy), male. Bacilliform sclerites, showing details of junction between cerci and junction between cercus and surstylus. Scale = 0.2 mm.

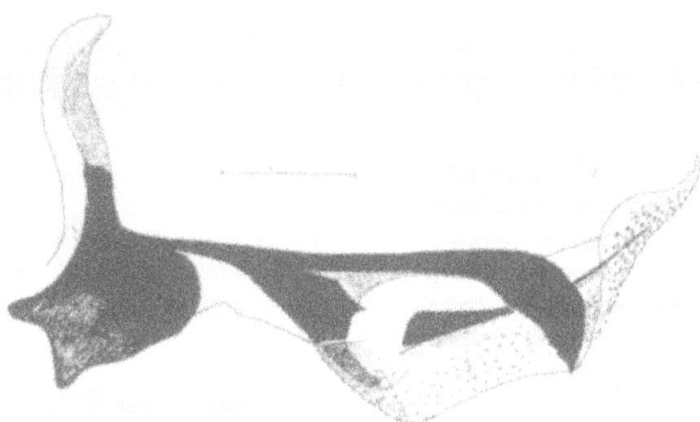

Fig. 71. *Bellardia vulgaris* (Robineau-Desvoidy), male. Aedeagus, left lateral view. Scale = 0.2 mm.

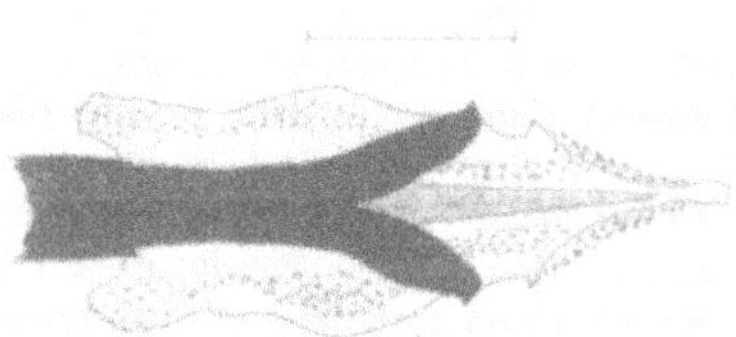

Fig. 72. *Bellardia vulgaris* (Robineau-Desvoidy), male. Distiphallus, dorsal view. Scale = 0.2 mm.

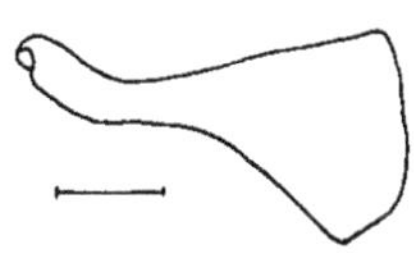

Fig. 73. *Bellardia vulgaris* (Robineau-Desvoidy), male. Ejaculatory sclerite. Scale = 0.1 mm.

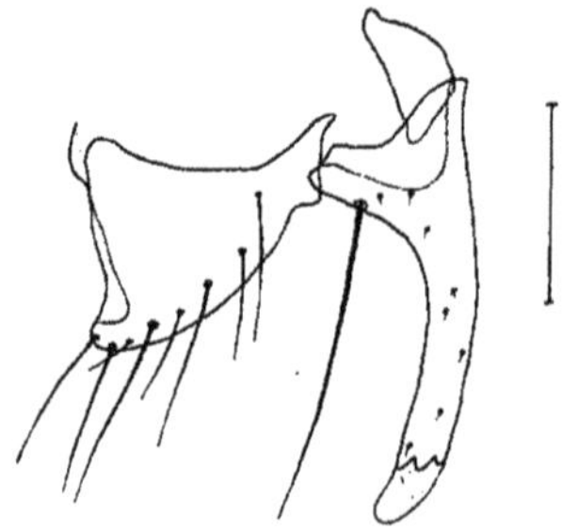

Fig. 74. *Bellardia vulgaris* (Robineau-Desvoidy), male. Pre- and postgonites. Scale = 0.2 mm.

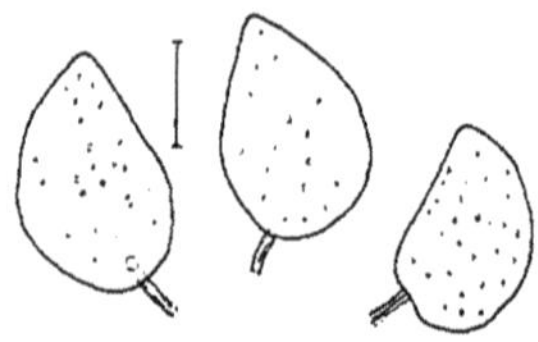

Fig. 75. *Bellardia vulgaris* (Robineau-Desvoidy), female. Spermathecae. Scale = 0.1 mm. G.pr. 220.

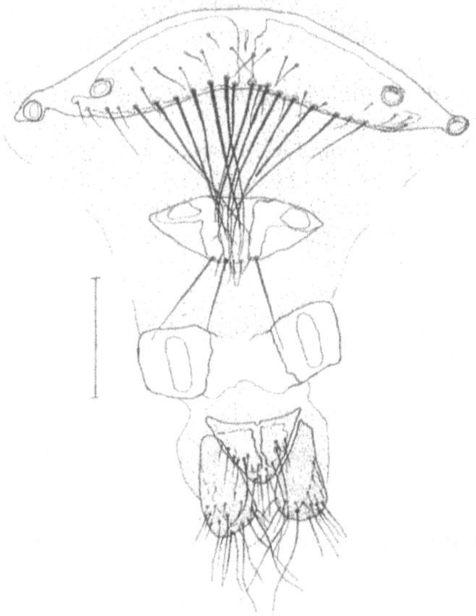

Fig. 76. *Bellardia vulgaris* (Robineau-Desvoidy), female. Ovipositor, dorsal sclerites. Stipple indicates extent of microtrichiae. Scale = 0.4 mm. G.pr. 220.

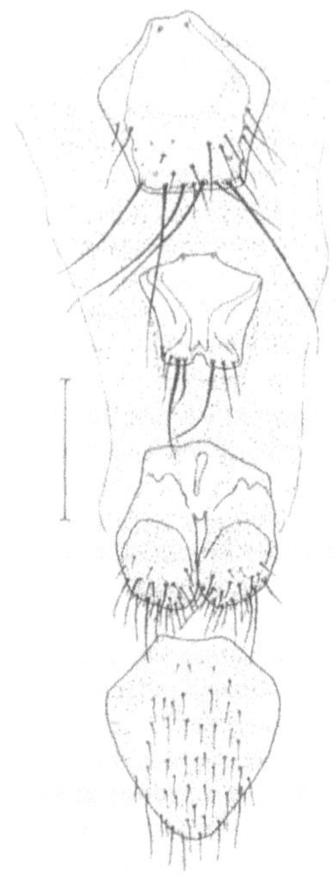

Fig. 77. *Bellardia vulgaris* (Robineau-Desvoidy), female. Ovipositor, ventral sclerites. Stipple indicates extent of microtrichiae. Scale = 0.4 mm. G.pr. 220.

♂. Frons 0.091-0.109 times head-width (n=5), frontal vitta narrower than fronto-orbital plates. Ventral epandrial corner acute, slightly less than a right angle in profile (arrow in Fig. 68). Surstyli massive, weakly curved in profile, longer than cerci, with an external groove which does not reach the apex.

♀. Frons 0.329-0.365 times head-width (n=5). Ovipositor with T6 shaped as a narrow transverse band, most often with a narrow zone of weak sclerotisation mid-dorsally. Marginal setae present along most of hind edge, some discal setae present, but discal setae absent from anterior part of T6. T7 with marginal setae. T8 bare. Epiproct triangular, sometimes even longer than broad, divided along mid-line. ST6 and ST7 about as long as broad. ST8 with prominent lobes apically; basally with a conspicuous mid-ventral sclerotisation in front of lobes. Cerci wholly covered with microtrichiae. Cerci and hypoproct rather large. In dried specimens ovipositor conspicuously protruding behind margins of T5. [9 ovipositor slides examined.]

Length. 5-11 mm.

Distribution. Common and widely distributed in Fennoscandia and Denmark. North to TRY (Norway). Ascends to the upper parts of the subalpine birch forest in South Norway (about 1000 m a.s.l.). Captured above tree-line in Sweden at Storlien (Jmt.). - Widely distributed in Europe (Schumann, 1974b, 1986). Also in eastern U.S.A. where it occurs in New Jersey and perhaps adjoining states (Shewell, 1987).

Biology. Earthworm host(s) unknown. First and third instar larvae have been described and figured by Lobanov (1971), Schumann (1974b) and Shewell (1987). The species overwinters in the third instar larval stage (Lobanov, 1971). Adults have been collected in Fennoscandia and Denmark from April to October, with maximum numbers in June, July and August.

Nomenclature. *Musca agilis* was described from ♂♀ syntypes, both sexes being figured on Meigen Plate 118, fig. 9 (Morge, 1976a). In Meigen's collection in MNHN are 2♂2♀ under *agilis* (labelled "Meigen" and "2141/40"). They are probably syntypes. Even though *agilis* was described as having "palpis nigris", Meigen's figure 9d clearly shows a yellow palp. 1♂, which carries Schumann's determination label and has the genitalia visible, is the present species, 1♂ = *viarum* Robineau-Desvoidy, 2♀ are *Bellardia* identifiable only after dissection. To fix the identity of the name *agilis* in the current sense (Schumann, 1974b), the ♂ with Schumann's determination label should be designated as lectotype. *Onesia vulgaris* was described by Robineau-Desvoidy (1830) as the fifth species under *Onesia*, and was reported to be "très-commune". It is likely to be a species of *Bellardia*. No specimens are left in his collection in Paris, and the name was listed as a nomen dubium by Schumann (1986). I have brought forward the unused name *vulgaris* of Robineau-Desvoidy in the interests of nomenclatural stability and in keeping with Recommendation 75E of the ICZN. I have selected and here designate a recent specimen in good condition as neotype for *Onesia vulgaris* to fix its identity. It is labelled (1) "Rambouillet / 20.vii.17", (2) "MUSEUM PARIS / Coll. E.SÉGUY 1919". I have placed it in Robineau-Desvoidy's collection in MNHN, box 12 under *Onesia*. Schumann synonymised *Calliphora amplectens* Pandellé, 1896: 210 with the present species. In Pandellé's collection in MNHN are 3♂3♀ under *amplectens*. 1♂ (with genitalia visible) and 1♀ with very short aristal hairs are probably *Bellardia vespillo* (Fabricius). The other males I find definitely not to be the present species, and, like the remaining females, not identifiable with certainty unless dissected. I have therefore refrained from designating a lectotype, and have removed *amplectens* from the synonymy of the present species for the time being. I have examined the holotype ♂ of *Onesia bisetosa* Hall together with 1♂4♀ paratypes (all in USNM). Both males have 1+3 *ia*, but none of the females do. I consider this an aberration without taxonomic importance, as the male genitalia agree exactly with European material. Both *agilis* and *bisetosa* are already established synonyms of the present species, even if formally new synonyms of the name *vulgaris*.

3. ***Bellardia bayeri*** (Jacentkovský, 1937b)
Figs 50, 78-87.

Onesia sicillensis Villeneuve (lapsus for *siciliensis*): Jacentkovský, 1936b: 115 (misidentification).

Onesia bayeri Jacentkovský, 1937b: 6, 20 (Czech), 49 (French).

Onesia townsendi Hall, 1948: 324. SYN.N.

Bellardia osetica Khitsova, 1979:1245. SYN.N.

Melinda pruinosa: Grunin, 1970a: 611; Lobanov, 1976: 25, 26, fig.25 (misidentifications, not Enderlein, 1933).

Melinda pusilla: Lehrer, 1972: 117, fig. 39 (misidentification, not Meigen, 1826).

Bellardia pusilla: Schumann, 1974b: 275, fig. 21 (mouthparts of third instar larva) (misidentification, not Meigen, 1826).

♂♀. *Bellardia bayeri* is a rather small species with a conspicuous brown spot on the upper part of the parafacials. The spot does not disappear when the parafacials are viewed tangentially from above. Arista (including hairing) 1.5-2.0 times width of first flagellomere. Palpi yellow. Upper half of occiput with 3-4 rows of black hairs in upper part, pale hairs below. Posterior spiracle large, much longer than katepimeron (barette). Wings infuscated, paler in the ♀. Costa almost always bare below between tips of Sc and R_1. Cell r_{4+5} widely open. Distance from bend of M to dm-cu is 2 times distance from bend of M to wing-margin. Upper and lower calypter strongly and evenly infuscated. Abdomen conspicuously dusted. T3 with strong marginals. T4 and T5 with discal setae.

♂. Frons narrow, 0.040-0.071 times head-width (n=5). Fronto-orbital plates touching. Parafacialia in profile narrower than length of first flagellomere. Mid tibia with 1 *ad*, often with 1-2 small ones above. Terminalia strikingly small, much smaller than in *viarum*. Cerci wedge-shaped, bent slightly ventrally in profile. Surstyli with parallel upper and lower edge, tip blunt, diverging from

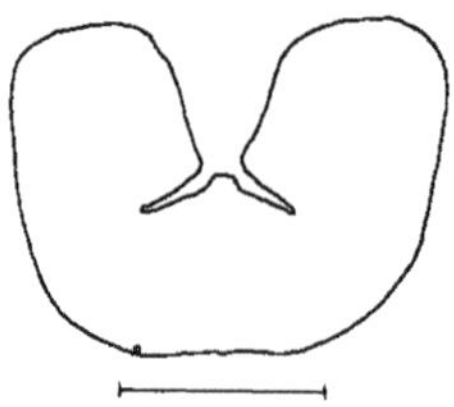

Fig. 78. *Bellardia bayeri* (Jacentkovský), male. ST5. Scale = 0.2 mm.

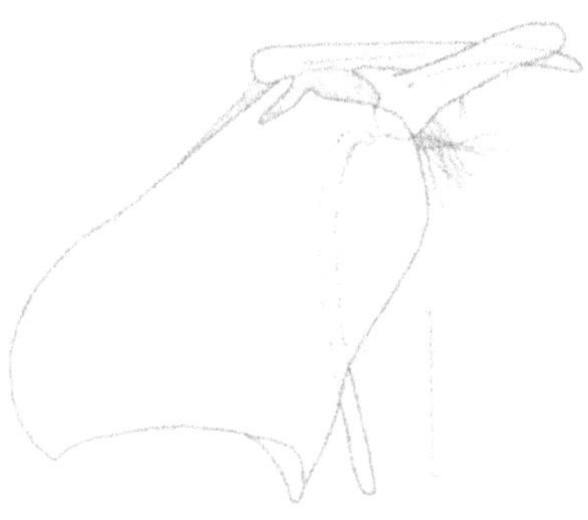

Fig. 79. *Bellardia bayeri* (Jacentkovský), male. Epandrium, cerci, surstylus and bacilliform sclerite, left lateral view. Scale= 0.2 mm.

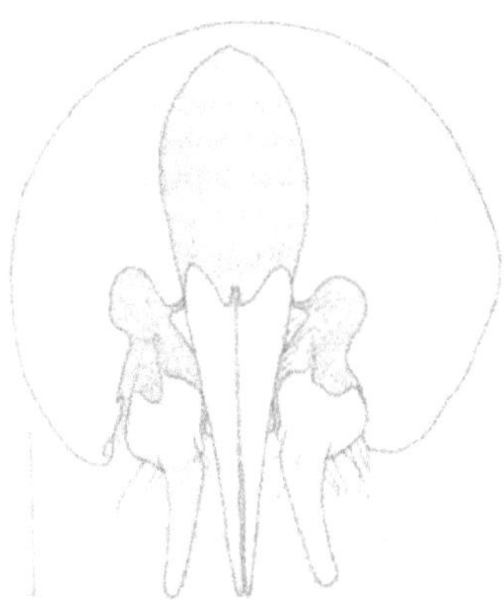

Fig. 80. *Bellardia bayeri* (Jacentkovský), male. Epandrium, cerci and surstyli, dorsal view. Scale = 0.2 mm.

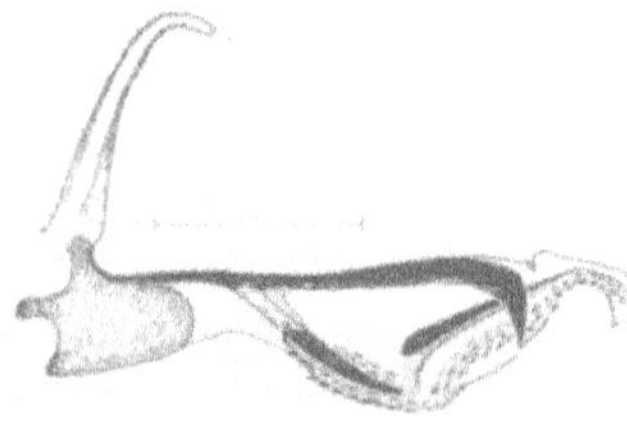

Fig. 81. *Bellardia bayeri* (Jacentkovský), male. Aedeagus, left lateral view. Scale = 0.2 mm.

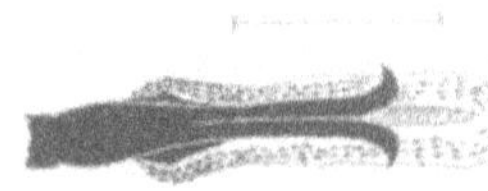

Fig. 82. *Bellardia bayeri* (Jacentkovský), male. Distiphallus, dorsal view. Scale = 0.2 mm.

Fig. 83. *Bellardia bayeri* (Jacentkovský), male. Ejaculatory sclerite. Scale = 0.2 mm.

Fig. 84. *Bellardia bayeri* (Jacentkovský), male. Post- and pregonites. Scale = 0.2 mm.

Fig. 85. *Bellardia bayeri* (Jacentkovský), female. Spermathecae. Scale = 0.2 mm. G.pr. 260.

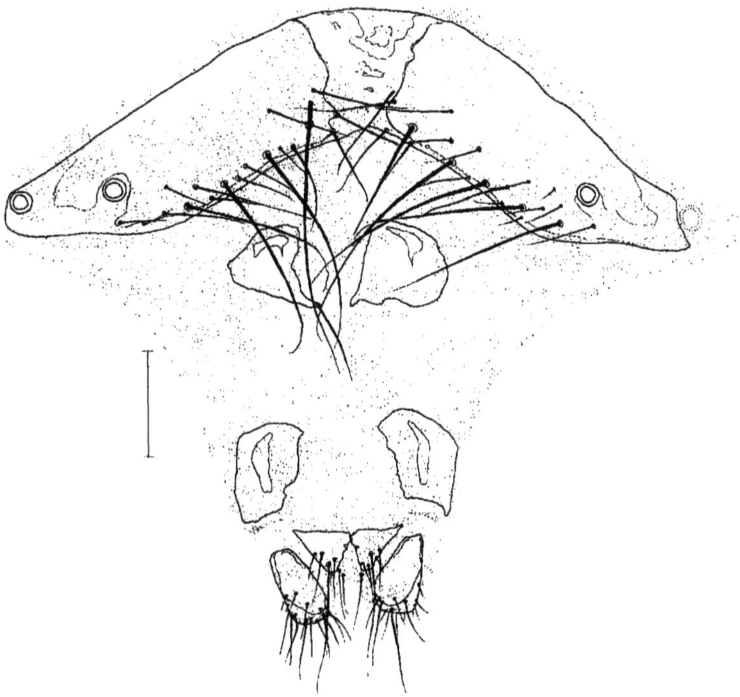

Fig. 86. *Bellardia bayeri* (Jacentkovský), female. Ovipositor, dorsal sclerites. Stipple indicates extent of microtrichiae. Scale = 0.2 mm. G.pr. 260.

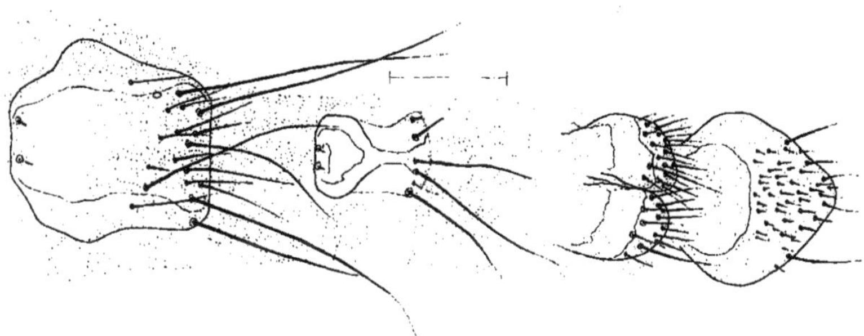

Fig. 87. *Bellardia bayeri* (Jacentkovský), female. Ovipositor, ventral sclerites. Stipple indicates extent of microtrichiae. Scale = 0.2 mm. G.pr. 260.

cerci in posterior view. Basally an almost transverse row of long setae on the underside. Epiphallus long and slender. Paraphalli not very strong, distally evenly curved, dorsally separated by a thin line reaching far basad.

♀. Frons 0.280-0.324 times head-width (n=5). An additional spot in some views at the middle of the fronto-orbital plates. Mid tibia with 2 *ad*. Ovipositor: T6 with a weakly pigmented central band. T8 covered with microtrichiae. Epiproct divided. Cerci wholly covered with microtrichiae. ST6 as long as broad or slightly longer. ST8 sclerotised only at tip of lobes. Hypoproct with a large circular very weakly pigmented basal area. Hypoproct as broad as long. [1 ovipositor slide examined.]

Length. 5-8 mm.

Distribution. Very rare in Fennoscandia and Denmark: 1♂2♀ are known from VAY (Norway) and 2♂ from NEZ (Denmark). - Also known from West Germany (Markgröningen), East Germany (Eberswalde (Jacentkovský, 1944b: 118)), Romania, Bulgaria, U.S.S.R. (Carpathians, Leningrad, Moscow, east to Tobol'sk and Tyumen' and south to Caucasus, Armenia and Azerbaijan) and U.S.A (Massachusetts). In the Carpathians it ascends to 700m a.s.l.

Biology. A ♂ from Denmark (in ZMC) was reared by Lundbeck from a pupa found beneath elm bark on December 25, 1910. It emerged on March 31, 1911. It was misidentified by Schumann as "*Bellardia pusilla*" and is the basis for his figures of the mouthparts of the third instar larva (thus erroneously assigned to "*pusilla*"). 1♀ from Tyumen' (U.S.S.R.) was found as a larva on September 20, 1975. It emerged October 10, i.e. 21 days later. 3♀ from Voronezhskaya province (U.S.S.R.) have been bred from the lumbricid *Eisenia foetida* (Savigny). Adults have been collected in all months of the year except February, March and November. The highest numbers have been found in August. In Fennoscandia and Denmark it has been taken on the wing in July and August.

Nomenclature. *Onesia bayeri* was described from 2♂ from "Bulharsko, Prešlav, Bačkovo", collected September 8 and 30, 1935. The syntype dated "1935.IX.30" and with Jacentkovský's "siciliensis" determination label, has been found in coll. Mesnil (ZIKIEL). It was designated as lectotype by Schumann (1974b: 249), and I have provided it with a label to this effect. It fits the description and has the terminalia exposed. Unfortunately the very distinct spot on the parafacials is not mentioned in Schumann's description of the species. *Onesia townsendi* was described from a single ♂ captured by Townsend in the Melrose Highlands, Mass., U.S.A. September 13, 1914. I have examined the holotype (in USNM) which has the dissected genitalia in a vial on the pin. I have also examined numerous specimens of Grunin's material of "*pruinosa*" from various parts of the U.S.S.R. (including the Caucasus). They are labelled variously as "*schumanni* sp.n." or "*pruinosa*". I have not seen the holotype of *Bellardia osetica* from the Caucasus but the published figures leave little doubt as to its identity. The peculiar angle between the surstyli and cerci shown by Khitsova (1979: figs. 3, 4) is most likely an artifact and does not reflect the normal position of these parts.

4. ***Bellardia pandia*** (Walker, 1849)
Figs 51, 88-98.

Musca pandia Walker, 1849: 898.
Musca unxia Walker, 1849: 899.
Onesia biseta Kramer, 1917: 284.

♂♀. *Bellardia pandia* is very similar to *vulgaris*, but the front tibia almost always has 2 *pv* setae and the costa often has small hairs on the underside between tips of Sc and R_1. Palpi rarely darkened in basal half. T4 often with discal setae.

♂. Frons 0.074-0.077 times head-width (n=5). Fronto-orbital plates almost touching. Parafacial shimmering white, without spots. Ventral epandrial corner projecting and rounded, forming about a right angle. Cerci very characteristically heart-shaped basally, with long, strong, shining black and bare prongs. Surstyli strong, curved ventrally in profile, with an external groove ending far from apex; in posterior view surstyli curved inwards distally, basally with a medial projection. Basal sclerotisation of distiphallus (ventral plates) rather long in profile.

♀. Frons 0.314-0.357 times head-width (n=5). Ovipositor: T6 evenly sclerotised, with discal and marginal setae. Discal setae absent from anterior parts of disc. T8 with or without setae. Epiproct much shorter than broad, with 2 strong and numerous additional medium-sized setae. ST6 1.5-2.0 times as long as broad. ST7 elongate. ST8 as two warts, without conspicuous sclerotisation in front of lobes. [5 ovipositor slides examined.]

Length. 5-11 mm.

Distribution. Widespread in Denmark, south Sweden and south Finland. Very rare in Norway where it is known only from 1♂2♀ from TEI, near Haukelisæter, 990 m a.s.l., near the upper limit of the subalpine birch zone. - Throughout Europe (Schumann, 1986).

Biology. Immature stages and host unknown. Adults have been collected in Fennoscandia and Denmark from April to September, with maximum numbers from May to July.

Nomenclature. I consider Schumann (1986) as the first reviser in selecting *pandia* as the valid name for this species. The holotype ♀ is in BMNH. In BMNH are also 4 syntypic males of *unxia*. I consider Schumann's (1974b) reference to the "Holotype" in BMNH as a valid lectotype fixation for *unxia*. The remaining 3♂ are paralectotypes. The name *Onesia biseta* was introduced by Kramer (1917: 284) for a species he had misidentified as "*sepulcralis*" in a previous work (thus: "Onesia biseta Villen. in litt. Sepulcralis in d. T.d.O.", i.e. "Die Tachiniden der Oberlausitz"

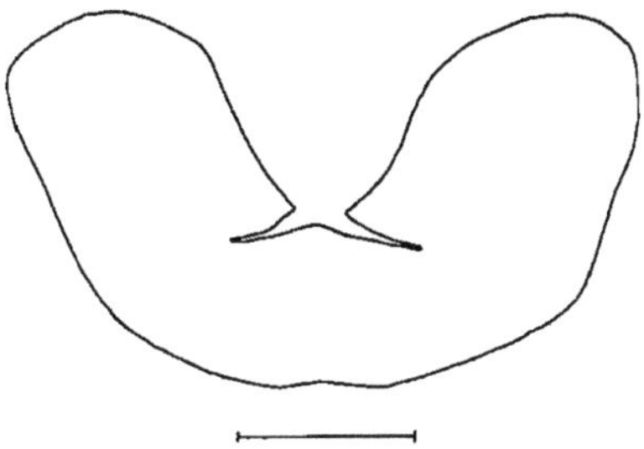

Fig. 88. ***Bellardia pandia*** **(Walker), male. ST5. Scale = 0.4 mm.**

Fig. 89. *Bellardia pandia* (Walker), male. Epandrium, cerci and surstylus, left lateral view. Scale = 0.2 mm.

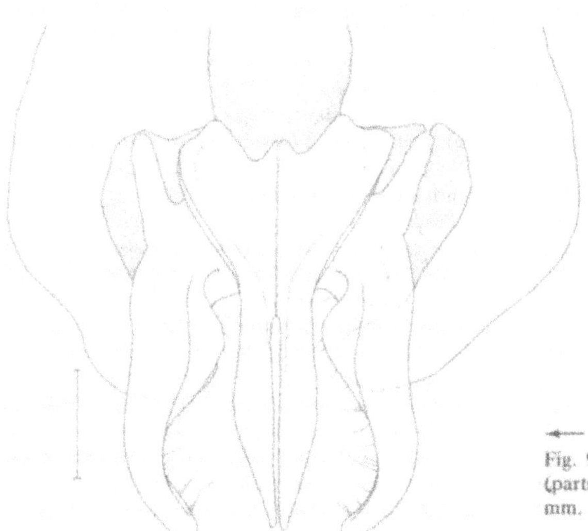

Fig. 90. *Bellardia pandia* (Walker), male. Epandrium (parts), cerci and surstyli, posterior view. Scale = 0.2 mm.

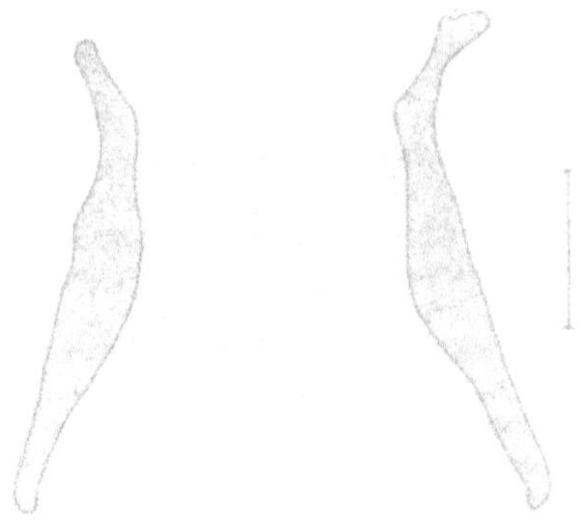

Fig. 91. *Bellardia pandia* (Walker), male. Bacilliform sclerites and membrane between them. Scale = 0.2 mm.

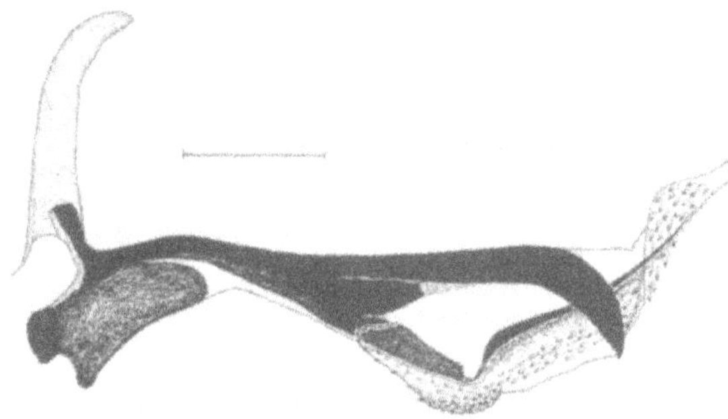

Fig. 92. *Bellardia pandia* (Walker), male. Aedeagus, left lateral view. Scale = 0.2 mm.

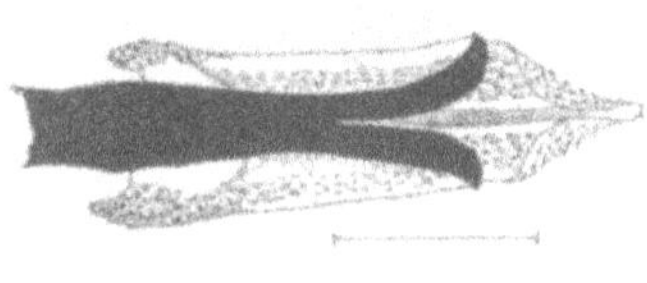

Fig. 93. *Bellardia pandia* (Walker), male. Distiphallus, dorsal view. Scale = 0.2 mm.

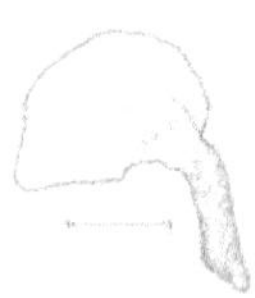

Fig. 94. *Bellardia pandia* (Walker), male. Ejaculatory sclerite. Scale = 0.1 mm.

Fig. 95. *Bellardia pandia* (Walker), male. Pre- and postgonites. Scale = 0.2 mm.

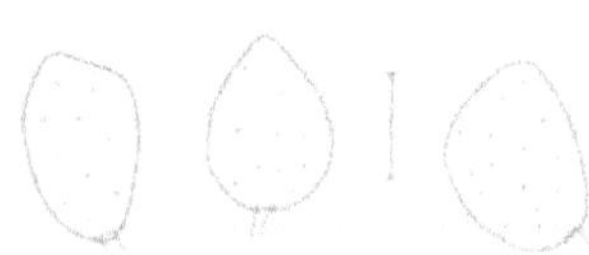

Fig. 96. *Bellardia pandia* (Walker), female. Spermathecae. Scale = 0.1mm. G.pr. 147.

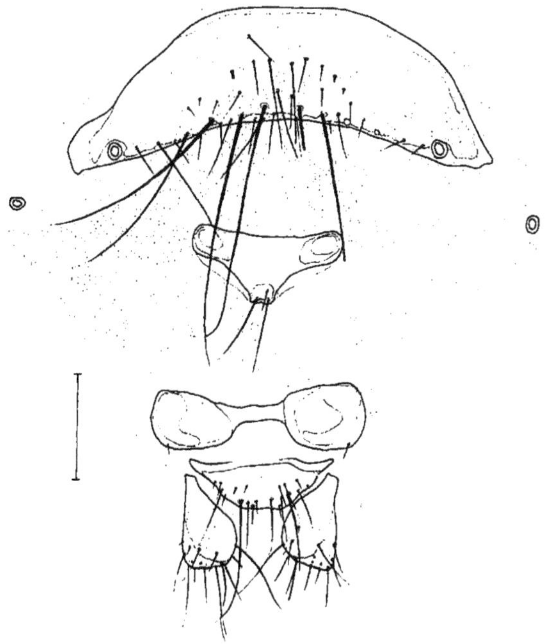

Fig. 97. *Bellardia pandia* (Walker), female. Ovipositor, dorsal sclerites. Stipple indicates extent of microtrichiae. Scale = 0.2 mm. G.pr. 13.

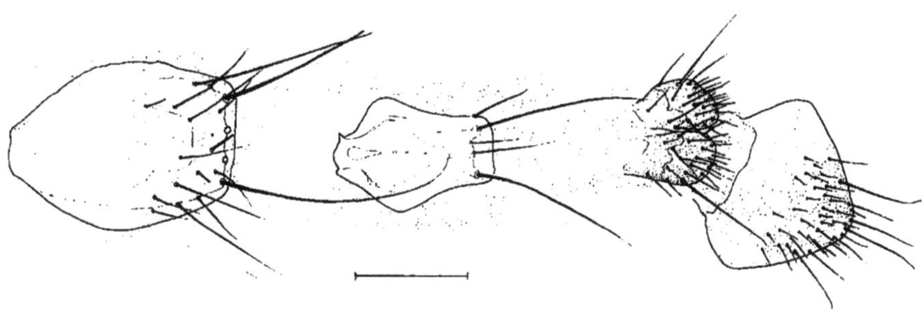

Fig. 98. *Bellardia pandia* (Walker), female. Ovipositor, ventral sclerites. Stipple indicates extent of microtrichiae. Scale = 0.2 mm. G.pr. 13.

= Kramer, 1911; in the 1911 work is a figure of the male cerci and surstyli in profile assignable to the present species). Mueller (1922: 59) construed this as indicating that Kramer regarded *biseta* and *sepulcralis* Meigen as the same species, but this is not so: Kramer lists *sepulcralis* 12 lines below *biseta*. Schumann (1986) incorrectly lists Kramer's name as a replacement name ("new name"). No lectotype has been selected for *biseta*. Kramer's collection is in Stadtmuseum Bautzen (Schumann, 1974b: 253).

5. ***Bellardia pubicornis*** (Zetterstedt, 1838), COMB.N.

Figs 47, 52, 99-109.

Tachina pubicornis Zetterstedt, 1838: 639.
Musca puberula Zetterstedt, 1838: 654.

Musca dolens Zetterstedt, 1838: 654.
Dexia curvipes Zetterstedt, 1859: 6175.

♂♀. Arista pubescent dorsally in basal two-thirds, practically bare below. Parafacialia densely silvery dusted, with sparse hairs. Lower part of face and lower facial margin strongly protruding, prementum correspondingly elongated. 1+1-2 *acr*, 0-1+2 *ia*. Outer *ph* often absent. Anterior spiracle strikingly small. Anterior part of anepimeron most often bare in lower half. r_{4+5} open or closed at margin. Section of M beyond bend straight. Costal spine strong. Lower calypter with some hairs or bare above. Front tibia with 1 *pv*. Abdomen black in ground-colour, dusted with conspicuous tessellations, sometimes with weakly bronze metallic sheen.

♂. Frons 0.099-0.127 times head-width (n=5). Fronto-orbital plates rather broad, slightly separated. Surstyli with parallel edges and weakly curved in profile. In posterior view, surstyli strongly curved inwards. Acrophallus with conspicuous lateral wings.

♀. Frons 0.352-0.388 times head-width (n=5). T8 halves large and close together, with numerous marginal setae. Epiproct much shorter than broad. Cerci completely microtrichiose. ST6 and ST7 broad. ST8 two broad warts covered with microtrichiae. [2 ovipositor slides examined.]

Length. 4.5-7.5 mm.

Distribution. Absent from Denmark. Subalpine and low alpine regions of Norway, Sweden, North Finland and the Kola Peninsula in the U.S.S.R.. In Western Norway also in the lowlands (MRY, SFI). Scotland, St. Kilda (Emden, 1954), Austria (Schumann, 1986). - Boreoalpine distribution in Europe.

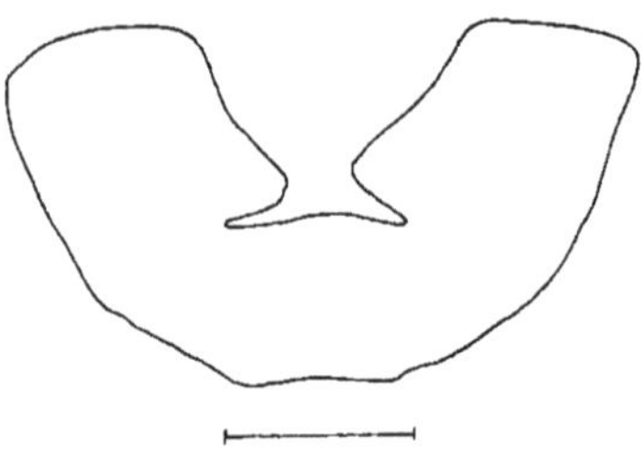

Fig. 99. *Bellardia pubicornis* (Zetterstedt), male. ST5. Scale = 0.4 mm.

Biology. Immature stages and hosts unknown. In North Europe collected in June-August, with maximum in June and July.

Nomenclature. Ringdahl (1945a) first established the above synonymy which has been overlooked by later authors (Zumpt, 1956a; Schumann, 1986). He was also the first author to discover and cite all the synonyrns from Zetterstedt's Insecta Lapponica work applicable to the present species. I therefore consider him as the first reviser when selecting *pubicornis* as the valid name for this species. *Tachina pubicornis* was described from 1♂ captured July 10, 1821 in the subalpine region "ad lacum insignem Torneåträsk" in T.Lpm. The holotype is in ZML and I have labelled it as such. *Musca puberula* was described from 1♂1♀ from "ad Tromsoe" July 24, 1821. Both specimens (in ZML) are on the same pin. I have labelled and here designate the ♂ as lectotype and the ♀ as paralectotype. *Musca dolens* was described from 1♂ captured July 9, 1821 in the subalpine region "ad lacum insignem Torneåträsk" in T.Lpm. The holotype is in ZML and I have labelled it as such. *Dexia curvipes* was described from an unspecified number of ♂♀ captured by "Proff. Boheman & Andersson" in "Lappon. Lulensi". 2♂2♀ are under this name in a list of determinations (in NHRM) prepared by Zetterstedt after examining specimens Boheman sent to him for identification, but only 1♀ has been found in Boheman's collection in NHRM. It fits the description. I have labelled and hereby designate this female as lectotype.

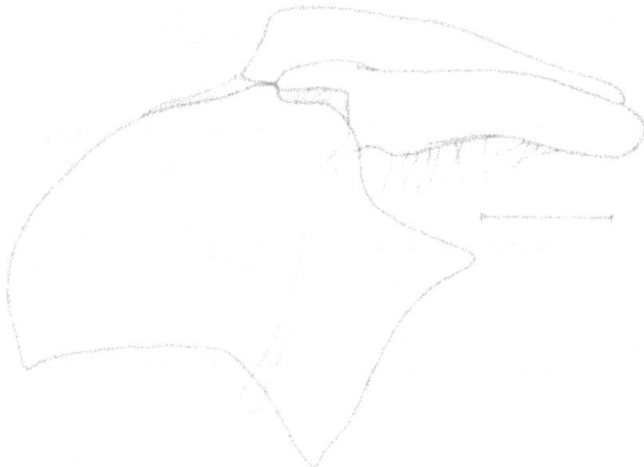

Fig. 100. *Bellardia pubicornis* (Zetterstedt), male. Epandrium, cerci, surstylus and bacilliform sclerite, left lateral view. Scale= 0.2 mm.

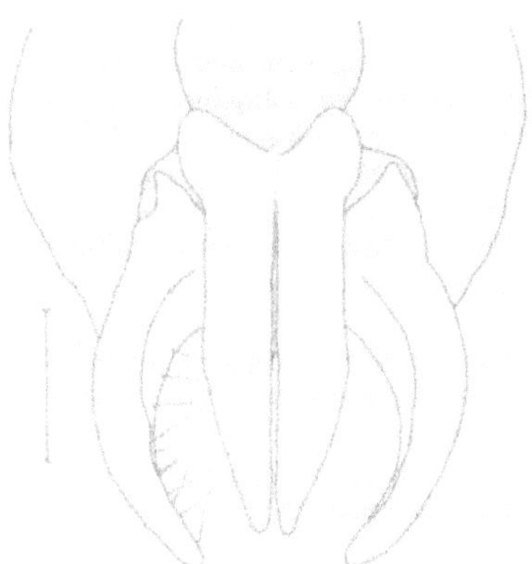

Fig. 101. *Bellardia pubicornis* (Zetterstedt), male. Epandrium (parts), cerci and surstyli, posterior view. Scale = 0.2 mm.

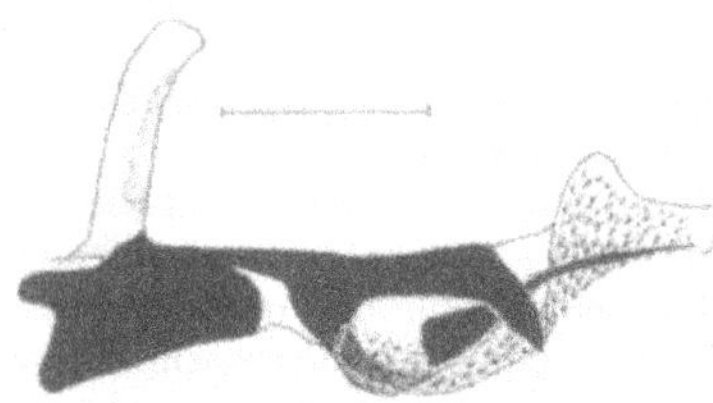

Fig. 103. *Bellardia pubicornis* (Zetterstedt), male. Aedeagus, left lateral view. Scale = 0.2 mm.

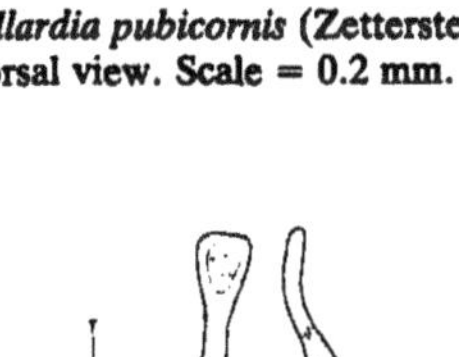

Fig. 104. *Bellardia pubicornis* (Zetterstedt), male. Distiphallus, dorsal view. Scale = 0.2 mm.

Fig. 105. *Bellardia pubicornis* (Zetterstedt), male. Ejaculatory sclerite, various views. Scale = 0.1 mm.

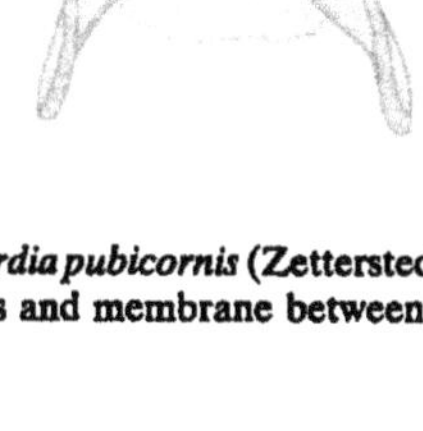

Fig. 102. *Bellardia pubicornis* (Zetterstedt), male. Bacilliform sclerites and membrane between them. Scale = 0.2 mm.

Fig. 106. *Bellardia pubicornis* (Zetterstedt), male. Pre- and postgonites. Scale = 0.1 mm.

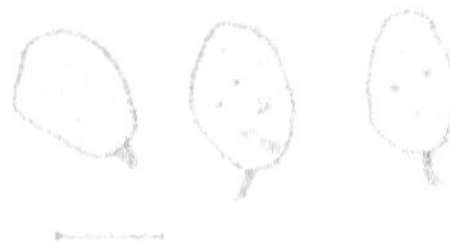

Fig. 107. *Bellardia pubicornis* (Zetterstedt), female. Spermathecae. Scale = 0.1 mm. G.pr. 152.

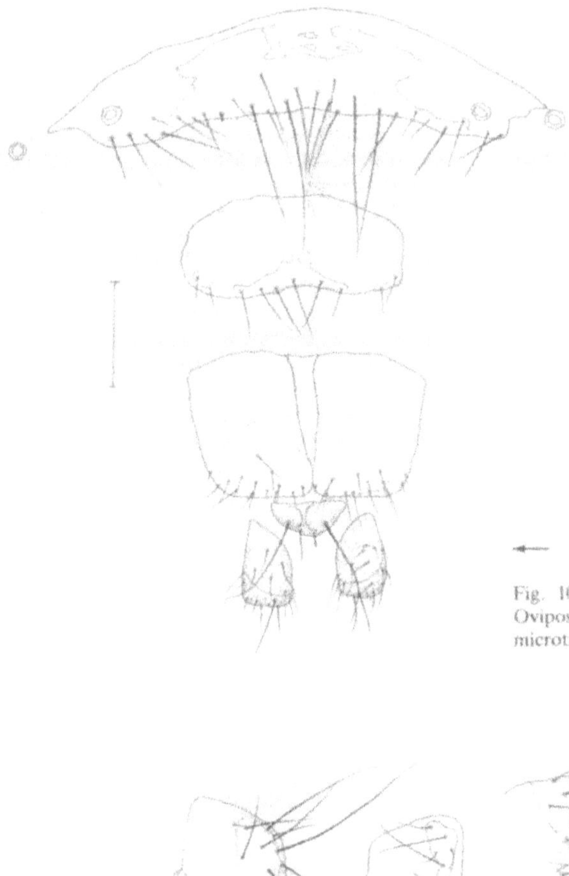

Fig. 108. *Bellardia pubicornis* (Zetterstedt), female. Ovipositor, dorsal sclerites. Stipple indicates extent of microtrichiae. Scale = 0.2 mm. G.pr. 153.

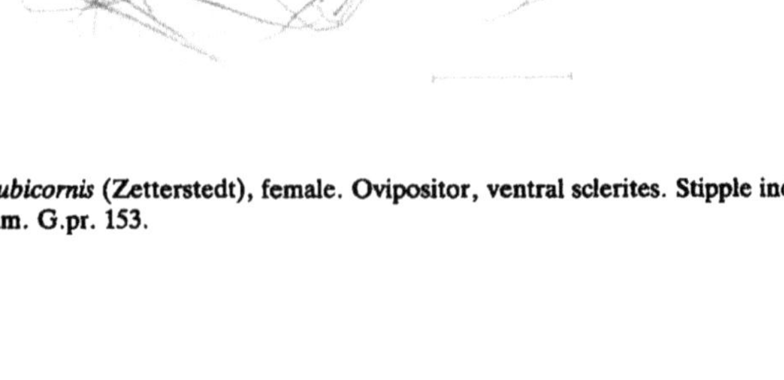

Fig. 109. *Bellardia pubicornis* (Zetterstedt), female. Ovipositor, ventral sclerites. Stipple indicates extent of microtrichiae. Scale = 0.2 mm. G.pr. 153.

6. ***Bellardia stricta*** (Villeneuve, 1926)
Figs 53, 110-117.

Onesia stricta Villeneuve, 1926: 131.
Xerophilophaga pumicata Enderlein, 1933: 122.
Ambodicria minutissima Enderlein, 1933: 125.
Bellardia xinganensis Chen, Fan & Cui, 1982. SYN.N.

♂♀. Arista short plumose in basal half. Parafacialia evenly and very densely dusted. Parafacial hairs sparse, almost absent along eye margin. r_{4+5} open, very rarely closed at margin. Lower calypter hairy above, but sometimes with very few hairs. Front tibia with 1 (rarely 2) *pv*. Costal spine conspicuous. Abdomen shining green, without or with very thin dusting. Abdominal ground vestiture sparse, especially on T4 and T5. Abdomen with numerous discal setae on T4 and T5.

♂. Frons 0.058-0.083 times head-width (n=5). Parafacialia with silvery dusting. Cerci wedge-shaped, longer than surstyli. Surstyli weakly curved with an external groove, in profile with a bulge on dorsal edge and an irregular ventral edge. Paraphallic hook right-angled, tip weakly dentate.

♀. Frons 0.321-0.398 times head-width (n=5). Parafacialia with silvery to golden dusting. T6 sclerotised in two widely separated halves, with 2-3 strong and a few weaker marginal setae in extreme median part. Epiproct with 2 strong and at most a few additional setae. Cerci wholly microtrichiose. ST6 and ST7 short and broad. [4 ovipositor slides examined.]

Length. 4-8 mm.

Distribution. Widely distributed in Fennoscandia and Denmark, from the lowlands to the subalpine region, but very few records from Norway. Absent from the British Isles (Emden, 1954). - Palaearctic Region, apparently with the exception of its most western parts (Schumann, 1986).

Biology. Immature stages and hosts unknown. One puparium in coll. Lundbeck (ZMC) found on June 6, 1911; the adult emerged the day after. In Fennoscandia and Denmark adults have been collected from May to September, with maximum numbers in June and July.

Note. Schumann (1974b, 1986) erroneously listed *polita* Mik from Denmark. The specimens on which the record is based (from Læsø) belong to the present species. Hackman (1980) erroneously recorded *polita* from Finland. All specimens under this label in ZMH belonged to *stricta*.

Nomenclature. I have not seen the types of *xinganensis*. The characters by which Chen, Fan & Cui (1982) differentiate *xinganensis* from *stricta* apply equally well to *stricta*.

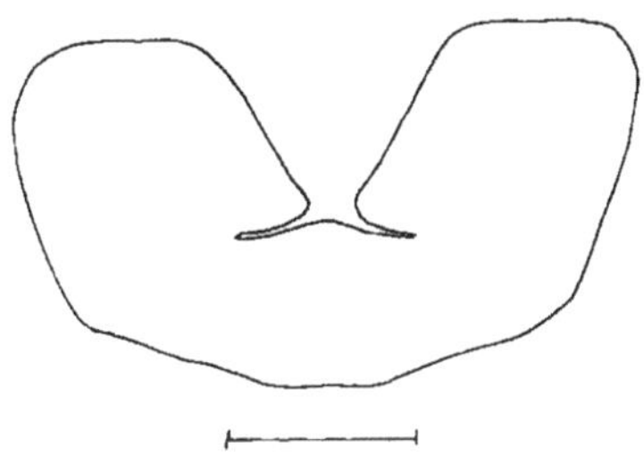

Fig. 110. ***Bellardia stricta*** (Villeneuve), male. ST5. Scale = 0.4 mm.

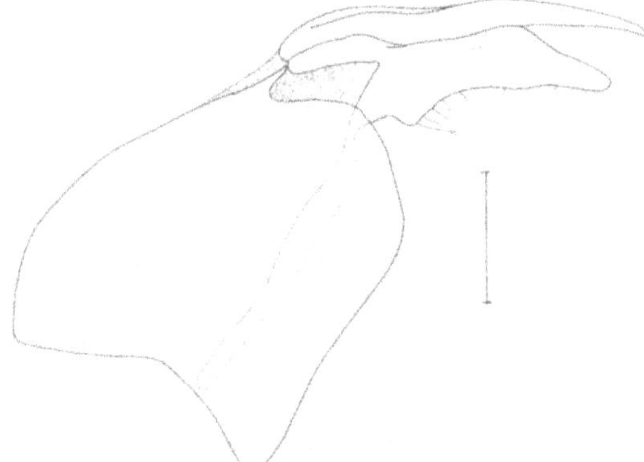

Fig. 111. ***Bellardia stricta*** (Villeneuve), male. Epandrium, cerci, surstylus and bacilliform sclerite, left lateral view. Scale = 0.2 mm.

Fig. 112. *Bellardia stricta* (Villeneuve), male. Epandrium (parts), cerci and surstyli, posterior view. Scale = 0.2 mm.

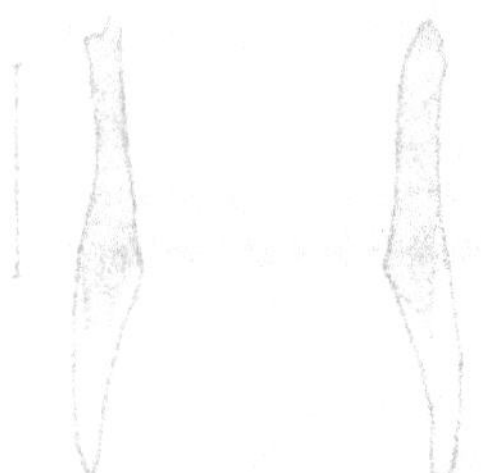

Fig. 113. *Bellardia stricta* (Villeneuve), male. Bacilliform sclerites and membrane between them. Scale = 0.2 mm.

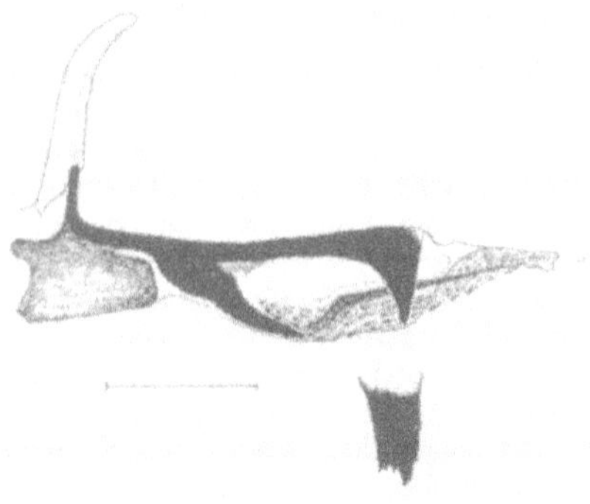

Fig. 114. *Bellardia stricta* (Villeneuve), male. Aedeagus, left lateral view. Inset: tip of paraphallus (enlarged). Scale = 0.2 mm.

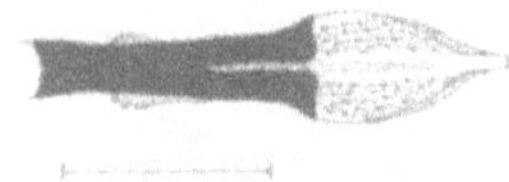

Fig. 115. *Bellardia stricta* (Villeneuve), male. Distiphallus, dorsal view. Scale = 0.2 mm.

Fig. 116. *Bellardia stricta* (Villeneuve), male. Ejaculatory sclerite. Scale = 0.1 mm.

Fig. 117. *Bellardia stricta* (Villeneuve), male. Pre- and postgonites. Scale = 0.2 mm.

Fig. 118. *Bellardia stricta* (Villeneuve), female. Spermathecae. Scale = 0.1 mm. G.pr. 224.

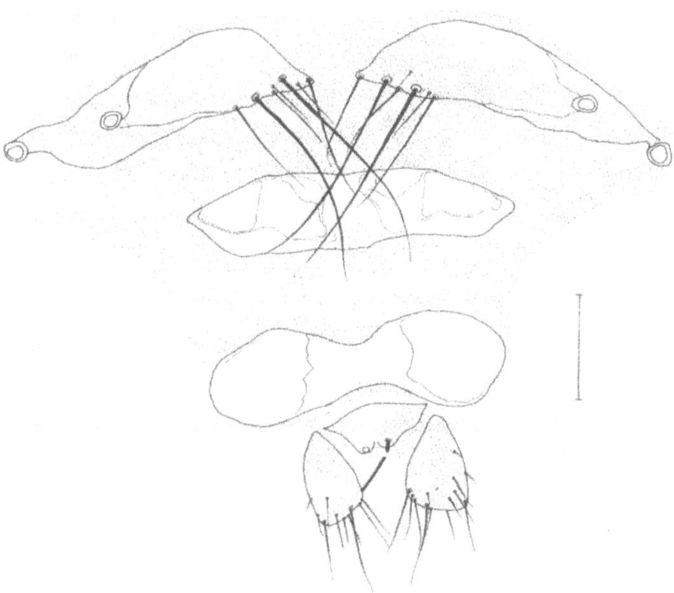

Fig. 119. *Bellardia stricta* (Villeneuve), female. Ovipositor, dorsal sclerites. Stipple indicates extent of microtrichiae. Scale = 0.2 mm. G.pr. 224.

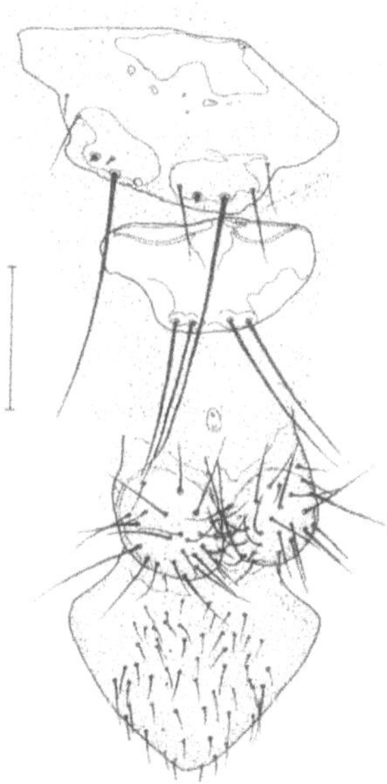

Fig. 120. *Bellardia stricta* (Villeneuve), female. Ovipositor, ventral sclerites. Stipple indicates extent of microtrichiae. Scale = 0.2 mm. G.pr. 224.

7. ***Bellardia vespillo*** (Fabricius, 1794), COMB.N.
Figs 54, 121-131.

Musca vespillo Fabricius, 1794: 318.
Onesia borotinensis Jacentkovský, 1944a: 45 (Czech), 47 (German). SYN.N.

♂♀. Arista (including hairing) about half as wide as first flagellomere, hairs short above, almost absent below. Palpi usually rather dark for a *Bellardia*, especially basally. r_{4+5} usually open, rarely closed at margin. Front tibia with 1 *pv* seta. Abdomen olive green, dusted, with tessellations.

♂. Frons 0.068-0.095 times head-width (n=5). Fronto-orbital plates almost touching. T4 with discal setae. Cerci resembling those of *vulgaris*, but somewhat grooved dorsally along the midline, and longer than surstyli. Surstyli strongly curved in profile, with an external groove which continues all the way to the tip. A distinct median projection basally and ventrally in posterior view.

♀. Frons 0.344-0.367 times head-width (n=5). Ovipositor with numerous discal setae on T6, also anteriorly. Most dried females can be recognised by this character without dissection. T6 without areas with weak sclerotisation. Epiproct much shorter than broad, with 2 strong and at most 2-3

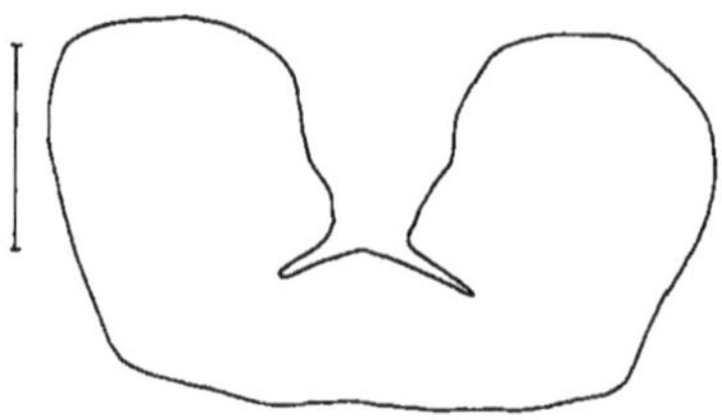

Fig. 121. *Bellardia vespillo* (Fabricius), male. ST5. Scale = 0.4 mm.

Fig. 122. *Bellardia vespillo* (Fabricius), male. Epandrium, cerci, surstylus and bacilliform sclerite, left lateral view. Scale = 0.2 mm.

Fig. 123. *Bellardia vespillo* (Fabricius), male. Epandrium (parts), cerci and surstyli, posterior view. Scale = 0.2 mm.

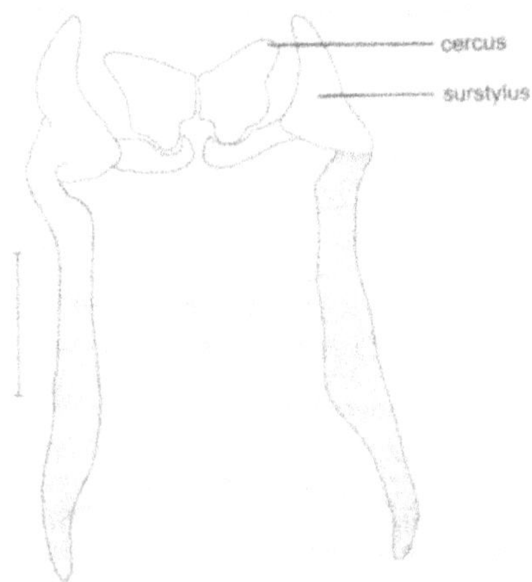

Fig. 124. *Bellardia vespillo* (Fabricius), male. Bacilliform sclerites, showing details of junction between cerci and junction between cercus and surstylus. Scale = 0.2 mm.

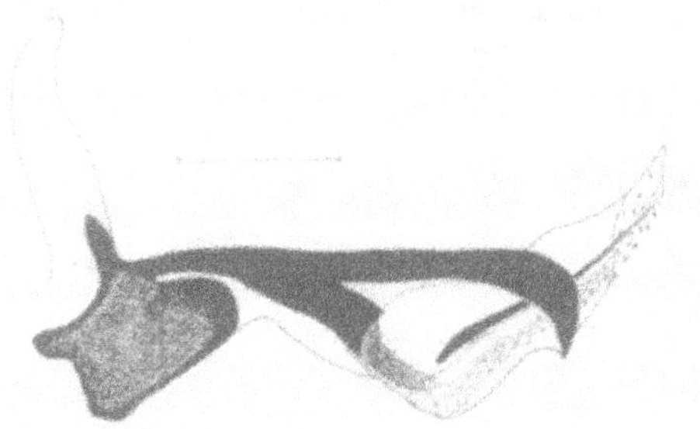

Fig. 125. *Bellardia vespillo* (Fabricius), male. Aedeagus, left lateral view. Scale = 0.2 mm.

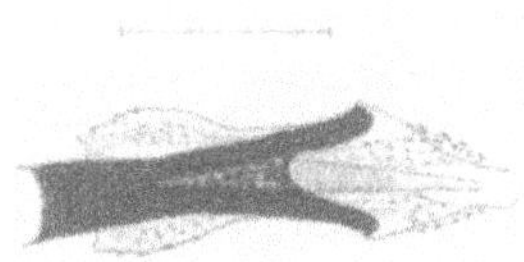

Fig. 126. *Bellardia vespillo* (Fabricius), male. Distiphallus, dorsal view. Scale = 0.2 mm.

Fig. 127. *Bellardia vespillo* (Fabricius), male. Ejaculatory sclerite, various views. Scale = 0.1 mm.

Fig. 128. *Bellardia vespillo* (Fabricius), male. Pre- and postgonites. Scale = 0.2 mm.

Fig. 129. *Bellardia vespillo* (Fabricius), female. Spermathecae. Scale = 0.1 mm. G.pr. 214.

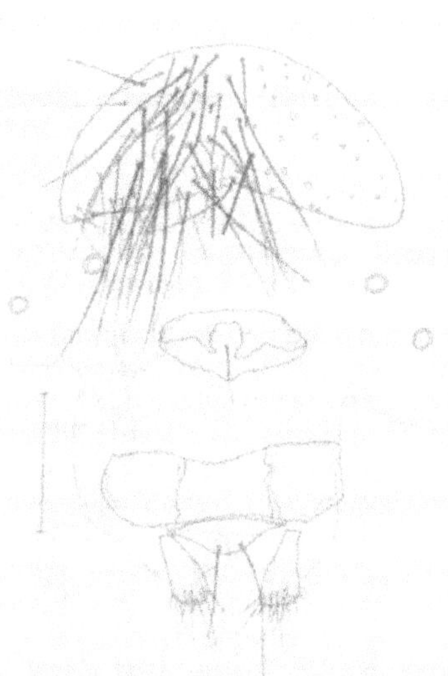

Fig. 130. *Bellardia vespillo* (Fabricius), female. Ovipositor, dorsal sclerites. Stipple indicates extent of microtrichiae. Scale = 0.4 mm. G.pr. 214.

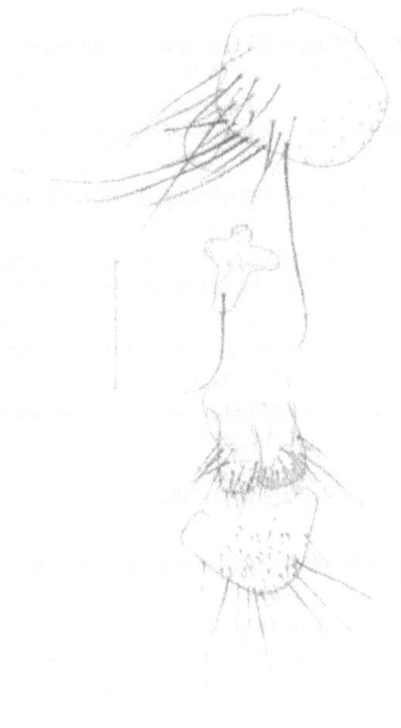

Fig. 131. *Bellardia vespillo* (Fabricius), female. Ovipositor, ventral sclerites. Stipple indicates extent of microtrichiae. Scale = 0.4 mm. G.pr. 214.

additional small setae. Cerci without microtrichiae in basal half, thus appearing polished black basally and dusted apically in dried specimens. ST6 as long as broad. [3 ovipositor slides examined.]

Length. 6.5-9.5 mm.

Distribution. Apparently very rare in Fennoscandia and Denmark. Known from Denmark (NEZ, NWZ, SZ) and Sweden (Sk.). Absent from Finland and Norway. - Outside Fennoscandia and Denmark I have seen specimens from Germany (the Fabrician syntypes, see above), Czechoslovakia (lectotype of *borotinensis*, Borotin in Moravia (see below); 1♂, Veltrusy-obora, M. Barták leg.), and Hungary (1♂, Dohnan, in ZMBERL, cf. Schumann, 1974b: 247, as "*agilis*"). It is probably mixed with *Bellardia vulgaris* in most collections.

Biology. Immature stages and hosts unknown. Apparently an early spring species, as all adults with known dates of capture (53 specimens) are from May or June.

Note. All credit for the discovery of the separate status of this species should be given to my friend Stig Andersen, ZMC, who collected a large number of specimens in Denmark and also did much background work concerning the synonymy. For various reasons we have agreed that all matters pertaining to this species should be treated by me in this volume. I am very grateful to him for his generosity in making this possible.

Nomenclature. *Musca vespillo* was described from an unspecified number of specimens from Kiel. In Fabricius' collection (ZMC, Kiel collection) there are two specimens under *vespillo* (not 1 as given by Zimsen, 1964). Both are male *Bellardia* in bad condition. They fit Fabricius' description tolerably well and I accept their status as syntypes. The first specimen lacks the abdomen and bears a label reading "vespillo" in Fabricius' hand. It has two *pv* setae on the front tibiae and the costa is bare below between tips of Sc and R_1. It may belong to *vulgaris* (Robineau-Desvoidy) or *pandia* (Walker). The second specimen, which is unlabelled, lacks the head. The abdomen has been dissected by Stig Andersen, ZMC. Its genitalia, though partly destroyed, are clearly referable to the present species. I have labelled and here designate this second specimen as lectotype and the first as paralectotype. *Onesia borotinensis* was described from a number of males. A single male present in coll. Mesnil, now in ZIKIEL, is probably the only extant syntype (cf. Schumann, 1974b). I have labelled and here designate this specimen as lectotype. Since the days of Meigen the name *vespillo* has been used for a species in the genus *Pollenia* Robineau-Desvoidy, and *Musca vespillo* Fabricius is the type-species of *Nitellia* Robineau-Desvoidy. For a discussion of the nomenclatural problems involved with the new status of the name *vespillo*, see the section Nomenclature under *Pollenia* below (p. 215).

Genus *Calliphora* Robineau-Desvoidy, 1830

Calliphora Robineau-Desvoidy, 1830: 433.
Type-species: *Musca vomitoria* Linnaeus, 1758: 595, by original designation.

Steringomyia Pokorny, 1889: 568.
Type-species: *Steringomyia stylifera* Pokorny, 1889: 569, by original designation and monotypy. SYN.N.

Neocalliphora Brauer & Bergenstamm, 1891: 87.
Type-species: *Calliphora dasyphthalma* Macquart, 1843: 287, by original designation; = *Musca quadrimaculata* Swederus, 1787.

Acrophaga Brauer & Bergenstamm, 1891: 367 [67].
Type-species: *Acrophaga stelviana* Brauer &

Bergenstamm, 1891: 367 [67], by designation of Brauer, 1893: 500.

Eucalliphora Townsend, 1908: 118.

Type-species: *Calliphora latifrons* Hough, 1899: 286, by original designation and monotypy.

Paracalliphora Townsend, 1916a: 151.

Type-species: *Calliphora oceaniae* Robineau-Desvoidy, 1830: 438, by original designation; = *Musca augur* Fabricius, 1775.

Abonesia Villeneuve, 1927: 357.

Type-species: *Musca genarum* Zetterstedt, 1838: 658, by original designation and monotypy.

Triceratopyga Rodendorf, 1931: 175.

Type-species: *Triceratopyga calliphoroides* Rodendorf, 1931: 175, by original designation and monotypy. SYN.N.

Aldrichiella Rodendorf, 1931: 177.

Type-species: *Calliphora grahami* Aldrich, 1930: 1, by original designation and monotypy. Junior homonym of *Aldrichiella* Vaughan, 1903, and *Aldrichiella* Hendel, 1911.

Stobbeola Enderlein, 1933: 126.

Type-species: *Stobbeola norwegica* Enderlein, 1933: 126, by original designation and monotypy; = *Acrophaga stelviana* Brauer & Bergenstamm, 1891.

Aldrichina Townsend, 1934: 111, replacement name for *Aldrichiella* Rodendorf, 1931.

Type-species: *Calliphora grahami* Aldrich, 1930: 1, automatic. SYN.N.

Acronesia Hall, 1948: 272.

Type-species: *Steringomyia aldrichia* Shannon, 1923: 112, by original designation.

Australocalliphora Kurahashi, 1971: 195, as subgenus of *Calliphora* Robineau-Desvoidy.

Type-species: *Calliphora onesioidea* Kurahashi, 1971: 200, by original designation.

Papuocalliphora Kurahashi, 1971: 167, as subgenus of *Calliphora* Robineau-Desvoidy.

Type-species: *Calliphora toxopeusi* Theowald, 1957: 158, by original designation.

In Fennoscandia and Denmark the genus *Calliphora* contains the familiar bluebottle flies, which are easily recognised by their large size, black ground-colour, broad compact body, blue metallic abdomen with silvery dusting and large first flagellomere. Scapes ("first antennal segments") touching. Cell r_{4+5} open.

Male frons narrow, usually without lateroclinate and proclinate orbital setae. Female frons broad, 1 lateroclinate and 2 proclinate orbital setae. Parts of head and antennae with reddish ground-colour. First flagellomere much more than 2 times as long as pedicel. Occiput often with mostly pale hairs. 2+3 *acr*, 3+3 *dc*, (0-)1+2 *ia* setae, 3 *sa*, 3+1 *h*, 1-2+1 *ph*, 3-5 *sc*. Posterior spiracle with large anterior lappet (Figs. 132-138). Mid tibia with 2-4 *ad* setae (but male *subalpina* usually 1 *ad*), most species also with 1-2 *v* setae. Hind tibia with strong *ad* and *d* preapicals of same size, any *pd* setae present in preapical position much weaker or hardly differentiated. Costa hairy below to junction with R_1 (Fig. 170). Lower calypter with hairs on most of disc. Abdomen metallic blue, dusted, and with more or less conspicuous tessellations. ST5 in male normal or enlarged with projecting lobes. Bacilliform sclerites separated, or more or less fused along the mid-line. Aedeagus with hypophallus serrated posteriorly, usually ending anteriorly in a more or less pronounced process (cornu) supporting the tip of paraphallus. Paraphallus often serrated apically and with a small projection subapically. Acrophallus varying in length, ventrally with posteriorly pointing denticles and often with a lateral duct, which is open along its whole length laterally and is roofed over by a shelf projecting from the medial part of the dorsal surface of acrophallus (Fig. 18, 19, 211).

Females with a primitive cylindrical uterus without lateral incubatory pouches and with unsclerotised lateral sacs (Hori, 1961; Kurahashi, 1971). In ovipositor, T6 with marginal setae only, without discal setae. T7 with rather long marginal setae. ST8 of full length, apically bilobed, even completely divided in one species. Cerci and epiproct without microtrichiae.

All species are oviparous, although delayed oviposition may result in larviposition. Larvae saprophagous in decaying animal matter. Adults visit flowers, faeces and dead animals. Many species are involved in human and animal myiasis. The genus is of great medical, hygienic and forensic importance (Schumann, 1965; Zumpt, 1965; Smith, 1973; 1986). The immature stages of 6 British species, which also occur in Fennoscandia and Denmark, have recently been described by Erzinçlioğlu (1985).

The genus is known from all zoogeographical regions, but is best represented in the Holarctic and Australian Regions. Its taxonomic limits are not well understood. Pending a phylogenetic analysis of *Calliphora* which takes full account of genital characters of both sexes, I favour (like Zumpt, 1956a) a broad concept of the genus. Contrary to Zumpt, I have also abandoned the monotypic genera *Steringomyia*, *Triceratopyga*, *Aldrichina*, with very peculiar male postabdomen but rather normal aedeagus and ovipositor (cf. Thomas, 1951). The two latter lack the *prst ia*, but this is also often the case with North American specimens of *Calliphora stelviana*. Unfortunately, the Manual of Nearctic Diptera. vol. 2 (Shewell, 1987) revives the genera *Acrophaga* and *Acronesia*, which had already been sunk into the synonymy of *Calliphora* by Zumpt (1956a), by splitting up *Acronesia* in the sense of Hall (1948; 1965). Furthermore, three of the species are combined in a separate genus (*Acrophaga*) for the first time to the exclusion of all the other *Calliphora*, i.e. *genarum* Zetterstedt, *stelviana* Brauer & Bergenstamm (called "*Acrophaga* sp." in the Manual) and *subalpina* Ringdahl. I can find no sound basis for this action in the structure of the male and female genitalia, and reject this assemblage as wholly artificial.

7 species in Fennoscandia and Denmark.

Note on synonymy. For the status of the names *Mya* Rondani and *Somomya* Bertoloni, listed by Schumann (1986) as synonyms of *Calliphora*, see Pont (1983). My suggestion (Rognes, 1986b) that Brauer & Bergenstamm (1891: 367) validly designated *Sarcophaga alpina* Zetterstedt as type-species of their new genus *Acrophaga* was based on a naive understanding of Brauer & Bergenstamm's use of the word "Type". The ovipositor of the subgenera *Paracalliphora*, *Australocalliphora* and *Papuocalliphora* (figured by Kurahashi, 1971) differs from other members of *Calliphora* in having very short ST8, and is rather similar to that of *Onesia* and *Bellardia*. Females of *Paracalliphora* have a short uterus with lateral incubatory pouches. They may not belong in *Calliphora*.

Key to species of *Calliphora*

1 Both calypters wholly white or almost wholly white; upper calypter with almost white rim (hairs along rim white or infuscated); basicosta black; basal *h* setae on a broken line, as if at the corners of an almost right-angled triangle; between *prst acr* often a conspicuous dark undusted stripe; occiput with 3 or more rows of black setulae behind postocular row of setae

2

– Both calypters dark brown or at least very conspicuously infuscated; upper calypter with infuscated rim and dark hairs along rim; lower calypter with white rim and white hairs along rim; basal *h* setae in a straight or gently curved line; most often 4 pairs of marginal scutellar setae

4

2(1) 2 inner and 1 outer *ph* setae; arista with long hairs above and below (Fig. 42), about as wide (including hairs) as first flagellomere; usually 4 pairs of marginal scutellar setae; face and parafacialia with yellow ground-colour and yellow dusting; undusted stripe between *prst acr* setae always broad and reaching setal bases; upper calypter with weakly infuscated hairs along rim; lower calypter sometimes with a slight darkening, especially across middle; strong median marginals present on T3 but these are usually shorter than half the length of T4; ♂: fronto-orbital plates contiguous, usually without a lateroclinate seta; TST7+8 convex in profile (Fig. 184); postabdomen swollen and prominent; lobes of ST5 large and projecting (Fig. 184); t_2 with 1 *ad* and almost always without a *v* seta; t_3 usually without, rarely with 1-2 small *av* seta; ♀: frons narrower than an eye when seen from above; t_2 with 2-3 *ad*; T5 longer than T4, marginal setae of T5 usually not strikingly strong nor very densely set so that no comb is formed (Fig. 193), posterior incision not extensive; T4 with decumbent ground-setulae

12. *subalpina* (Ringdahl)

– 1 inner and 1 outer *ph* setae; usually only 3 pairs of marginal scutellar setae (as in Fig. 169); lower calypter pure white except very close to median edge; ♂: fronto-orbital plates separated by a frontal vitta about as broad as or broader than each fronto-orbital plate, usually with a lateroclinate seta; t_2 with 2-3 *ad* and a *v* seta; t_3 with 2-4 *av* setae

3

3(2) Arista with short hairs above and below (Fig. 43), hairs very short on lower side of proximal half, their length here not exceeding width of arista itself (without hairs); arista (including hairs) about half as wide as first flagellomere; parafacialia and face usually with black ground-colour and silvery dusting; undusted stripe between *prst acr* broad and reaching setal bases; upper calypter posteriorly with pure white hairs along rim, hairs dark laterally near junction with rim of lower calypter; T3 with strong, erect median marginals, usually more than half as long as T4; ♂: postabdomen not swollen, normal in appearance; lobes of ST5 small, not projecting; ♀: T4 with rather sparse and mostly decumbent ground-setulae; T5 shorter than T4, marginal setae of T5 weak and inconspicuous, posterior incision hardly present

9. *genarum* (Zetterstedt)

– Arista with long hairs above and below (as in Fig. 42), those on lower side of proximal half longer than width of arista itself (without hairs); arista (including hairs) about as wide as first flagellomere; parafacialia and face usually with yellow ground-colour and yellowish dusting; undusted stripe between *prst acr* often indistinct and usually not reaching setal bases; *prst ia* very rarely absent; upper calypter posteriorly with infuscated hairs along rim; T3 usually without strong median marginals, these always less than half as long as T4; ♂: frons broad; fronto-orbital plates narrower than frontal vitta; a lateroclinate seta usually present; TST7+8 straight in profile (Fig. 171); postabdomen strongly swollen; lobes of ST5 large and strongly projecting (Fig. 171); ♀: frons broader than an eye when seen from above; T4 with densely set and rather erect ground-setulae; T5 about as long as T4, posterior margin with a very conspicuous mid-dorsal incision reaching forwards almost halfway to anterior edge and with numerous closely-set very strong marginal setae, forming a conspicuous comb at the hind end of the preabdomen, the combs of each side overlapping with each other when the abdomen is in the relaxed position (Fig. 180); T5 in profile strongly convex

11. *stelviana* (Brauer & Bergenstamm)

4(1) Basicosta yellow or yellowish-brown, but if darker then never all black; anterior thoracic spiracle orange; frons, upper half of parafacialia, most of face, posterior third of jowls and occiput with dark ground-colour; facial ridges, mouth-edge, and anterior two-thirds of jowls with orange ground-colour; occiput mostly with pale hairs, 1-2 irregular rows of black setulae behind postocular row of setae; hairs of gena all black; at most a faint and variable dark line between *prst acr* setae, often absent; infuscation of lower calypter not reaching rim; abdomen with conspicuous shifting tessellations, also on T3; ♂: width of frons at narrowest part 1.5-2 times distance between posterior ocelli inclusive; fronto-orbital plates here separated by a frontal vitta about as broad as front ocellus and about as broad as each fronto-orbital plate; ♀: dusting in upper half of parafacialia with a shifting pattern of dirty yellow and brownish-grey colour

8. *vicina* Robineau-Desvoidy

– Basicosta black; anterior thoracic spiracle usually infuscated

5

5(4) Postgena and lower parts of genal dilation clothed with orange hairs; parafacialia, face and jowls with mostly dark ground-colour; occiput with pale hairs only, at most with 1 irregular row of black setulae behind postocular row; a faint and variable dark line between *prst acr* setae; infuscation of lower calypter reaching rim; abdomen with less conspicuous shifting tessellations, not evident on T3; ♂: frons about as broad as distance between posterior ocelli inclusive

14. *vomitoria* (Linnaeus)

– Postgena and lower parts of genal dilation with black hairs only; occiput with several rows of black setulae behind postocular row of setae

6

6(5) Dark, conspicuous and undusted stripe present between *prst acr* and usually reaching setal bases; parafacialia, face and gena with black ground-colour; in upper part of parafacialia a shining white spot in some lights; facial membrane black; sometimes only 3 marginal scutellar setae; lower calypter sometimes with rather faint infuscation, this usually not quite reaching rim; abdomen with shifting tessellations on T3; ♂: frons narrower than distance between posterior ocelli inclusive; fronto-orbital plates contiguous; ♀: first flagellomere very large, its tip almost reaching lower facial margin, separated from it by less than maximum width of palp; parafacialia at narrowest point narrower than half the distance between inner margins of pores at base of large vibrissae; T5 shorter than T4, with a conspicuous incision in posterior edge almost reaching middle of tergite

10. *loewi* Enderlein

– Usually no undusted stripe between *prst acr* setae; parafacialia, face and often anterior parts of gena mostly with yellow ground-colour; upper part of parafacialia with uniform whitish-yellow to golden-yellow dusting, hardly shifting in different lights and no shining white spot; infuscation of lower calypter reaching rim; abdomen without shifting tessellations on T3; ♂: frons 1.5 times distance between posterior ocelli inclusive; ♀: first flagellomere rather small, its tip separated from lower facial margin by about its own width; parafacialia at narrowest point broader than half the distance between pores at base of large vibrissae; T5 about as long as T4, with a less conspicuous incision in posterior edge

13. *uralensis* Villeneuve

8. ***Calliphora vicina*** Robineau-Desvoidy, 1830
Figs 132, 139-148; Plate - fig 1.

Musca carnivora Fabricius, 1794: 313. [An application to the Commission proposing the suppression of this name has been published by Rognes & Blackith (1990).]

Musca erythrocephala Meigen, 1826: 62. Junior primary homonym of *Musca erythrocephala* De Geer, 1776: 146; *Musca erythrocephala* Fabricius, 1787: 351; *Musca erythrocephala* Villers, 1789: cxxxvii.

Calliphora vicina Robineau-Desvoidy, 1830: 435.

Calliphora littoralis Robineau-Desvoidy, 1830: 435.

Calliphora spitzbergensis Robineau-Desvoidy, 1830: 435.

Calliphora monspeliaca Robineau-Desvoidy, 1830: 436.

Calliphora musca Robineau-Desvoidy, 1830: 436.

Calliphora nana Robineau-Desvoidy, 1830: 436.

Calliphora scutellata Macquart, 1834: 161.

Musca thuscia Walker, 1849: 897.

Calliphora rufifacies Macquart, 1851: 216 [243].

Musca aucta Walker, 1853: 334.

Calliphora insidiosa Robineau-Desvoidy, 1863: 695.

Calliphora turanica Rodendorf, 1926: 90 (cf. Rodendorf, 1928: 336-337). SYN.N.

♂♀. This species is surprisingly often confused with *uralensis* in collections. It is easily recognised by the characters given in the key, especially the yellow basicosta, conpicuously tessellated abdomen, and yellowish anterior spiracle. Note, however, that occasional ♀ *uralensis* may have rather pale anterior spiracles, so this character should not be used on its own. Occiput weakly concave (especially in ♂), with 1-2 rows of small black setae behind the postocular setae, otherwise with pale hairs. 2 inner and 1 outer *ph* setae. f_2 without *a* preapical seta. t_1 with 1 *pv* seta. t_3 with 2-3 short *av* setae which are about as long as tibial diameter.

♂. Frons 0.080-0.120 times head-width (n=5), frontal vitta distinct, about as broad as each fronto-orbital plate. ST5 normal, not projecting. Bacilliform sclerites widely separated, membrane between them microtrichiose in ventral half and dorsally along mid-line. Cerci straight; surstyli rather broad, very weakly bent, upper and lower edge parallel, with long hairs. Acrophallus with lateral duct. Paraphallus with small subapical projection and apical denticles. Ventral plates not fused mid-ventrally. Hypophallus ending apically in a strongly projecting cornu. Epiphallus situated basally on the basiphallus.

♀. Frons 0.344-0.372 times head-width (n=5). Frontal vitta about as broad as or slightly broader than fronto-orbital plates. T6 rather long, also laterally. T8 with two small patches of short microtrichiae mid-dorsally. ST6 and ST7 rather narrow. Pleural membrane 8 microtrichiose on more than distal half. Spermathecae oval and rather short. [2 ovipositor slides examined.]

Length. 5-12 mm.

Distribution. Very common all over Fennoscandia and Denmark, from the lowlands to above the tree-line. Ascends to high alpine region in South Norway (1658 m a.s.l., TEI, Vinje: Vassdalseggi). Also Iceland (Nielsen, Ringdahl & Tuxen, 1954), but apparently not Greenland. - Cosmopolitan, in temperate and subtropical regions.

Biology. In Finland it is dependent on conditions created by man, except in the most southwestern parts (Nuorteva, 1963). Nuorteva (1977) gives a mean development time of 31.4 to 34.7 days under field conditions during July and August at 15°C at different sites in southwestern Finland. In a nest of *Parus major* a female *Protocalliphora azurea* was reared together with 19 specimens of *Calliphora vicina* (Lennart Magni leg. et cult, j.nr. M51, ZML). It has also been reared from a nest of *Hirundo rustica* with 4 dead nestlings (R. Mehl leg. et cult., j.nr. 268/67, RM). Involved in a case of intestinal myiasis in a baby in Finland (Nuorteva & Auvinen, 1968). Reported from a case of blowfly myiasis in the hedgehog (Nielsen, Nielsen & Walhovd, 1978) and in cases

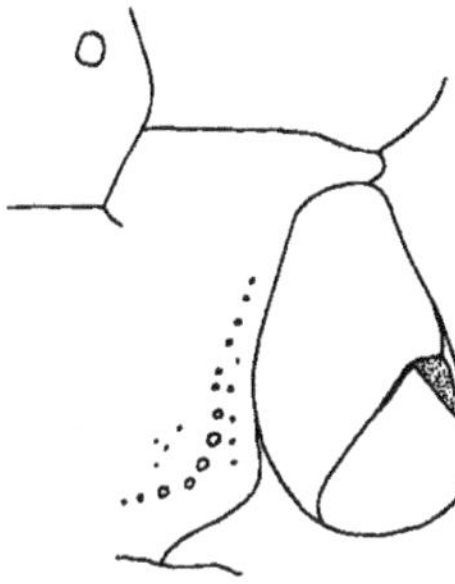

Fig. 132. *Calliphora vicina* (Robineau-Desvoidy), posterior spiracle.

Fig. 133. *Calliphora genarum* (Zetterstedt), posterior spiracle.

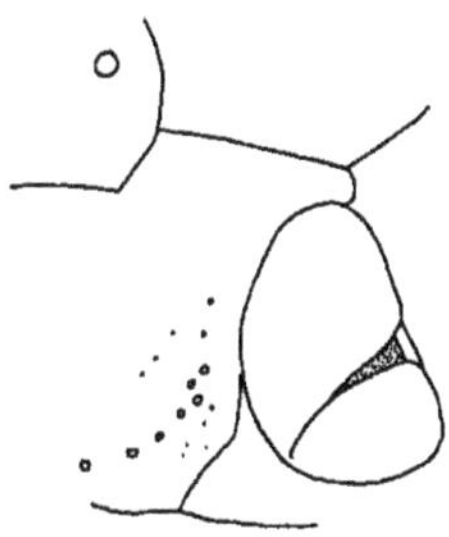

Fig. 134. *Calliphora loewi* (Enderlein), posterior spiracle.

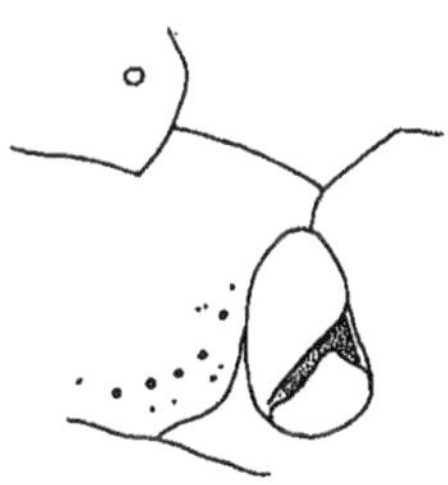

Fig. 135. *Calliphora stelviana* (Brauer & Bergenstamm), posterior spiracle.

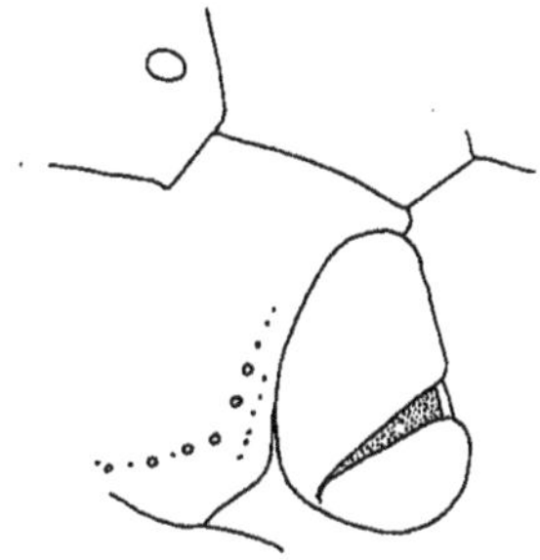

Fig. 136. *Calliphora subalpina* (Ringdahl), posterior spiracle.

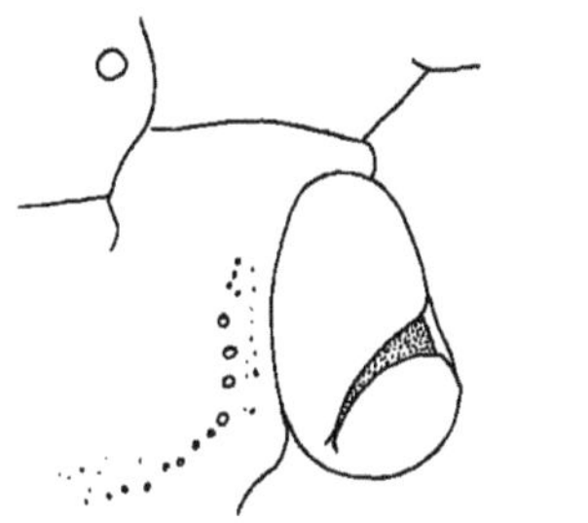

Fig. 137. *Calliphora uralensis* (Villeneuve), posterior spiracle.

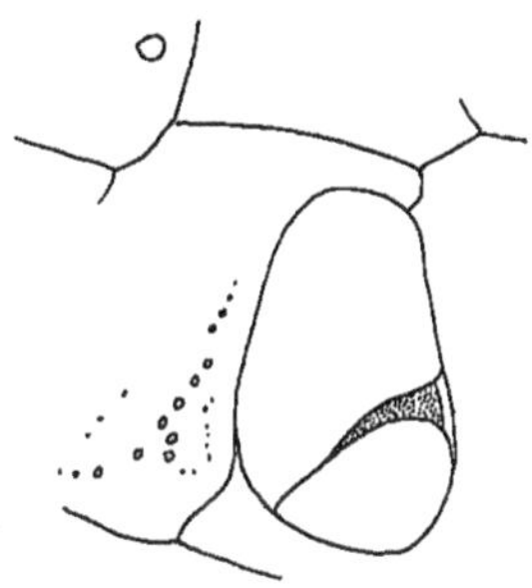

Fig. 138. *Calliphora vomitoria* (Linnaeus), posterior spiracle.

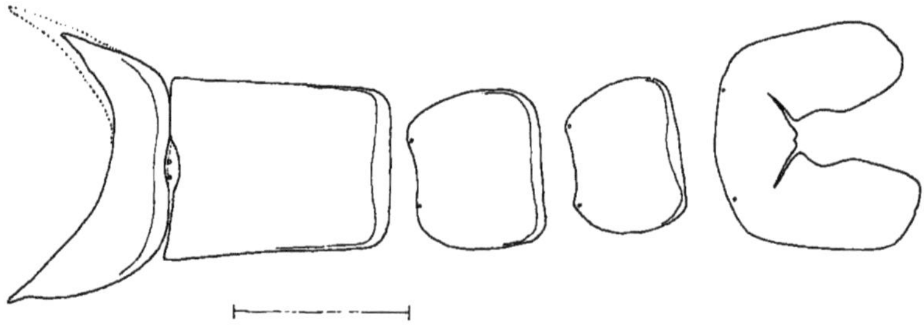

Fig. 139. *Calliphora vicina* Robineau-Desvoidy, male. ST 1-5. Scale = 1.0 mm.

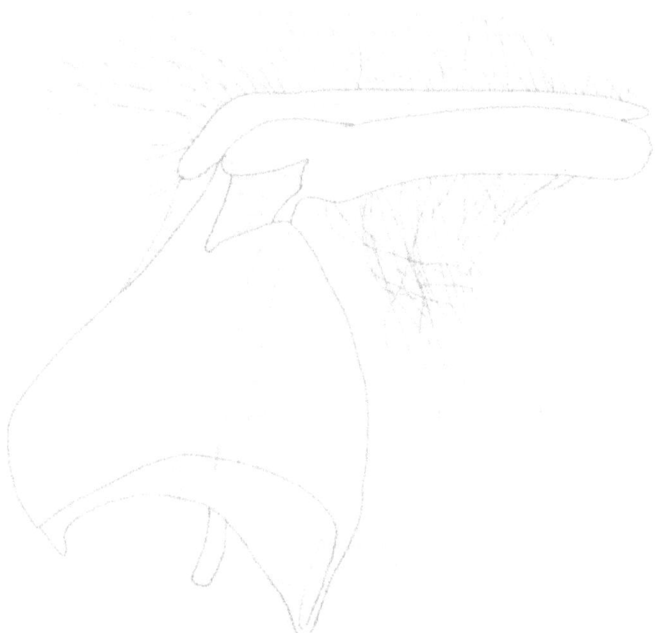

Fig. 140. *Calliphora vicina* Robineau-Desvoidy, male. Epandrium, cerci, surstylus and bacilliform sclerite, left lateral view. Scale= 0.2 mm.

Fig. 142. *Calliphora vicina* Robineau-Desvoidy, male. Bacilliform sclerites and vestiture on membrane between them. Scale = 0.2 mm.

Fig. 141. *Calliphora vicina* Robineau-Desvoidy, male. Epandrium, cerci and surstyli, posterior view. Scale = 0.2 mm.

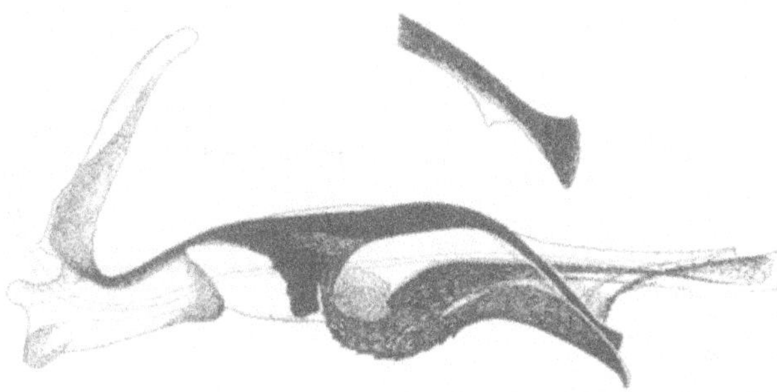

Fig. 143. *Calliphora vicina* Robineau-Desvoidy, male. Aedeagus, left lateral view. Inset: tip of paraphallus (enlarged). Scale = 0.2 mm.

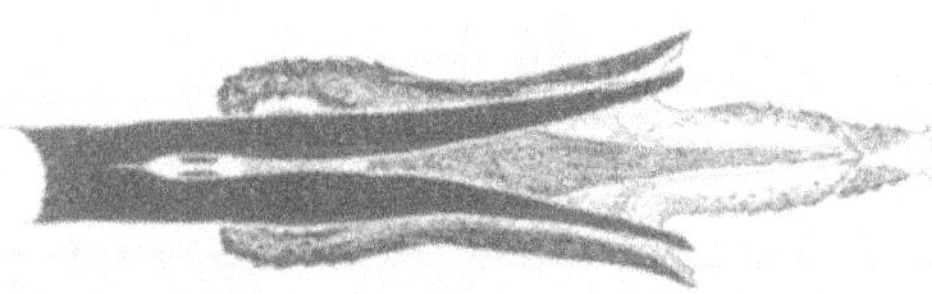

Fig. 144. *Calliphora vicina* Robineau-Desvoidy, male. Distiphallus, dorsal view. Scale = 0.2 mm.

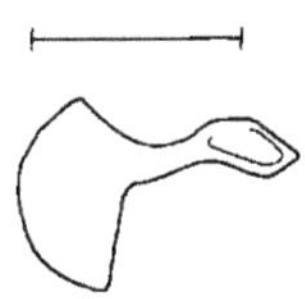

Fig. 145. *Calliphora vicina* Robineau-Desvoidy, male. Ejaculatory sclerite. Scale = 0.2 mm.

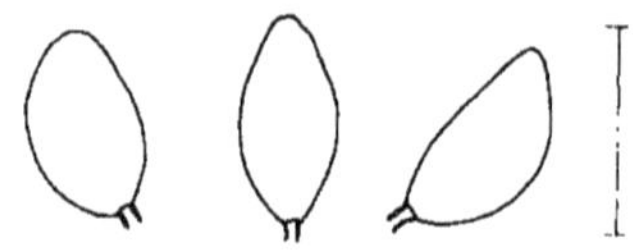

Fig. 147. *Calliphora vicina* Robineau-Desvoidy, female. Spermathecae. Scale = 0.2 mm. G.pr. 273.

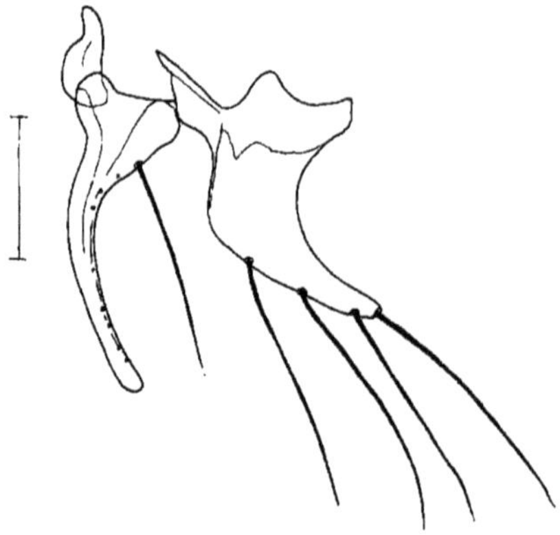

Fig. 146. *Calliphora vicina* Robineau-Desvoidy, male. Post- and pregonites. Scale = 0.2 mm.

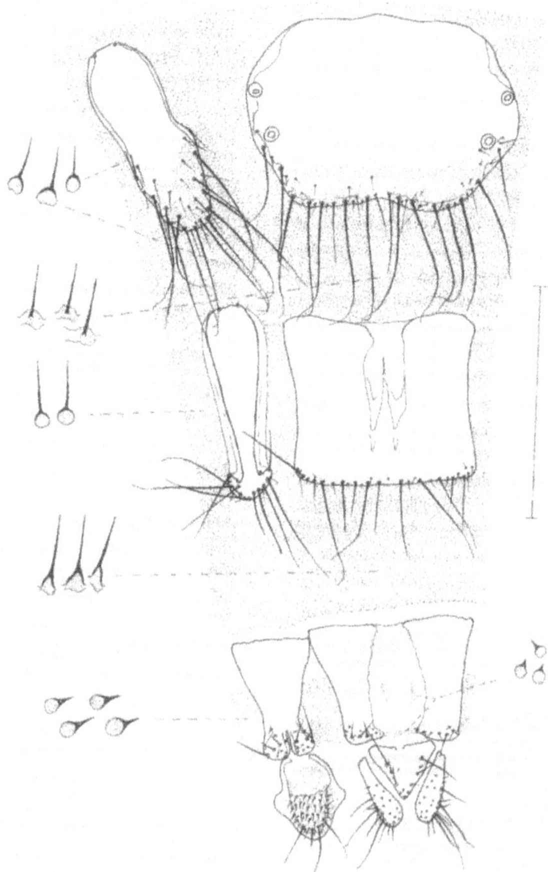

Fig. 148. *Calliphora vicina* Robineau-Desvoidy, female. Ovipositor. Stipple indicates extent of microtrichiae. Insets: shape of individual microtrichiae in areas indicated (enlarged). Scale = 1.0 mm. G.pr. 273.

of sheep-strike in Denmark (Nielsen, 1984a, 1984b). Attracted to the stinkhorn fungus (*Phallus impudicus* Pers.) in Denmark (Nielsen, 1963, 1967, 1968). In Fennoscandia it is on the wing mostly from April to October, but it has also been collected during the winter months which indicates that it overwinters in the adult stage (cf. also data in Nuorteva, Isokoski & Laiho, 1967). This is definitely the case in Norway. In England it is reported as overwintering as a larva (Norris, 1965; Bourke, 1975).

Nomenclature. *Musca carnivora* Fabricius, listed as a synonym of *Calliphora vomitoria* (Linnaeus) by Schumann (1986), was described from an unspecified number of specimens of unspecified sex, "Habitat in Germaniae carne putrescente". In Fabricius' collection in ZMC (Kiel collection) are 2 specimens under *carnivora*. The first carries a label reading "carnivor" in Fabricius' hand. It is in very bad condition, consisting of fragments of a wing, lower calypter and scutellum. However, it is clearly identifiable as *vicina* Robineau-Desvoidy as currently understood. The second is a fragment of a specimen of *Sarcophaga* (s.lat.). I regard the first specimen as the holotype of *carnivora* and have labelled it as such. The second specimen is probably the later addition mentioned by Fabricius (1805). (This treatment of the specimens was first suggested by A. C. Pont on a label fixed beside the specimens in the Kiel collection). *Musca erythrocephala* was described from a syntypic series of both sexes, which is evident from Meigen's Plate 248, fig. 10 (Morge, 1976b). In Meigen's collection in MNHN are 1♂2♀ under *Musca erythrocephala*. 1♂1♀ are the present species, 1♀ = *vomitoria*. To fix the identity of the name *erythrocephala* in the current sense, the male should be designated as lectotype. Dear (1986) has recognised and labelled the holotype of *Calliphora vicina* Robineau-Desvoidy (in Bigot's Diptera Exotica collection in Oxford). I have seen the ♂ holotype of *Calliphora rufifacies* Macquart in MNHN. Zumpt (1956a) and Schumann (1986) treat *Calliphora turanica* as a synonym of *uralensis*. This is an error. Rodendorf (1926; 1928) figures the genitalia, and they definitely belong to *vicina*.

9. ***Calliphora genarum*** (Zetterstedt, 1838)
Figs 7, 43, 133, 149-158.

Musca genarum Zetterstedt, 1838: 658.
Musca laticornis Zetterstedt, 1838: 658.
Sarcophaga alpina Zetterstedt, 1838: 651.
Sarcophaga vomitorina Zetterstedt, 1845: 1305. Unavailable, published as a junior synonym of *Sarcophaga alpina* Zetterstedt, 1838: 651, and never adopted. SYN.N.
Calliphora popoffana Townsend, 1908: 117.
Acronesia collini Hall, 1948: 279.
Calliphora erectiseta Fan, 1957: 347. SYN.N.

♂♀. Easily recognised by the short haired arista, white calypters, strong marginal setae on T3, normal non-projecting postabdomen in the ♂, and weak and inconspicuous marginal setae of T5 in the ♀. Head with black ground-colour except for lower facial margin, vibrissal corner and lower parts of facial ridge. First flagellomere large and black. Facial membrane black. Parafacials black and densely silvery dusted. Occiput strongly convex with black setae. 1 inner and 1 outer *ph* setae. 3 pairs of scutellar marginals. Costal spine conspicuous. f_2 with an *a* preapical seta in both sexes. t_1 with 1 *pv* seta. t_3 with 2-3 *av* setae, slightly longer than tibial diameter.

♂. Frons rather broad, 0.168-0.212 times head-width (n=5). Lateroclinate seta present. Frontal vitta 1.5-2 times width of fronto-orbital plates. Postocular setae long, slender and gently curved. Bacilliform sclerites widely separated, membrane between them microtrichiose ventrally. Cerci slender. Surstylus moderately broad, with soft hairs only. Acrophallus strikingly short, with a lateral duct. Paraphallus dentate distally. Hypophallus distally with a conspicuous ventral projection. Epiphallus situated basally on the basiphallus.

♀. Frons 0.344-0.358 times head-width (n=5). Frontal vitta almost 3 times width of fronto-orbital plates. T8 without microtrichiae mid-dorsally. ST6 with rather strong marginals. ST6 and ST7 broad. ST8 broad proximally and cleft distally. Pleural membrane 8 microtrichiose on more than distal half. Spermathecae oval and rather short. (The bifid apex of the one shown in Fig. 157 is an aberration.) [2 ovipositor slides examined.]

Length. 6.5-10 mm.

Distribution. Rather infrequent in subalpine and low alpine regions of Norway, Finland and Sweden. Absent from Denmark. - Palaearctic: Alps (Austria, West Germany, Switzerland) (material in ZMC, Zumpt, 1956a); North European territory of the U.S.S.R.; northern West Sibiria, northern East Siberia, Kirgizia, Far East (Chernov, 1959, 1961, 1965; Sýchevskaya, 1970; Ozerov, 1986; Schumann, 1986); China (Inner Mongolia) (Fan, 1957). Nearctic: Alaska to Quebec and Labrador, south to north Manitoba and Labrador (Hall, 1965).

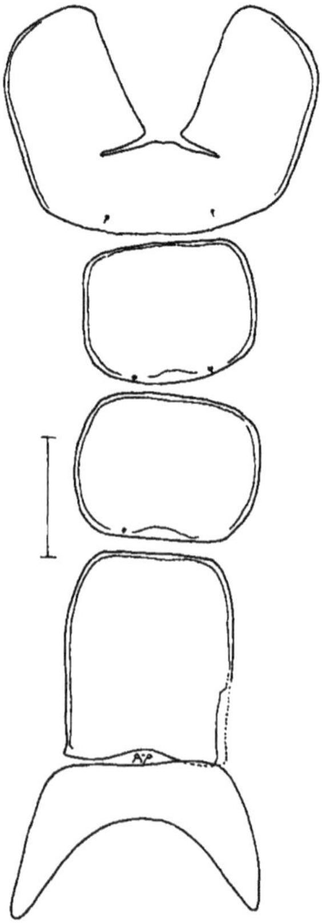

Fig. 149. *Calliphora genarum* (Zetterstedt), male. ST 1-5. Scale = 0.4 mm.

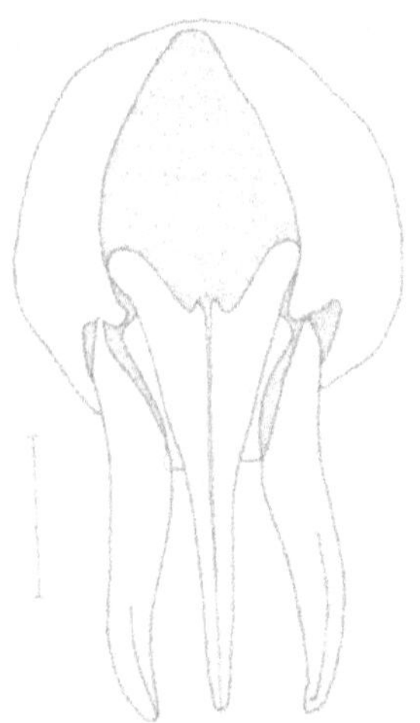

Fig. 151. *Calliphora genarum* (Zetterstedt), male. Epandrium, cerci and surstyli, posterior view. Scale = 0.2 mm.

Fig. 150. *Calliphora genarum* (Zetterstedt), male. Epandrium, cerci, surstylus and bacilliform sclerite, left lateral view. Scale= 0.2 mm.

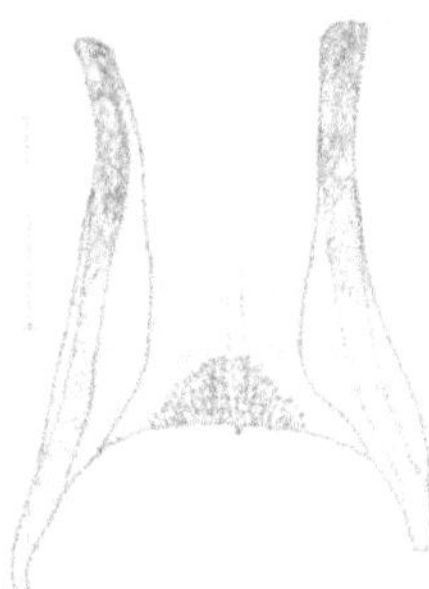

Fig. 152. *Calliphora genarum* (Zetterstedt), male. Bacilliform sclerites and vestiture on membrane between them. Scale = 0.2 mm.

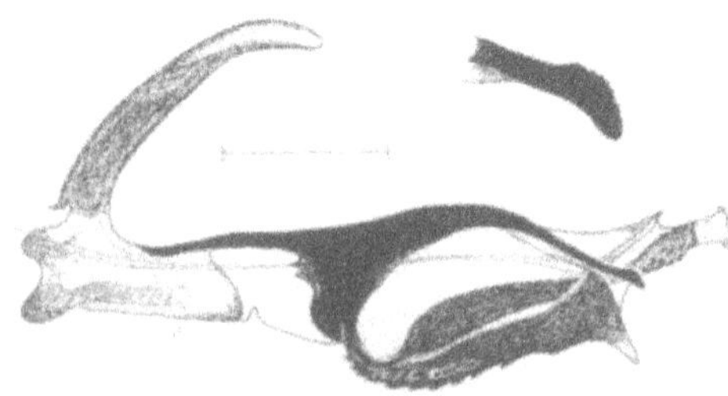

Fig. 153. *Calliphora genarum* (Zetterstedt), male. Aedeagus, left lateral view. Inset: tip of paraphallus (enlarged). Scale = 0.2 mm.

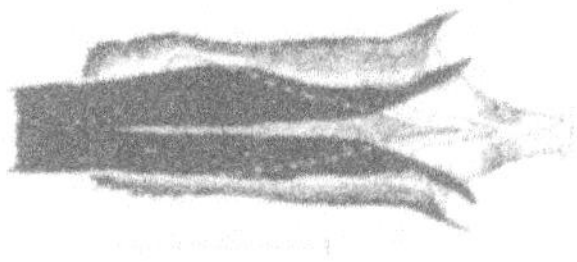

Fig. 154. *Calliphora genarum* (Zetterstedt), male. Distiphallus, dorsal view. Scale = 0.2 mm.

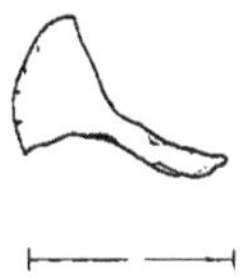

Fig. 155. *Calliphora genarum* (Zetterstedt), male. Ejaculatory sclerite. Scale = 0.2 mm.

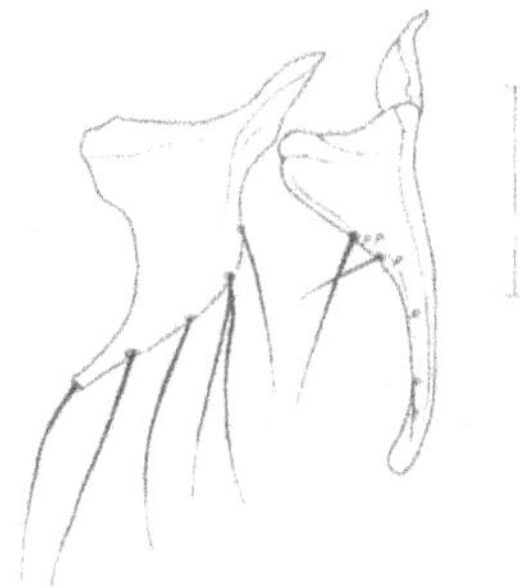

Fig. 156. *Calliphora genarum* (Zetterstedt), male. Pre- and postgonites. (Only a single basal seta present on the postgonite of other side). Scale = 0.2 mm.

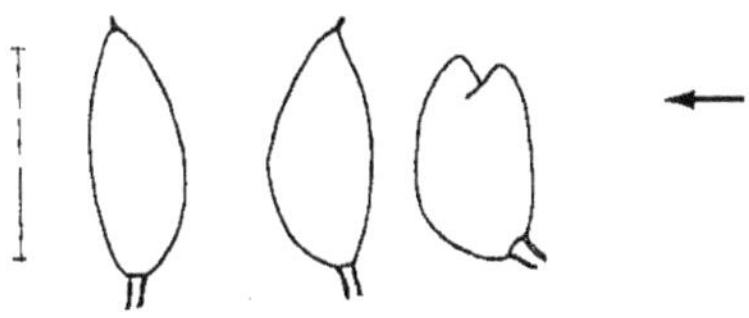

Fig. 157. *Calliphora genarum* (Zetterstedt), female. Spermathecae. (The bifid tip of one them is an abnormality.) Scale = 0.2 mm. G.pr. 263.

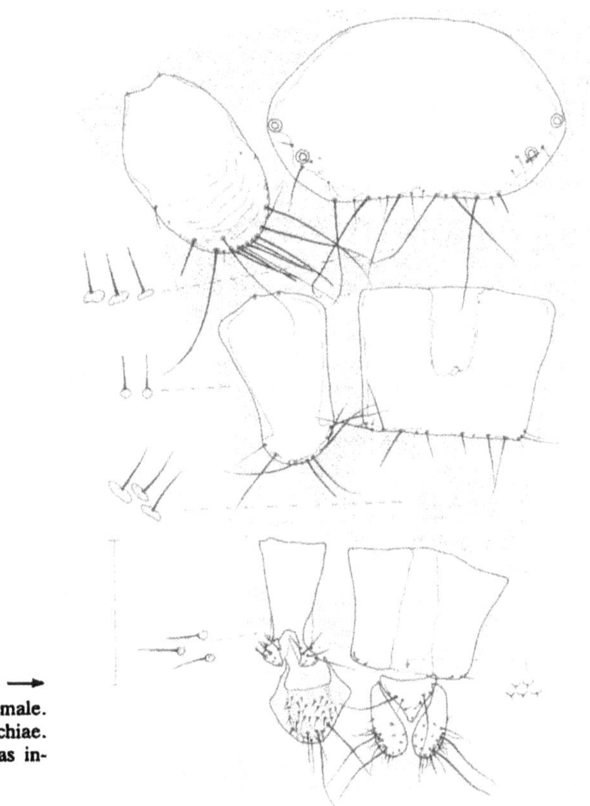

Fig. 158. *Calliphora genarum* (Zetterstedt), female. Ovipositor. Stipple indicates extent of microtrichiae. Insets: shape of individual microtrichiae in areas indicated (enlarged). Scale = 0.5 mm. G.pr. 263.

Biology. In Finland largely independent of conditions created by man (Nuorteva, 1963, 1964c). Reared from carcasses of small rodents in Northern Siberia (Chernov, 1965). Larval stages undescribed. In Sweden (Jmt.) taken in copula July 10 (O.Ringdahl leg). In Fennoscandia adults have been collected from June to August.

Note. Its closest relative may be *Calliphora rohdendorfi* Grunin, 1970b, from Tadzhik S.S.R. (U.S.S.R.) and Nan Shan Mountains (China), to judge from the structure of the aedeagus. Both have very short acrophallus and rather similar cerci and surstyli. Note that *C. rohdendorfi* has dark calypters and yellowish-grey occipital vestiture. The Manual of Nearctic Diptera, vol. 2 (Shewell, 1987) unites *genarum* with *stelviana* and *subalpina* in a separate genus (*Acrophaga*) on the basis of white lower calypters, sometimes large postabdomen in ♂, and posteriorly cleft T5 in the ♀. The latter character does not apply to *subalpina* or *genarum*. The structure of the ♂ and ♀ genitalia of *genarum* does not support this classification and it is not accepted here.

Nomenclature. I have not seen types of *Calliphora erectiseta*, but I suggested the synonymy to Dr. Fan, Shanghai Institute of Entomology, who agrees with it (in litt. August 5, 1987). For the identity of *Sarcophaga alpina* Zetterstedt, see Rognes (1986b).

10. ***Calliphora loewi*** Enderlein, 1903
Figs 134, 159-168.

Calliphora loewi Enderlein, 1903: 254.
Onesia germanorum Villeneuve, 1907: 398.
Calliphora morticia Shannon, 1923: 116. SYN.N.

♂♀. 3-4 pairs of marginal scutellars. Head with black ground-colour, except for lower facial margin, vibrissal angle and lower parts of facial ridge. 2 inner and 1 outer *ph* setae. Occiput concave (especially in the ♂) with 3-5 rows of black setae, centrally with pale hairs. t_2 with 1 *pv* seta. t_3 with 2-3 *av* setae.

♂. Frons narrow, 0.040-0.054 times head-width (n=5). Fronto-orbital plates contiguous. f_2 usually without *a* preapical seta. Bacilliform sclerites elongate, well-separated. Membrane between them microtrichiose ventrally. Cerci long and narrow. Surstyli long and slender, apically pointed, ventrally with rather regular moderately long vestiture. Acrophallus of normal length, with lateral duct. Paraphallus apically dentate with subapical projection. Hypophallus with well developed distal cornu. Epiphallus situated basally on basiphallus. Postgonite with distal flattened widening.

♀. Frons 0.341-0.363 times head-width (n=5). Frontal vitta about 3 times width of fronto-orbital plates. First flagellomere very large. f_2 with *a* preapical seta. T5 with large posterior incision. T6 long, almost divided into two halves by a weak mid-dorsal sclerotisation. T7 halves narrowly joined behind. T8 microtrichiose mid-dorsally between strongly sclerotised parts, bare on anterior fourth. ST7 very narrow. Pleural membrane 8 microtrichiose on posterior three-fourths. [2 ovipositor slides examined.]

Length. 7-12 mm.

Distribution. Widespread in Fennoscandia and Denmark, but rare in south Sweden and Denmark. Ascends to subalpine and low alpine zone (Ringdahl, 1951; Nuorteva, 1964c). - Palaearctic: From North and Central Europe east to Amur Oblast (U.S.S.R.), Mongolia and Japan (Ozerov, 1986; Schumann, 1986). Nearctic: Alaska (Shannon, 1923; Hall, 1965), British Columbia, Yukon Territory (Shewell, in litt. Oct. 31, 1983), Newfoundland (Øystein Rognes leg., in KR).

Biology. In Finland it is asynanthropic and avoids human settlements (Nuorteva, 1963). Collected on the stinkhorn fungus (*Phallus impudicus* Pers.) in Denmark (Nielsen, 1967, 1968). In Fennoscandia and Denmark adults have been collected from June to October; a female has been taken indoors (N, TEI: Notodden).

Nomenclature. I have examined all the type-material of *Calliphora morticia* (from Alaska, in USNM), including the holotype ♂, which has the genitalia dissected.

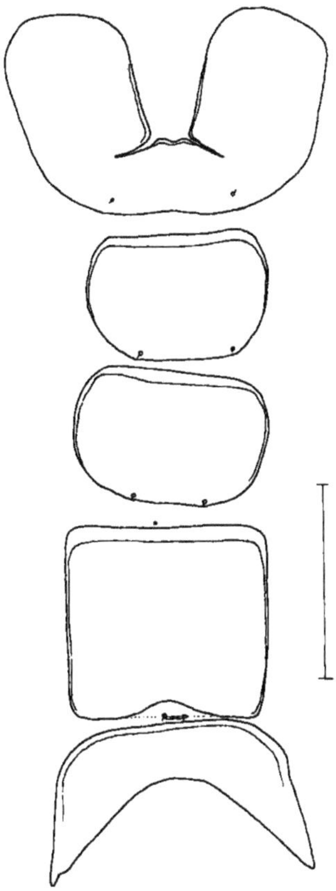

Fig. 159. *Calliphora loewi* (Enderlein), male. ST 1-5. Scale = 1.0 mm.

Fig. 160. *Calliphora loewi* (Enderlein), male. Epandrium, cerci, surstylus and bacilliform sclerite, left lateral view. Scale = 0.2 mm.

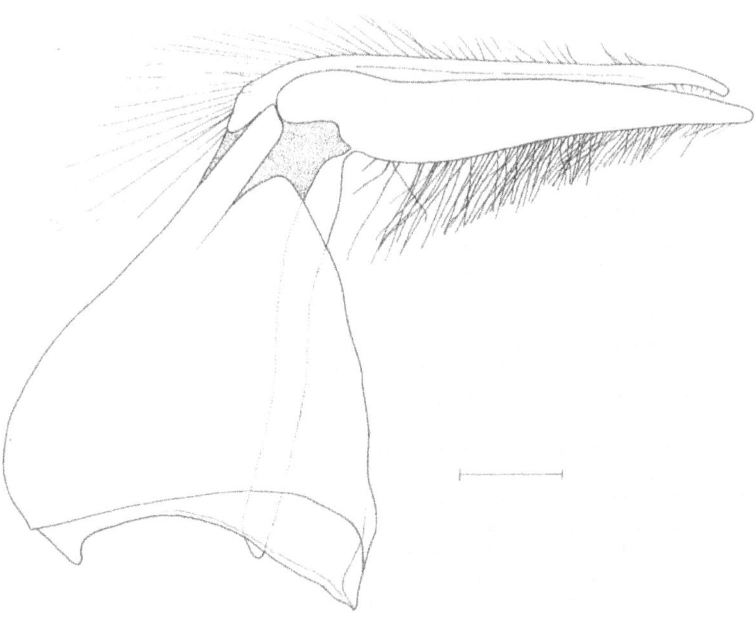

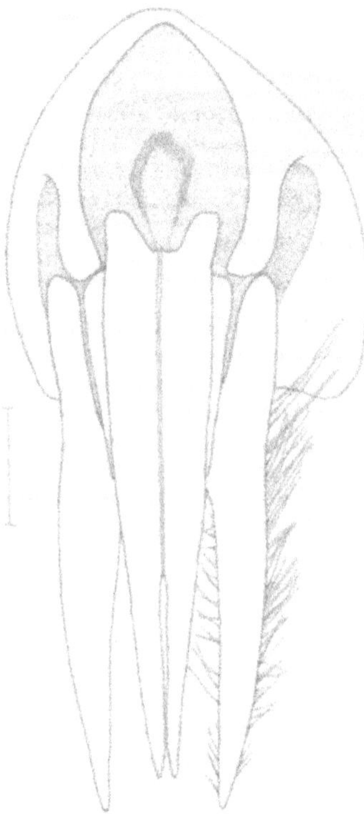

Fig. 161. *Calliphora loewi* (Enderlein), male. Epandrium, cerci and surstyli, posterior view. Scale = 0.2 mm.

Fig. 162. *Calliphora loewi* (Enderlein), male. Bacilliform sclerites and vestiture on membrane between them. Scale = 0.4 mm.

Fig. 163. *Calliphora loewi* (Enderlein), male. Aedeagus, left lateral view. Inset: tip of paraphallus (enlarged). Scale = 0.2 mm.

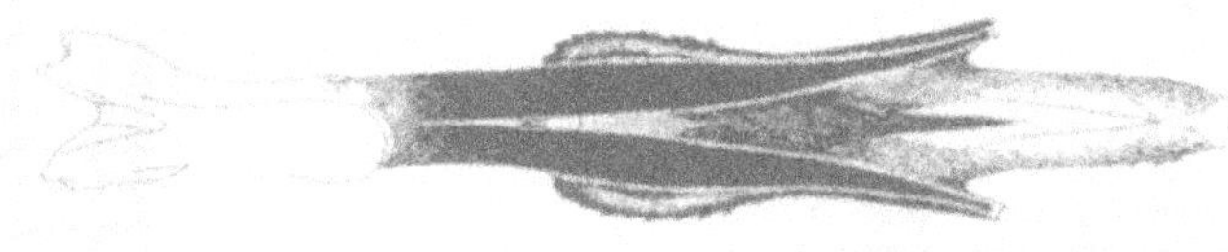

Fig. 164. *Calliphora loewi* (Enderlein), male. Aedeagus, dorsal view. Scale = 0.2 mm.

Fig. 165. *Calliphora loewi* (Enderlein), male. Ejaculatory sclerite. Scale = 0.2 mm.

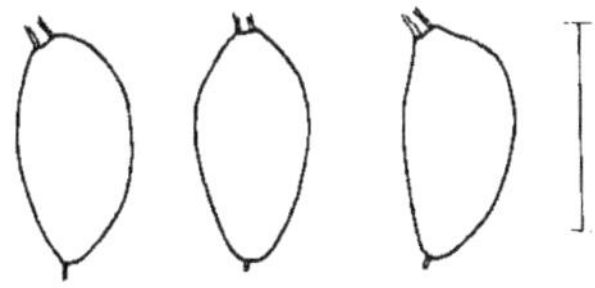

Fig. 167. *Calliphora loewi* (Enderlein), female. Spermathecae. Scale = 0.2 mm. G.pr. 269.

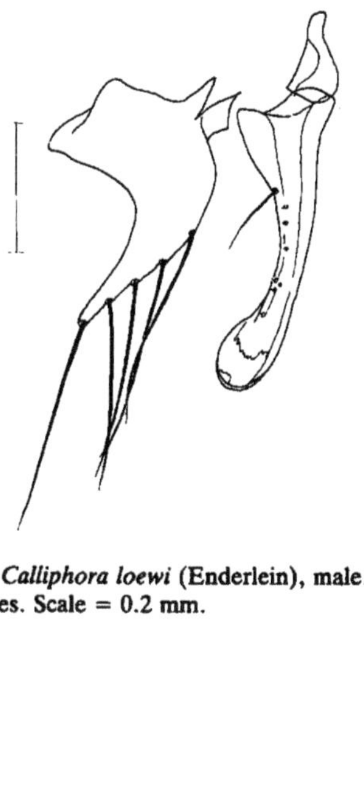

Fig. 166. *Calliphora loewi* (Enderlein), male. Pre- and postgonites. Scale = 0.2 mm.

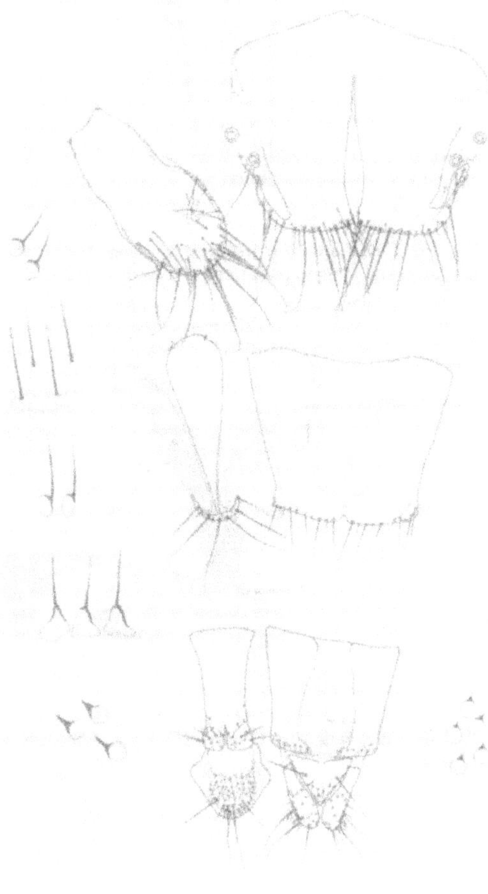

Fig. 168. *Calliphora loewi* (Enderlein), female. Ovipositor. Stipple indicates extent of microtrichiae. Insets: shape of individual microtrichiae in areas indicated (enlarged). Scale = 1.0 mm. G.pr. 269.

11. ***Calliphora stelviana*** (Brauer & Bergenstamm, 1891)
Figs 135, 169-183.

Acrophaga stelviana Brauer & Bergenstamm, 1891: 367 [63].
Stobbeola norwegica Enderlein, 1933: 126.
Acronesia abina Hall, 1948: 274. SYN.N.
Acronesia anana Hall, 1948: 278.
alpina: authors (e.g. Zumpt, 1956a; Schumann, 1986), not Zetterstedt, 1838; misidentifications.

♂♀. Easily recognised by the characters given in the key, especially presence of 1 inner and 1 outer *ph* seta, strongly projecting postabdomen and ST5 in the ♂, and dense marginal row of heavy setae on T5 in the ♀. 1 *prst ia* (Scandinavian material) or 0-1 *prst ia* (North American material). Costal spine conspicuous. Occiput convex with mostly black setae. t_1 with 2 *pv* setae. f_2 with *a* preapical seta. t_3 with 2-3 rather strong *av* setae. Scutellum with 3 strong marginal pairs of setae.

♂. Frons 0.155-0.172 times head-width (n=5). Frontal vitta as broad as or slightly broader than fronto-orbital plates. Lateroclinate setae usually (but not always) present. Postabdomen large and projecting. TST7+8 straight in profile. Epandrium long. ST5 large and projecting. Bacilliform sclerites fused along mid-line. Cerci slightly longer than the surstyli, both clothed with short straight spines, which are absent along a narrow zone on the external surface of the surstylus. Paraphallus unarmed apically. No shelf or projection on the hypophallus forming a support for the movement of the paraphallus. Acrophallus apparently without a lateral duct. Pregonite of striking form, very narrow, with about 7 apical setae. Postgonite without or with a small basal seta.

♀. Frons 0.382-0.406 times head-width (n=5). Frontal vitta more than 2 times as wide as fronto-orbital plates. T5 margin with a row of very strong spines. T6 with a peculiar transverse deep broad groove. Spiracles not included in T6 sclerotisation. T7 and T8 without microtrichiae. ST6 short and broad. ST7 broad. ST8 divided in two separate sclerotisations, apically with a pair of very long setae. Pleural membrane 8 wholly without microtrichiae. Spermathecae very long and narrow with apical tubercle. [2 ovipositor slides examined.]

Length. 6-11 mm.

Distribution. Subalpine to high alpine parts of Norway, Sweden and Finland. Absent from Denmark. In South Norway taken at 1841 m a.s.l. (ON, Lom: Juvvasshytta). - Palaearctic: North European territory (Vaigach Is and Yugorskiy Peninsula), and northern East Siberia (Agapa) of the U.S.S.R. (Chernov, 1959, 1965), northern Great Britain (Emden, 1954; Dear, 1981; Horsfield, 1984), Alps (Austria (Zumpt, 1956a), Italy (type locality of *Acrophaga stelviana*)), Czechoslovakia (Gregor, 1987), Carpathians (Poland) (Draber-Mońko, 1978), Tien-Shan, Pamir (Grunin, 1970a). Nearctic: Alaska to northern Quebec and northern Labrador, Colorado (Hall, 1965); Greenland (Henriksen, 1939).

Biology. Widespread in uninhabited areas, also taken in dense settlements in northern Finland (Nuorteva, 1963, 1964c). This is also the case in northern U.S.S.R. (Chernov, 1965). It has a single generation each year in northern Finland (Nuorteva, 1971b). I have reared it from carcasses of *Microtus oeconomus* (Pallas) (TEI, Vinje; VAI, Sirdal) and *Sorex araneus* Linnaeus (TEI, Vinje) from the mountains of southern Norway, alone or together with *Cynomya mortuorum*. In Fennoscandia adults have been collected from June to August.

Note. *Calliphora stelviana* appears to be extremely close to *Calliphora chinghaiensis* Van & Ma from China, which also has a very narrow pregonite, but in *chinghaiensis* it has only three apical setae (Van & Ma, 1978). Furthermore, *chinghaiensis* has 4 marginal scutellars. The aedeagus appears indistinguishable from that of *stelviana*. *C. stelviana* and *chinghaiensis* may form a monophyletic group on the basis of the structure of the pregonite and aedeagus. *C. stelviana* and *chinghaiensis* share with *alaskensis* Shannon, *echinosa* Grunin and *subalpina* Ringdahl the dense cover of short straight spines on the cerci and surstyli and the distal position of the epiphallus on the basiphallus. Possibly these characters are synapomorphies, indicating the monophyly of this group. Note that *alaskensis* and *echinosa* both have dark lower calypters. *C. stelviana* shares with *subalpina* the absence of a lateral duct from the acrophallus and the absence of microtrichiae from pleural membrane 8 in the ovipositor.

Nomenclature. Bezzi & Stein (1907) cite *Cynomyia mohileviana* Portschinsky, 1875: 37 as a separate species. Schumann (1986) regards it as a questionable synonym of *stelviana* ("*alpina*"). The type locality is "circa Mohileviam ad Borysthenem" [= near Mogilev by the river Dnepr, Byelorussian S.S.R.]. As *C. stelviana* is not known to occur there (Schumann, 1986), the name must surely apply to another species. I have therefore deleted it from the synonymy of *stelviana*. *Acronesia abina* and *A. anana* are based on the absence or presence of the *prst ia*. I have examined specimens

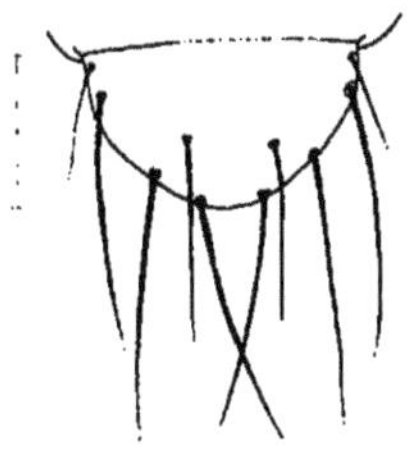

Fig. 169. *Calliphora stelviana* (Brauer & Bergenstamm), male. Scutellum. Scale = 1.0 mm.

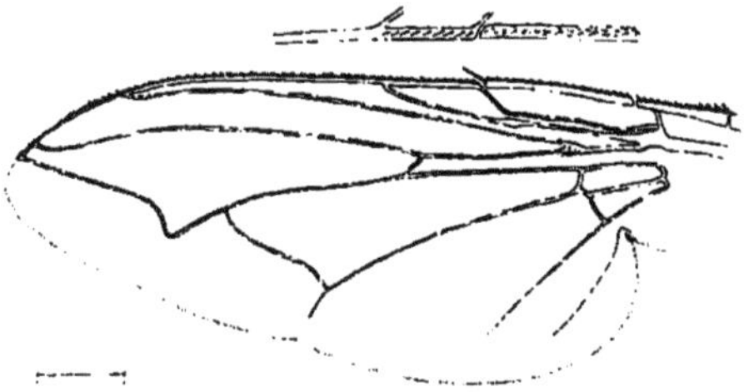

Fig. 170. *Calliphora stelviana* (Brauer & Bergenstamm), male. Left wing from above. Inset: costa from below showing distribution of small hairs on its underside. Scale = 1.0 mm.

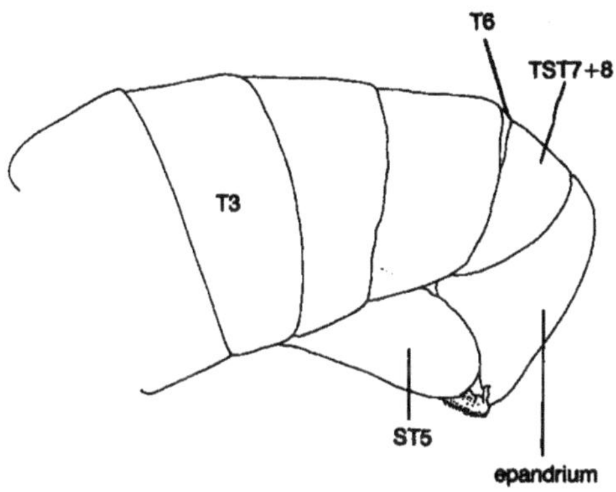

Fig. 171. *Calliphora stelviana* (Brauer & Bergenstamm), male. Abdomen, left lateral view. Arrow points to proximal end of anal membrane.

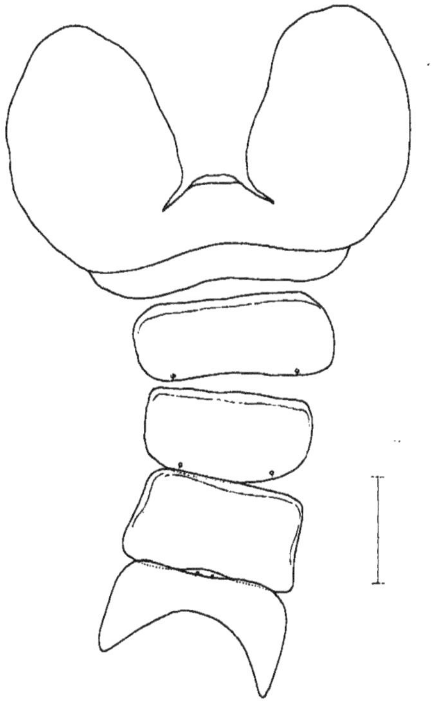

Fig. 172. *Calliphora stelviana* (Brauer & Bergenstamm), male. ST 1-5. Scale = 0.6 mm.

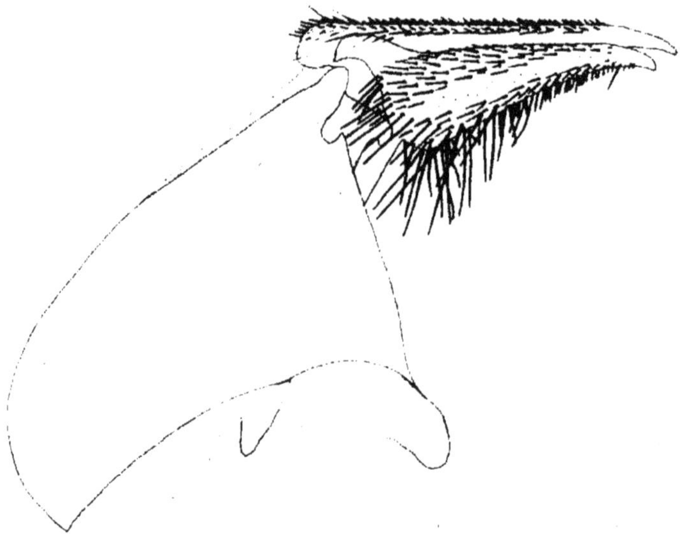

Fig. 173. *Calliphora stelviana* (Brauer & Bergenstamm), male. Epandrium, cerci, surstylus and bacilliform sclerite, left lateral view. Scale = 0.5 mm.

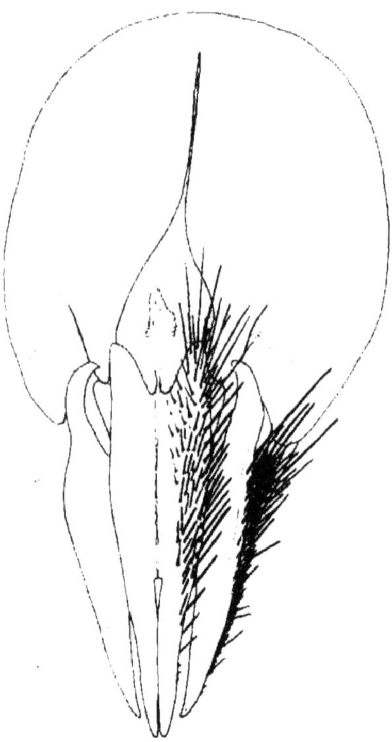

Fig. 174. *Calliphora stelviana* (Brauer & Bergenstamm), male. Epandrium, cerci and surstyli, posterior view. Scale = 0.5 mm.

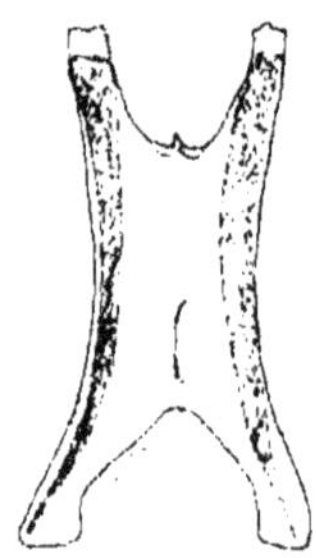

Fig. 175. *Calliphora stelviana* (Brauer & Bergenstamm), male. Bacilliform processes. Scale = 0.4 mm.

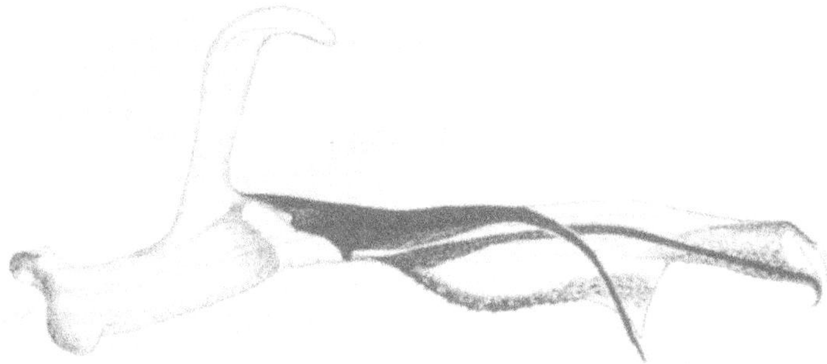

Fig. 176. *Calliphora stelviana* (Brauer & Bergenstamm), male. Aedeagus, left lateral view. Scale = 0.2 mm.

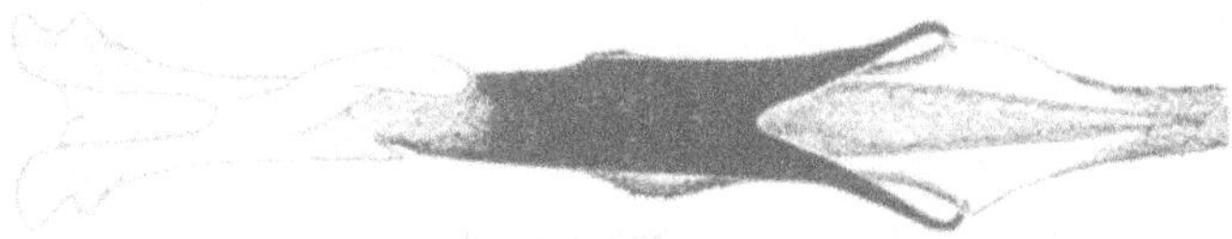

Fig. 177. *Calliphora stelviana* (Brauer & Bergenstamm), male. Aedeagus, dorsal view. Scale = 0.2 mm.

Fig. 178. *Calliphora stelviana* (Brauer & Bergenstamm), male. Ejaculatory sclerite. Scale = 0.2 mm.

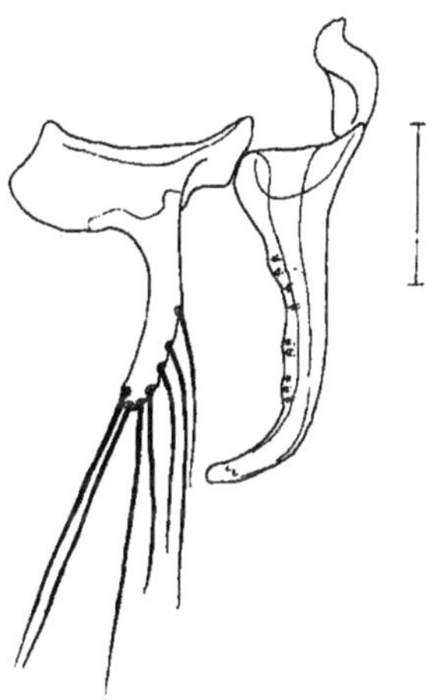

Fig. 179. *Calliphora stelviana* (Brauer & Bergenstamm), male. Pre- and postgonites. Scale = 0.2 mm.

Fig. 180. *Calliphora stelviana* (Brauer & Bergenstamm), female. T5 with marginal setae, dorsolateral view. Ovipositor protruding behind setae.

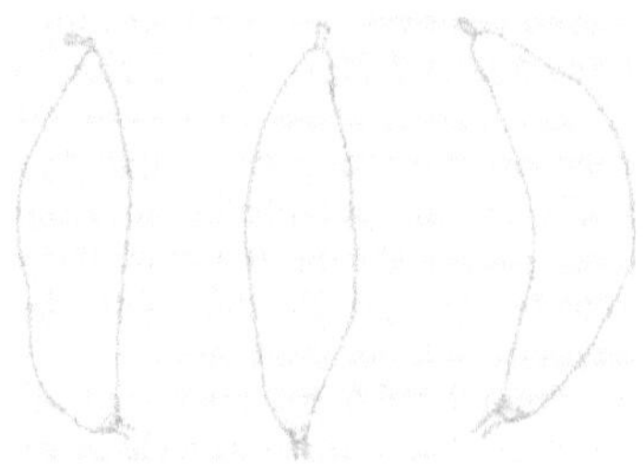

Fig. 181. *Calliphora stelviana* (Brauer & Bergenstamm), female. Spermathecae. Scale = 0.2 mm. G.pr. 265.

Fig. 183. *Calliphora stelviana* (Brauer & Bergenstamm), female. Proximal segment of ovipositor (T6 and ST6), left lateral view. G.pr. 265.

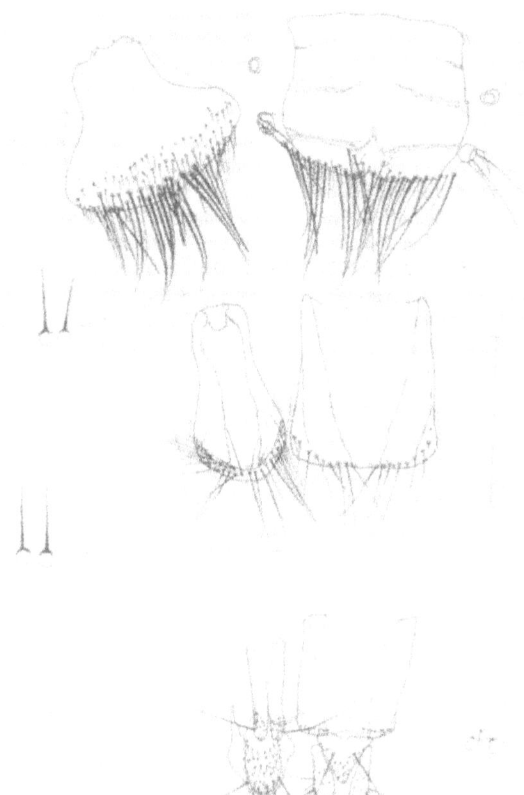

Fig. 182. *Calliphora stelviana* (Brauer & Bergenstamm), female. Ovipositor. Stipple indicates extent of microtrichiae. Insets: shape of individual microtrichiae in areas indicated (enlarged). Scale = 1.0 mm. G.pr. 265.

of both species in BMNH, and can find no differences in the genitalia from *stelviana* (note especially the shape and vestiture of the pregonite). I have also seen a specimen from North America which is asymmetric in the differentiation of the *prst ia*. I agree with Shewell (in litt. November 6th, 1984) in synonymising *abina* with *stelviana*.

12. ***Calliphora subalpina*** (Ringdahl, 1931)
Figs 1-5, 136, 184-195.

Steringomyia subalpina Ringdahl, 1931: 172.
Calliphora franzi Zumpt, 1956a: 17. SYN.N.

♂♀. Easily recognised by the characters mentioned in the key, especially the combination of white (occasionally faintly infuscated) lower calypter and presence of 2 inner *ph* setae. Parafacialia usually with yellow dusting, sometimes silvery. Occiput convex, with 3-4 rows of black setulae, centrally with pale hairs. f_2 with an *a* preapical seta in both sexes. 1 *prst ia* seta. 4 pairs of marginal scutellars. t_1 with 1 *pv*.

♂. Frons 0.086-0.093 times head-width (n=5), fronto-orbital plates touching. A lateroclinate seta usually absent. t_2 with 1 *ad* and usually without *v* seta (occasionally present even in Scandinavian material). t_3 usually without *av* setae. Cerci slightly shorter than the surstyli. Both covered with short straight spines, absent along a stripe on the outer surface of the surstylus. Bacilliform sclerites separated for most of their length, close together or fused ventrally. Acrophallus long, apparently lacking a lateral duct. Ventral plates closed midventrally. Hypophallus with a conspicuous distal projection supporting the subapically sigmoid and apically dentate paraphallic tip. Epiphallus very strong, joining the distal part of basiphallus. Pregonite broad, apically blunt with about 9 setae. Postgonite without basal seta.

♀. Frons 0.319-0.360 times head-width. Frontal vitta more than 2 times as wide as fronto-orbital plates. t_2 with 2-3 *ad* and a *v* seta. t_3 with 1 *av* seta. Margin of T5 with setae, but not as strong as in *stelviana*. T6 very large, one spiracle pair enclosed within it, its posterior margin heavily setose. T7 microtrichiose in posterior two-thirds between the rod-like main parts. T8 similarly microtrichiose in posterior half. ST6 short and broad. ST7 narrow subapically. ST8 complete, divided apically. Pleural membrane 8 devoid of microtrichiae. Spermathecae oval. [2 ovipositor slides examined.]

Length. 7-13 mm.

Distribution. Widely distributed in Fennoscandia and Denmark from low altitude to the subalpine region (in South Norway to about 1000 m a.s.l.), where it occurs together with *Calliphora stelviana*. - Widespread in Palaearctic Region: Austria, Bulgaria, Czechoslovakia, Germany, Great Britain, Hungary, Poland, Romania, most of U.S.S.R., Mongolia, Japan (Jacentkovský, 1944b; Emden, 1954; Dear, 1981; Ozerov, 1986; Schumann, 1986).

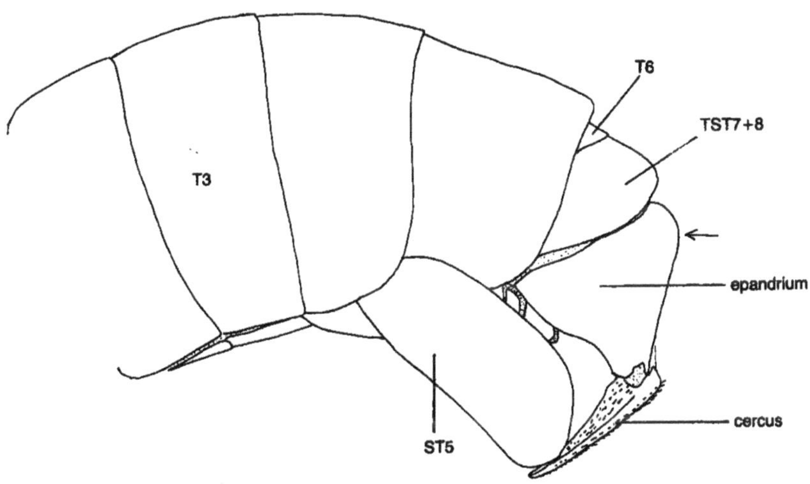

Fig. 184. *Calliphora subalpina* (Ringdahl), male. Abdomen, left lateral view. Arrow points to proximal end of anal membrane.

Fig. 185. *Calliphora subalpina* (Ringdahl), male. ST 1-5. Scale = 0.6 mm.

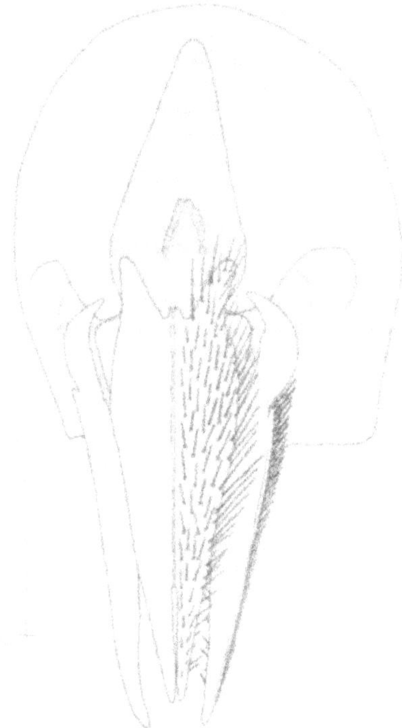

Fig. 187. *Calliphora subalpina* (Ringdahl), male. Epandrium, cerci and surstyli, posterior view. Scale = 0.5 mm.

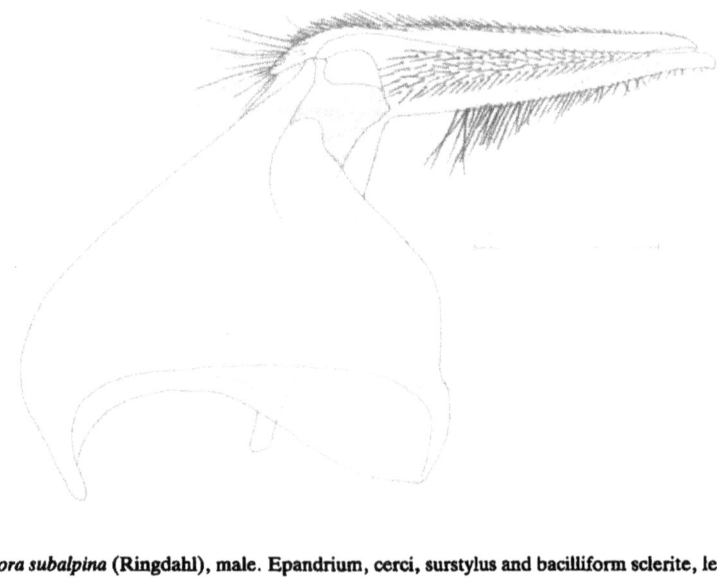

Fig. 186. *Calliphora subalpina* (Ringdahl), male. Epandrium, cerci, surstylus and bacilliform sclerite, left lateral view. Scale = 0.5 mm.

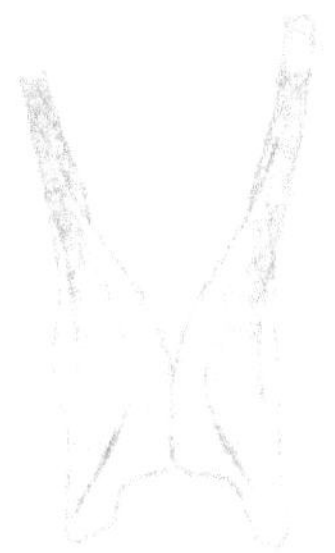

Fig. 188. *Calliphora subalpina* (Ringdahl), male. Bacilliform sclerites. Scale = 0.4 mm.

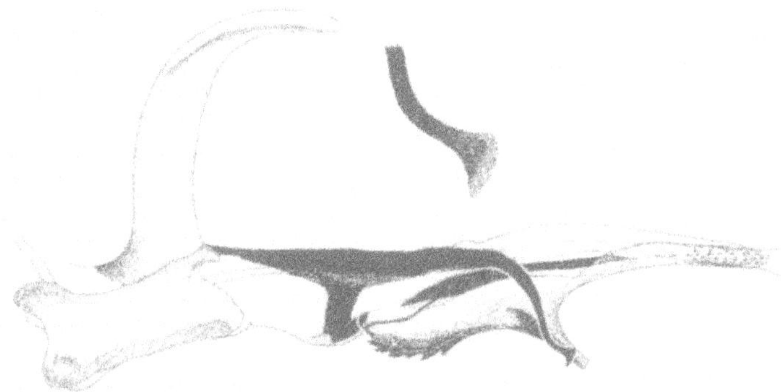

Fig. 189. *Calliphora subalpina* (Ringdahl), male. Aedeagus, left lateral view. Note that right paraphallic tip has slipped out of its groove. Inset: tip of paraphallus (enlarged). Scale = 0.2 mm.

Fig. 190. *Calliphora subalpina* (Ringdahl), male. Aedeagus, dorsal view. Paraphallic tips partly displaced. Scale = 0.2 mm.

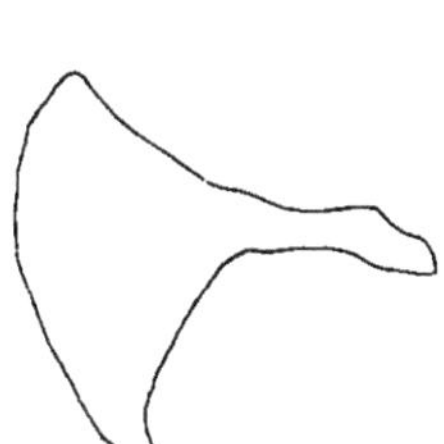

Fig. 191. *Calliphora subalpina* (Ringdahl), male. Ejaculatory sclerite. Scale = 0.2 mm.

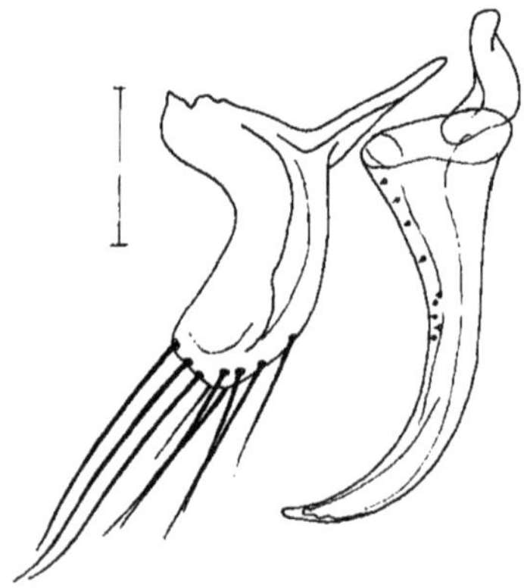

Fig. 192. *Calliphora subalpina* (Ringdahl), male. Pre- and postgonites. Scale = 0.2 mm.

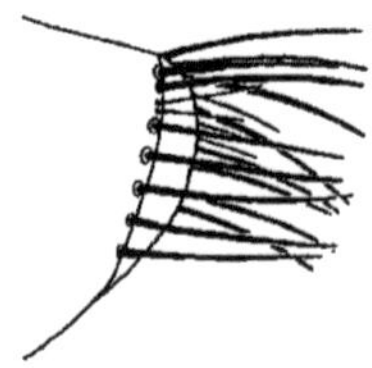

Fig. 193. *Calliphora subalpina* (Ringdahl), female. Distal part of T5 with marginal setae.

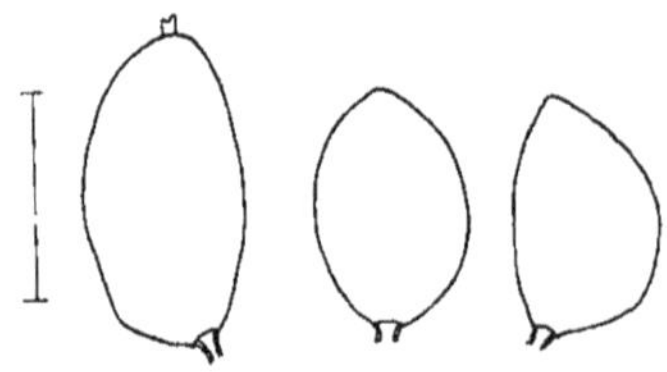

Fig. 194. *Calliphora subalpina* (Ringdahl), female. Spermathecae. Scale = 0.2 mm. G.pr. 267.

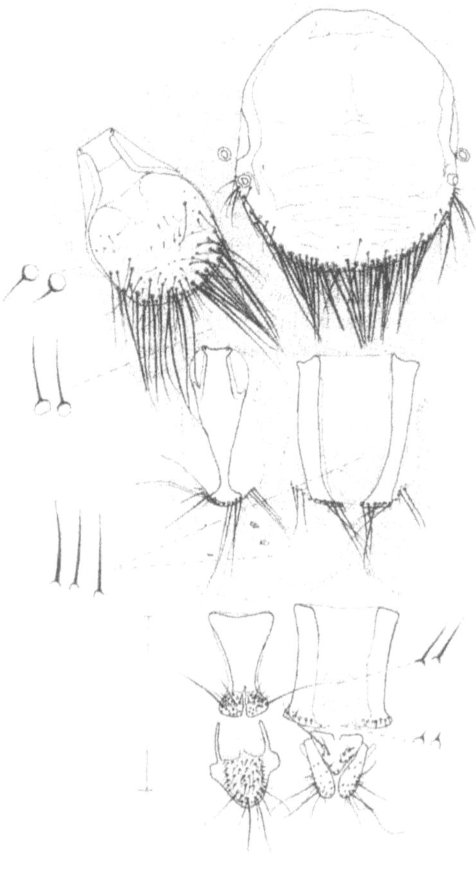

←

Fig. 195. *Calliphora subalpina* (Ringdahl), female. Ovipositor. Stipple indicates extent of microtrichiae. Insets: shape of individual microtrichiae in areas indicated (enlarged). Scale = 1.0 mm. G.pr. 267.

Biology. In Finland it avoids human settlements (Nuorteva, 1963). Collected on the stinkhorn fungus (*Phallus impudicus* Pers.) in Denmark (Nielsen, 1967, 1968). It has been reared from shrew carcasses in Great Britain (Disney, 1973). In Fennoscandia and Denmark adults have been collected from June to October.

Note. Japanese ♂♂ examined (in KR, BMNH) have a *v* seta on t_2, and 1 *av* seta on t_3, and this fits the description by Kano & Shinonaga (1968). Note however that Kano, Okazaki, Utsumi & Hagi (1959) do not mention these setae and that Japanese material may be variable as regards the chaetotaxy of the mid and hind legs. A Japanese ♂ in KR also has slightly separated fronto-orbital plates and both this ♂ and a ♀ from Japan in KR lack an *a* preapical seta on the mid femur. *C. subalpina* shares with *stelviana* the apparent absence of a lateral duct on the acrophallus and the absence of microtrichiae on pleural membrane 8 in the ovipositor. The pregonite of *subalpina* is rather characteristic. A similarly shaped pregonite occurs in *alaskensis* Shannon and *echinosa* Grunin, and this may be a synapomorphy defining a monophyletic group consisting of these species. See also the notes on *stelviana* above.

Nomenclature. I have designated a ♂ labelled "Åsljunga / 1.8 -31" in coll. Ringdahl (ZML) as lectotype of *subalpina*, 7♂ 3♀ (dated not later than September 1931) from coll. Ringdahl and 1♂ from coll. Zetterstedt (labelled "Sarcophaga / vomitorina ♀ / Gus. Wb. n.sp [Gusum Wahlberg]") as paralectotypes (Ringdahl's 1931 paper was printed on December 10, 1931). I have examined the holotype ♂ of *Calliphora franzi* (in BMNH) with accompanying slides of the genitalia. The frons has a lateroclinate setae, but in other external features and terminalia it agrees exactly with *subalpina*. Note that there is an error in Zumpt's (1956a: Taf. II, fig. 11) figure of the aedagus of *subalpina*: the epiphallus is shown as joining the basal part of the basiphallus, whereas in fact it joins it distally (cf. Fig. 189). The figure of *franzi* shows this feature correctly.

13. ***Calliphora uralensis*** Villeneuve, 1922
Figs 42, 137, 196-206.

Calliphora uralensis Villeneuve, 1922: 515.
Calliphora pseuderythrocephala Kramer, 1928: 63.

♂♀. In spite of its wholly black basicosta this species is surprisingly often confused with *vicina* in collections, possibly because of its yellowish gena and sometimes pale anterior spiracle (♀). It is easily recognised by the characters given in the key. Separable from *loewi* by the different pattern of dusting on the presutural area, the colour and dusting of the face and parafacialia (♂♀), and the width of the parafacialia and size of the first flagellomere (♀). Separable from *vomitoria* by the colour of the postgenal and lower genal vestiture. Occiput concave (especially in the ♂), with 2-3 rows of black setae behind postocular setae, otherwise with pale hairs. Genae with black hairs, also posteriorly. 2 inner and 1 outer *ph* seta. t_1 with 1 *pv* seta. f_2 without *a* preapical seta. t_3 with 2-4 short *av* setae.

♂. Frons 0.063-0.083 times head-width (n=5), fronto-orbital plates touching or slightly separated. Cerci and surstyli very long and slender, polished black with rather sparse long setae, both with a slight dilation apically. Cerci about as long as surstyli, slender, bent upwards in profile. Surstylus slightly curved downwards in profile, straight (not bending towards mid-line) in dorsal view. Bacilliform sclerites rather short, separated throughout their whole length. Acrophallus normal, with a lateral duct. Ventral plates closed mid-ventrally. Projection on distal part of hypophallus rather weak. Paraphallus subapically with a projection, apically dentate. Epiphallus joining basal part of basiphallus. Pregonite acute. Postgonite slender, without apical dilation.

♀. Frons 0.353-0.388 times head-width (n=5). Frontal vitta slightly more than 2 times width of fronto-orbital plates. Discal ground-vestiture erect on mid-dorsal parts of T3 and whole of T4. T6 diamond-shaped, sclerotisation strengthened in mid-line near posterior edge. T7 also with strengthened sclerotisation posteriorly near midline, microtrichiae absent. T8 without microtrichiae, often joined with ST8 through a band of sclerotisation encircling the ovipositor at the base of T8 and ST8. ST6 and ST7 broad. Pleural membrane 8 with a small patch of microtrichiae distally. Spermathecae oval. [2 ovipositor slides examined.]

Length. 8-13 mm.

Distribution. Widely distributed in Fennoscandia. Absent from Denmark and south Sweden. Common on sandy beaches on the coast of southwestern Norway. Very common in mountainous districts where it ascends to the high alpine region. In South Norway to 1841 m a.s.l. (ON, Lom: Juvasshytta). - Palaearctic: Faroes (Lyneborg, 1968), Iceland (Nielsen, Ringdahl & Tuxen, 1954), Great Britain (Emden, 1954; Dear, 1981), Ireland (Irwin, 1976), North and Central Europe, U.S.S.R. east to Amur Oblast, Mongolia, China

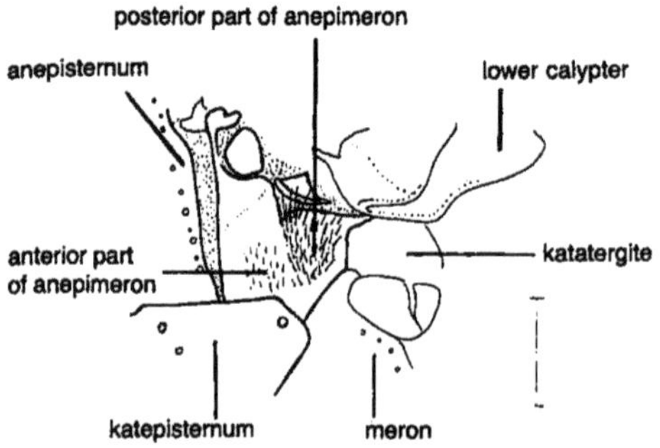

Fig. 196. *Calliphora uralensis* (Villeneuve). Thorax, left lateral view of anepimeral area. Scale = 1.0 mm.

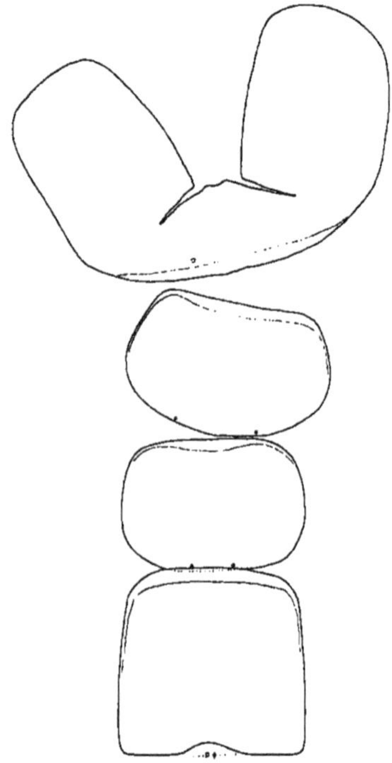

Fig. 197. *Calliphora uralensis* (Villeneuve), male. ST2-5. Scale = 1.0 mm. →

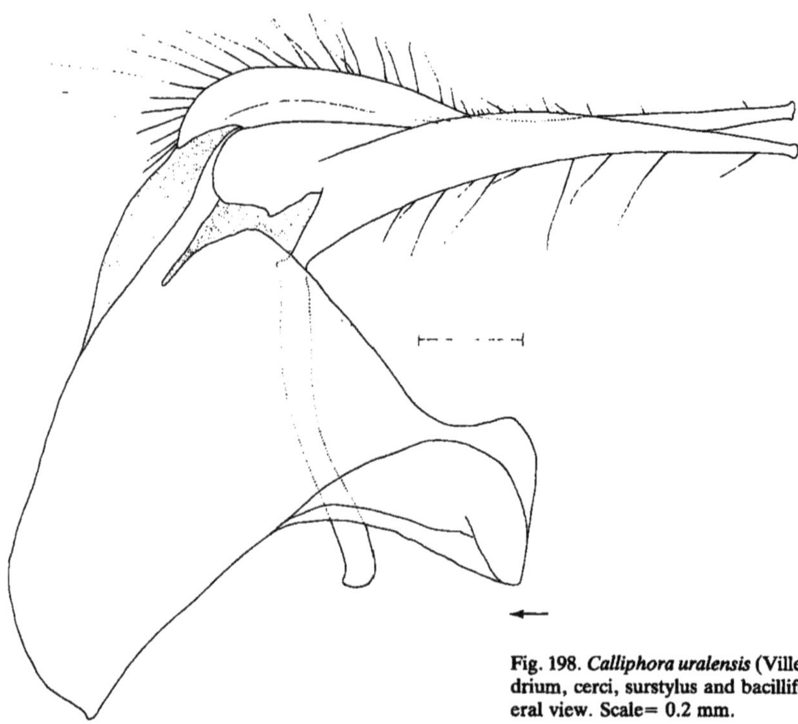

Fig. 198. *Calliphora uralensis* (Villeneuve), male. Epandrium, cerci, surstylus and bacilliform sclerite, left lateral view. Scale= 0.2 mm.

Fig. 199. *Calliphora uralensis* (Villeneuve), male. Epandrium, cerci and surstyli, posterior view. Scale = 0.2 mm.

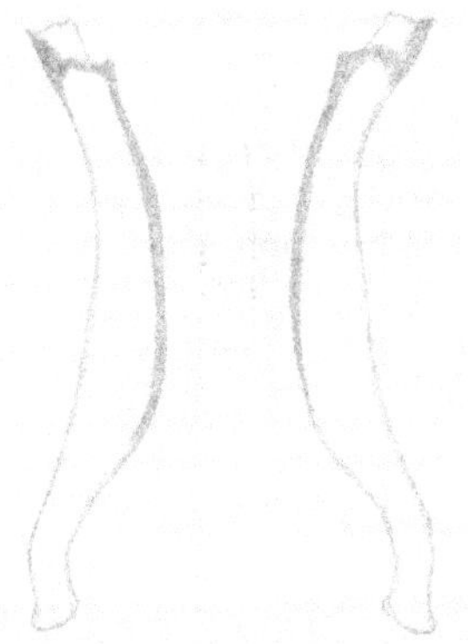

Fig. 200. *Calliphora uralensis* (Villeneuve), male. Baciliform sclerites and membrane between them. Scale = 0.2 mm.

Fig. 201. *Calliphora uralensis* (Villeneuve), male. Aedeagus, left lateral view. Inset: tip of paraphallus (enlarged). Scale = 0.2 mm.

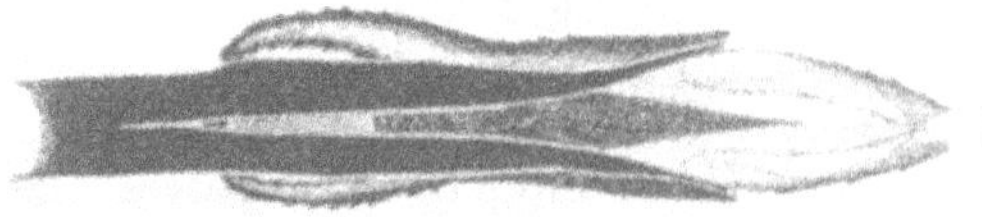

Fig. 202. *Calliphora uralensis* (Villeneuve), male. Distiphallus, dorsal view. Scale = 0.2 mm.

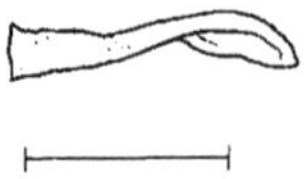

Fig. 203. *Calliphora uralensis* (Villeneuve), male. Ejaculatory sclerite. Scale = 0.2 mm.

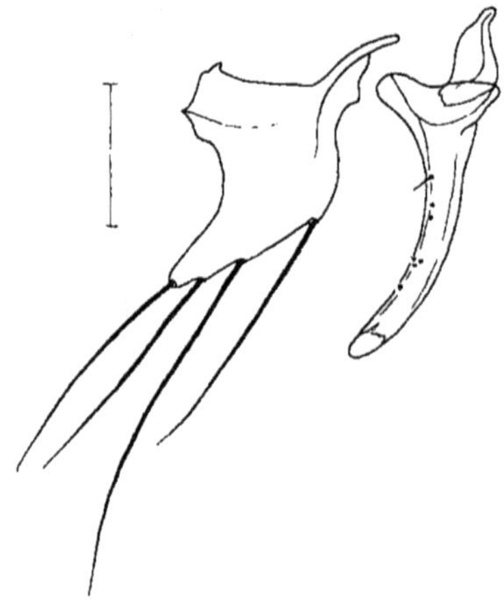

Fig. 204. *Calliphora uralensis* (Villeneuve), male. Pre- and postgonites. Scale = 0.2 mm.

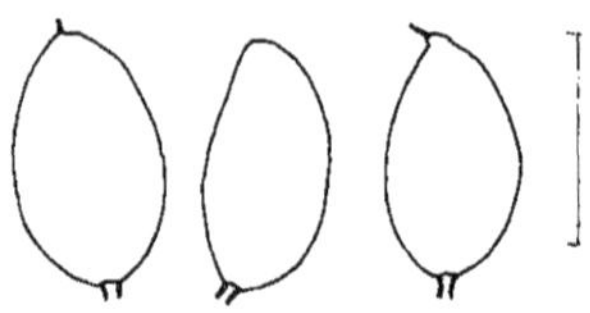

Fig. 205. *Calliphora uralensis* (Villeneuve), female. Spermathecae. Scale = 0.2 mm. G.pr. 272.

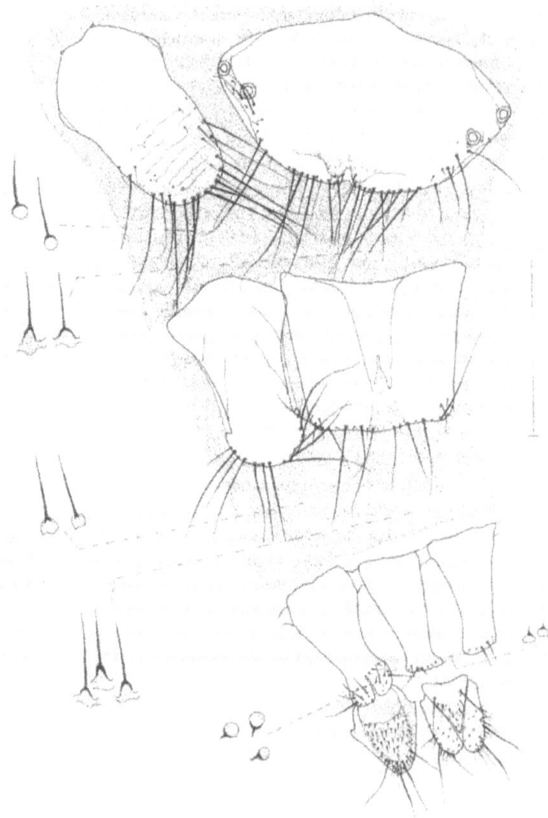

Fig. 206. *Calliphora uralensis* (Villeneuve), female. Ovipositor. Stipple indicates extent of microtrichiae. Insets: shape of individual microtrichiae in areas indicated (enlarged). Scale = 1.0 mm. G.pr. 271.

(Ozerov, 1986; Schumann, 1986). Nearctic: Greenland (north to 71°15'N) (material in USNM, ZMC; also Henriksen, 1939).

Biology. Nuorteva (1977) gives a mean development time of 37.1 and 39 days under field conditions during July and August at 15°C at different sites in southwestern Finland. In northern Finland (both coniferous and subarctic zones) it is highly synanthropic, but apparently not in the oak zone in the southwest (Nuorteva, 1963). According to Chernov (1965) it is maintained in the arctic zone of the U.S.S.R. by purely synanthropic populations. In the mountainous zones of Norway, its abundance may be maintained by slaughter refuse (skulls, viscera) left behind in the field during the reindeer hunting season in the autumn. In the U.S.S.R. it is also closely associated with latrines and faecal matter (Nuorteva, 1963; Chernov, 1965). In Fennoscandia adults have been collected from May to October.

Note. *Calliphora terraenovae* Macquart, 1851, which is widely distributed in the Nearctic from Alaska to Newfoundland and south to California, Colorado and northern Florida, and which also occurs in the Far East of the U.S.S.R. (Grunin, 1970b; Ozerov, 1986), is very similar in external appearance. However, it has dirty, dark brownish-grey areas on the upper part of the parafacialia that alternate with shining spots according to incidence of light, an area which in *uralensis* is evenly coloured without changing hue, and darker facial membrane and antennae. In the ♂ the surstyli curve towards each other apically. The cerci are almost straight in profile, and pointed apically. The distal projection on the hypophallic lobe is much more prominent than in *uralensis*. The ♀ has decumbent, not erect, discal ground-vestiture on T3 and T4. *C. terraenovae* is listed from Greenland by Hall (1965), but not by Hall (1948), Henriksen (1939) or Danks (1981). Hall (1948) suspected that Collin's (1931) material of *uralensis* from Greenland belonged to *terraenovae*, and this may be his basis for recording *terraenovae* from Greenland in the Nearctic catalogue. From Collin's description I find this highly unlikely. I have seen *uralensis* but not *terraenovae* from Greenland in ZMC and USNM. *C. uralensis* and *terraenovae* may be sister-species, judging from their vicariant distribution (overlap only in the easternmost Palaearctic).

Nomenclature. Zumpt (1956a) and Schumann (1986) also list *turanica* Rodendorf as a synonym. This is an error: see above under *vicina*.

14. ***Calliphora vomitoria*** (Linnaeus, 1758)
Figs 18, 19, 31, 138, 207-216.

Musca vomitoria Linnaeus, 1758: 595.
Musca carnaria Scopoli, 1763: 325. Junior primary homonym of *Musca carnaria* Linnaeus, 1758: 596.
Musca carnaria caerulea De Geer, 1776: 57. Unavailable; published in synonymy with *Musca vomitoria* Linnaeus, 1758, and never adopted.
Musca vomitoria minimus Harris, 1780. SYN.N.
Musca obscoena Eschscholz, 1823: 113.
Calliphora brunibarbis Robineau-Desvoidy, 1830: 434.
Calliphora fulvibarbis Robineau-Desvoidy, 1830: 434.
Calliphora affinis Macquart, 1835: 263.
Calliphora rubrifrons Townsend, 1908: 116.
Calliphora pseudovomitoria Baranov, 1943: 62. SYN.N.

♂♀. Easily recognised in both sexes by the reddish colour of the postgenal and lower genal vestiture. Occiput concave (especially in the ♂), at most a single row of black setae behind postocular setae, otherwise clothed with pale hairs which become gradually reddish ventrally. 2 inner and 1 outer *ph* setae. Posterior thoracic spiracle large and invading the meron to a greater extent than any of the other *Calliphora* species treated here. t_1 with 1 *pv* seta. f_2 without *a* preapical seta. t_3 with 2-4 short *av* setae.

♂. Frons 0.033-0.056 times head-width (n=5). Fronto-orbital plates touching. Posterior half of katepimeron (barette) a narrow stripe above the spiracle. Cerci straight, long. Surstylus curved in profile, bending slightly towards the mid-line apically in dorsal view. Bacilliform sclerites long, widely separated throughout their whole length. Acrophallus normal, with a lateral duct. Ventral plates not fused mid-ventrally. Projection on distal part of hypophallus very strong. Epiphallus joining basal part of basiphallus. Pregonite acute.

♀. Frons 0.332-0.363 times head-width (n=5). Frontal vitta slightly more than 2 times width of fronto-orbital plates. Ground-vestiture on T4 erect. T6 diamond-shaped, sclerotisation strengthened on mid-line, especially posteriorly. This is also the case with T7. T8 with some microtrichiae near posterior edge. ST6 narrower than in *uralensis*. ST7 broad. Pleural membrane 8 microtrichiose posteriorly. Spermathecae oval. [3 ovipositor slides examined.]

Length. 9-14 mm.

Distribution. Very common all over Fennoscandia and Denmark, also above the tree-line.

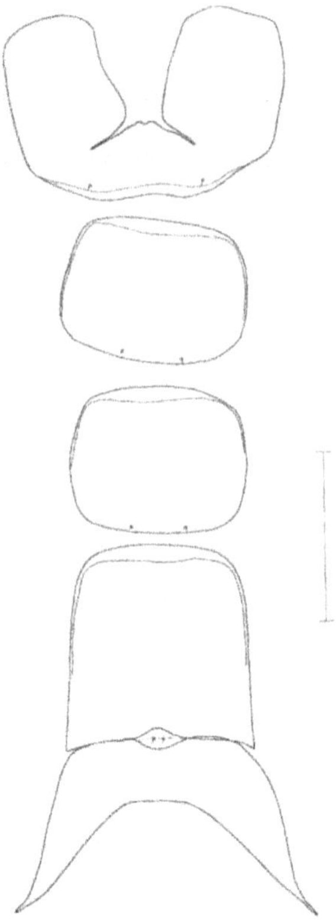

Fig. 207. *Calliphora vomitoria* (Linnaeus), male. ST 1-5. Scale = 1.0 mm.

Fig. 208. *Calliphora vomitoria* (Linnaeus), male. Epandrium, cerci, surstylus and bacilliform sclerite, left lateral view. Scale= 0.2 mm.

Fig. 209. *Calliphora vomitoria* (Linnaeus), male. Epandrium, cerci and surstyli, posterior view. Scale = 0.2 mm.

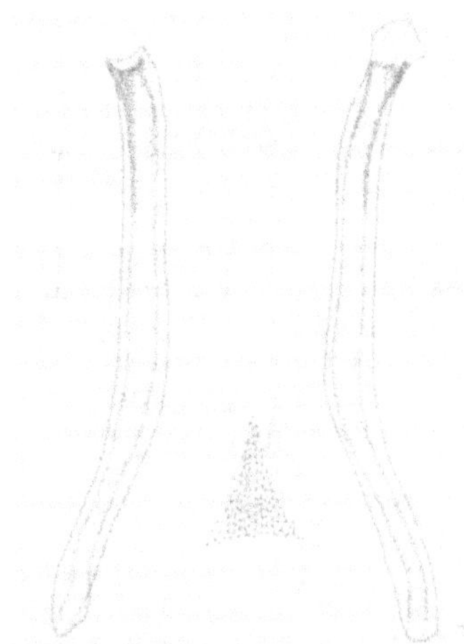

Fig. 210. *Calliphora vomitoria* (Linnaeus), male. Bacilliform sclerites and vestiture on membrane between them. Scale = 0.2 mm.

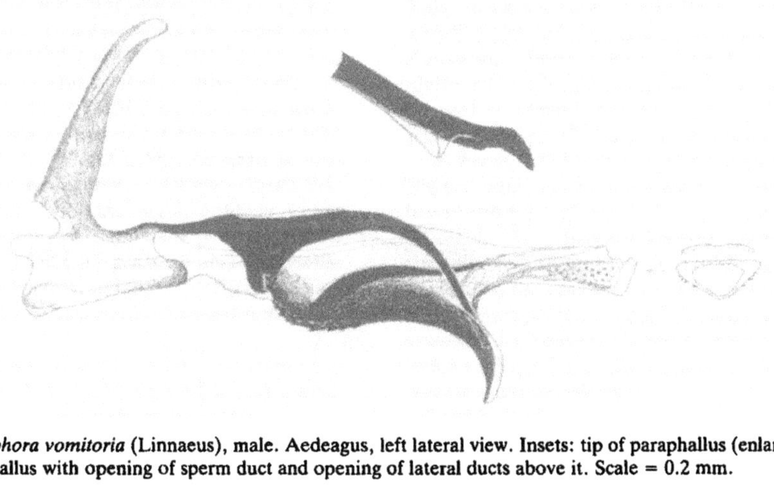

Fig. 211. *Calliphora vomitoria* (Linnaeus), male. Aedeagus, left lateral view. Insets: tip of paraphallus (enlarged) and apex of acrophallus with opening of sperm duct and opening of lateral ducts above it. Scale = 0.2 mm.

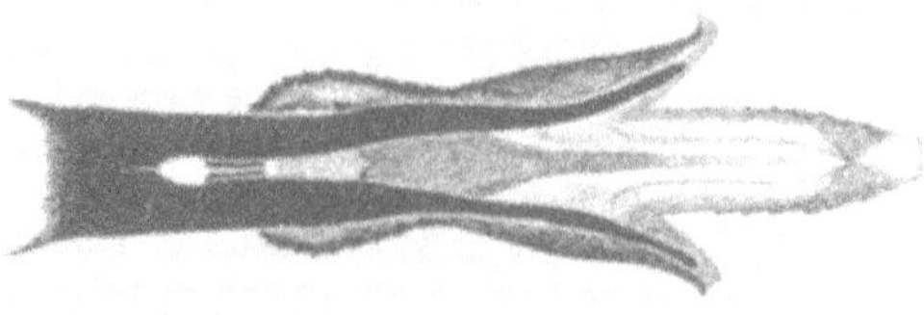

Fig. 212. *Calliphora vomitoria* (Linnaeus), male. Distiphallus, dorsal view. Scale = 0.2 mm.

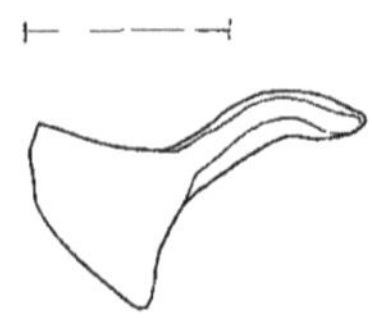

Fig. 213. *Calliphora vomitoria* (Linnaeus), male. Ejaculatory sclerite. Scale = 0.2 mm.

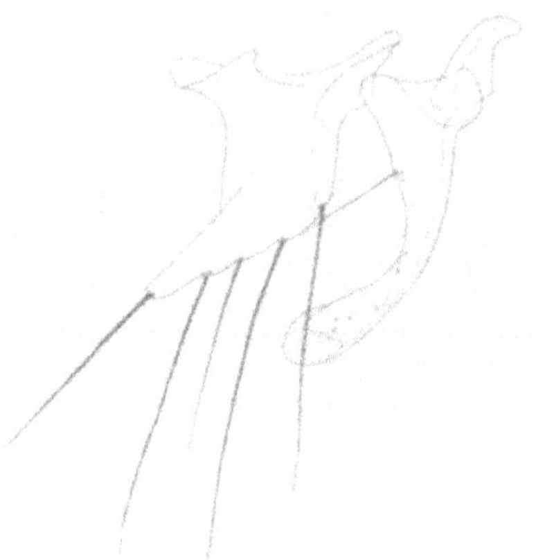

Fig. 214. *Calliphora vomitoria* (Linnaeus), male. Pre- and postgonites. Scale = 0.2 mm.

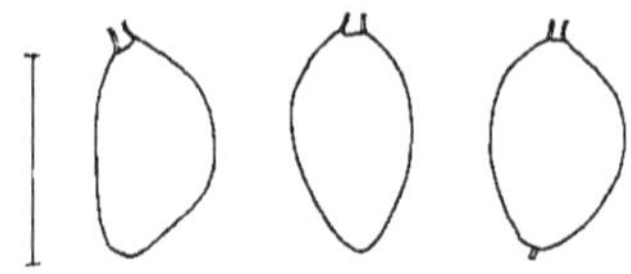

Fig. 215. *Calliphora vomitoria* (Linnaeus), female. Spermathecae. Scale = 0.2 mm. G.pr. 277.

Ascends to the high alpine zone. In South Norway taken at 1841 m a.s.l. (ON, Lom. Juvasshytta). - Widely distributed in Palaearctic and Nearctic Regions (Schumann, 1986; Hall, 1965). Also Oriental Region (Delfinado & Hardy, 1977).

Biology. Nuorteva (1977) gives a development time of 38 days under field conditions during July and August at 15°C in southwestern Finland. Attracted to the stinkhorn fungus (*Phallus impudicus* Pers.) in Denmark (Nielsen, 1963, 1967, 1968). Bred from cases of sheep-strike in Norway (Brinkmann, 1976) and Denmark (Nielsen, 1984a, 1984b), probably as a secondary myiasis agent. No synanthropic tendencies in most of Europe, or even Finland (Nuorteva, 1963). In England it overwinters as a larva (Norris, 1965). In Fennoscandia and Denmark collected from May to October.

Note. Recorded from Greenland by Hall (1948, 1965), Danks (1981) and Messersmith (1982), but not by Henriksen (1939). No material of this species from Greenland is present in ZMC or USNM. Recorded from Iceland by Messersmith (1982), but not by Nielsen, Ringdahl & Tuxen (1954).

Nomenclature. I interpret *Musca vomitoria minimus* Harris, 1780 as a dwarf form of *vomitoria* and have removed it from the synonymy of *vicina* where it was placed by Pont (1976). I have examined the holotype ♂ (in USNM) of *Calliphora peudovomitoria* from Zagreb, Yugoslavia (cf. Sabrosky & Crosskey, 1970: 427). It has the genitalia exposed.

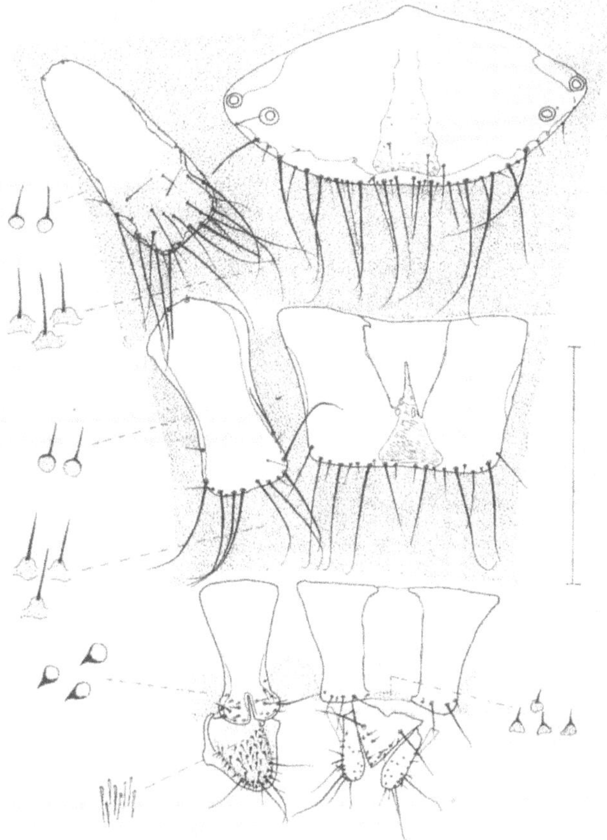

Fig. 216. *Calliphora vomitoria* (Linnaeus), female. Ovipositor. Stipple indicates extent of microtrichiae. Insets: shape of individual microtrichiae in areas indicated (enlarged). Scale = 1.0 mm. G.pr. 277.

Genus *Cynomya* Robineau-Desvoidy, 1830

Cynomya Robineau-Desvoidy, 1830: 363.
Type-species: *Musca mortuorum* Linnaeus, 1761: 452, by designation of Macquart, 1834: 176.
Cynophaga Lioy, 1864: 890.
Type-species: *Musca mortuorum* Linnaeus, 1761: 452, by monotypy.
Carcinomyia Townsend, 1915a: 21.
Type-species: *Cynomya hirta* Hough, 1898b: 166, by original designation; = *Musca mortuorum* Linnaeus, 1761.
Cynomyopsis Townsend, 1915b: 118.
Type-species: *Cynomya cadaverina* Robineau-Desvoidy, 1830: 365, by original designation.
Cynomyia, incorrect subsequent spelling.

A distinct genus of large species, with a shining green abdomen, which is practically without or with a very thin layer of dust, and bluish thorax with at least three dark stripes. Occiput with pale hairs except for 2-3 rows of black setae behind postocular setae. Head profile as in Fig. 45. Basal humeral setae on a weakly curved or straight line. 0+2 *ia*, 2+1-2 *acr*, 2-4+3 *dc*, 1+1 *ph*, (aberrations frequent). Scutellum with 3 pairs of marginals. Upper and lower calypter white, rim of upper dark. Basicosta black. Cell r_{4+5} in wing open (Fig. 218). t_1 with 2 *pv* setae; f_2 with *a* preapical seta in both sexes; t_3 with strong *ad* and *d* preapical setae, *pd* preapical seta weak or absent. Preabdominal sternites short and broad. A complete row of strong marginal setae on T4. Marginal and discal setae present on T5, the latter weak or absent in the male, very strong in the female.

The male is strongly hirsute on abdominal sternites and ventral surface of abdominal tergites and on femora. T5 much shorter than T4. Postabdomen in male greatly enlarged. TST7+8 and epandrium shining black. Cerci short. Surstyli in the form of very long shining black curved rods. Aedeagus with reduced paraphallus, long meso-hypophallus and short tube-like sclerotisation encircling opening of ejaculatory duct. Hypophallus without a projecting cornu like the one present in *Calliphora*. Acrophallus without a lateral duct. Ventral surface of acrophallus on each side with broad bands of posteriorly-projecting denticles. Postgonites without basal seta.

In the female, T5 with very strong discal seta, extremely stout in *mortuorum*, less so in *cadaverina*. T7 with rather long marginal setae. Pleural membrane 7, cerci and epiproct without microtrichiae. Spermathecae very elongate.

Uterus of normal cylindrical type. Oviparous. Larvae saprophagous, occasionally involved in animal myiasis (Itämies & Koskela, 1980).

Widely distributed in the Holarctic Region. 2 species, showing vicariant distribution. Enderlein (1933) quotes some old references asserting that *C. cadaverina* occurs on the Vaigach Peninsula (Northern U.S.S.R.), but it is not listed by Grunin (1970a). Both species of *Cynomya* are keyed out below.

1 species in Fennoscandia and Denmark.

Key to Holarctic species of *Cynomya*

1 Facial membrane and antenna with bright yellow ground-colour and golden dusting; anterior part of first flagellomere darkened; fronto-orbital plates and frontal vitta with bright yellow ground-colour in anterior half, black ground-colour in posterior half; parafacialia bright yellow with golden dusting, no shifting spots according to incidence of light; gena with yellow ground-colour and golden dusting except for posterior 1/4 which is black with greyish dusting, the boundary very sharp; usually 1 *post acr*; abdomen practically without dusting; ♂: a weak lateroclinate seta usually present; surstyli 3 times as long as cerci (Fig. 221); aedeagus with hairs on ventral membrane in front of basiphallus (Fig. 223); postgonite without a lateral wing-like projection; ♀: T5 in profile most often concave, discal setae very strong; T5 much longer than T4.

15. *mortuorum* (Linnaeus)

– Facial membrane dark, antennae brownish-black except for junction between pedicel and first flagellomere which is rather yellow; fronto-orbital plates and frontal vitta with dark ground-colour all the way to lunula; gena with black ground-colour and greyish dust, ground-colour somewhat lighter anteriorly, but no sharp boundary present; usually 2 *post acr*; abdomen with a very weak layer of dust; ♂: lateroclinate setae absent; parafacialia in upper part with dark ground-colour and dense silvery or golden dust, with shifting spots according to direction of light; silvery dusting also on fronto-orbital plates; surstyli 5-6 times as long as the cerci; aedeagus without hairs on ventral membrane in front of basiphallus; postgonite with lateral wing-like projection (well illustrated by Grunin, 1970b: fig. 40); ♀: parafacialia dark with dirty blackish-golden brown shifting spots in upper part, similar shifting spots also on fronto-orbital plates; T5 in profile usually straight, discal setae moderately strong; T5 as long as T4.

cadaverina Robineau-Desvoidy

15. ***Cynomya mortuorum*** (Linnaeus, 1761)
Figs 45, 217-228.

Musca mortuorum Linnaeus, 1761: 452.
Musca chrysocephala De Geer, 1776: 60. Unavailable; published in synonymy of *Musca mortuorum* Linnaeus, 1761, and never adopted.
Cynomya hirta Hough, 1898b: 166. SYN.N.
Cynomya gregorpovolnyi Čepelák & Čepelák, 1978: 1. SYN.N.

♂♀. Easily recognised by the characters given in the key, especially by the bright yellow colour of the anterior half of the head and the shining green abdomen. Occasionally a second *post acr* is present in front of the normal pair.

♂. Frons 0.196-0.233 times head-width (n=5). A weak lateroclinate seta present. Frontal vitta 1.5-2 times as wide as fronto-orbital plates, golden dusted when seen from in front. Bacilliform sclerites long, fused for their whole length.

♀. Frons 0.364-0.371 times head-width (n=5). Frontal vitta a little more than twice as wide as each fronto-orbital plate. Disc of T5 usually strik-

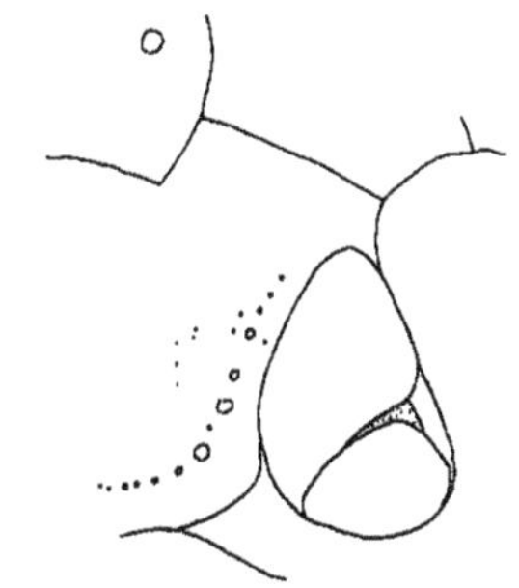

Fig. 217. *Cynomya mortuorum* (Linnaeus). Posterior spiracle.

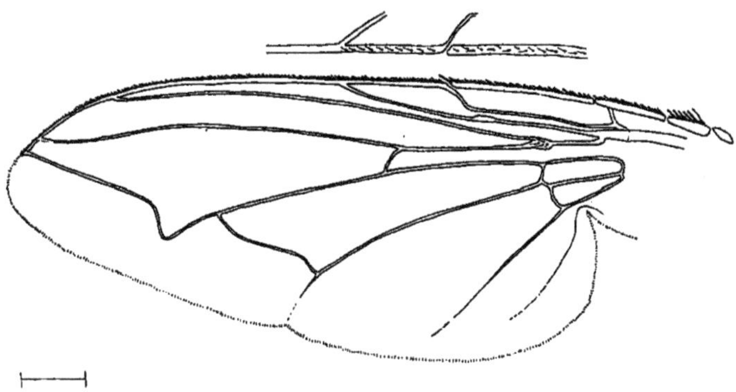

Fig. 218. *Cynomya mortuorum* (Linnaeus), male. Left wing from above. Inset: costa from below showing ditribution of small hairs on its underside. Scale = 1.0 mm.

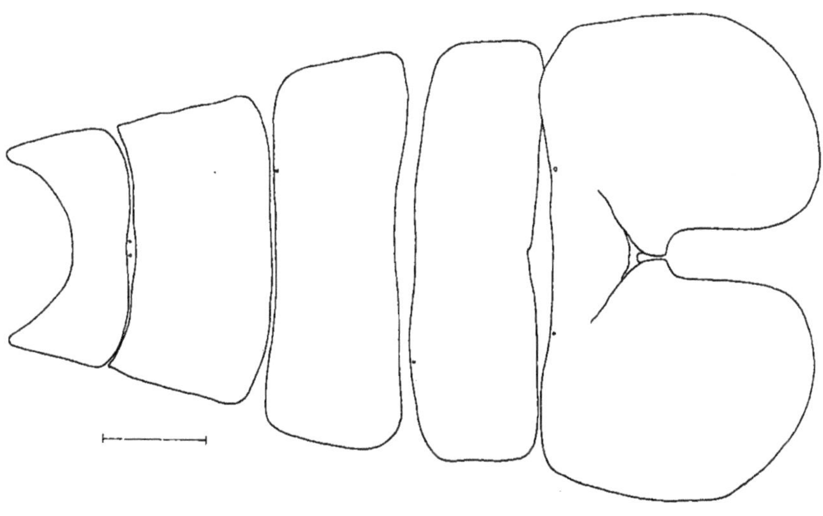

Fig. 219. *Cynomya mortuorum* (Linnaeus), male. ST1-5. Scale = 0.8 mm.

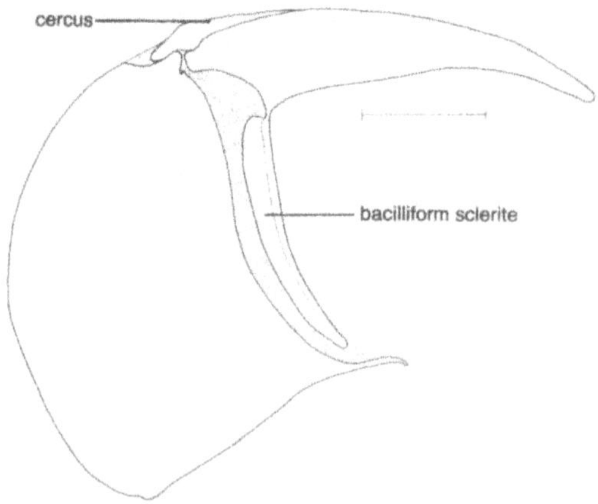

Fig. 220. ***Cynomya mortuorum*** **(Linnaeus), male. Epandrium, cerci, surstylus and bacilliform sclerite, left lateral view. Scale = 0.8 mm.**

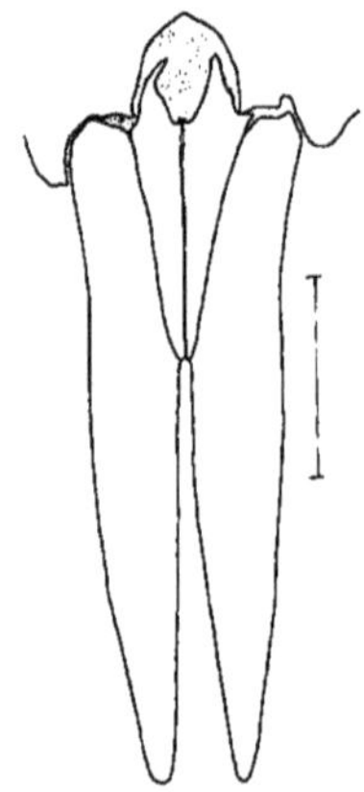

Fig. 221. ***Cynomya mortuorum*** **(Linnaeus), male. Cerci and surstyli, posterior view. Scale = 0.8 mm.**

Fig. 222. ***Cynomya mortuorum*** **(Linnaeus), male. Bacilliform sclerites. Scale = 0.8 mm.**

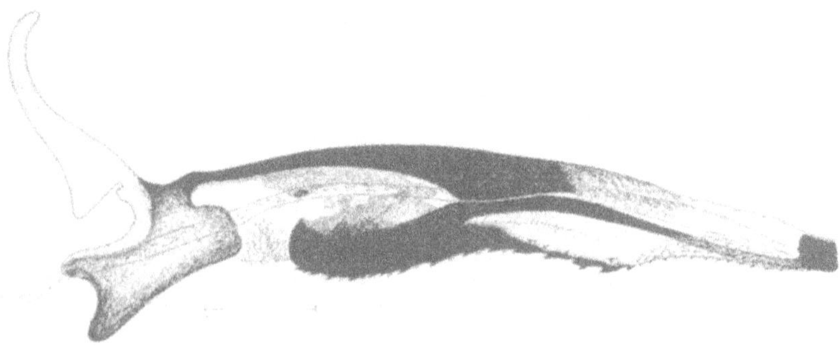

Fig. 223. ***Cynomya mortuorum*** **(Linnaeus), male. Aedeagus, left lateral view. Scale = 0.2 mm.**

Fig. 224. ***Cynomya mortuorum*** **(Linnaeus), male. Distiphallus, dorsal view. Scale = 0.2 mm.**

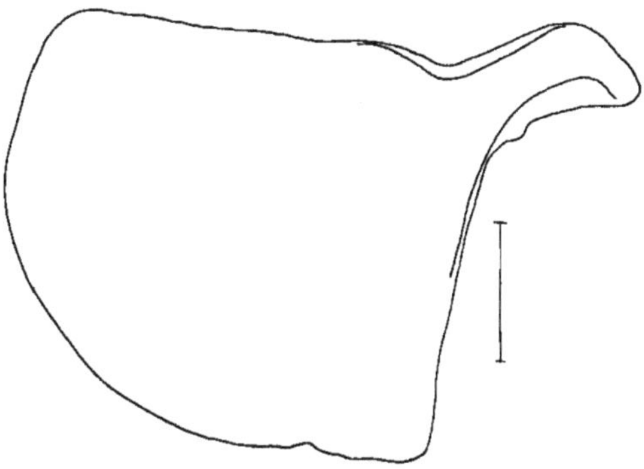

Fig. 225. ***Cynomya mortuorum*** **(Linnaeus), male. Ejaculatory sclerite. Scale = 0.2 mm.**

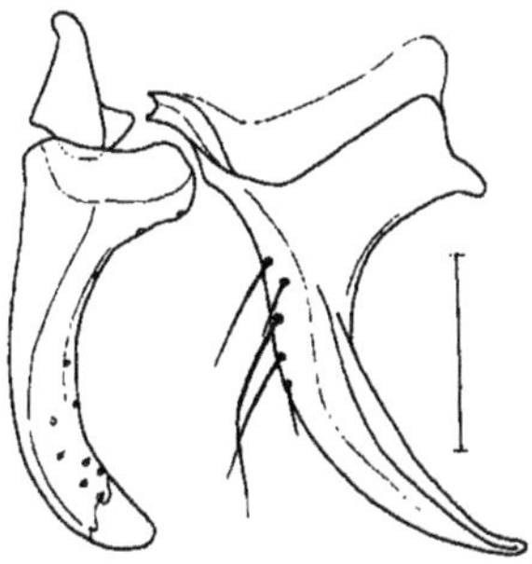

Fig. 226. ***Cynomya mortuorum*** **(Linnaeus), male. Post- and pregonites. Scale = 0.2 mm.**

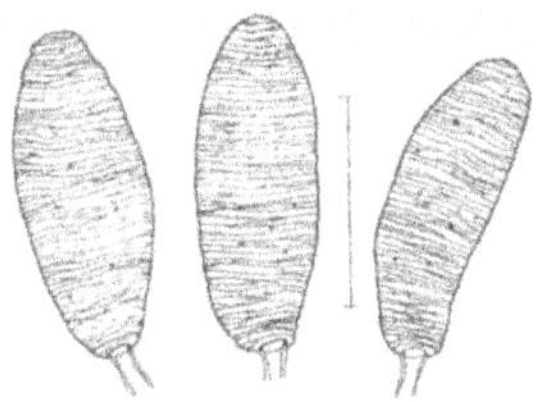

Fig. 227. *Cynomya mortuorum* (Linnaeus), female. Spermathecae. Scale = 0.2 mm. G.pr. 213.

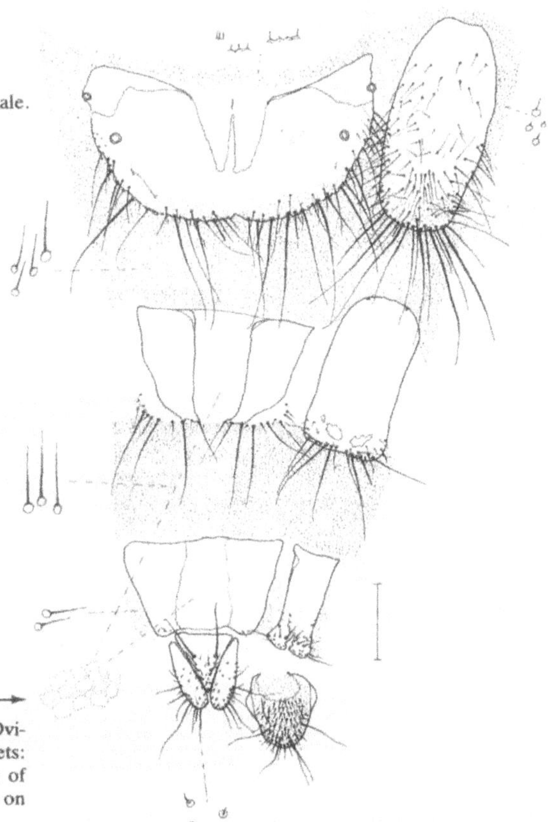

Fig. 228. *Cynomya mortuorum* (Linnaeus), female. Ovipositor. Stipple indicates extent of microtrichiae. Insets: shape of microtrichiae in areas indicated, pattern of sclerotisation mid-dorsally on T7 and T8, sensillae on cerci (all enlarged). Scale = 0.4 mm. G.pr. 213.

ingly concave. T6 with a triangular area of weak sclerotisation on mid-line anteriorly. T8 without microtrichiae. ST6 very long, disc covered with weak setae. ST7 broad. Spermathecae elongate. [2 ovipositor slides examined.]

Length. 7-18 mm.

Distribution. Very common all over Fennoscandia and Denmark from the lowlands to the high alpine zone (in Norway collected at 1658 m a.s.l.). - All over the Palaearctic Region (Schumann, 1986; Ozerov, 1986), also Iceland and Faroes (Nielsen, Ringdahl & Tuxen, 1954; Lyneborg, 1968) and Jan Mayen (Ringdahl, 1933). Nearctic: Greenland; Alaska. No *C. mortuorum* has been found in the Nearctic Region between Alaska and Greenland, where it is replaced by *cadaverina* Robineau-Desvoidy, 1830. Its distribution overlaps with *C. cadaverina* in Alaska and the Far East of the U.S.S.R. (cf. Grunin, 1970b). Hall (1965, but not 1948) and Danks (1981) state that *cadaverina* occurs in Greenland, but this appears to be an error. Henriksen (1939) does not list it from there. All material of *Cynomya* from Greenland in ZMC and USNM is *mortuorum* (K.R. det.). Shewell (in litt. November 6, 1984) also says that he does not know *cadaverina* from Greenland. The source for the reference in the Nearctic Catalogue to its occurrence in Greenland may be Hough (1898a), who said that he had 4 specimens from Greenland which he referred with some doubt to "this species" ["*americana*" = *cadaverina*], or Sailer & Lienk (1951) who list it from West Greenland.

Biology. Immature stages have been described by Erzinçlioğlu (1985). Collected on the stinkhorn fungus (*Phallus impudicus* Pers.) in Denmark (Nielsen, 1968). No synanthropic preferences demonstrated in Finland (Nuorteva, 1963). Nuorteva (1977) gives a mean development time of 25.4 to 31.5 days under field conditions during July and August at 15°C at different sites in southwestern Finland. In subarctic Finland, according to experiments performed by Nuorteva (1972), the average duration of its development is between 46.7 and 72.3 days. It has a low grade of heliophily and is apparently the commonest blowfly visiting burrows of small mammals (Hackman, 1963; Nuorteva & Vesikari, 1966). It has been reported as involved in cases of myiasis in the hare *Lepus timidus* L. (Itämies & Koskela, 1980). I have reared it from carcasses of *Lemmus lemmus* (Linnaeus) (SFI, Vik), *Microtus oeconomus* (Pallas) (TEI, Vinje) and *Sorex araneus* Linnaeus (TEI, Vinje) from the mountains of southern Norway, alone or together with *Calliphora stelviana*. In Fennoscandia and Denmark adults have been collected from April to October.

Nomenclature. I have examined the ♂ and ♀ syntypes of *Cynomya hirta* from St Paul Is, Alaska (in FMC) and have labelled and here designate the ♂ as lectotype and the ♀ as paralectotype. A third ♂ (from St George Is) in FMC is not a syntype. I find all three specimens conspecific with *mortuorum*, although both sexes have smaller eyes (♂ frons 0.240-0.260 times head-width, n=2; ♀ frons 0.386 times head-width, n=1) and the males have much longer and denser hairing on the legs, on the thoracic pleura and dorsum, and on the abdomen than does the material of *mortuorum* from Europe available to me. The specimens are also very large (♂♂ 17mm, ♀ 13mm). The genitalia of the lectotype are exposed and indistinguishable from those of European *mortuorum*. Even the hairing on the ventral membrane just in front of the basiphallus is present. I consider it to be a northern form (small eyes, dense hairing) of *mortuorum*, not deserving full specific status. I have not been able to examine the holotype ♂ of *gregorpovolnyi*, but it appears from the description to be based on an aberrant ♂ of *mortuorum*.

Genus *Onesia* Robineau-Desvoidy, 1830

Onesia Robineau-Desvoidy, 1830: 365.
Type-species: *Onesia floralis* Robineau-Desvoidy, 1830: 366, by designation of Townsend, 1916b: 8.

Marsilia Robineau-Desvoidy: 1863: 535.
Nomen nudum. [No taxonomic characters.]

Thelesina Robineau-Desvoidy: 1863: 535.
Nomen nudum. [No taxonomic characters.]

Macrophallus Mueller, 1922: 62.
Type-species: *Onesia austriaca* Villeneuve, 1920: 204, by designation of Townsend, 1935: 170.

Macronesia Villeneuve, 1926: 130.
Type-species: *Onesia hendeli* Villeneuve, 1926: 130, by designation of Townsend, 1935: 170; = *Onesia kowarzi* Villeneuve, 1920.

Onesioides Schumann, 1974a: 338.
Type-species: *Melinda hokkaidensis* Baranov, 1939: 112, by original designation.

Pellonesia Lu & Fan in Wang, Lu, Chen & Fan, 1981: 255, 258.
Type-species: *Onesia pterygoides* Lu & Fan in Wang, Lu, Chen & Fan, 1981: 255, 258, by monotypy.

In Fennoscandia and Denmark, a distinct genus with olive-greenish tessellated abdomen and striped thorax, resembling *Bellardia* but having 3 *post ia*, a more protruding face and broader parafacialia, and a first flagellomere which is about 3 times as long as pedicel. Head profile as in Fig. 44, ground-colour black. Occiput weakly concave in ♂, with pale hairs except for 1-3 irregular rows of small black setae behind postocular row. 2+3 *acr*, 3+3 *dc*, 2+1 *ph*. Scutellum with 3 pairs of marginal setae. Upper and lower calypters pale. Cell r_{4+5} in wing open (Fig. 230). t_1 with 1 *pv* seta. f_2 without *a* preapical in both sexes. t_3 with strong *ad*, *d* and *pd* preapicals. Posterior spiracle as in Fig. 229. Frons in male without lateroclinate seta. Aedeagus with reduced paraphallus with an apex not free from wall of distiphallus. (In the Eastern Palaearctic subgenus *Onesioides* Schumann the paraphalli are slender rods apparently free apically). Paraphalli never forming massive hooks as in *Bellardia*. Mesohypophallus long. Ventral surface of distiphallus covered with posteriorly-projecting denticles. Acrophallus long, without lateral duct. Ovipositor with shortened sclerites. All intersegmental and pleural membranes fully microtrichiose. T6 strongly invaded by microtrichiae. Epiproct and cerci microtrichiose, the latter with some bare areas basally. ST8 apically bilobed in all species, except *floralis* (Schumann, 1964). Uterus short with large lateral incubatory pouches. All species are viviparous and numerous first instar larvae can be recovered from mature females. Larvae are thought to be parasites or predators of earthworms or snails (Schumann, 1964), but no firm evidence appears to be available. Krivosheina (1961) reports that "*Onesia sepulcralis*" develops within the bodies of earthworms, but this species does not occur in the U.S.S.R. (Grunin, 1970a; Schumann, 1986) and Krivosheina's record probably refers to *Onesia austriaca* Villeneuve or a species of *Bellardia*.

Widely distributed in the Palaearctic. 11 species were listed by Schumann (1986). 1 species in Fennoscandia and Denmark.

Notes on synonymy. Schumann (1986) states that Hendel (1901) designated *Musca sepulcralis* Meigen as type-species of *Onesia* Robineau-Desvoidy. However, this nominal species was not among the sixteen originally included species, neither does Hendel cite any of the included species as synonyms of *sepulcralis* (ICZN, Article 69, (a) (v)), and the designation is therefore invalid under the current Code.

16. ***Onesia floralis*** Robineau-Desvoidy, 1830
Figs 26, 44, 229-241.

Musca sepulcralis Meigen, 1826: 71. Junior primary homonym of *Musca sepulchralis* Linnaeus, 1758: 596 (Syrphidae) (ICZN, Article 58 (6)).
Onesia floralis Robineau-Desvoidy, 1830: 366.
Onesia claripennis Robineau-Desvoidy, 1830: 367.
Onesia subapennina Rondani, 1862: 182.
Calliphora retrocurva Pandellé, 1896: 211.

♂♀. Recognisable from other European *Onesia* by the lack of *prst ia* (occasionally a weak one present), open cell r_{4+5} and a curved distal section of vein M on wing, lack of a *v* seta on the t_2 in the ♂ and a large, complete and undivided ST8 in the ♀ ovipositor. Parafacialia broad with greyish-yellow dusting, weakly shifting spots. No alphasetae on anterior edge of abdominal sternites.

♂. Frons 0.099-0.127 times head-width (n=5). Fronto-orbital plates separated by a frontal vitta about as broad as anterior ocellus. 3-4 *post ia*. t_3 with 1-2 *av*. Lobes of ST5 with concave median edges. Bacilliform sclerites strong, short, curving medially and separated. Cerci straight. Surstylus strongly curved, concavity facing cerci. Pregonite without apical setae, a bundle of 2-5 setae present at the basal plate. Postgonite without basal seta.

♀. Frons 0.318-0.329 times head-width (n=5). Frontal vitta with parallel edges. t_2 with *v* seta. Intrapostocular cilia absent. T6 completely divided, in dried specimens the two parts standing almost vertically next to each other within the opening formed by the posterior margin of T5, with the rest of the ovipositor protruding between them. T7 reduced, without setae. T8 larger, with a few marginal setae. ST6 square. ST7 strongly reduced but marginal setae strong. ST8 a large undivided convex plate, distally rounded and covered with setae. Spermathecae oval. [2 ovipositor slides examined.]

Length. 6-12 mm.

Distribution. Common in Denmark and South Sweden. A few captures from Ø in Norway. Absent from Finland. - Europe. Also Spain (Barcelona, 1♀, in ZMC). Absent from British Isles and U.S.S.R. (Pont, 1976; Schumann, 1986).

Biology. Immature stages described by Schumann (1964). In Denmark the species has been taken in copula on the following dates: May 25, May 30, July 26, August 30. It has been reared from a puparium found in soil in Denmark. In Fennoscandia and Denmark adults have been taken from May to October.

Nomenclature. *Musca sepulcralis* was described from ♂♀ syntypes, both sexes being figured on Meigen Plate 118, fig. 8 (Morge, 1976a). In Mei-

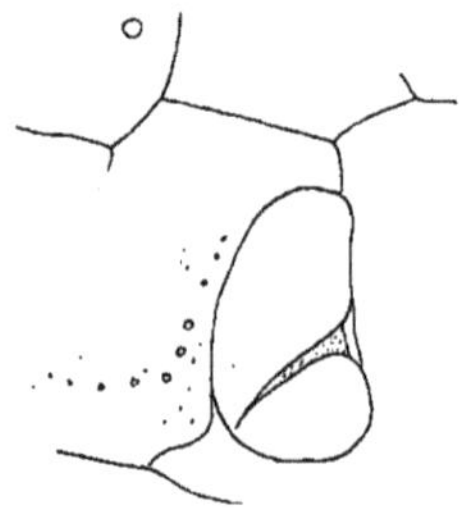

Fig. 229. *Onesia floralis* (Robineau-Desvoidy). Posterior spiracle.

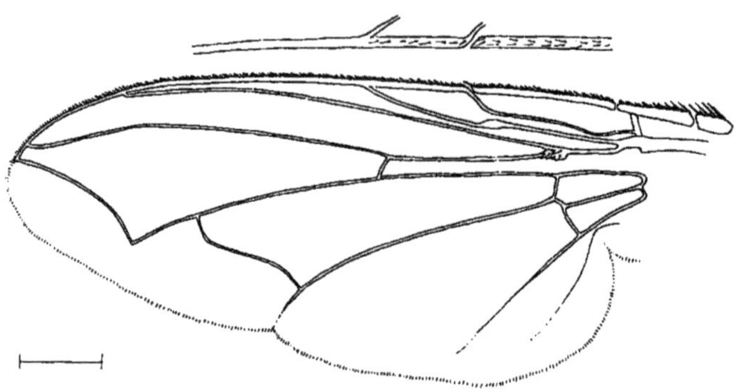

Fig. 230. *Onesia floralis* (Robineau-Desvoidy). Left wing from above. Inset: costa from below showing distribution of small hairs on its underside. Scale = 1.0 mm.

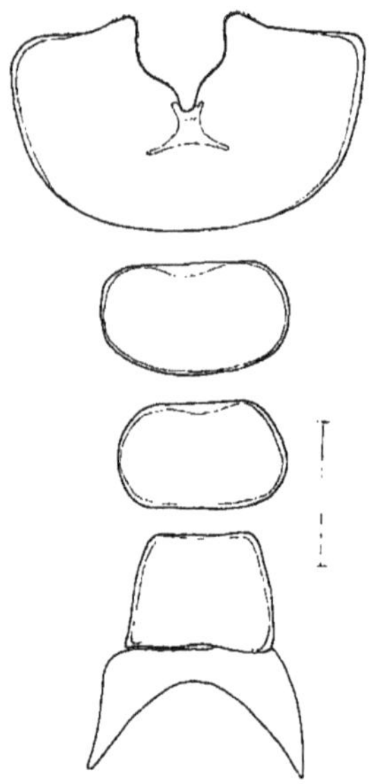

Fig. 231. *Onesia floralis* (Robineau-Desvoidy), male. ST1-5. Scale = 0.8 mm.

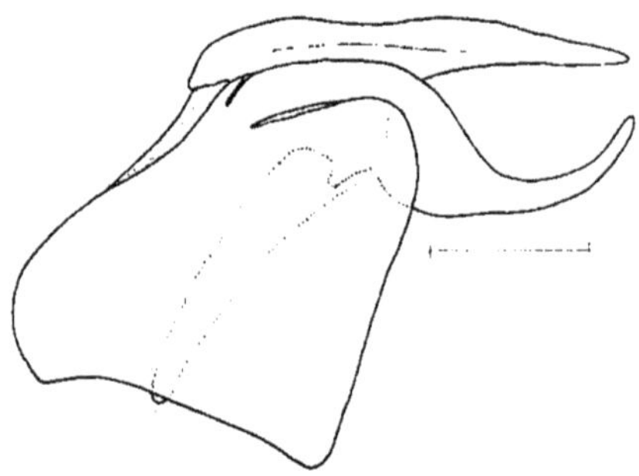

Fig. 232. *Onesia floralis* (Robineau-Desvoidy), male. Epandrium, cerci, surstylus and bacilliform sclerite, left lateral view. Scale= 0.4 mm.

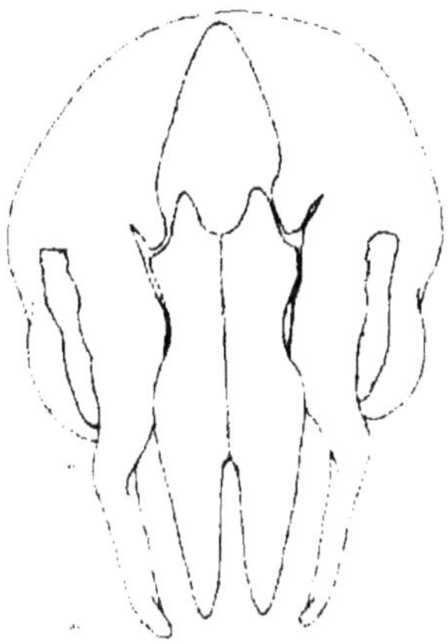

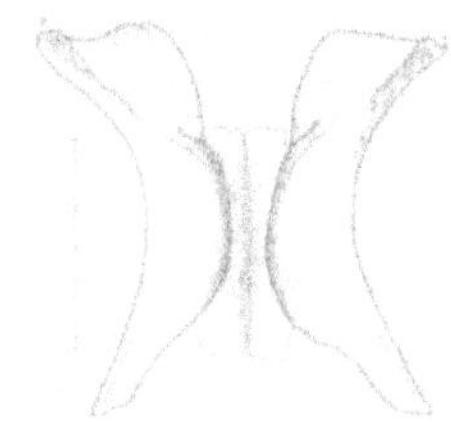

Fig. 234. *Onesia floralis* (Robineau-Desvoidy), male. Bacilliform sclerites. Scale = 0.4 mm.

Fig. 233. *Onesia floralis* (Robineau-Desvoidy), male. Epandrium, cerci and surstyli, posterior view. Scale = 0.4 mm.

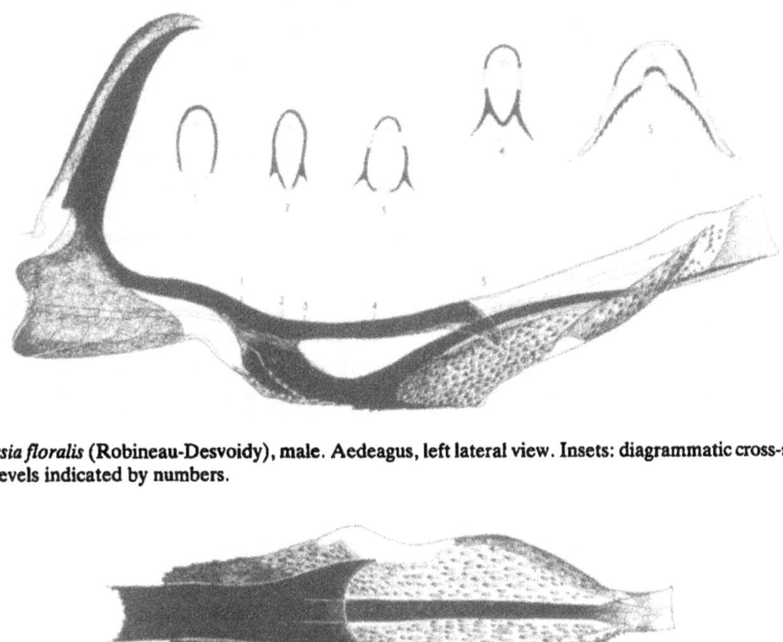

Fig. 235. *Onesia floralis* (Robineau-Desvoidy), male. Aedeagus, left lateral view. Insets: diagrammatic cross-sections of aedeagus at levels indicated by numbers.

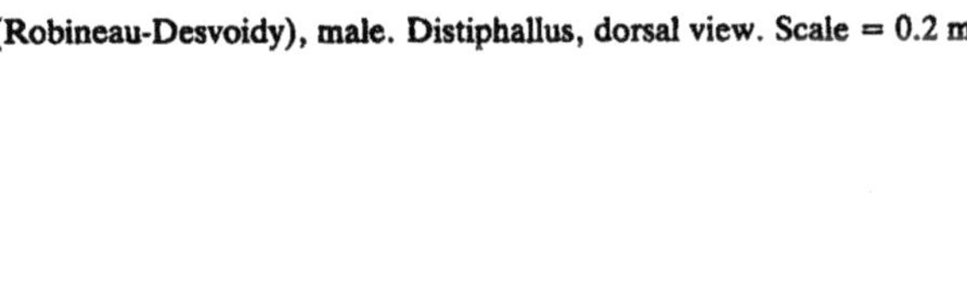

Fig. 236. *Onesia floralis* (Robineau-Desvoidy), male. Distiphallus, dorsal view. Scale = 0.2 mm.

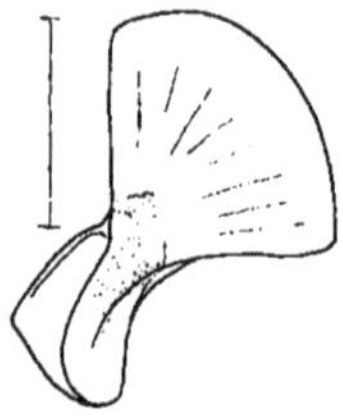

Fig. 237. *Onesia floralis* (Robineau-Desvoidy), male. Ejaculatory sclerite. Scale = 0.2 mm.

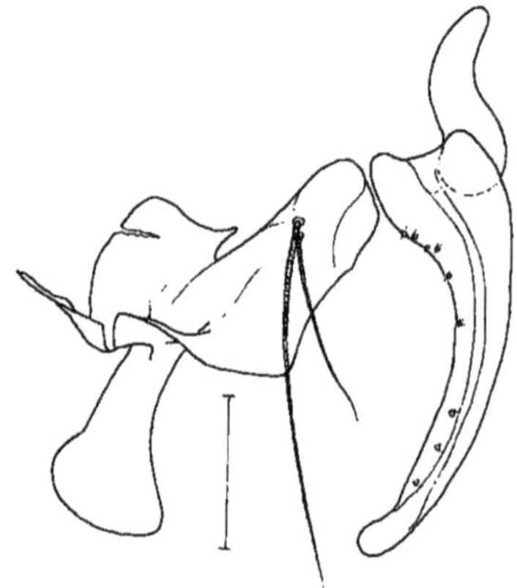

Fig. 238. *Onesia floralis* (Robineau-Desvoidy), male. Pre- and postgonites. Scale = 0.2 mm.

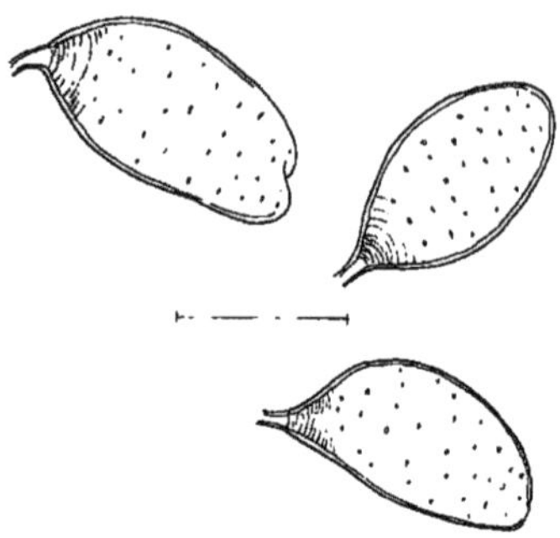

Fig. 239. *Onesia floralis* (Robineau-Desvoidy), female. Spermathecae. (Bilobed apex of one of the spermatheca shown is an abnormality.) Scale = 0.2 mm. G.pr. 102.

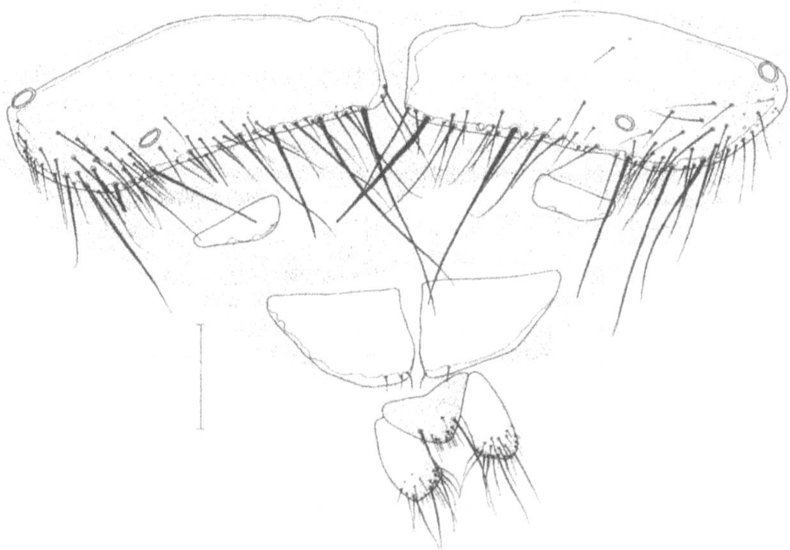

Fig. 240. *Onesia floralis* (Robineau-Desvoidy), female. Ovipositor, dorsal sclerites. Stipple indicates extent of microtrichiae. Scale = 0.4 mm. G.pr. 101.

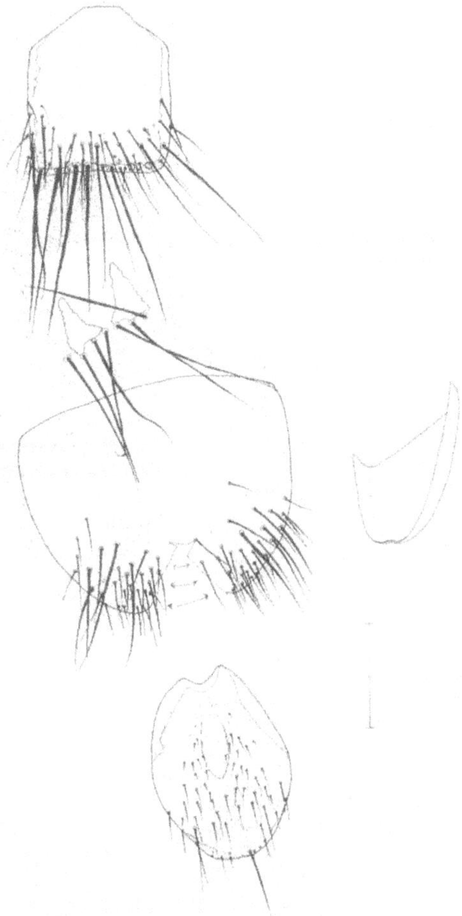

Fig. 241. *Onesia floralis* (Robineau-Desvoidy), female. Ovipositor, ventral sclerites. Stipple indicates extent of microtrichiae. ST8 has ruptured distally. Inset: approximate shape of intact ST8. Scale = 0.4 mm. G.pr. 101.

gen's collection in MNHN are 1♂2♀ and 1 empty pin under *Musca sepulcralis*, all labelled "Meigen" and "2143/40", a ♀ also with a label reading "sepulcralis/♀" in Meigen's hand. All are definitely the present species. None of them appears to have been seen by Schumann (1964). *Onesia floralis* was described by Robineau-Desvoidy (1830) as the first species under *Onesia*. No specimen is left in Robineau-Desvoidy's collection in MNHN (*Onesia*, box 12). A neotype should be designated. *Onesia claripennis* was described from females only, so the single male (= *Bellardia pandia* (Walker)) in Robineau-Desvoidy's collection (*Onesia*, box 12, no. 533) cannot be a syntype. In the MNHN General collection under *Onesia* are 2 females, labelled "192" and "claripennis" respectively in Macquart's hand (cf. Séguy, 1928a: 134, footnote). I do not believe that these are syntypes. In Pandellé's collection in MNHN (under no. 2669) are 3♂3♀ under *Calliphora retrocurva* all of which are the present species (labelled as *sepulcralis* by Schumann). All the males have the genitalia visible. I have selected and here designate a male in good condition labelled "♂ / 2072" in Pandellé's hand, as lectotype, and the other specimens as paralectotypes. Schumann (1986) lists *Melinda albiceps* Robineau-Desvoidy, 1830: 440 as a synonym. No specimen of this species is present in the Robineau-Desvoidy collection in Paris. However, in Box 12, under *Onesia* no. 531, are two specimens above a label reading "*Ones. albiceps* R.D. / Sp. inedita" in Robineau-Desvoidy's hand, an unpublished manuscript name. One of them has been identified as *Onesia sepulcralis* by Schumann, and the synonymy published by Schumann may be based on this specimen. I therefore believe that a confusion of names has occurred and have removed *Melinda albiceps* from the synonymy of *floralis*.

Subfamily Chrysomyinae

The subfamily Chrysomyinae in the sense adopted here is probably a monophyletic group justified by the following ground-plan apomorphies:

(1) Coxopleural streak absent.
(2) Numerous fine setae in a row along anteroventral edge of posterior (metathoracic) spiracle.
(3) Hind coxa with hairs on posterior surface.
(4) Stem-vein with row of setae on dorsal side of wing.
(5) Subcostal sclerite with black setulae.
(6) Postgonite with a more or less distinct process bearing the basal seta.
(7) Anterior half of ST8 in ovipositor unsclerotised or very weakly sclerotised.
(8) Third instar larva with incomplete peritreme medially, ecdysial scar tending to become indistinct or absent (Fig. 39).

The following genera probably belong to this subfamily: *Chloroprocta* Wulp, *Chrysomya* Robineau-Desvoidy, *Chrysopyrellia* Séguy, *Cochliomyia* Townsend, *Compsomyiops* Townsend, *Hemilucilia* Brauer, *Paralucilia* Brauer & Bergenstamm, *Phormia* Robineau-Desvoidy, *Phormiata* Grunin, *Protocalliphora* Hough, *Protophormia* Townsend and *Trypocalliphora* Peus.

Hall (1948) divided this assemblage into two tribes, Phormiini and Chrysomyini, on the basis of body colour, alleged presence (Chrysomyini) or absence (Phormiini) of hairs on the posterior surface of the hind coxa, and presence (Phormiini) or absence (Chrysomyini) of a well-defined "button" (ecdysial scar) in the posterior spiracle of the third instar larva, the phormiines being a northern group, the chrysomyines mainly a tropical and subtropical one. Schumann (1986) gave Hall's tribes separate subfamily rank. However, as most phormiine genera except *Protocalliphora* and *Trypocalliphora* most often have the hind coxa hairy behind to about the same extent as the South American chrysomyine genus *Chloroprocta* Wulp, and as the presence of a distinct button in the posterior spiracle is most likely to be plesiomorphous at this level and therefore cannot be used to define a monophyletic group, I do not accept Hall's classification. Zumpt (1956a) based the Phormiini on the shape of the lower calypter, which is narrow with an inner edge diverging backwards from the longitudinal axis of the fly. This system works for the Old World (the genus *Chrysomya* Robineau-Desvoidy has broad lower calypter) but breaks down when the Neotropical fauna is considered (Dear, 1985). Dear (1985) retained Hall's tribes using various features of the head to distinguish them, even though he found *Phormia regina* Robineau-Desvoidy a really "non-conformist Phormiine" and difficult to separate from less typical New World chrysomyines. Rognes (1985a) suggested that *Phormia* Robineau-Desvoidy might be more closely related to some of the "chrysomyines" than to any of its tribal fellows on the basis of the structure of the aedeagus. Pending a phylogenetic analysis of the genera making up this group, I prefer to abandon this two-tribe (or two-subfamily) division and to deal with a single subfamily Chrysomyinae (s. lat.). This avoids the nomenclatural problems involved in removing *Phormia* from the tribe named after it. It seems possible to establish homogenous subgroups on the basis of the structure of the aedeagus, but these are best given informal species-group names. Boyes & Shewell (1975) and Shewell (1987) use the name Chrysomyinae in an even wider sense, i.e. for all calliphorid genera with stem-vein hairy above (except *Pollenia atramentaria* (Meigen)) and give separate tribal rank to the chrysomyines, phormiines (part of the Neotropical toxotarsines is included here), sarconesiines (the remainder of the Toxotarsinae) and the rhiniines (= Rhiniinae in the present work). This system is not adopted here.

In Fennoscandia and Denmark large to medium sized flies with black ground-colour and metallic blackish blue-green, blue, green, olive-green or bronze sheen. Head profiles as in Figs. 242-246. Eyes bare. Male frons narrow, without proclinate orbital setae. Female frons usually with 2-4 proclinate orbitals. Arista long plumose to short pubescent in alpine forms, hairs always present on underside (Fig. 242-246). Facial carina weakly developed. Parafacialia hairy, without strong setae. Mesonotum often conspicuously flattened dorsal-

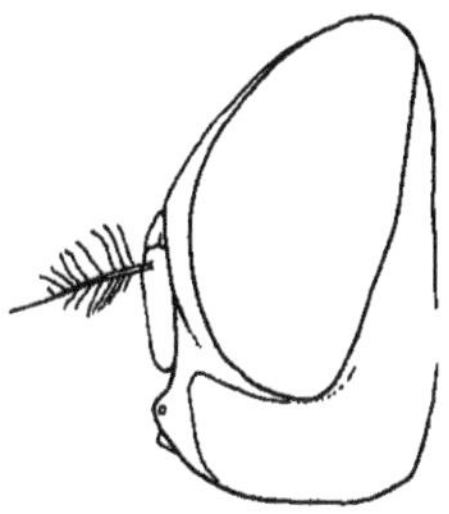

Fig. 242. *Phormia regina* (Meigen), head profile, male.

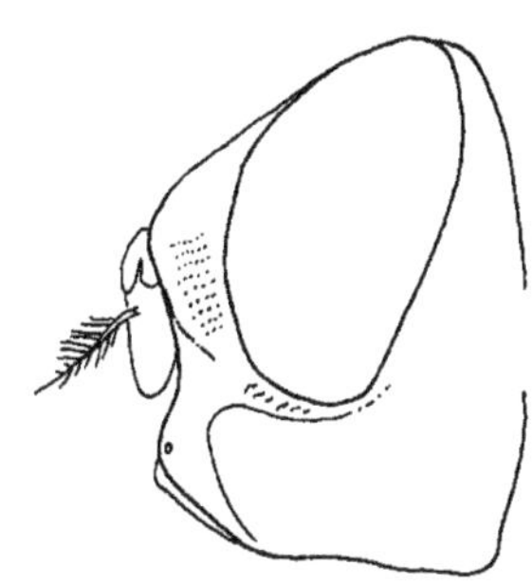

Fig. 243. *Protocalliphora chrysorrhoea* (Meigen), head profile, male.

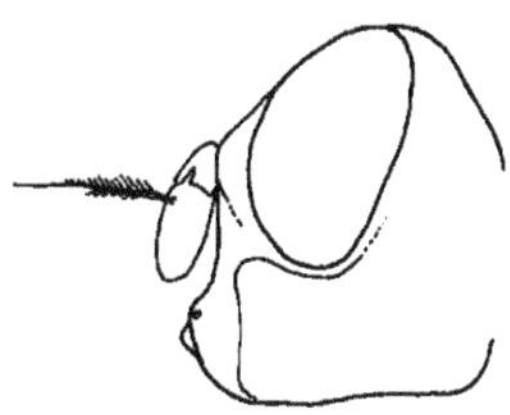

Fig. 244. *Protophormia atriceps* (Zetterstedt), head profile, male.

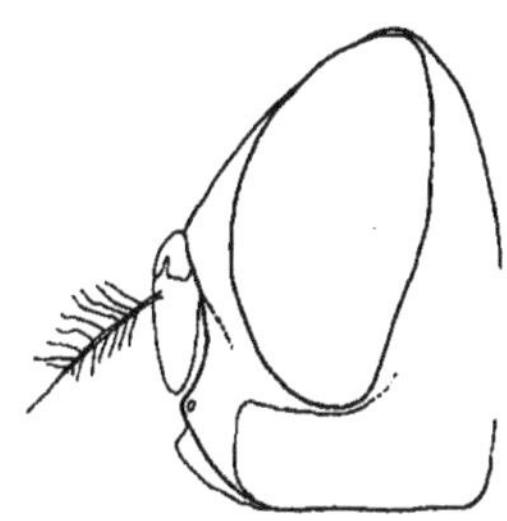

Fig. 245. *Protophormia terraenovae* (Robineau-Desvoidy), head profile, male.

Fig. 246. *Trypocalliphora braueri* (Hendel), head profile, male.

ly between suture and scutellum. Proepisternal depression, prosternum, metasternal area and usually also postalar wall hairy. Hindmost *prst acr*, if present, close to suture. 2-8 *h*. Outer *ph* slightly outside or in line with *prs* seta. 2+1 *kepst* setae. Lower half of anterior part of anepimeron bare, at most a few hairs just in front of border between anterior and posterior parts. Greater ampulla only pubescent (the Old World genus *Chrysomya* has the greater ampulla and the entire anterior half of the anepimeron covered with conspicuous erect black or pale hairs). Katatergite pubescent. Anatergite with short hairs below lower calypter. Coxopleural streak absent. Posterior spiracle rather small with subequal lappets, a row of short setae present along lower margin (Figs. 247, 259-263, 312-313, 339) (these are also present in Neotropical, Afrotropical and Oriental forms which have very large posterior spiracles). Area above hind coxa (metakatepisternum) bare. Stem-vein hairy on dorsal surface of wing (Figs. 8, 248, 281, 317). Costa hairy below to junction with Sc, bare beyond this point (Figs. 8, 248, 281, 317). Subcostal sclerite setulose. R_{4+5} hairy halfway to r-m. Cell r_{4+5} opens well in front of apex of wing. Lower calypter bare above, narrow, inner edge diverging backwards from longitudinal axis of fly. Upper calypter bare or hairy on upper side. Mid tibia with *v* seta. t_3 with strong *ad* and *d* preapical, *pd* preapical much weaker. Hind coxa with bundle of small hairs laterally on posterior surface, or bare (*Protocalliphora*, *Trypocalliphora*). Lateral margins of ST2 overlapping ventral margins of T2. Basal part of ST5 rather large compared with lobes, dusted and covered with small hairs. Aedeagus variable. Paraphallus long or short, its distal parts always situated medial to hypophallic lobes. Mesohypophallus present or absent. Hypophallus variable. Acrophallus armed with small denticles, without lateral duct (I have neither been able to find one in *Protophormia terraenovae*, as Lewis & Pollock, 1975, did). Distalmost seta on pregonite subapical in position, not apical. Marginal setae of T7 rather long. ST8 of ovipositor apically entire and rounded, basal half unsclerotised or very weakly sclerotised. Epiproct and cerci of ovipositor shining black without microtrichiae. Oviparous.

Widespread, with representatives in all faunal regions. Four genera (all "phormiines") in Fennoscandia and Denmark.

Key to genera of Chrysomyinae

1 Both calypters dark brown; upper calypter with black hairs on upper surface (viewed when wings are in resting position) (Fig. 316); hind coxa with a few hairs laterally on posterior surface; *prst acr* absent; 2 *post ia*; very dark bluish-green metallic body without dusting

Protophormia Townsend (p. 128)

\- Both calypters white or yellow, at most weakly infuscated; upper calypter bare or with white hairs laterally on upper surface (not to be confused with hairs along margin); *prst acr* conspicuous; body blue or green metallic with conspicuous dusting, at least on anterior slope of mesonotum

2

2(1) Upper calypter with white hairs on lateral half of upper surface; *prst acr* hardly twice as long as ground-vestiture (thorax examined in profile); 1+2 *ia*; postalar wall with numerous long black hairs; basicosta and anterior thoracic spiracle yellow; hind coxa with a few hairs or bare on posterior surface

Phormia Robineau-Desvoidy (p. 105)

\- Upper calypter bare on upper surface; *prst acr* longer, about 4 times as long as ground-vestiture (thorax examined in profile); 1+3 *ia*; hind coxa always bare behind

3

3(2) Basicosta, both thoracic spiracles and both calypters bright yellow; postalar wall practically bare (a few short inconspicuous hairs sometimes present); ♂: fronto-orbital plates contiguous, frons 2 times as wide as anterior ocellus (Fig. 338); ♀: frons strikingly narrow; lateroclinate seta absent (Fig. 338).

Trypocalliphora Hendel (p. 137)

\- Basicosta and both thoracic spiracles brown, dark brown or blackish; both calypters white or lower one sometimes infuscated brown (*azurea* ♂); postalar wall with or without long hairs; ♂: fronto-orbital plates never contiguous, frontal vitta broader than each fronto-orbital plate; ♀: lateroclinate seta present

Protocalliphora Hough (p. 106)

Genus *Phormia* Robineau-Desvoidy, 1830

Phormia Robineau-Desvoidy, 1830: 465.
Type-species: *Musca regina* Meigen, 1826: 58, by designation of Robineau-Desvoidy, 1849: v.
Euphormia Townsend, 1919: 542.
Type-species: *Musca regina* Meigen, 1826: 58, by original designation.

A monotypic genus easily recognised by the characters given in the key. Similar in general appearance to the genus *Trypocalliphora* but separable in both sexes by the long hairs on the postalar wall, mostly white not orange-yellow calypters, and presence of white hairs on outer part of upper surface of the upper calypter. Green metallic. Mesonotum with little dust except on anterior slope and not flattened between suture and scutellum. Chaetotaxy variable and difficult to describe as setae are often weakly differentiated from ground-vestiture. 4+2-3 *acr*. 3-6+5-6 *dc*. 3-4 *ph*. 4-7 *h*. 1+2-3 *ia*. 2+1 *npl*. t_1 with 1 *pv* seta. 3-4 pairs of marginal, 2-3 pairs of discal scutellar setae. Oviparous.

Phormia shares 3 synapomorphies with the genus *Chrysomya* (cf. Rognes, 1985a):

(1) white hairs laterally on upper surface of upper calypter,
(2) ventral surface of distiphallus with a dense covering of hairs proximal to hypohallic lobes,
(3) hypophallic lobes circular.

It may therefore be more closely related to this genus than to any other chrysomyine.

17. ***Phormia regina*** (Meigen, 1826)
Figs 242, 247-258.

Musca regina Meigen, 1826: 58.
Musca thalassina Meigen, 1826: 54.
Musca accincta Wiedemann, 1830: 407.
Phormia philadelphica Robineau-Desvoidy, 1830: 466.
Phormia cuprea Robineau-Desvoidy, 1830: 467.
Phormia fulvifacies Robineau-Desvoidy, 1830: 467.
Phormia vittata Robineau-Desvoidy, 1830: 467.
Phormia squalens Robineau-Desvoidy, 1830: 468.
Phormia mollis Walker, 1849: 892.
Musca proxima Walker, 1853: 241.
Somomyia lucens Rondani, 1862: 189.
Lucilia rufipalpis Jaennicke, 1867: 340.
Lucilia stigmaticalis Thomson, 1869: 544.
Lucilia nigrina Bigot, 1877: 247.
Somomyia rufigena Bigot, 1888a: 598; 1888b: clxxxi
Somomyia rupicola Bigot, 1888a: 603; 1888b: clxxx.
Phormia aurisquama Villeneuve, 1928: 151.

♂♀. Parafacialia densely silvery dusted. Palpi yellow. Antennae brownish, yellow at tip of pedicel and at base and along hind edge of first flagellomere. Facial membrane dark. Occiput concave with 0-2 rows of black hairs behind postocular row of setae, otherwise with pale hairs. Gena with black hairs. Genal dilation black with a very thin layer of white dust. Hind coxa without (70%) or with (30%) small hairs posteriorly. Ground-vestiture decumbent on T1+2-T4, erect on T5.

♂. Frons very narrow, 0.030-0.039 times head-width (n=4). Fronto-orbital plates contiguous, shining black in posterior half. f_2 without *a* preapical seta. t_2 with 1 *ad* seta. Cerci short. Surstyli short and triangular in profile. Aedeagus without mesohypophallus. Paraphallus almost straight, proceeding apically medial to the hypophallus. Hypophallus circular, with small denticles along lower and hind edge. A fold in the proximal part of the hypophallus where it joins the main stem of the distiphallus mimics a "lobe hypophallique médian" when the aedeagus is seen in exact profile (cf. Rognes, 1988). Acrophallus long with a few inconspicuous denticles, widened in middle in dorsal view. Ventral plates not joining each other ventrally. Proximal part of distiphallus covered ventrally with hair-like denticles. Pregonite blunt, with a few setae along posterior edge. Postgonite with basal process bearing the basal seta.

♀. Frons 0.283-0.316 times head-width (n=3). Fronto-orbital plates shining black in posterior half. f_2 with *a* preapical seta. t_2 with 3-4 *ad* setae. Ovipositor with rather large spiracles in T6. T7 and T8 halves broad. ST6 and ST7 broad. Narrow anterior zone of pleural membrane 7 devoid of microtrichiae. Microtrichiae on intersegmental membrane 7-8 and on posterior half of pleural membrane 8 very short, peg-like. Spermathecae elongate. [1 ovipositor slide examined.]

Length. 6-10 mm.

Distribution. In Fennoscandia and Denmark known only from a single ♀ captured by Ringdahl in Hälsingborg (Sk.) August 12, 1945. - Widely distributed in the Nearctic and Palaearctic Regions. Also Hawaii.

Biology. Immature stages described by Hall (1948). *Phormia regina* has not been reported as a myiasis-producer in the Old World (Zumpt, 1965), but commonly so in the Nearctic (Hall, 1948). A case of human myiasis has been reported by Hall, Anderson & Clark (1986).

Nomenclature. *Musca regina* was described from a syntypic series, "... im Jahr 1814 hier [= West Germany, probably Stolberg]... sehr gemein ...". Specimens were not sexed, but Meigen Plate 120, fig. 12 (Morge, 1976a) shows both sexes. It was assigned to *Lucilia* by Meigen (1838: 292). In Meigen's collection in MNHN is a male labelled (1) "Meigen" and "2112/40"; (2) "Lucilia / regina / ♂ [Meigen's hand]" with the left front and mid legs lost but otherwise in good condition. I have labelled and here designate this specimen as lectotype of *Musca regina* Meigen. *Musca thalassina* was decribed from a holotype ♀, locality unstated [West Germany, probably Stolberg] with "...Taster schwarz...", figured on Meigen Plate 120, fig. 6 (Morge, 1976a). It was assigned to *Lucilia* by Meigen (1838: 292). In Meigen's collection in MNHN is a female in moderate condition, which has the yellow palps hidden. The dark mouthparts are slightly projecting and may have been mistaken for palps. It is labelled (1) "Meigen" and "2109/40"; (2) "Lucilia / thalassina / ♀ [Meigen's hand]". I have given it a holotype label. Séguy (1928a: 169) first established the synonymy of *regina* and *thalassina* (confirmed by Aubertin, 1933). I consider him the first reviser in selecting *regina* as the valid name for this species.

Genus *Protocalliphora* Hough, 1899

Protocalliphora Hough, 1899b: 66.
Type-species: *Musca azurea* Fallén, 1817: 245, by original designation.
[Suppression of Hennig's (1939) designation of

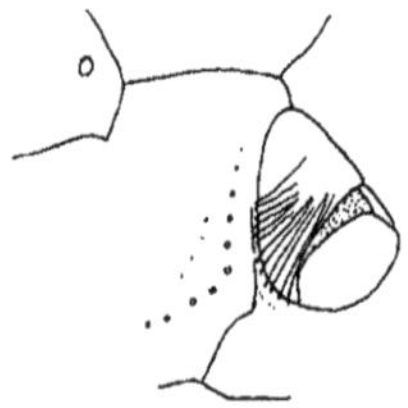

Fig. 247. *Phormia regina* (Meigen), posterior spiracle.

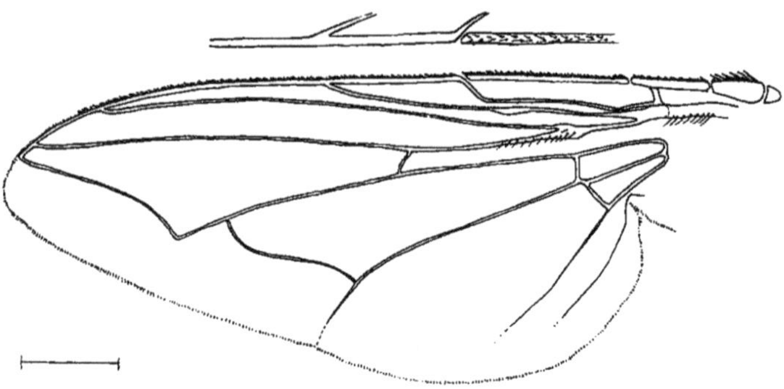

Fig. 248. *Phormia regina* (Meigen). Left wing from above. Inset: costa from below showing distribution of small hairs on its underside. Scale = 1.0 mm.

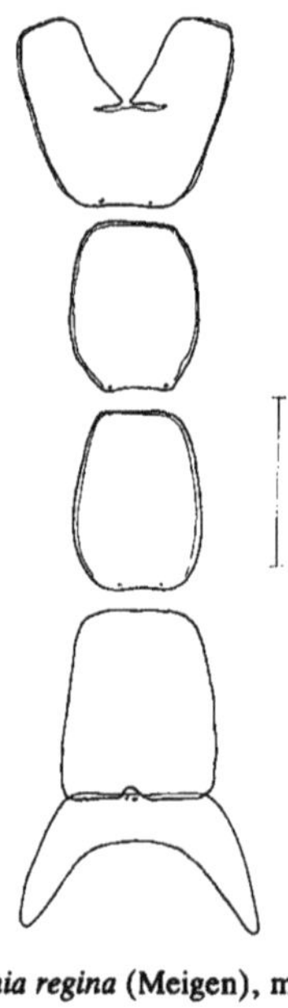

Fig. 249. *Phormia regina* (Meigen), male. ST1-5. Scale = 0.8 mm.

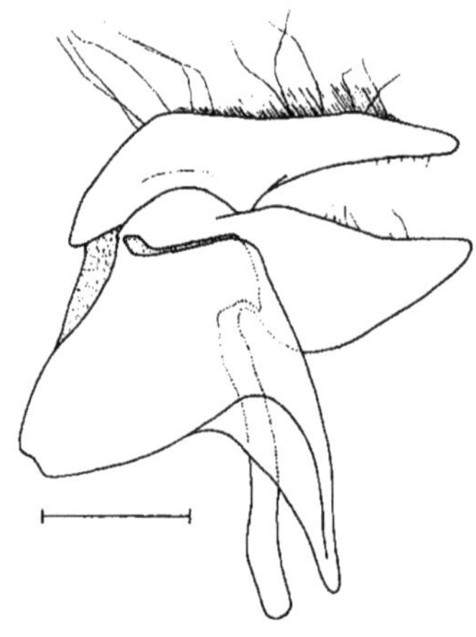

Fig. 250. *Phormia regina* (Meigen), male. Epandrium, cerci, surstylus and bacilliform sclerite, left lateral view. Scale = 0.2 mm.

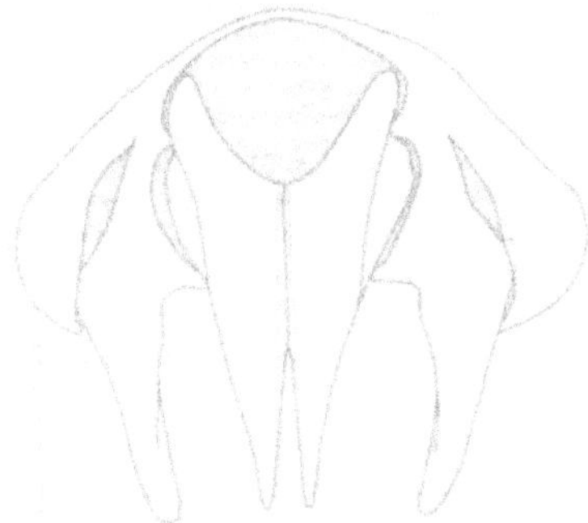

Fig. 251. *Phormia regina* (Meigen), male. Epandrium, cerci and surstyli, posterior view. Scale = 0.2 mm.

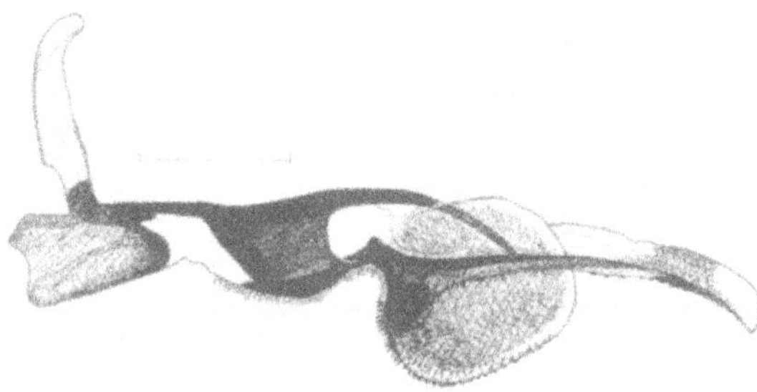

Fig. 252. *Phormia regina* (Meigen), male. Aedeagus, left lateral view. Scale = 0.2 mm.

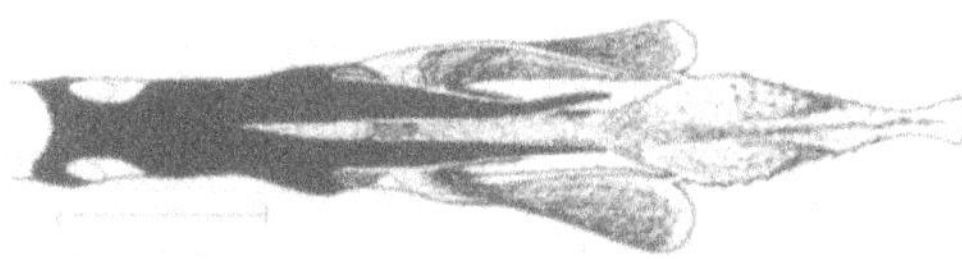

Fig. 253. *Phormia regina* (Meigen), male. Distiphallus, dorsal view. Scale = 0.2 mm.

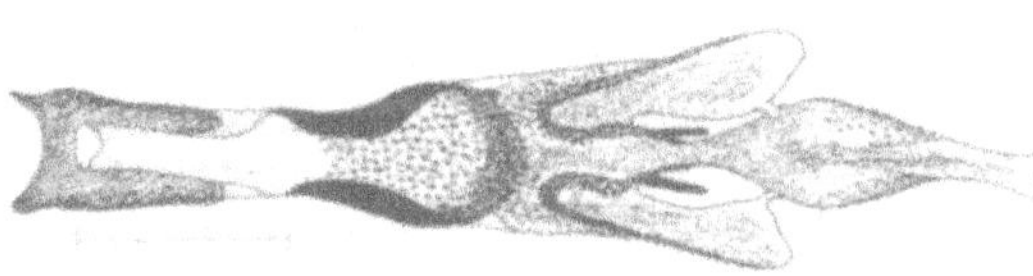

Fig. 254. *Phormia regina* (Meigen), male. Aedeagus, ventral view. Scale= 0.2 mm.

Fig. 255. *Phormia regina* (Meigen), male. Ejaculatory sclerite. Scale =0.2 mm.

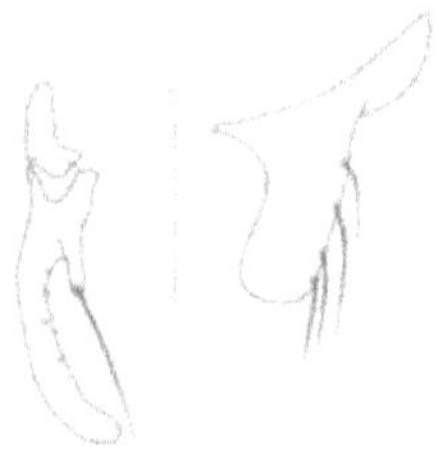

Fig. 256. *Phormia regina* (Meigen), male. Post- and pregonites. Scale =0.2 mm.

Fig. 257. *Phormia regina* (Meigen), female. Spermathecae. Scale = 0.2 mm. G.pr. 178.

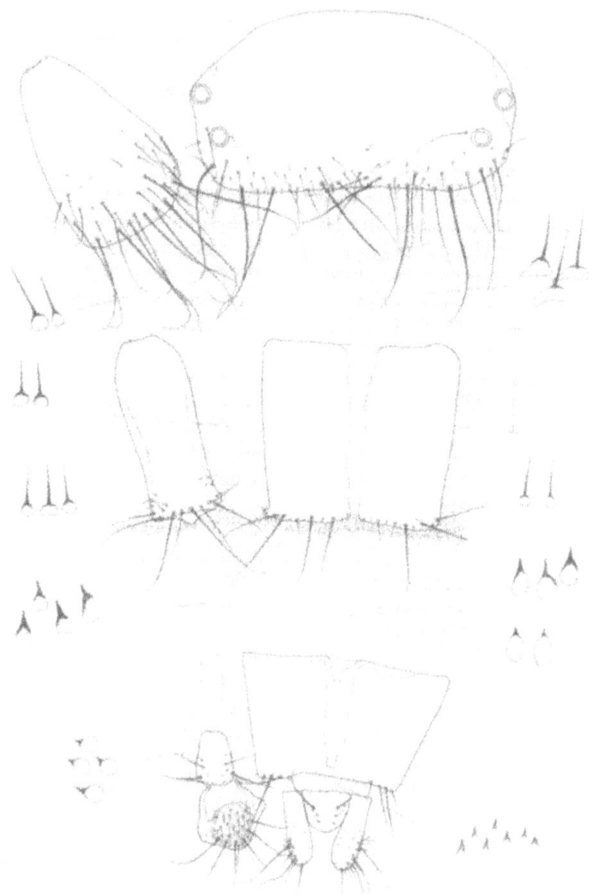

Fig. 258. *Phormia regina* (Meigen), female. Ovipositor. Stipple indicates extent of microtrichiae. Insets: shape of individual microtrichiae in areas indicated (enlarged). Scale = 0.4 mm. G.pr. 178.

an original Fallén ♂ specimen of *Protophormia terraenovae* as lectotype of *Musca azurea* is necessary to preserve the name *azurea* in the current sense. An application to the Commission has been prepared by Sabrosky (1989) to this effect. A possible Fallén syntype conforming to *azurea* in the present sense (see below) is available in ZML, and has been designated as lectotype by Sabrosky (1956).]

Avihospita Hendel, 1901: 29.

Type-species: *Musca azurea* Fallén, 1817: 245, by original designation.

Apaulina Hall, 1948: 179.

Type-species: *Protocalliphora avium* Shannon & Dobroscky, 1924: 250, by original designation.

Orneocalliphora Peus, 1960: 198.

Type-species: *Musca chrysorrhoea* Meigen, 1826: 60, by original designation.

A Holarctic genus of large, blue or green metallic flies whose larvae live obligatorily as external blood-sucking parasites on nestling birds. The genus is easily recognised by the characters given in the key. The taxonomy at the species level, however, especially of the females, is very difficult.

Head black except sometimes for vibrissal corner and tip of pedicel and base of first flagellomere. Frontal vitta always broader than each fronto-orbital plate. Latter always with a lateroclinate seta in the female. First flagellomere short, 1-2 times as long as pedicel. Arista with long hairs. Parafacialia densely dusted with characteristic colour, often with numerous transverse wave-like ridges (rugae) and irregular grooves of varying depth. These rugae continue to a varying extent onto the genal groove below the eye. Facial carina present, not high but sharp in upper part. Lunula with a more or less transverse ridge which meets the upper end of the facial carina in a "T", or ridge not transverse but dipping ventrally at mid-line to join the facial carina in a "Y". Vertex in female with the area between the posteromedian corner of eye, outer vertical, inner vertical and lateroclinate seta sometimes polished black, without dusting. Variable number (1-4) of but most often 2 proclinate orbitals in female. Occiput with pale hairs except for about 3 irregular rows of black setae behind the postocular row. Gena with black hairs. Genal dilation shining, very thinly dusted. Palpi yellow, often more or less infuscated. Mesonotum dusted with conspicuous stripes. Conspicuous flattening present between suture and scutellum. Chaetotaxy difficult to describe and often very variable. 3-4+3-6 *ac*, 3-4+3-4 *dc*, 3-5 *ph*, outer *ph* in line with *prs* or slightly outside, 4-6 *h*, 1+3 *ia*, 3-6 pairs of marginal and 1-3 discal pairs of scutellar setae. 2 *npl*. Postalar wall with varying vestiture, from almost bare to covered with numerous black setae. Both spiracles brown, lappets of posterior spiracle as in Figs. 259-263. t_1 with 2 *pv* setae. f_2 with weak *a* preapical in males, stronger in females. t_2 with *v* seta and several *ad* setae. t_3 with 3-4 *av* setae. Hind coxa always bare behind. Wing-veins dark. Basicosta of varying colour. Abdomen shining, weakly dusted. Ground-vestiture on the dorsal surface of the tergites decumbent on all segments except on T5 where it is erect. Aedeagus very uniform in structure. Long torpedo-shaped mesohypophallus present in a ventral groove. Hypophallus pointing ventrally, short and strong, without spinules apically. Paraphallus strong, apex often weakly serrate, in lateral view hidden behind hypophallus. Two apical clusters of spinules on acrophallus. Apical opening of ejaculatory duct narrow. Pregonite small and triangular with about 3 setae. Postgonite with a basal seta on a weak projection. Ovipositor with small spiracles in T6. Sclerites long and narrow. Broad anterior zone of pleural membrane 7 devoid of microtrichiae. Microtrichiae on intersegmental membrane 7-8 and on posterior part of pleural membrane 8 long and similar to those elsewhere on the ovipositor. Spermathecae oval, one slightly longer than the other two.

Third instar larva with a crown of setae on the cephalic segment forming a kind of sucker with which it anchors itself to the host during feeding.

All species are obligatory external blood-suckers in the larval stages, feeding intermittently on nestling birds. In the Palaearctic up to 155 and 250, in the Nearctic up to 970, larvae have been reported from a single nest (Lindner, 1957; Sabrosky & Bennett, 1958) with a mean number of 47 (Lindner, 1957; Gregor & Povolný, 1959).

For a discussion of the relationship of *Protocalliphora* to *Trypocalliphora*, see below under the latter genus.

Hall (1965) catalogued 9 species from the Nearctic. A revision of the Nearctic species has been completed by Sabrosky, Bennett & Whitwort (1989). They recognise 26 species, of which 2 are Holarctic. Like Sabrosky (1967) they treat *Trypocalliphora* (with *braueri*) as a subgenus of *Protocalliphora*. Grunin (1971) thought that about 50 species were present in the Palaearctic, many still undescribed. Schumann (1986) catalogued 14 Palaearctic species. 5 species are known in Fennoscandia and Denmark.

The key below allows identification of speci-

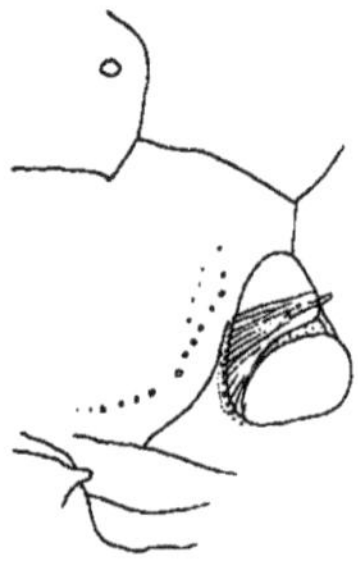

Fig. 259. *Protocalliphora azurea* (Fallén), posterior spiracles.

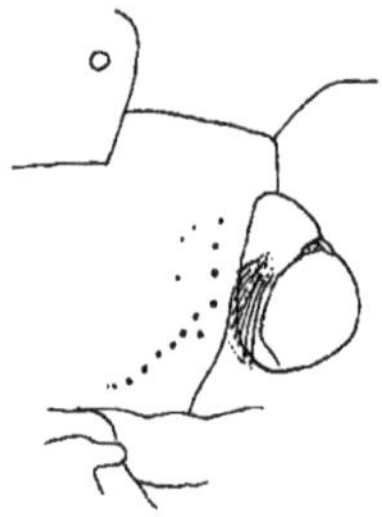

Fig. 261. *Protocalliphora nuortevai* (Grunin), posterior spiracles.

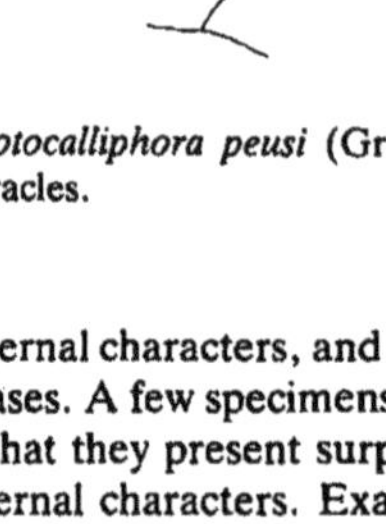

Fig. 263. *Protocalliphora peusi* (Gregor & Povolný), posterior spiracles.

mens on external characters, and appears to work in 90% of cases. A few specimens are notoriously difficult in that they present surprising combinations of external characters. Examination of the genitalia of both sexes is essential for a reliable identification. Only 2 females of *proxima* have been available so its inclusion in the key is rather tentative.

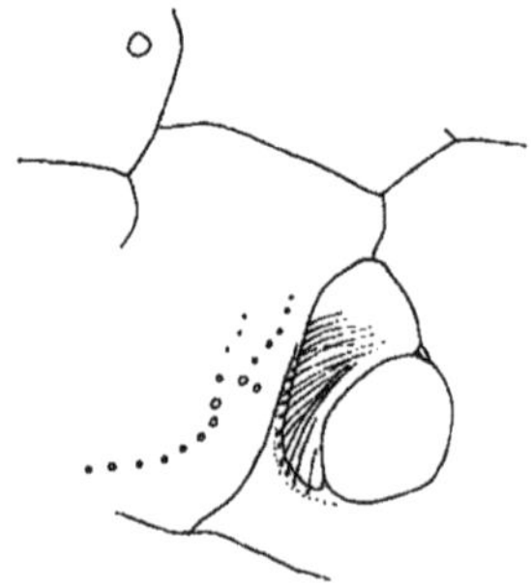

Fig. 260. *Protocalliphora chrysorrhoea* (Meigen), posterior spiracles.

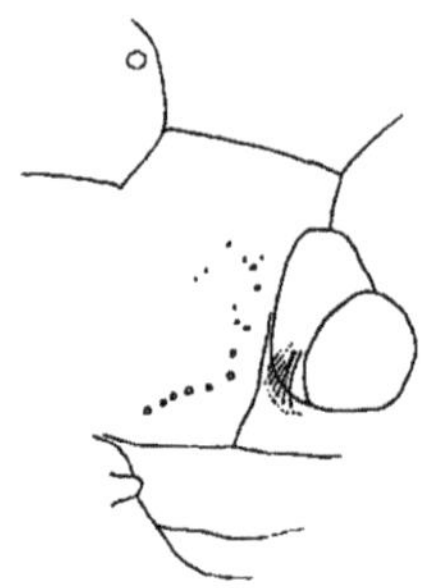

Fig. 262. *Protocalliphora proxima* (Grunin), posterior spiracles.

Key to species of *Protocalliphora*

Males

1 Parafacialia and usually fronto-orbital plates dusted light brown to dark brown; frons broad, almost 2 times as wide as distance between posterior ocelli inclusive; parafacialia densely rugose, rugae also present anteriorly in genal groove; vibrissal angle reddish; both calypters, including margin, white; cerci and surstyli as in Figs. 273, 274.

19. *chrysorrhoea* (Meigen)

– Parafacialia dusted white, grey or dirty grey 2

2(1) Calypters, especially margin of lower one, infuscated; frons narrow, 1-1.5 times as wide as distance between posterior ocelli inclusive; usually no rugae in genal groove; lunula with a transverse rib, joining facial carina in a

"T" shaped structure; cerci and surstyli as in Figs. 265, 266.

18. *azurea* (Fallén)

– Calypters, including margins, rather pure white

3

3(2) Frons broad, 2 times as wide as distance between posterior ocelli inclusive; lunula without a transverse rib, junction with facial carina shaped as a "Y"; parafacialia distinctly rugose, rugae also present anteriorly in genal groove; vibrissal angle black; basicosta almost black; cerci and surstyli as in Figs. 301, 302.

21. *peusi* Gregor & Povolný

– Frons narrower, 1-1.5 times as wide as distance between posterior ocelli inclusive; lunula with a transverse rib, junction with facial carina shaped as a "T"; parafacialia with few and inconspicuous rugae, usually no rugae in genal groove; vibrissal angle usually reddish; basicosta most often brown

4

4(3) Dusted band on fronto-orbital plates strikingly broad at middle (i.e. at 'centre' of concentric circles through eye facets), as broad as or broader than anterior ocellus; parafacialia dusted densely, intensely white, and sparsely haired; cerci and surstyli as in Figs. 283, 284.

20. *nuortevai* Grunin

– Dusted band on fronto-orbital plates narrow as usual at middle (i.e. at 'centre' of concentric circles through eye facets), narrower than anterior ocellus; parafacialia more greyish-white dusted, sometimes with a very slight golden sheen, and more conspicuously haired; cerci and surstyli as in Figs. 292, 293.

22. *proxima* Grunin

Females

1 Parafacialia and fronto-orbital plates dusted light brown to dark brown; thorax and abdomen blue or greenish-blue metallic; calypters white

2

– Parafacialia and fronto-orbital plates dusted white, usually no rugae in genal groove; lunula with a transverse rib, junction with facial carina "T" shaped; basicosta yellowish-brown to brown

4

2(1) Area on vertex between posteromedian corner of eye, lateroclinate seta and outer and inner vertical setae evenly dusted; parafacialia densely rugose, about 10 rugae present; rugae also present anteriorly in genal groove; vibrissal corner reddish; lunula usually with a transverse rib, junction with facial carina most often "T" shaped; basicosta dark brown; t_2 with 4-5 *ad* setae; size large

19. *chrysorrhoea* (Meigen)

– Area on vertex between posteromedian corner of eye, lateroclinate seta and outer and inner vertical setae polished black, sometimes irregularly and thinly dusted with dark colour shining through

3

3(2) Lunula without a transverse rib, junction with facial carina "Y" shaped; vibrissal angle black; parafacialia and fronto-orbital plates greyish-brown; rugae present anteriorly in genal groove; basicosta almost black; thorax and abdomen blue; t_2 with 5-6 *ad* setae; size large

21. *peusi* Gregor & Povolný

– Lunula usually with a transverse rib, junction with facial carina "T" shaped; vibrissal angle reddish; parafacialia golden glistening brown, fronto-orbital plates matt brown; usually a very pronounced ruga in upper part of parafacialia, in some lights appearing as a dark transverse band; no rugae in genal groove; basicosta brown; thorax and abdomen often with a conspicuous greenish tinge; t_2 with 3-4 *ad* setae; size small

20. *nuortevai* Grunin

4(1) Abdomen and thorax usually green; calypters sometimes infuscated

18. *azurea* (Fallén)

– Abdomen and thorax blue; calypters pure white

22. *proxima* Grunin

18. ***Protocalliphora azurea*** (Fallén, 1817)
Figs 259, 264-271.

Musca azurea Fallén, 1817: 245.
Phormia caerulea Robineau-Desvoidy, 1830: 466.
Musca violacea Meigen, 1838: 301. Junior primary homonym of *Musca violacea* Scopoli, 1763: 340 (Stratiomyidae) and *Musca violacea* Fabricius, 1805: 288 (Calliphoridae, not Palaearctic, cf. Michelsen, 1979; Dear, 1985). SYN.N.
Lucilia dispar Dufour, 1845: 205.
Protocalliphora surcoufi Séguy, 1928a: 166. SYN.N.

♂♀. Usually sexually dimorphic body-colour, males blue and females green.

♂. Frons narrow, 0.067-0.080 times head-width (n=5). Cerci in profile almost straight and rather short and broad. Ventral plates of distiphallus not extensively sclerotised.

♀. Frons narrow, 0.234-0.247 times head-width (n=5). Area on vertex between posteromedian corner of eye, lateroclinate seta, and outer and inner vertical setae dusted. T6 almost always weakly sclerotised mid-dorsally, thus appearing subdivided. Pleural membrane 7 microtrichiose for posterior two-thirds. T7 relatively short, T7 halves well sclerotised to hind margin. T8 sclerotised to margin, marginal setae usually present along whole hind margin of each T8 half. [4 ovipositor slide-mounts examined.]

Length. 7-12 mm.

Distribution. The most common *Protocalliphora* species and widely distributed in Fennoscandia and Denmark. Ascends to subalpine region. North to NSY and Ly.Lpm. - Palaearctic Region east to Japan (Kano & Shinonaga, 1968; Ozerov, 1986; Schumann, 1986; Hori & Iwasa, 1988).

Biology. Immature stages described by Engel (1920, as *sordida*), Rodendorf (1957, as *chrysorrhoea*) and Ishijima (1967). The larvae are fully grown within a week, pupariation takes place in the nest and the adults emerge from 9 to 23 days afterwards (cf. Zumpt, 1965). In Fennoscandia and Denmark it has been bred from the following hosts: *Erithacus rubecula*, *Ficedula hypoleuca*, *Parus caeruleus*, *Parus ater*, *Parus major*, *Phylloscopus trochilus*, *Turdus iliacus* and even a mouse-nest. Zumpt (1965) should be consulted for a complete host-list; Hori & Iwasa (1988) for host-birds in Japan. Mixed infestations with *Trypocalliphora braueri* are known from two nests of *Phylloscopus trochilus* (R. Mehl leg. et cult. j.nr. 66/71, 16/85, RML, KR). A female has been reared from a nest of *Parus major* together with 19 specimens of *Calliphora vicina* (Lennart Magni leg. et cult. j.nr. M51, ZML). The nestling was probably dead or almost dying at the time of infection, the *vicina* specimens emerging 4 days after the *azurea* specimen.

Adults have been collected from April to September, mostly on flowers, sunny walls, etc. They are not attracted to carcasses or decomposing organic matter. They overwinter in the adult stage (Zumpt, 1965). Erzinçlioğlu (1984) has reported puparia to be parasitised by *Nasonia vitripennis* (Walker) (Hym., Pteromalidae).

Note. A few males may have a totally white lower calypter, and they may run to *proxima* in the key. A few rugae may occasionally be present in the anterior part of the genal groove. Kano & Shinonaga (1968) do not describe Japanese female *azurea* as having a green abdomen.

Nomenclature. Sabrosky (1956, 1989) and Sabrosky, Bennett & Whitworth (1989) have discussed in detail the problem of the identity of *Musca azurea* Fallén and I here follow their views. The main points are as follows: Fallén's original syntypic material ("♂♀") of *azurea* (sent to him from Västergötland by Gyllenhal) is now located both in Fallén's collection in NHRM and in Zetterstedt's collection in ZML. It was mixed, a fact pointed out by Fallen's contemporaries Meigen (1826: 53-54) and Zetterstedt (1838: 657). The latter segregated Fallén's material into the saprophagous *groenlandica* (now = *Protophormia terraenovae*) and the bird-parasitic *azurea*. Hennig (1939) borrowed a ♂ from NHRM and, by referring to it as "der Typus" (inadvertently), is deemed under the ICZN (Art. 74 (a)) to have designated this specimen as lectotype; Hennig was apparently unaware of the existence of further syntypic material in NHRM and unaware of any Fallén material in ZML. He dissected and figured the genitalia and identified it as a specimen of *Protophormia*, thus barring the use of the name *azurea* for any bird blowfly. Sabrosky (1956) studied all the *Protocalliphora* material in the Fallén and Zetterstedt collections in NHRM and ZML. In NHRM he located 2♂1♀ under Fallén's label "*azurea*", all = *Protophormia terraenovae*. In ZML he located 5 specimens (4♂1♀) in Zetterstedt's collection under *azurea*. 3♂1♀ are now in the Diptera Scandinavica collection (drawer 24) and 1♂ in the Gothenburg collection (drawer 3). The ♀ is labelled "M.azurea /♀. Svecia"; 1♂ bears a small white square label with three black stripes on the underside and is also labelled "M. azurea /♂.Lärket," [= Ög.: Lärketorp]; 2 ♂ bear a white small paper square (also white on under-

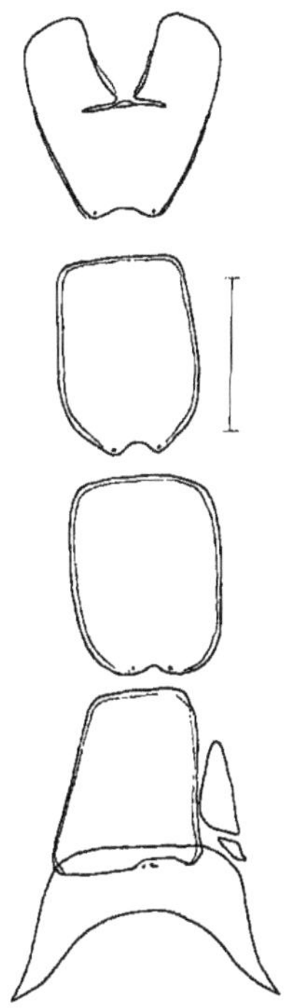

Fig. 264. *Protocalliphora azurea* (Fallén), male. ST1-5. Scale = 0.8 mm.

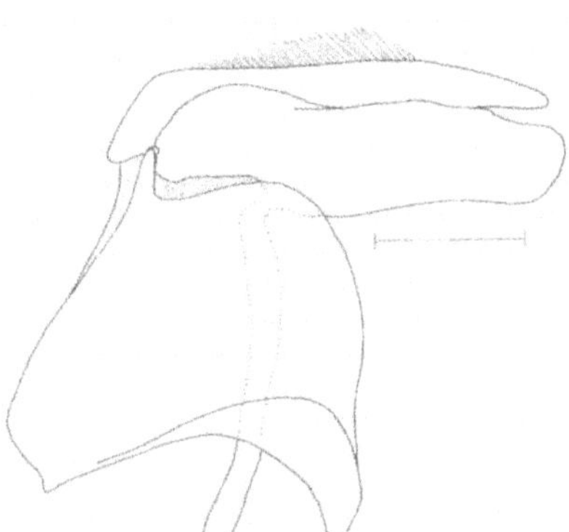

Fig. 265. *Protocalliphora azurea* (Fallén), male. Epandrium, cerci, surstylus and bacilliform sclerite, left lateral view. Scale= 0.2 mm.

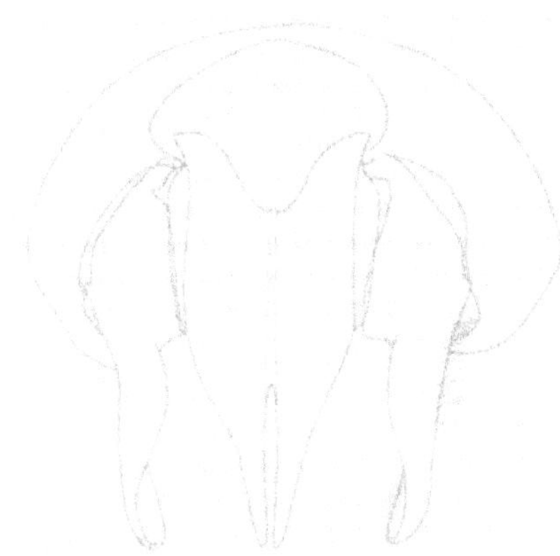

Fig. 266. *Protocalliphora azurea* (Fallén), male. Epandrium, cerci and surstyli, posterior view. Scale = 0.2 mm.

Fig. 267. *Protocalliphora azurea* (Fallén), male. Aedeagus, left lateral view. Inset: tip of paraphallus (enlarged). Scale = 0.2 mm.

Fig. 268. *Protocalliphora azurea* (Fallén), male. Ejaculatory sclerite. Scale = 0.2 mm.

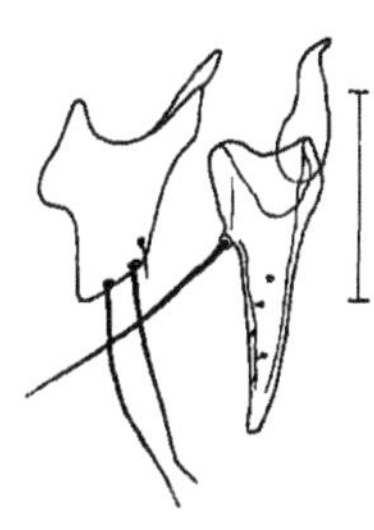

Fig. 269. *Protocalliphora azurea* (Fallén), male. Pre- and postgonites. Scale = 0.2 mm.

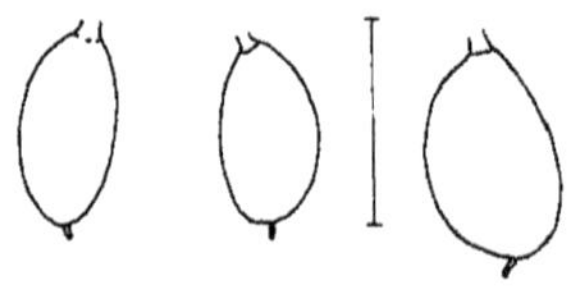

Fig. 270. *Protocalliphora azurea* (Fallén), female. Spermathecae. Scale = 0.2 mm. G.pr. 121.

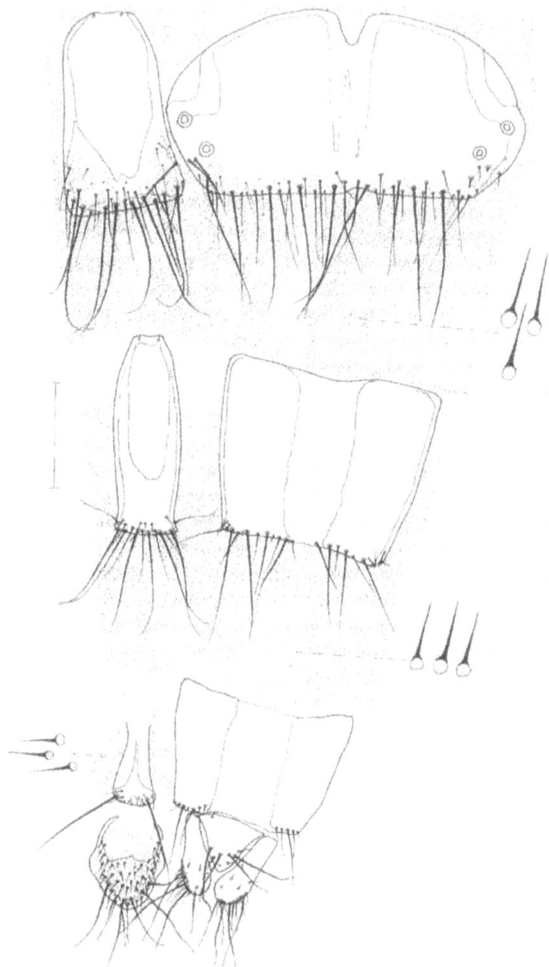

Fig. 271. *Protocalliphora azurea* (Fallén), female. Ovipositor. Stipple indicates extent of microtrichiae. Insets: shape of individual microtrichiae in areas indicated (enlarged). Scale = 0.4 mm. G.pr. 121.

side) which indicates that they were collected by Zetterstedt during his first Lapland journey in 1821 [one of these, without further labels, is *proxima* Grunin; the second, which carries a second label reading "M.azurea / Zett. Fll., ♂", is *azurea*]; and finally 1♂ is wholly without labels. Sabrosky (1956: 179) suggested that this latter specimen might be an original Fallén specimen and believed "it should be considered the lectotype". If this is indeed an original syntype one would have expected it to carry a label reading "Mus. Fall." or "F." in Zetterstedt's hand, which Zetterstedt usually put on the Fallén material he revised (cf. Michelsen, 1983; Pont, 1984; R. Danielsson, pers.comm.). I have dissected it and find that it agrees with the current concept of *azurea* (Peus, 1960; Grunin, 1970a). Under the current Code it appears that the specimen Hennig (1939) dissected and figured is a valid lectotype. Sabrosky (1956) considered Hennig's designation invalid since inconsistent with the actions of the first reviser (Meigen or Zetterstedt), and he designated the unlabelled and presumed Fallén specimen as lectotype to preserve the use of the name *azurea* for a bird parasite blowfly. Under the current Code, lectotype designations take precedence over restrictions by revisers (Article 74 (a) (ii)) and Sabrosky's (1956) lectotype designation is therefore invalid. However, apart from Ringdahl and Hennig, no author has used *azurea* for the saprophagous *Protophormia terraenovae*. Strict application of the current rules of the Code would upset the well-established nomenclature not only of the bird blowfly but also of the saprophagous species. I agree with Sabrosky (1989) and Sabrosky, Bennett & Whitworth (1989) that the best course is to suspend the rules, suppress the lectotype designation by Hennig and fix the name *azurea* for the bird blowfly species. I therefore accept the unlabelled specimen as a syntype, and treat it as the lectotype of *Musca azurea* Fallén. I have labelled it as such. *Musca violacea* was described by Meigen from an unspecified number of males from "Hiesige Gegend" [= West Germany, Stolberg]. In Meigen's collection in MNHN is a single ♂ labelled (1) "Meigen" and "2138 / 40"; (2) "Musca / violacea / ♂ [Meigen's hand]". This fits the description and I have labelled it as holotype. It was identified by Rodhain & Bequaert (1916) as *sordida* Zetterstedt, but listed as a nomen dubium in Calliphorinae by Schumann (1986). Meigen Plate 299, fig. 16 (Morge, 1976b), however, shows a female with metallic green fronto-orbital plates, probably a *Neomyia* Walker (Muscidae). It is named "violacea ♀ (Var. nitida)" by Meigen in his "Erklärung der Figuren" (Morge, 1975: 459), but simply as "violacea" in his "Supplement" to the "Alphabetisches Verzeichnis" (Morge, 1975: 497). It seems that Meigen mistakenly believed that the female he illustrated was the same species as his male *violacea*. The holotype of *surcoufi* is in MNHN. The infuscated lower calypter, the narrow frons and the genitalia (Séguy slide 376) indicate that it belongs to *azurea*. I have therefore removed *surcoufi* from the synonymy of *chrysorrhoea* Meigen, where it was placed by Zumpt (1956a) (cf. Schumann, 1986).

19. ***Protocalliphora chrysorrhoea*** (Meigen, 1826)

Figs 243, 260, 272-279.

Musca chrysorrhoea Meigen, 1826: 60.
Musca sordida Zetterstedt, 1838: 657.

♂♀. Body colour not sexually dimorphic, blue in both sexes.

♂. Frons 0.116-0.148 times head-width (n=5). Surstyli in profile similar to those of *azurea*, but a little longer. Ventral plates of distiphallus extensively sclerotised when seen in profile.

♀. Frons 0.271-0.303 times head-width (n=7). T6 usually (in 6 of 7 cases) undivided except for an area of weak sclerotisation in fore margin. T7 halves rather narrow. T8 with a full row of marginals behind strongly sclerotised T8 halves. Pleural membrane 7 without microtrichiae in anterior 1/2 to 2/3. [7 ovipositor slides examined.]

Length. 9-14 mm.

Distribution. Rare. Absent from Denmark, scattered all over Norway and Sweden. Northern Finland. - Germany, Austria, France, Poland (Draber-Mońko, 1978). Also Nearctic (Sabrosky, Bennett & Whitworth, 1989).

Biology. Immature stages briefly described by Engel (1920, as *azurea*). In Fennoscandia reared from nests of *Riparia riparia*, which appears to be its only known host. Developmental time not known. Adults collected in the wild from June to August, sometimes on flowers. Probably overwinters in the adult stage.

Nomenclature. *M. chrysorrhoea* Meigen was described from specimens "...Aus Oesterreich, von hrn. Megerle v. Mühlfeld...". This collection is assumed to have been destroyed (cf. Pont, 1986). A neotype ♀ has been designated for *Musca chrysorrhoea* Meigen by Peus (1960) (not examined). *M. sordida* was described from 2 ♀ from Lycksele (Ly.Lpm.). Sabrosky (1956) designated a ♀ labelled "M.sordida / ♀ Lycksele" in Zetterstedt's hand as lectotype. The second syntype,

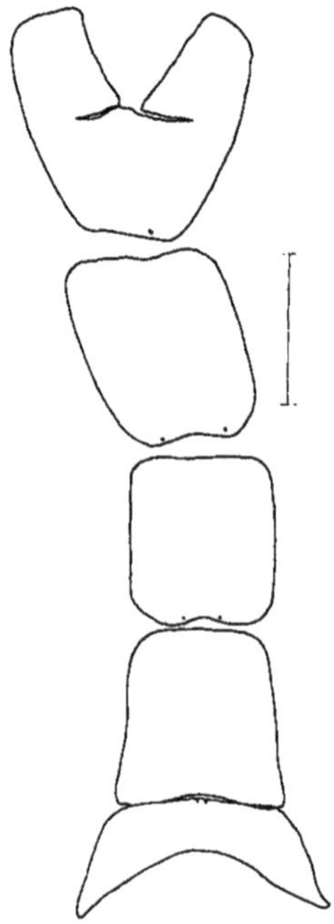

Fig. 272. *Protocalliphora chrysorrhoea* (Meigen), male. ST1-5. Scale = 0.8 mm.

Fig. 274. *Protocalliphora chrysorrhoea* (Meigen), male. Epandrium, cerci and surstyli, posterior view. Scale = 0.2 mm.

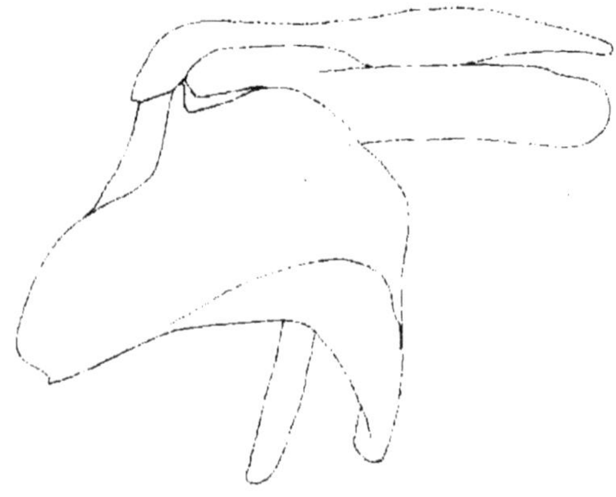

Fig. 273. *Protocalliphora chrysorrhoea* (Meigen), male. Epandrium, cerci, surstylus and bacilliform sclerite, left lateral view. Scale = 0.2 mm.

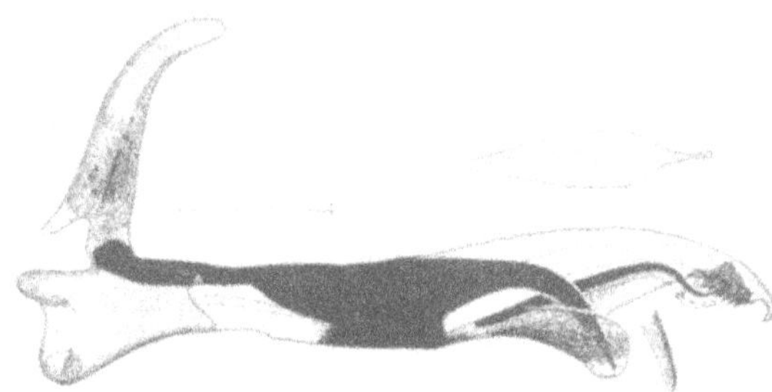

Fig. 275. *Protocalliphora chrysorrhoea* (Meigen), male. Aedeagus, left lateral view. Insets: tip of paraphallus (enlarged), shape of mesohypophallus (dorsal view). Scale = 0.2 mm.

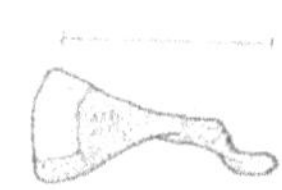

Fig. 276. *Protocalliphora chrysorrhoea* (Meigen), male. Ejaculatory sclerite. Scale = 0.2 mm.

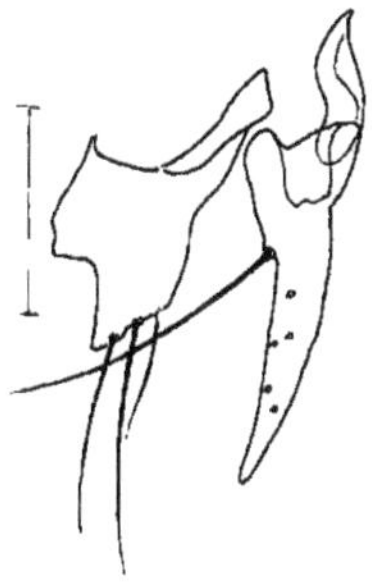

Fig. 277. *Protocalliphora chrysorrhoea* (Meigen), male. Pre- and postgonites. Scale = 0.2 mm.

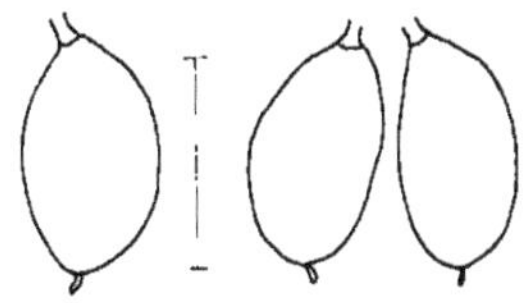

Fig. 278. *Protocalliphora chrysorrhoea* (Meigen), female. Spermathecae. Scale = 0.2 mm. G.pr. 127.

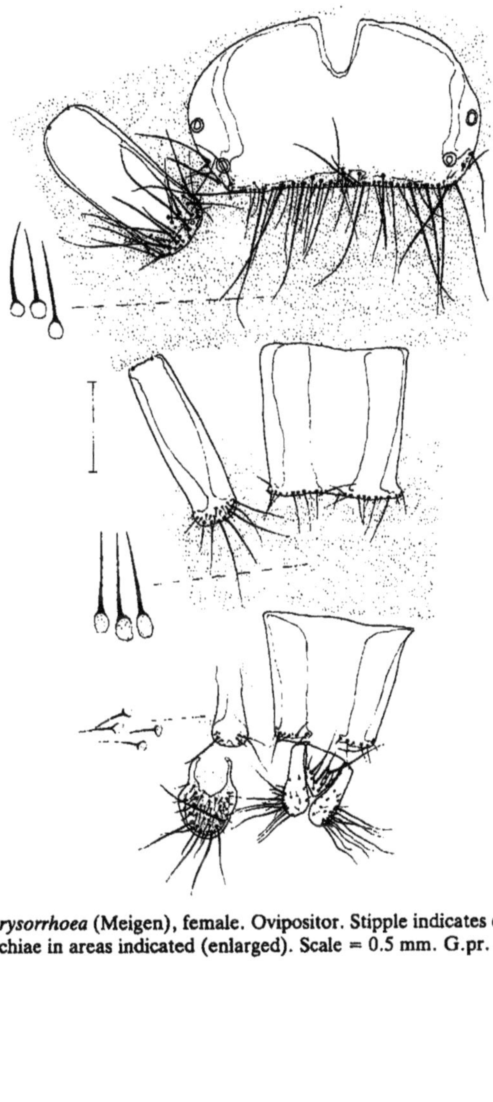

Fig. 279. *Protocalliphora chrysorrhoea* (Meigen), female. Ovipositor. Stipple indicates extent of microtrichiae. Insets: shape of individual microtrichiae in areas indicated (enlarged). Scale = 0.5 mm. G.pr. 127.

considered by Sabrosky as possibly lost, has since been found in the Wallengren collection (drawer labelled "Diptera, W. IV, 14") in ZML. It bears an identical label in Zetterstedt's hand. I have labelled it as paralectotype. Both specimens also have a black small paper square, indicating that they were collected by Zetterstedt on his second Lapland journey in 1832. t_2 has 3-4 *ad* setae, parafacialia are rather silvery for a *chrysorrhoea* (not in the paralectotype which appears conspecific with the lectotype), and vertex in front of vertical setae dulled. I have dissected the lectotype (G.pr. 226). The sclerites of the ovipositor are more slender than in *azurea*. T6 is divided, which would suggest that it is *azurea*, but the frons is too wide (LT 0.299 and PLT 0.303 times head-width), the genal groove rugose anteriorly, and the abdomen is blue, not green (G.pr. 226). *M. sordida* was synonymised with *chrysorrhoea* by Zumpt (1956a), but with *azurea* by Schumann (1986). Schumann (1986) also listed *Protocalliphora surcoufi* Séguy as a synonym. This is an error: see above under *azurea*.

20. ***Protocalliphora nuortevai*** Grunin, 1972
Figs 261, 280-290.

Protocalliphora nuortevai Grunin, 1972: 156.

♂♀. A rather small species. Body colour not sexually dimorphic. Rather strongly dusted. Most specimens easily recognised, the male by the very dense dusting on the parafacialia and the broadly dusted area on the fronto-orbital plates, the female by the dark brown colour of the parafacialia and fronto-orbital plates and the deep ruga in the upper part of the parafacialia.

♂. Frons 0.100-0.137 times head-width (n=11). Cerci rather long and narrow, weakly curved in profile. Ventral plates of distiphallus not extensively sclerotised.

♀. Frons 0.284-0.330 times head-width (n=19). The area on vertex between posteromedian corner of eye, lateroclinate seta and outer and inner vertical setae usually polished, but sometimes weakly dusted. T6 usually undivided, except for a small area on anterior edge, but sometimes more or less divided in the manner of *azurea*. T7 narrow and elongate. T8 with several marginal setae behind strongly sclerotised T8 halves. Pleural membrane 7 without microtrichiae in anterior 1/4 to 1/3. [9 ovipositor slides examined.]

Length. 6.5-10 mm.

Distribution. Scattered in Norway, northern Sweden and northern Finland. Absent from Denmark. Ascends to subalpine and low alpine regions. Appears to be rather frequent in the former. - Not known outside Fennoscandia.

Biology. Immature stages undescribed. Adults have been reared from *Calcarius lapponicus*, *Motacilla alba*, *Phylloscopus trochilus*, *Turdus torquatus*, and *Turdus iliacus*. Developmental time not known. Adults have been collected in the wild from June to August, also abundantly in Malaise-traps (cf. Rognes, 1982). It probably overwinters in the adult stage.

Note. A few females may have the parafacialia and fronto-orbital plates rather light brown in colour.

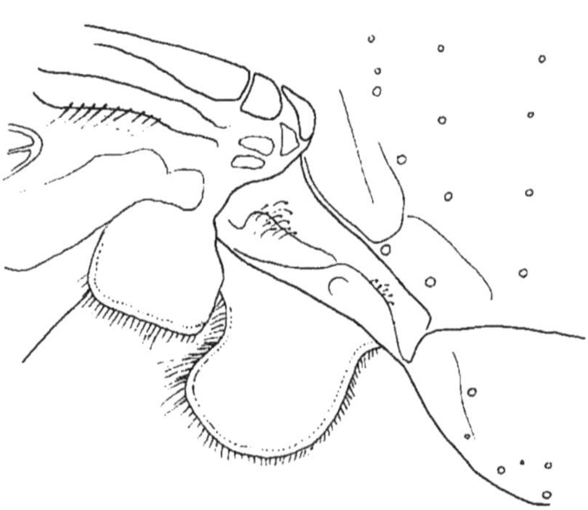

Fig. 280. *Protocalliphora nuortevai* Grunin, male. Area behind wing base, dorsoposterior view.

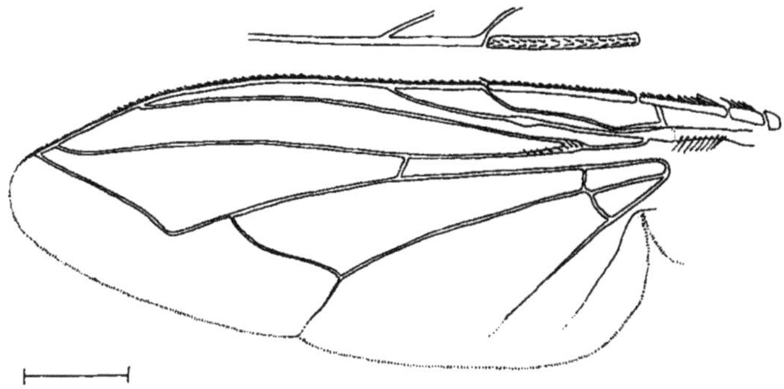

Fig. 281. *Protocalliphora nuortevai* Grunin, male. Left wing from above. Inset: costa from below showing distribution of small hairs on its underside. Scale = 1.0 mm.

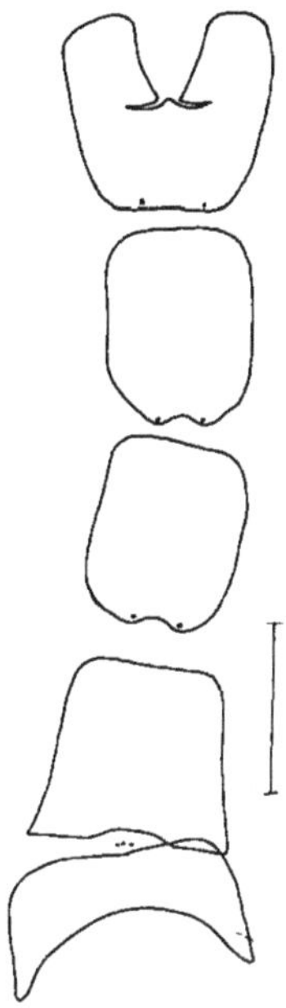

Fig. 282. *Protocalliphora nuortevai* Grunin, male. ST1-5. Scale = 0.8 mm.

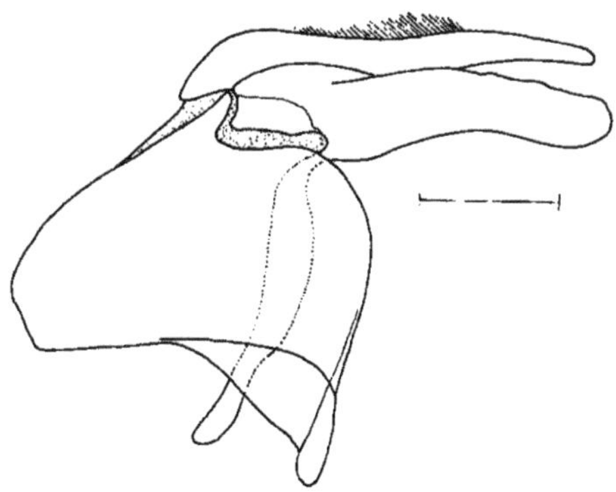

Fig. 283. *Protocalliphora nuortevai* Grunin, male. Epandrium, cerci, surstylus and bacilliform sclerite, left lateral view. Scale = 0.2 mm.

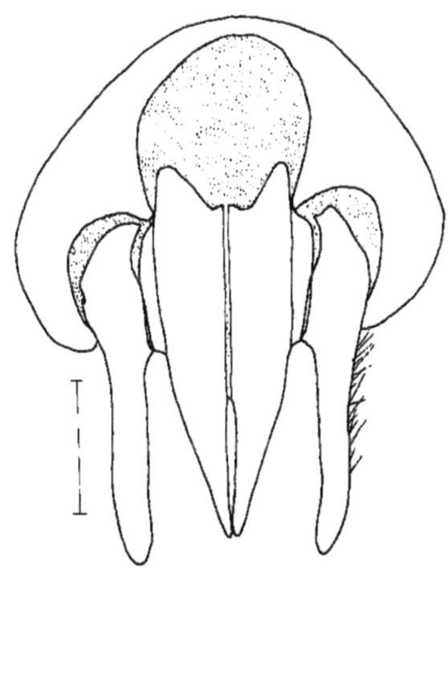

Fig. 284. *Protocalliphora nuortevai* Grunin, male. Epandrium, cerci and surstyli, posterior view. Scale = 0.2 mm.

Fig. 285. *Protocalliphora nuortevai* Grunin, male. Aedeagus, left lateral view. Inset: tip of paraphallus (enlarged). Scale = 0.2 mm.

Fig. 286. *Protocalliphora nuortevai* Grunin, male. Distiphallus, ventral view. Scale = 0.2 mm.

Fig. 287. *Protocalliphora nuortevai* Grunin, male. Ejaculatory sclerite. Scale = 0.2 mm.

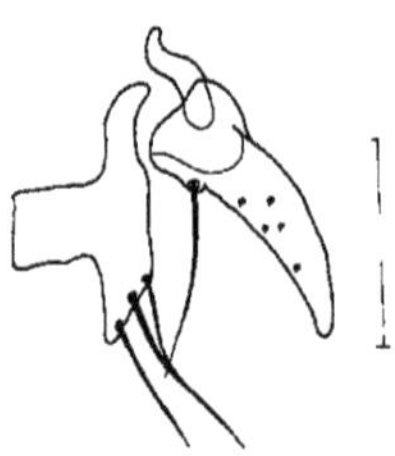

Fig. 288. *Protocalliphora nuortevai* Grunin, male. Pre- and postgonites. Scale = 0.2 mm.

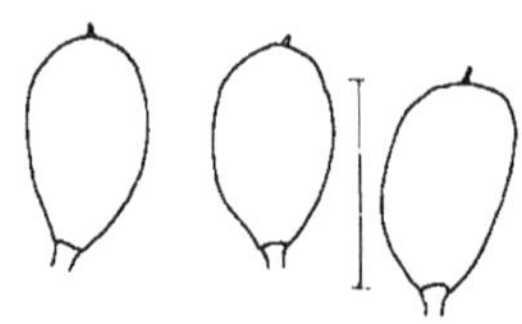

Fig. 289. *Protocalliphora nuortevai* Grunin, female. Spermathecae. Scale = 0.2 mm. G.pr. 124.

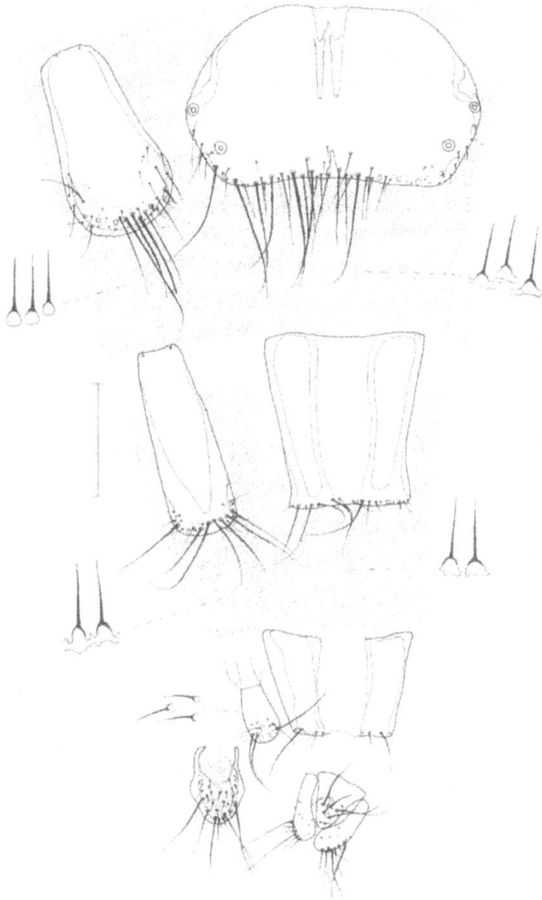

Fig. 290. *Protocalliphora nuortevai* Grunin, female. Ovipositor. Stipple indicates extent of microtrichiae. Insets: shape of individual microtrichiae in areas indicated (enlarged). Scale = 0.5 mm. G.pr. 120.

21. ***Protocalliphora peusi*** Gregor & Povolný, 1959

Figs 263, 300-309.

Protocalliphora peusi Gregor & Povolný, 1959: 45.

♂♀. Body colour not sexually dimorphic.

♂. Frons 0.144-0.152 times head-width (n=2). t_2 with 5-7 *ad* setae. Cerci long and strong, weakly curved in profile. Surstyli shorter than cerci, stout in profile, slightly narrowing, weakly curved, blunt apically. In dorsal view slightly curved inwards apically. Ventral plates of distiphallus long.

♀. Frons 0.274-0.289 times head-width (n=7). The area on vertex between posteromedian corner of eye, lateroclinate seta and outer and inner vertical setae rather polished. t_2 with 4-6 *ad* setae. T6 with weakly sclerotised lateral margins. T7 and T8 halves weakly sclerotised posteriorly. T8 with only 1-2 strong marginal setae on each side. ST7 very narrow. [4 ovipositor slides examined.]

Length. 9.5-12 mm.

Distribution. Very rare. Known from a single bred series of 2♂ 6♀ from Norway (ON, Sel: Otta, dated July 24, 1984, R. Mehl leg. et cult., RML, KR). A single ♀ from Sweden (Dlr.: Vansbro, July 29, 1936, O. Ringdahl leg.). - Widely distributed in Europe (Peus, 1960; Schumann, 1986).

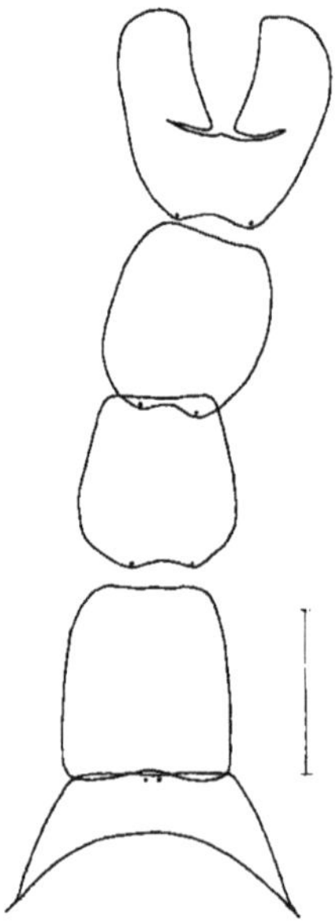

Fig. 291. *Protocalliphora proxima* Grunin, male. ST1-5. Scale = 0.8 mm.

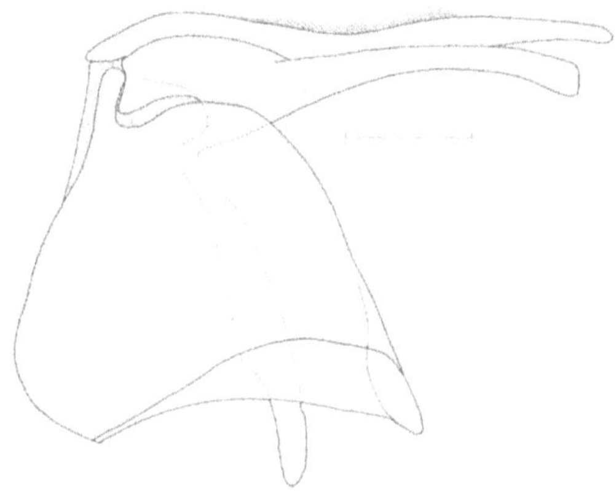

Fig. 292. *Protocalliphora proxima* Grunin, male. Epandrium, cerci, surstylus and bacilliform sclerite, left lateral view. Scale= 0.2 mm.

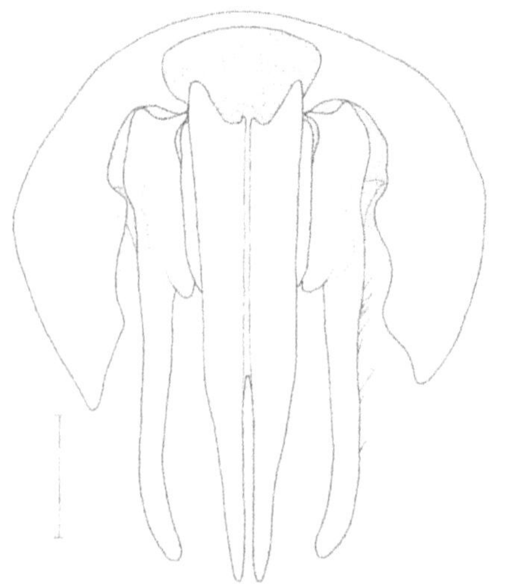

Fig. 293. *Protocalliphora proxima* Grunin, male. Epandrium, cerci and surstyli, posterior view. Scale = 0.2 mm.

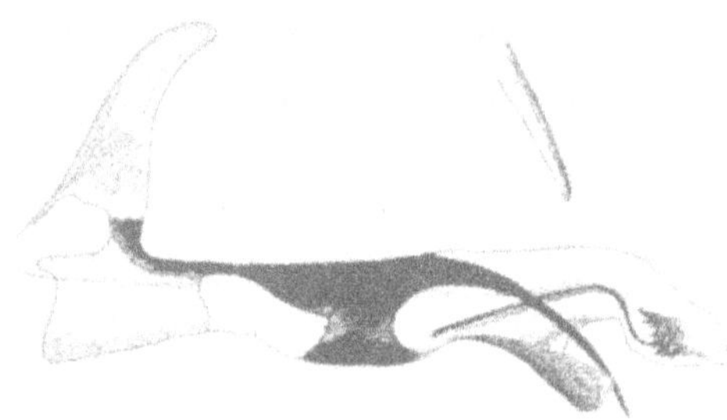

Fig. 294. *Protocalliphora proxima* Grunin, male. Aedeagus, left lateral view. Inset: tip of paraphallus (enlarged). Scale = 0.2 mm.

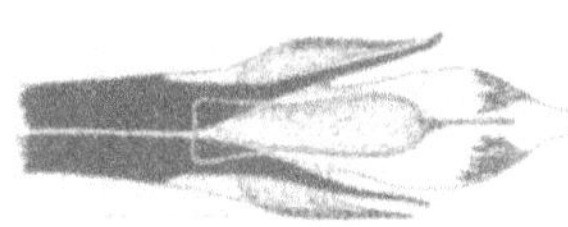

Fig. 295. *Protocalliphora proxima* Grunin, male. Distiphallus, dorsal view. Scale = 0.2 mm.

Fig. 296. *Protocalliphora proxima* Grunin, male. Ejaculatory sclerite, various views. Scale = 0.2 mm.

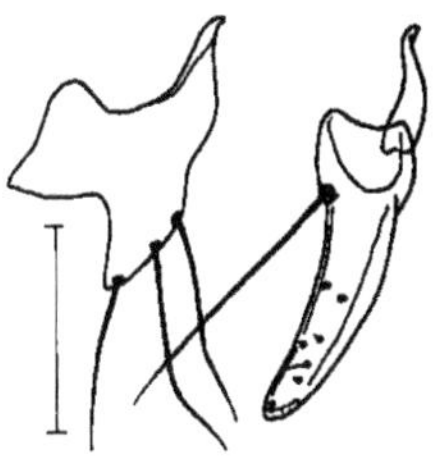

Fig. 297. *Protocalliphora proxima* Grunin, male. Pre- and postgonites. Scale = 0.2 mm.

Fig. 298. *Protocalliphora proxima* Grunin, female. Spermathecae. Scale = 0.2 mm. G.pr. 173.

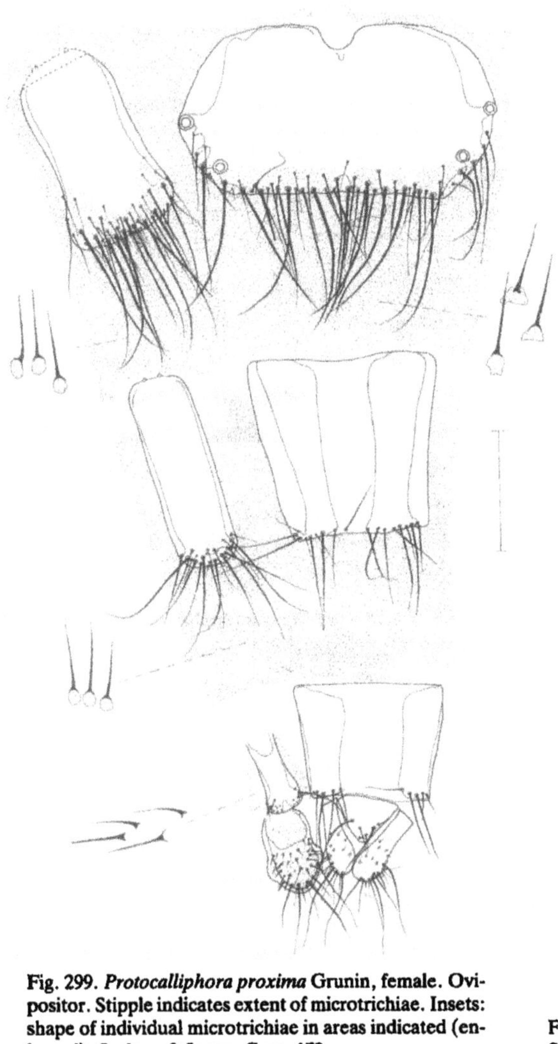

Fig. 299. *Protocalliphora proxima* Grunin, female. Ovipositor. Stipple indicates extent of microtrichiae. Insets: shape of individual microtrichiae in areas indicated (enlarged). Scale = 0.5 mm. G.pr. 173.

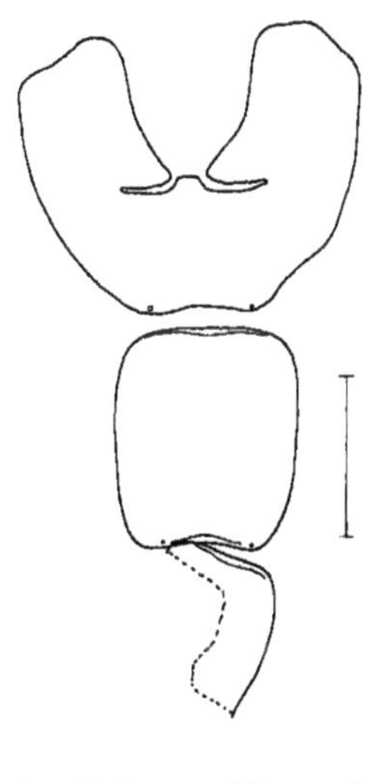

Fig. 300. *Protocalliphora peusi* Gregor & Povolný, male. ST4-5. Scale = 0.8 mm.

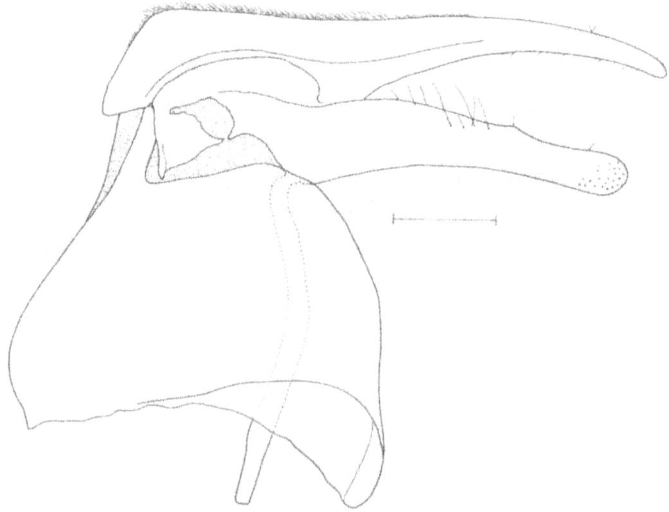

Fig. 301. ***Protocalliphora peusi*** **Gregor & Povolný, male. Epandrium, cerci, surstylus and bacilliform sclerite, left lateral view. Scale = 0.2 mm.**

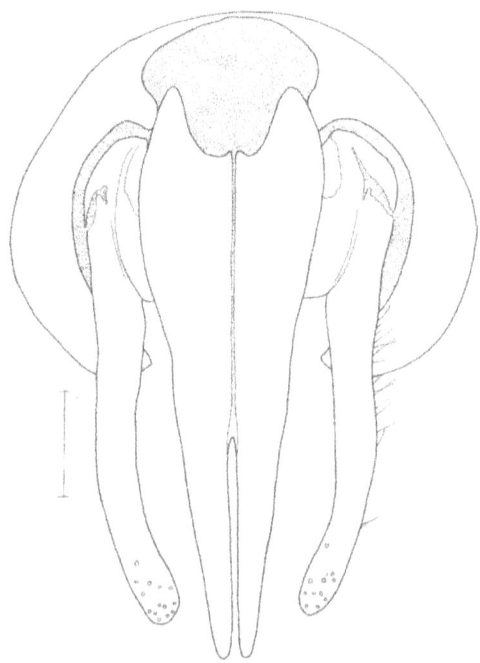

Fig. 302. *Protocalliphora peusi* Gregor & Povolný, male. Epandrium, cerci and surstyli, posterior view. Scale = 0.2 mm.

Fig. 303. *Protocalliphora peusi* Gregor & Povolný, male. Aedeagus, left lateral view. Inset: tip of paraphallus (enlarged). Scale = 0.2 mm.

Fig. 304. *Protocalliphora peusi* Gregor & Povolný, male. Aedeagus, dorsal view. Scale = 0.2 mm.

Fig. 305. *Protocalliphora peusi* Gregor & Povolný, male. Aedeagus, ventral view. Scale = 0.2 mm.

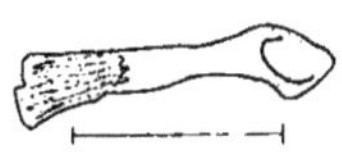

Fig. 306. *Protocalliphora peusi* Gregor & Povolný, male. Ejaculatory sclerite. Scale = 0.2 mm.

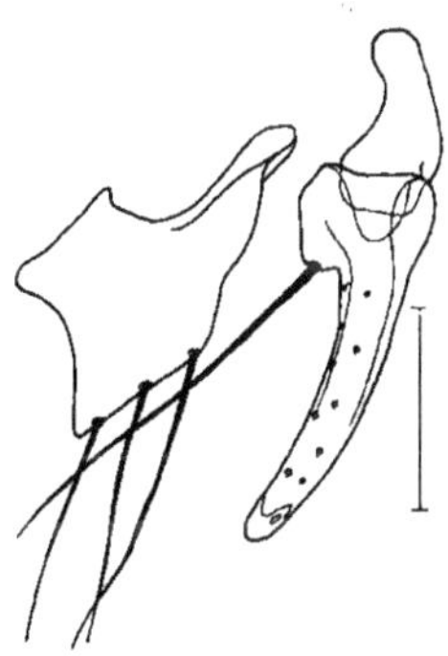

Fig. 307. *Protocalliphora peusi* Gregor & Povolný, male. Pre- and postgonites. Scale = 0.2 mm.

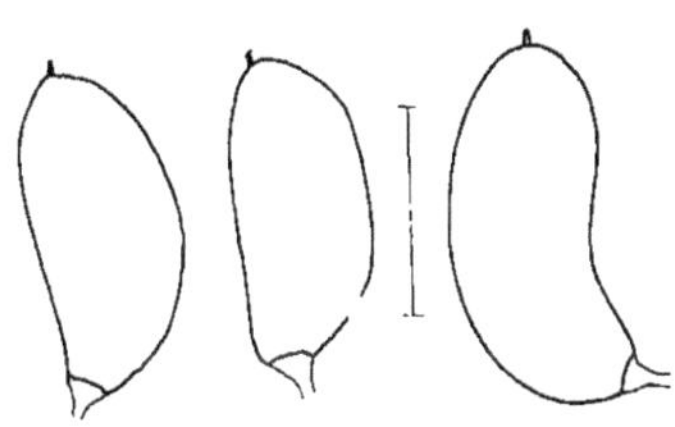

Fig. 308. *Protocalliphora peusi* Gregor & Povolný, female. Spermathecae. Scale = 0.2 mm. G.pr. 161.

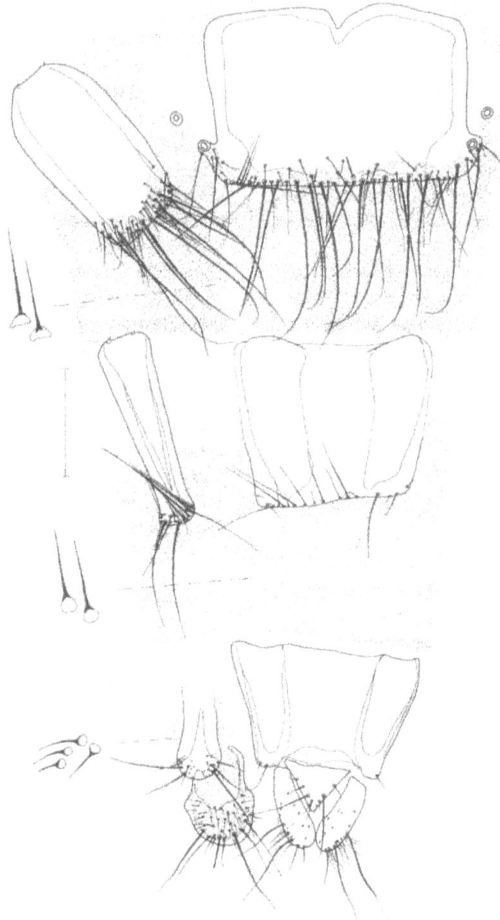

Fig. 309. *Protocalliphora peusi* Gregor & Povolný, female. Ovipositor. Stipple indicates extent of microtrichiae. Insets: shape of individual microtrichiae in areas indicated (enlarged). Scale = 0.5 mm. G.pr. 159.

Biology. Immature stages undescribed. Reared from nests of *Corvus cornix* L. and *Corvus corone* L. (Peus, 1960), and from *Pica pica* L. (series from Otta, above). Developmental time not known. Puparia parasitised by *Nasonia vitripennis* (Walker) (Hym., Pteromalidae) (Peus, 1960).

22. ***Protocalliphora proxima*** Grunin, 1966
Figs 262, 291-299.

Protocalliphora proxima Grunin, 1966: 899.

♂♀. Body-colour not sexually dimorphic. Costa occasionally hairy below between junction with Sc and R_1.

♂. Frons 0.108-0.133 times head-width (n=4). Cerci very long and narrow. Surstylus slightly shorter than cerci, very long and narrow, weakly curved in profile. Apex curved towards mid-line in dorsal view. Ventral plates of distiphallus not as extensively sclerotised as in *chrysorrhoea*.

♀. Frons 0.268 times head-width (n=1). T6 entire except a small area of weak sclerotisation on anterior edge. T7 halves very narrow. A full row of marginal setae behind each strongly sclerotised T8 half. Pleural membrane 7 without microtrichiae in anterior 1/4. [Only 2 partly teneral females from the same bred series as 3 males have been available; 1 ovipositor slide examined.]

Length. 9-11 mm.

Distribution. Very rare. A single record from Norway (1♂ from HEN, "Aamot", Siebke leg., in TM). A few scattered records from Sweden (including a specimen from Lapland in coll. Zetterstedt, see above under *azurea*) and Finland. Absent from Denmark. - A single record from Poland (Draber-Mońko, 1971). Widely distributed in the U.S.S.R., also the Far East (Grunin, 1966, 1970a).

Biology. Immature stages and biology unknown. 3♂ 2♀, all with puparia on pin, are present in ZML, but no breeding data are given. Adults have been collected from June to September. Collected "in a trap with a frankfurther [sausage] as bait" (Grunin & Nuorteva, 1969).

Genus *Protophormia* Townsend, 1908

Protophormia Townsend, 1908: 123.

Type-species: *Phormia terraenovae* Robineau-Desvoidy, 1830: 467, by monotypy.

Boreellus Aldrich & Shannon in Shannon, 1923: 107.

Type-species: *Boreellus aristatus* Aldrich & Shannon in Shannon, 1923: 107, by monotypy; = *Sarcophaga atriceps* Zetterstedt, 1845. SYN.N.

Mallochomyia Townsend, 1926: 25.

Type-species: *Mallochomyia johanseni* Townsend, 1926: 25, by original designation; = *Sarcophaga atriceps* Zetterstedt, 1845. SYN.N.

A Holarctic genus of very dark bluish-green flies with practically undusted thorax and abdomen. Head black. Frontal vitta broader than each fronto-orbital plate, latter always with a lateroclinate seta in female. Parafacialia dusted with shifting areas. Occiput convex. Palpi yellow or infuscated. 0+3 *acr*. 2+3 *dc*. 0-1+2 *ia*. 2 *ph*, outer on line with or very slightly external to *prs* seta. Posterior spiracles as in Figs. 312, 313. t_1 with 1 *pv* seta. t_2 with 2-3 *ad* setae. f_2 with *a* preapical seta in females. t_3 with 3-4 *av*. t_3 with subequal *ad* and *d* preapical setae, *pd* preapical much weaker but conspicuous. Hind coxa usually with a few short hairs laterally on posterior surface, these occasionally absent. Both calypters dark brown. Upper calypter with dark hairs above (Fig. 316). Basicosta black, wing-veins dark brown. Aedeagus with well-developed mesohypophallus. Acrophallus widened subapically, distal aperture narrow. Hypophallic lobes projecting anteroventrally, with denticles apically. Paraphallus well-developed. Pregonite with 2-3 setae. Postgonite with a projection bearing the basal seta. Ovipositor with wide T7 and T8 halves. A very narrow anterior zone of pleural membrane 7 devoid of microtrichiae. Posterior 1/2 to 2/5 of intersegmental membrane 7-8, thus a strikingly broad zone, devoid of microtrichiae. Epiproct and cerci without microtrichiae. Spermathecae oval.

The fully-grown larva remains on the surface of the substrate or makes only a short migration for pupariation (McAlpine, 1965; Nuorteva, 1977, 1987; Smith, 1986).

2 species, both in Fennoscandia and Denmark.

Notes on synonymy. The status of the monotypic nominal genera *Protophormia* and *Boreellus* has been discussed by Rognes (1985a). There can be no doubt about the monophyly of this group. Rodendorf (1924; 1926), Collin (1925), Séguy (1928a) and Villeneuve (1930) also treat the constituent species as congeneric.

Key to species of *Protophormia*

1 Lower part of face strongly protruding (Fig. 244); arista almost bare below, with very short hairs above (Fig. 244); antennae all black; parafacialia dusted dirty brownish-grey; occiput with only black setae; palpi infuscated; 3 *h* setae; *prst ia* absent; 2 pairs of marginal scutellar setae (Fig. 315); anterior spiracle very large, about as wide as half length of anepisternum (Fig. 311); cell r_{4+5} narrowly open or closed at margin (Fig. 317); ground-vestiture on dorsal surface of all abdominal tergites erect in both sexes

23. *atriceps* (Zetterstedt)

– Lower part of face normal (Fig. 245); arista plumose (Fig. 245); tip of pedicel and basal part of first flagellomere reddish; parafacialia dusted greyish-white; occiput with 6-7 rows of black setae, pale hairs in middle; palpi yellow; 5-6 *h*; *prst ia* present; 3-4 pairs of marginal scutellar setae (Fig. 314); anterior spiracle of normal size, about as wide as one-fifth of length of anepisternum (Fig. 310); cell r_{4+5} widely open (Fig. 328); ground-vestiture on dorsal surface of abdominal tergites decumbent in both sexes, except on T5 where it is erect

24. *terraenovae* (Robineau-Desvoidy)

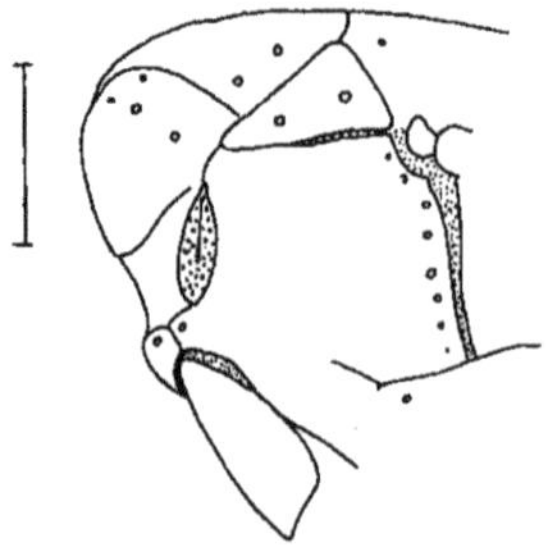

Fig. 310. *Protophormia terraenovae* (Robineau-Desvoidy), anterior spiracle and anepisternal area. Scale = 1.0 mm.

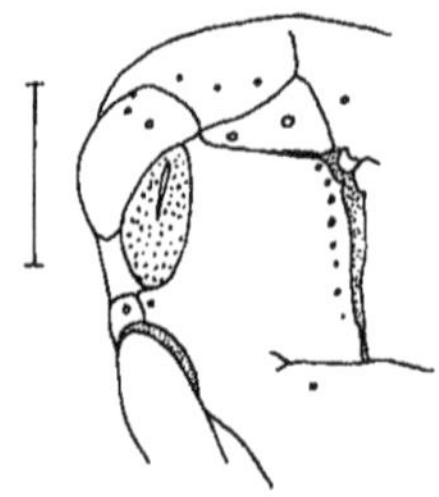

Fig. 311. *Protophormia atriceps* (Zetterstedt), anterior spiracle and anepisternal area. Scale = 1.0 mm.

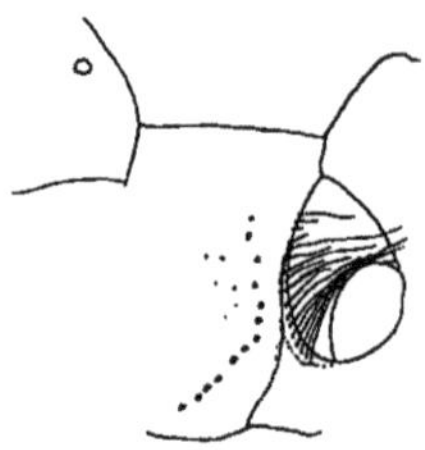

Fig. 312. *Protophormia terraenovae* (Robineau-Desvoidy), posterior spiracle. Scale = 1.0 mm.

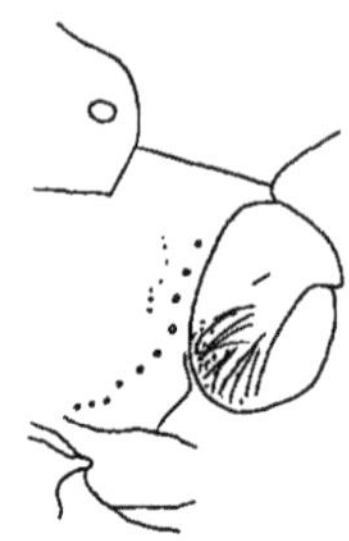

Fig. 313. *Protophormia atriceps* (Zetterstedt), posterior spiracle. Scale = 1.0 mm.

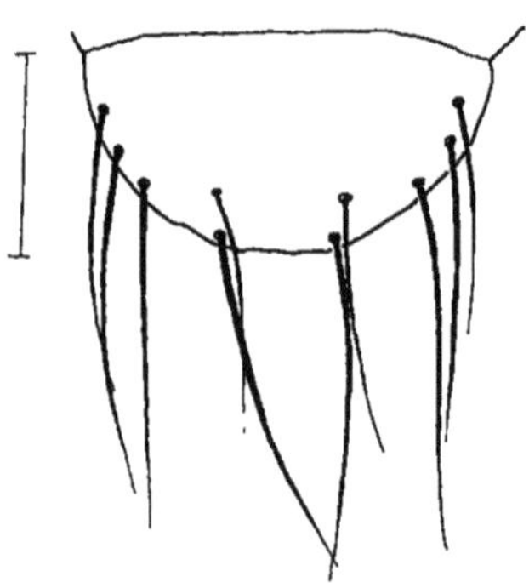

Fig. 314. *Protophormia terraenovae* (Robineau-Desvoidy), scutellum. Scale = 1.0 mm.

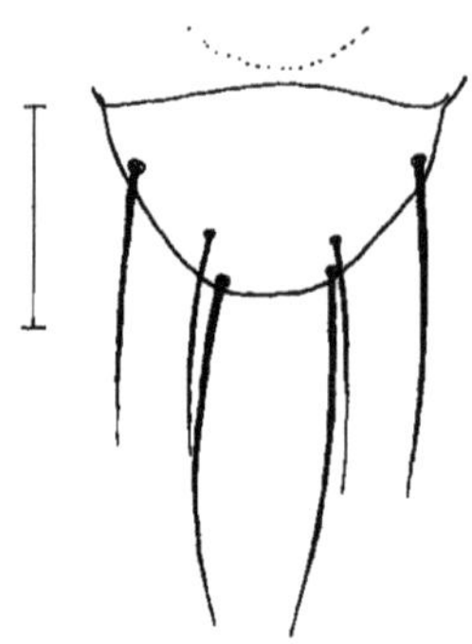

Fig. 315. *Protophormia atriceps* (Zetterstedt), scutellum. Scale = 1.0 mm.

23. ***Protophormia atriceps*** (Zetterstedt, 1845), COMB.N.

Figs 244, 311, 313, 315-326.

Sarcophaga atriceps Zetterstedt, 1845: 1311.
Musca boganidae Erichsson, 1851: 67.
Phormia caerulea Malloch, 1919: 59. Junior primary homonym of *Phormia caerulea* Robineau-Desvoidy, 1830: 466.
Boreellus aristatus Aldrich & Shannon in Shannon, 1923: 107, replacement name for *Phormia caerulea* Malloch, 1919.
Mallochomyia johanseni Townsend, 1926: 25, replacement name for *Phormia caerulea* Malloch, 1919.

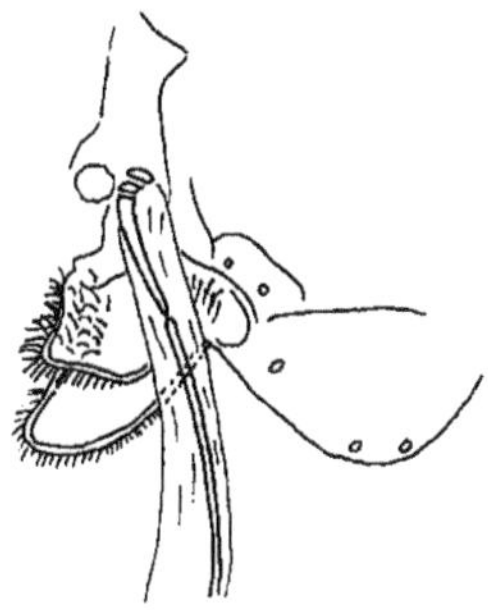

Fig. 316. *Protophormia atriceps* (Zetterstedt), area around wing base, dorsoposterior view.

Fig. 317. *Protophormia atriceps* (Zetterstedt). Left wing from above. Inset: costa from below showing distribution of small hairs on its underside. Scale = 1.0 mm.

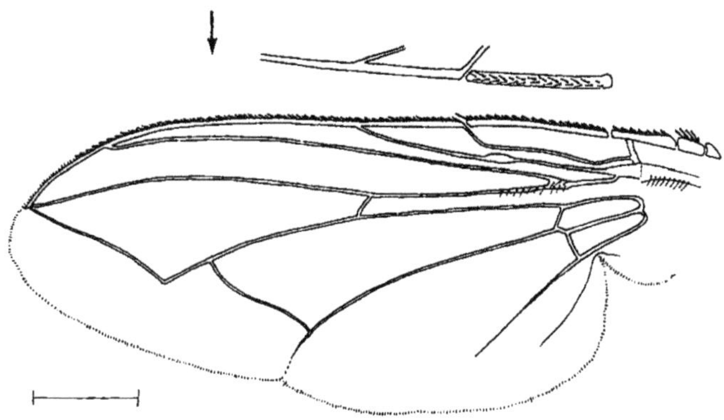

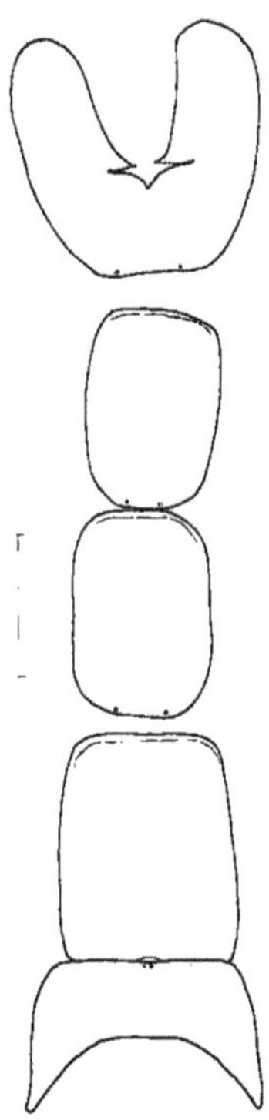

Fig. 318. *Protophormia atriceps* (Zetterstedt), male. ST1-5. Scale = 0.5 mm.

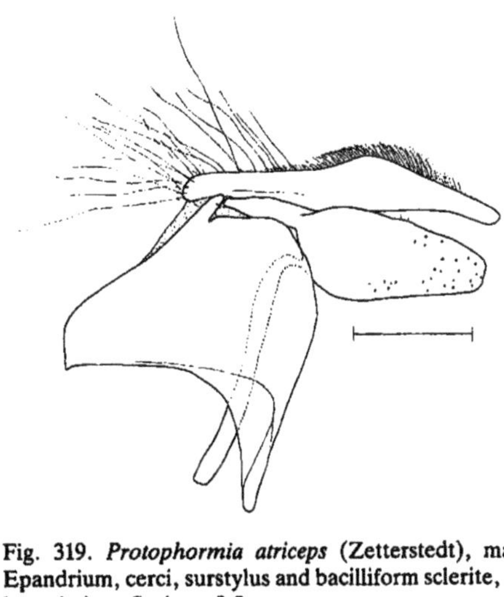

Fig. 319. *Protophormia atriceps* (Zetterstedt), male. Epandrium, cerci, surstylus and bacilliform sclerite, left lateral view. Scale = 0.2 mm.

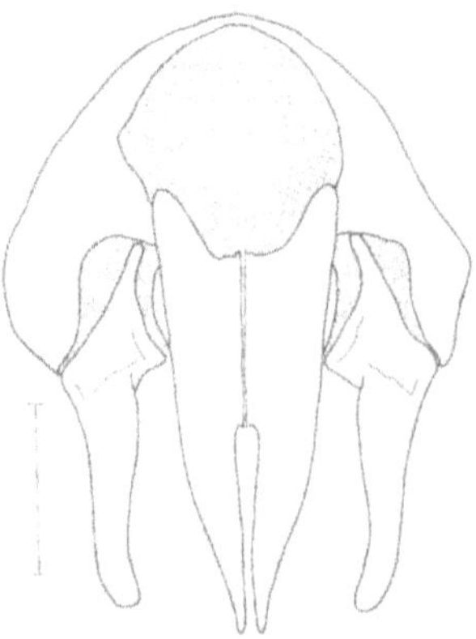

Fig. 320. *Protophormia atriceps* (Zetterstedt), male. Epandrium, cerci and surstyli, posterior view. Scale = 0.2 mm.

Fig. 321. *Protophormia atriceps* (Zetterstedt), male. Aedeagus, left lateral view. Inset: tip of paraphallus (enlarged). Scale = 0.2 mm.

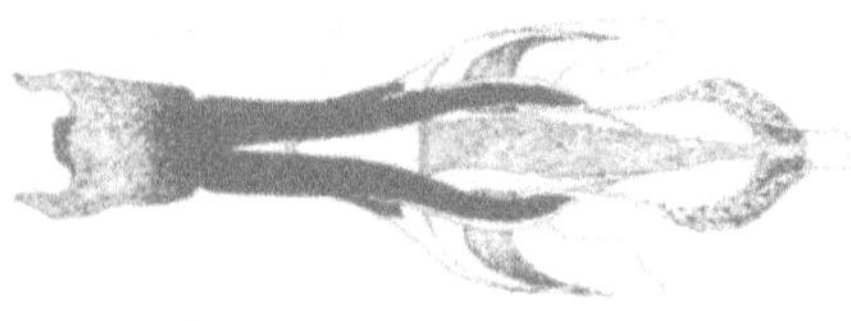

Fig. 322. *Protophormia atriceps* (Zetterstedt), male. Aedeagus, dorsal view. Scale = 0.2 mm

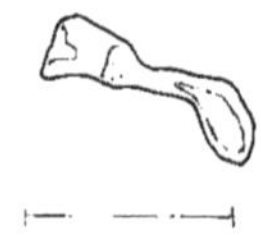

Fig. 323. *Protophormia atriceps* (Zetterstedt), male. Ejaculatory sclerite. Scale = 0.2 mm.

Fig. 324. *Protophormia atriceps* (Zetterstedt), male. Pre- and postgonites. Scale = 0.2 mm.

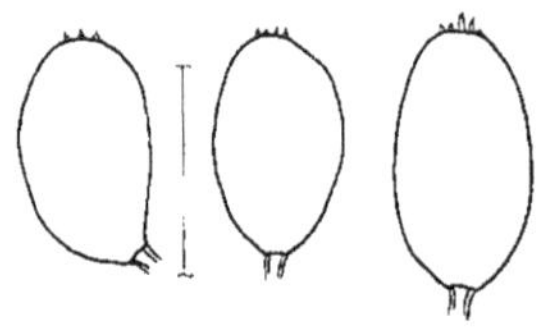

Fig. 325. *Protophormia atriceps* (Zetterstedt), female. Spermathecae. Scale = 0.2 mm. G.pr. 158.

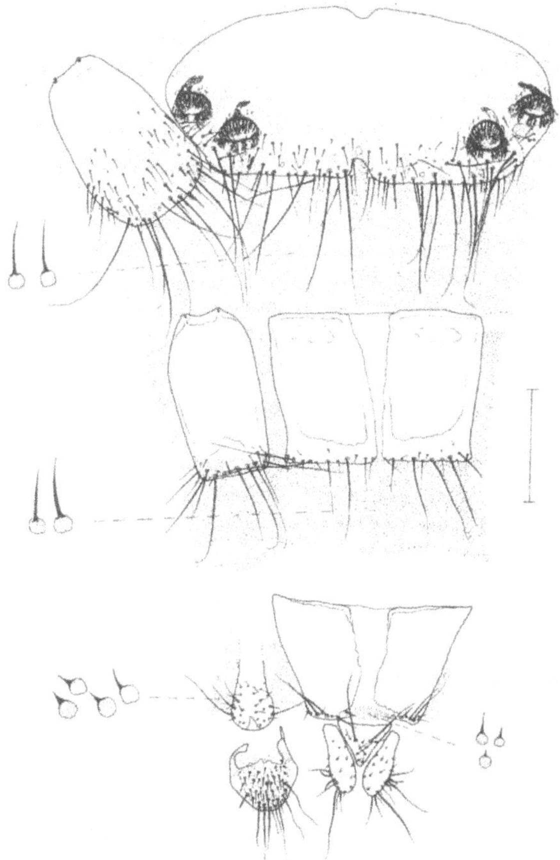

Fig. 326. *Protophormia atriceps* (Zetterstedt), female. Ovipositor. Stipple indicates extent of microtrichiae. Insets: shape of individual microtrichiae in areas indicated (enlarged). Scale = 0.5 mm. G.pr. 158.

♂♀. Similar in colour to *Protophormia terraenovae*, but smaller. Easily distinguished by the characters given in the key. Eyes small. Fronto-orbital plates with dense long hairs in front of the vertical seta. Postocular setae long, thin and gently curving forwards. Palpi dark brown.

♂. Frons 0.214-0.237 times head-width (n=5). A weak lateroclinate seta sometimes differentiated. f_2 usually with *a* preapical seta. Cerci short, curved in profile, dorsal hairing not quite reaching apex when seen in profile. Surstylus wide and short, almost triangular. Aedeagus in dorsal view broader at middle than acrophallus. Mesohypophallus evenly tapering. Ejaculatory sclerite small and narrow.

♀. Frons 0.402-0.443 times head-width, measured at posteromedian corner of eye (n=5). Fronto-orbital plates with one or two long lateroclinate setae just in front of the verticals, and a long lateroclinate and a long weakly inclinate seta in positions where the two proclinate orbitals are found in other blowfly females. Spiracles of T6 in ovipositor very large and with peculiar external ornamentation consisting of densely-set branched hairs along circumference. Microtrichiae on intersegmental membrane 7-8 long. Spermathecae oval with 3-4 apical papillae. [2 ovipositor slides examined.]

Length. 6-9.5 mm.

Distribution. Rare in Fennoscandia. In middle alpine to high alpine regions in Southern Norway (ON, Sognefjell; SFI, west of Horrungane; TEI,

Vassdalseggi). In Sweden, apart from type locality (Jmt., Mullfjällen), an old record from Gtl. (sic) (P. F. Wahlberg leg.) and a single capture from Ly.Lpm. North Finland (Li: Ailigas). Absent from Denmark. - Arctic circumpolar, well north of 80°N in Canada (Ellesmere Is). Also Svalbard (Nuorteva, 1967; Rognes, 1983b). In the account given by Rasmussen (1934: 89) of his first Thule expedition to northern Greenland is an entry for June 28-29 from "Stankelbenenes Kløft" [82° N, near Adam Biering's Land, just south of Peary Land]. He describes a piece of fresh meat which "var bleven een eneste sammenhængende Klump af Spyfluer. Der var ikke et Sted, hvor man kunne sætte en finger for Fluer" ["had been transformed into a large single mass of blowflies. Nowhere to put one's finger for flies."]. This probably refers to *Protophormia atriceps* and may be the northernmost locality known.

Biology. Immature stages undescribed, but anterior end of first instar larva figured by Shewell (1987). A single batch of eggs is laid in spring. Development time is 1-2 years (Danks, 1981: 281, 287). Larvae move to the exterior of the carcass to pupate in the hair exposed to the sun, to receive the fullest possible benefit from the insolational heat (McAlpine, 1965, see also discussion in Rognes, 1983b). It is reported to have bred in carcasses of the following mammals: walrus - *Odobaenus rosmarus* (L.), lemmings - *Dicrostonyx groenlandicus* (Traill), *D. torquatus* (Pallas), *Lemmus sibiricus* = *Lemmus obensis* (Brants), arctic hare - *Lepus arcticus* Ross, husky dog - *Canis familiaris* L., arctic fox - *Alopex lagopus* L. (see Rognes, 1983b for references). Dead *Lemmus lemmus* L. is probably the main substrate for larval development in Fennoscandia, although no direct evidence for this is available. Overwinters as larva, pupa or adult (Danks, 1981: 284, 285, 286). According to reports from north Siberia, where it has been collected in June and July, adults visit meat, faecal matter and flowers (Chernov, 1961). In Fennoscandia all adult specimens have been collected in July. All captures from Norway have been on fresh liver bait.

Nomenclature. I have examined the holotype ♀ of *S. atriceps* in ZML.

24. ***Protophormia terraenovae*** (Robineau-Desvoidy, 1830)

Figs 39, 245, 310, 312, 314, 327-337; Plate - fig 2.

Phormia terraenovae Robineau-Desvoidy, 1830: 467.

Musca groenlandica Zetterstedt, 1833: 34, nomen nudum.

Musca groenlandica Zetterstedt, 1838: 657.

Lucilia terraenovae Macquart, 1851: 224. Junior secondary homonym of *Phormia terraenovae* Robineau-Desvoidy.

Phormia nigripalpis Robineau-Desvoidy, 1863: 846.

Phormia coruscа Robineau-Desvoidy, 1863: 849.

♂♀. Easily recognised by the characters given in the key. Eyes normal. Postocular setae short and rather strong. Palpi yellow.

♂. Frons 0.118-0.140 times head-width (n=5). Fronto-orbital plates without lateroclinate seta. f_2 without *a* preapical seta. Cerci long, dorsal hairing reaching tip when seen in profile. Surstylus long, parallel-sided. Aedeagus in dorsal view about as broad at middle as acrophallus. Ejaculatory sclerite large, fan-shaped.

♀. Frons 0.365-0.386 times head-width (n=5). 2 normal proclinate orbital setae. Spiracles in T6 of ovipositor normal, no external ornamentation along circumference. Microtrichiae on intersegmental membrane 7-8 short peg-like, except in a narrow zone immediately behind T7. Spermathecae oval, with at most a single papilla. [2 ovipositor slides examined.]

Length. 6-11 mm.

Distribution. Very common all over Fennoscandia and Denmark. Collected up to the high alpine region in southern Norway (TEI, Vinje: Vassdalseggi, at 1658 m a.s.l.). - Common all over Holartic Region, also Greenland, Svalbard (TM, ZML, see also Nuorteva, 1967), Iceland (Nielsen, Ringdahl & Tuxen, 1954) and Faroes (Lyneborg, 1968).

Biology. Immature stages described by Hall (1948), Schumann (1954), Ishijima (1967). The development time is only 11 days (at 27°C.) which is relatively short compared to other blowflies (Kamal, 1958; Nuorteva, 1960b, 1977). The larvae pupate on the carcass if dry and not exposed to bright light (Nuorteva, 1977; Smith, 1986), and wander at most 0.5 m away from an infested dead human body, which is much less than other blowfly larvae (0.5-6 m) (Nuorteva, 1977, 1987). Reported to attack and kill living reindeer in Alaska (Hall, 1948). No similar records exist from Fennoscandia. Reported also in cases of myiasis in dehorned cattle and docked lambs in North America (Hall, 1948). Occasionally known as a sheep myiasis fly in the Old World (Zumpt, 1965), but no such records from Fennoscandia and Denmark. Adults have been collected from March to October. The first specimens to appear in the spring are overwintered adults, although in the north the immature stages may also overwinter

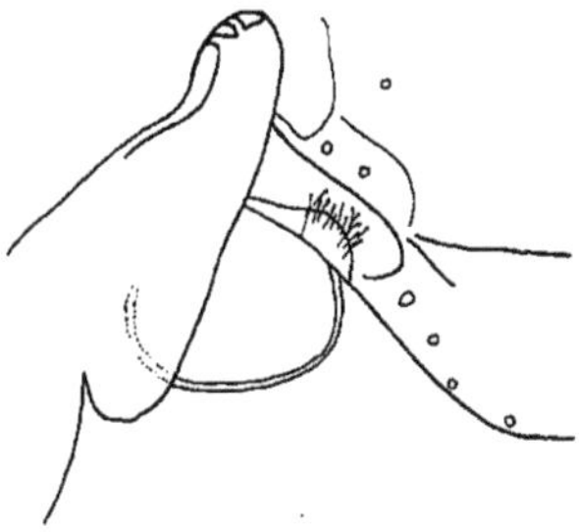

Fig. 327. *Protophormia terraenovae* (Robineau-Desvoidy), area behind wing base, dorsoposterior view.

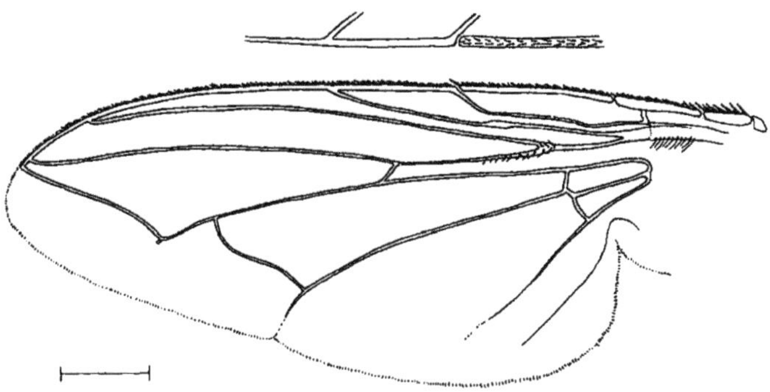

Fig. 328. *Protophormia terraenovae* (Robineau-Desvoidy). Left wing from above. Inset: costa from below showing distribution of small hairs on its underside. Scale = 1.0 mm.

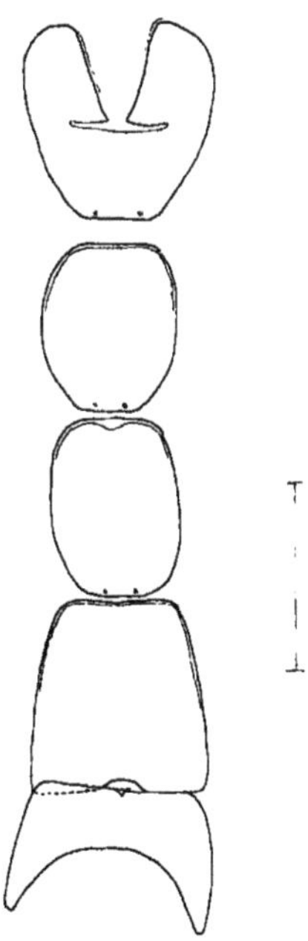

Fig. 329. *Protophormia terraenovae* (Robineau-Desvoidy), male. ST1-5. Scale = 1.0 mm.

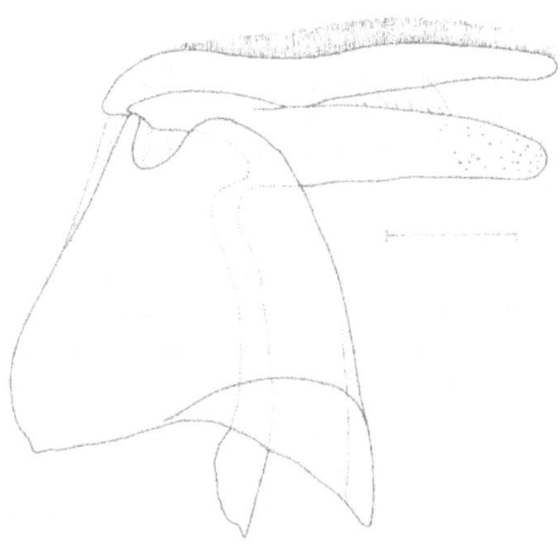

Fig. 330. *Protophormia terraenovae* (Robineau-Desvoidy), male. Epandrium, cerci, surstylus and bacilliform sclerite, left lateral view. Scale = 0.2 mm.

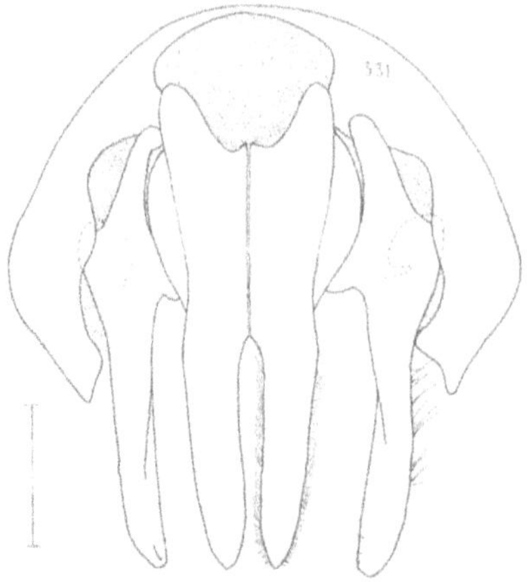

Fig. 331. *Protophormia terraenovae* (Robineau-Desvoidy), male. Epandrium, cerci and surstyli, posterior view. Scale = 0.2 mm.

Fig. 332. *Protophormia terraenovae* (Robineau-Desvoidy), male. Aedeagus, left lateral view. Scale = 0.2 mm.

Fig. 333. *Protophormia terraenovae* (Robineau-Desvoidy), male. Distiphallus, dorsal view. Scale = 0.2 mm.

Fig. 334. *Protophormia terraenovae* (Robineau-Desvoidy), male. Ejaculatory sclerite. Scale = 0.2 mm.

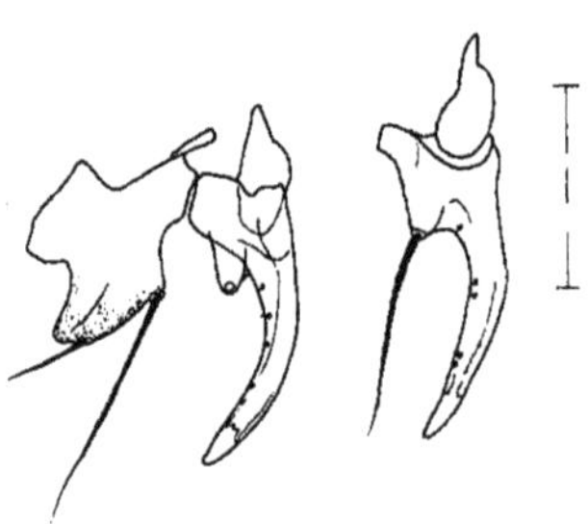

Fig. 335. *Protophormia terraenovae* (Robineau-Desvoidy), male. Pre- and postgonites (latter also from opposite side). Scale = 0.2 mm.

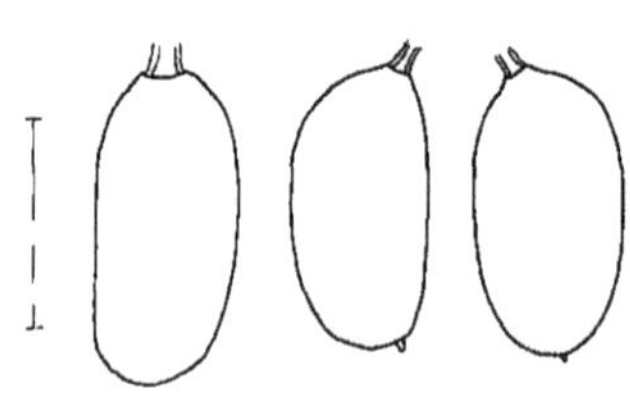

Fig. 336. *Protophormia terraenovae* (Robineau-Desvoidy), female. Spermathecae. Scale = 0.2 mm. G.pr. 162.

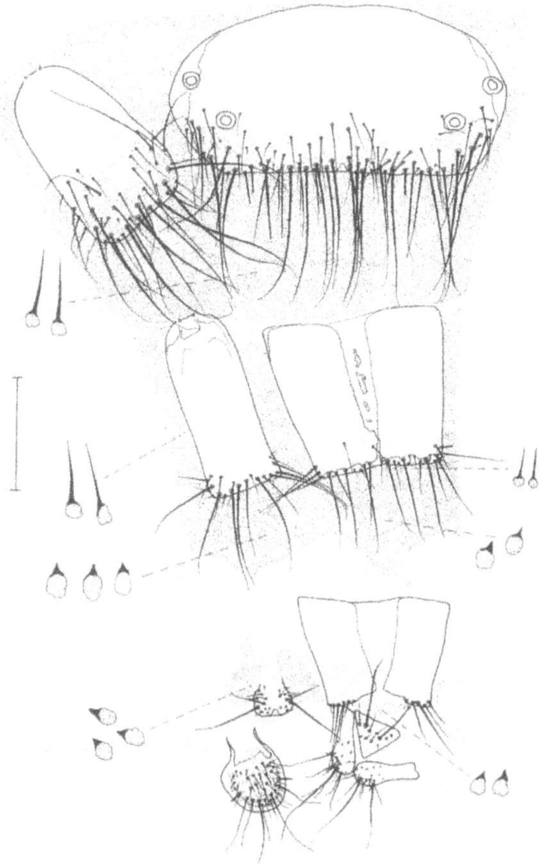

Fig. 337. ***Protophormia terraenovae*** **(Robineau-Desvoidy), female. Ovipositor. Stipple indicates extent of microtrichiae. Insets: shape of individual microtrichiae in areas indicated (enlarged). Scale = 0.5 mm. G.pr. 162.**

(Nuorteva, 1959e). All over Finland it is the dominant blowfly in spring, and north of 63°30'N it is the dominant blowfly in every season (Nuorteva, 1959e, 1960b). In Finland there are two generation each year. More details of its ecology will be found in Pulliainen (1974) and the papers by Nuorteva and coworkers (see bibliography). Its forensic importance in Finland is discussed by Nuorteva (1987), who maintains that human cadavers appear to be especially attractive to this species. It is very common on refuse dumps in southern Norway.

Nomenclature. *M. groenlandica* was described from a ♂ from Greenland, which was sent to Zetterstedt by B. W. Westermannn, and numerous specimens from various parts of Northern Norway and Sweden. I have labelled and here designate a ♂ labelled "Grönland" in the Insecta Lapponica collection in ZML (drawer 14) as lectotype. 1♂3♀ in the Insecta Lapponica collection (drawer 14), 2♂1♀ in the Diptera Scandinaviae collection (drawer 24), 1♂ in the Wallengren collection (drawer IV, 14) and 1♂2♀ in the Gothenburg collection (drawer 3) from localities mentioned by Zetterstedt (1838), all in ZML, have been labelled as paralectotypes.

Genus *Trypocalliphora* Peus, 1960

Trypocalliphora Peus, 1960: 199.
Type-species: *Avihospita braueri* Hendel, 1901: 30, by original designation.

A monotypic Holarctic genus, easily recognised by the characters given in the key. Medium sized metallic, weakly dusted blue-green (♂) or strongly dusted bronze-green (♀) flies. Arista with long hairs. Mesonotum sometimes flattened between suture and scutellum. 3+3-4 *acr.* 3+3-4 *dc.* 5 *h.* 4-5 *ph*, outer on line with or slightly outside of *prs* seta. 1+3 *ia.* 2 strong + 1-3 weak *npl.* Postalar wall bare, rarely with a few short hairs in middle. t_1 with 1 *pv* seta. t_2 with 2-4 *ad.* t_3 with 3-4 *av* setae. t_3 with rather strong *ad* and *d* preapical setae, *pd* preapical shorter and weaker. Hind coxa bare behind. Wings as in Fig. 8. Aedeagus with a broad apical opening of almost the same width as base of basiphallus. Hypophallic lobes small, pointing laterally, not ventrally. Ventral surface of distiphallus flat, not grooved as in *Protocalliphora*, evenly sclerotised, with spinules widely distributed but concentrated in two ventral subapical fields. Mesohypophallus absent. A stripe of hyaline membrane mid-ventrally. Acrophallus with a complete ring-like sclerotisation. A distinct dorsal sclerotisation proceeding basad from ring. Paraphallic processes strong, unarmed. Ventral plates not fused mid-ventrally. Postgonite with distinct process bearing basal seta.

Third stage larva without crown of seta on cephalic segment, and anterior spiracle with 2-4 branches.

Probably oviparous. Larvae burrowing subcutaneously on nestling birds.

Note. Rognes (1985a) argued that *Trypocalliphora* might be the sister-group of the composite group *Protocalliphora* + *Protophormia* + *Phormiata*. However, all species of *Protocalliphora* share with the genus *Trypocalliphora* the following apomorphies:

(1) Absence of hairs on the posterior surface of the hind coxa (secondary reduction from the assumed ground-plan condition of Chrysomyinae).
(2) Pleural membrane 7 of the ovipositor devoid of microtrichiae on a broad anterior zone.

These may both be interpreted as synapomorphies for a possibly monophyletic group *Protocalliphora* + *Trypocalliphora*. If this is true, there may be no need to give *Trypocalliphora* full generic rank (cf. Rognes, 1985a). However, the structure of the aedeagus, third instar larva, and the apparently different larval habits make me hesitate to sink *Trypocalliphora* as a synonym of *Protocalliphora*, as done by Sabrosky (1967), before a phylogenetic analysis has been made of all the genera of the Chrysomyinae.

25. *Trypocalliphora braueri* (Hendel, 1901)

Figs 8, 33, 246, 338-349.

Calliphora nidicola Nowicki, 1868: 44, nomen nudum.
Avihospita braueri Hendel, 1901: 30.
Protocalliphora hirudo Shannon & Dobroscky, 1924: 252.
Trypocalliphora lindneri Peus, 1960: 227.
Trypocalliphora lindneri compacta Grunin, 1966: 902.

♂♀. Head black, except for extensive reddish areas in front of genal dilation, at vibrissal angles, on lower part of facial ridges and on lower parts of parafacialia. Pedicel reddish in apical half. First flagellomere reddish basally and along hind edge. Occiput concave, with whitish hairs except for 1-3 rows of black setulae behind postocular row of setae. Palpi yellow. Parafacialia dusted greyish-white, with shifting reflections according to direction of light, especially in females which in certain lights show large triangular dark spots in upper part of the parafacialia. Prosternum bare in a few cases. Anterior spiracle yellow to light brownish-yellow. Posterior spiracle as in Fig. 339, brown to yellowish-brown. Basicosta, both calypters, subcostal sclerite and wing-veins yellow to yellowish-brown, veins darker apically.

♂. Frons 0.034-0.064 times head-width (n=26). Fronto-orbital plates touching or almost so. Parafacialia with parallel edges. f_2 without *a* preapical seta. Cerci straight in profile. Surstylus variable when seen in profile (cf. Rognes, 1985a), narrower in distal than in proximal half, apically blunt, anteroventral corner rather sharp.

♀. Frons 0.18-0.22 times head-width (n=19). Frontal vitta about 2 times as wide as fronto-orbital plates. Parafacialia narrowing below. Lateroclinate seta absent. f_2 with *a* preapical seta. Wing veins paler than in male. T6 with indications of subdivision mid-dorsally. Spiracles small. T8 with an almost complete row of marginals. Large area in anterior part of pleural membrane 7 devoid of microtrichiae. Intersegmental membrane 7-8 with normal, long microtrichiae. Spermathecae elongate with apical papilla. [2 ovipositor slides examined.]

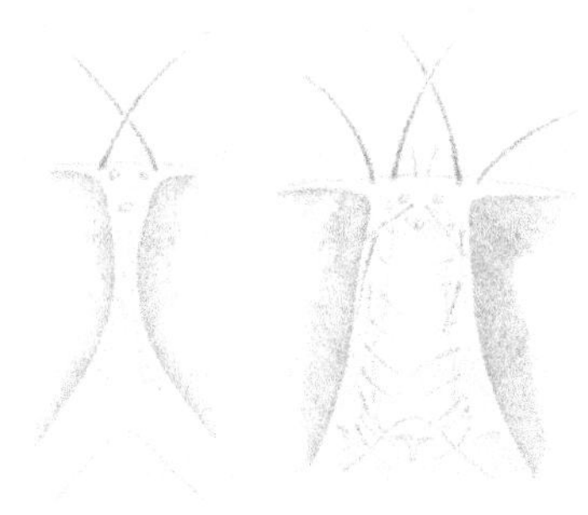

Fig. 338. *Trypocalliphora braueri* (Hendel). Male (left) and female frons. (From Rognes, 1985a: figs. 7, 8).

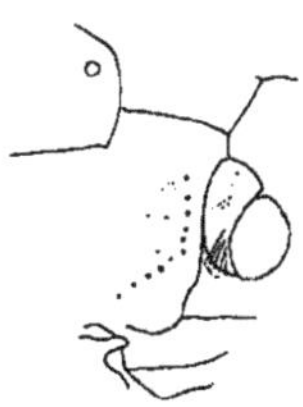

Fig. 339. *Trypocalliphora braueri* (Hendel). Posterior spiracle.

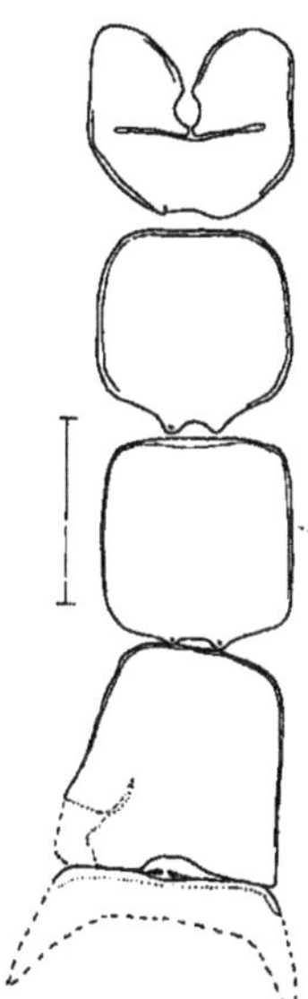

Fig. 340. *Trypocalliphora braueri* (Hendel), male. ST1-5. Scale = 0.8 mm.

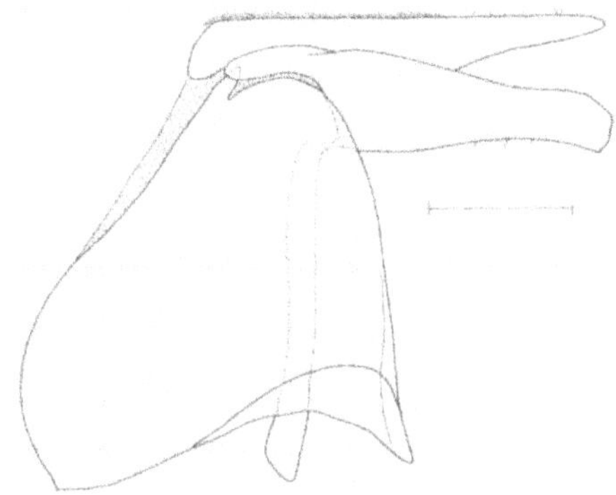

Fig. 341. *Trypocalliphora braueri* (Hendel), male. Epandrium, cerci, surstylus and bacilliform sclerite, left lateral view. Scale = 0.2 mm.

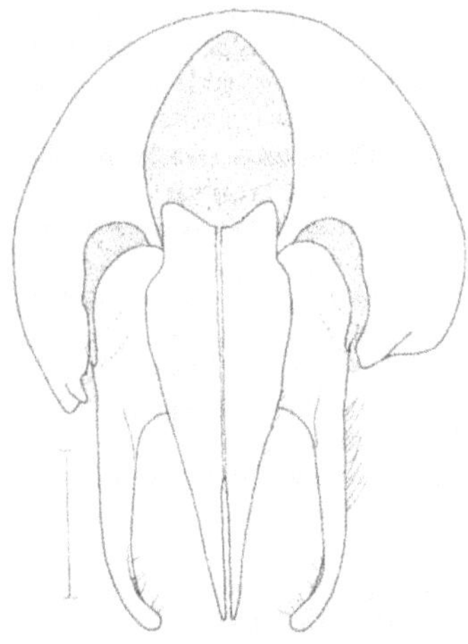

Fig. 342. *Trypocalliphora braueri* (Hendel), male. Epandrium, cerci and surstyli, posterior view. Scale = 0.2 mm.

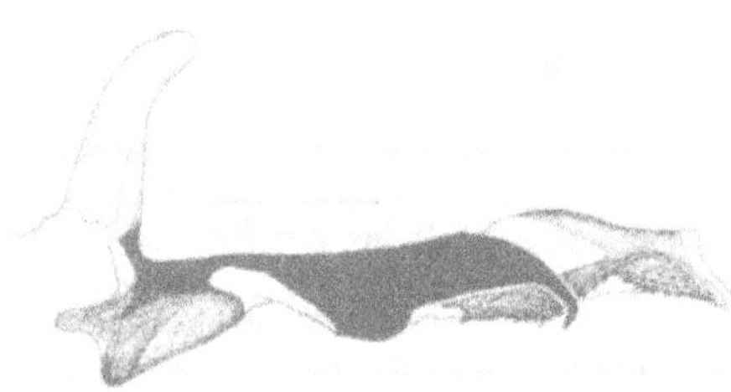

Fig. 343. *Trypocalliphora braueri* (Hendel), male. Aedeagus, left lateral view. Scale = 0.2 mm.

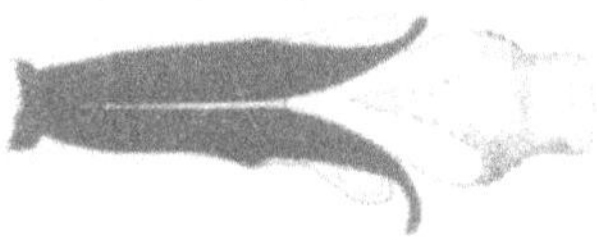

Fig. 344. *Trypocalliphora braueri* (Hendel), male. Distiphallus, dorsal view. Scale = 0.2 mm.

Fig. 345. *Trypocalliphora braueri* (Hendel), male. Aedeagus, ventral view. Scale = 0.2 mm.

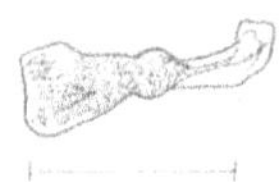

Fig. 346. *Trypocalliphora braueri* (Hendel), male. Ejaculatory sclerite. Scale = 0.2 mm.

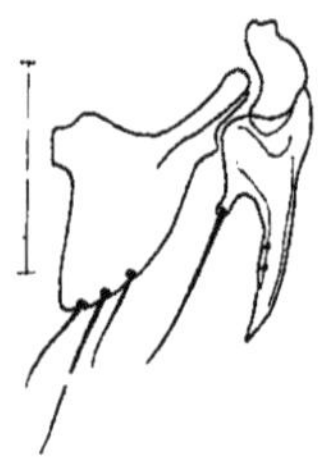

Fig. 347. *Trypocalliphora braueri* (Hendel), male. Pre- and postgonites. Scale = 0.2 mm.

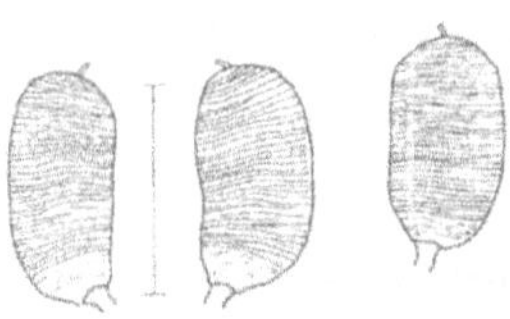

Fig. 348. *Trypocalliphora braueri* (Hendel), female. Spermathecae. Scale = 0.2 mm. G.pr. 157.

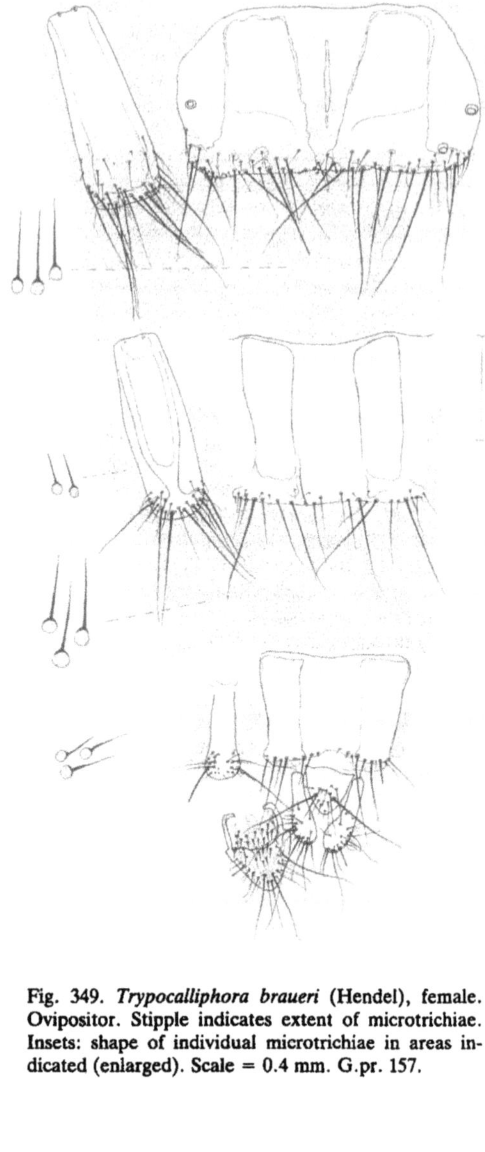

Fig. 349. *Trypocalliphora braueri* (Hendel), female. Ovipositor. Stipple indicates extent of microtrichiae. Insets: shape of individual microtrichiae in areas indicated (enlarged). Scale = 0.4 mm. G.pr. 157.

Length. 6.0-9.5 mm.

Distribution. Very rare in collections. Lennart Magni, a Swedish ornithologist doing field-work in Sm., has observed five cases of infestation during the years 1955-1957 but none during the subsequent 25 years of ring-marking. A few further records known from Sweden. Widespread in Finland and southern Norway. Catalogue entries for Ø and VAY are based on information given by Meinert (1890: 315, *Miliaria calandra*) and Collett (1921: 96, 401, *Oenanthe oenanthe*). Absent from Denmark. - Palaearctic: apart from Fennoscandia also Austria, Czechoslovakia, East Germany, Hungary, Poland, U.S.S.R (from Kr to the Far East) (Rognes, 1985a; Ozerov, 1986), also Japan (Iwasa & Hori, 1988). Nearctic: Alaska to Quebec, south to New Mexico and Georgia (references in Rognes, 1985a).

Biology. Immature stages have been described by Rodendorf (1957, as "Genus species"), Zumpt (1965), Cais (1965) and Rognes (1985a). Third instar larva with irregular spinulation on segments (although George & Mitchell, 1948, describe the larva as spineless), and without a crown of setae on the cephalic segment. A single infested nestling may support 1-57 larvae. The larvae make burrows up to 12 mm deep and 4.1 mm broad on various parts of the body from nostrils to feet, but mostly on the wings. Mature larvae leave the host at death or at the fledgling stage (thus outside the nest). Pupariation occurs in the nest or on the ground away from the nest. Reared material has emerged from June to August. Adults hibernate, but the sites are not known. No captures in the wild are known from Fennoscandia. The following birds are known to act as hosts in Fennoscandia: *Anthus pratensis, Carduelis flammea, Ficedula hypoleuca, Hirundo rustica, Luscinia luscinia, Miliaria calandra, Motacilla alba, Oenanthe oenanthe, Parus major, Phylloscopus sibilatrix, Phylloscopus trochilus*. Mixed infestations with *Protocalliphora azurea* are known from two nests of *Phylloscopus trochilus* in Norway (R. Mehl leg. et cult. j.nr. 66/71, 16/85, RML, KR). In Lr (U.S.S.R., Murmansk Oblast, Kandalashka Nature Reserve) Kovalev & Verves (1987) reported *Phoenicurus phoenicurus* and *Parus cinotus* as new hosts. They found mixed infestations with *Protocalliphora azurea* in nests of *Motacilla alba, Parus cinotus* and *Phoenicurus phoenicurus*.

Subfamily Helicoboscinae

The subfamily Helicoboscinae is probably a monophyletic group with the following groundplan apomorphies:

(1) Non-metallic body colour.
(2) Very strong setae in lower half of parafacialia.
(3) Postalar wall bare (secondary loss).
(4) 2+1+1 *kepst* setae.
(5) First instar larva with labrum separate from tentoropharyngeal sclerites
(6) Third instar larva with posterior spiracular plates completely sclerotised.

Members of this group are easily recognised by their large size, strong parafacial setae, black shining ground-colour with silvery dusting, and very strong setae. Head profile as in Fig. 350. Eyes bare. Male frons narrow or broad, always with a lateroclinate seta, in two species with 2 proclinate orbitals. Female frons broad with lateroclinate and 2 proclinate orbitals. Arista strongly pubescent in basal half (Fig. 350). Facial carina absent. Scape erect and projecting. First flagellomere large and broad, 2.5-3 times as long as pedicel. Palpi long, distally widened, yellow. 5-7 *h*. 3-4 *ph*, outer *ph* in line with *prs* or rarely slightly exterior to it. 5+3 *acr*. 4+4 *dc*. 0-1+3 *ia*. 2+1 *npl*. 2+1+1 *kepst* setae. Scutellum with 2 pairs of discal setae, and 5-7 pairs of marginals (Fig. 352). Prosternum, proepisternal depression and metasternal area hairy. Postalar wall and suprasquamal ridge bare (Fig. 351). Coxopleural streak present (Fig. 353). Posterior spiracle as in Fig. 353, anterior lappet small. Basicosta black. Subcostal sclerite bare. Stem-vein bare. Costa hairy below to tip of vein Sc. No costal spine. Bend of M rectangular and rather far from posterior wing-margin. Both calypters white, bare on disc. Lower calypter with inner edge converging backwards with longitudinal axis of fly (Fig. 351). t_1 with 2 *p* setae. t_2 with 3-5 *ad* setae. f_2 without *a* preapical seta. t_3 with 2-4 *av* setae and strong *ad*, *d* and *pd* preapical setae of uniform size. Hind coxa bare behind. T3 with 2-4 strong erect median marginals. T4 and T5 each with 12-16 very strong, erect and evenly spaced marginals. No discal setae on any tergite. ST2-4 widely exposed, lateral margins of ST2 overlapped by margins of T2. Sternites usually without alphasetae on anterior margin. TST7+8 in male with numerous strong setae. Aedeagus with paraphallic processes distally rounded and denticulate marginally and on outer side. Ventral plates fused mid-ventrally into a ring-shaped structure with triangular pointed flanges. Mesohypophallus parallel-sided. Hypophallic flanges well-developed. Aperture of ejaculatory duct apical, oval and very large, strengthened laterally by sclerotisations curving dorsally from apex of mesohypophallic sclerotisation. Lateral wall of acrophallus weakly sclerotised and armed with small denticles. Lateral duct absent. Pregonites with about 5 setae. Postgonite with a weak basal seta. Ovipositor with short sclerites. T6 large, divided. T7 small, divided. T8 partly divided, with a varying number of marginal setae. ST6 large. ST7 small. ST8 vestigial. Hypoproct large. Uterus large and cylindrical with anterior blind pouch, accessory gland ducts and spermathecal ducts entering dorsal side far from tip (Rognes, 1986c). Spermathecae oval.

Macrolarviparous (unilarviparous) reproduction, the single first instar larva developing on dead snails.

Widespread in Western Palaearctic Region. A single genus *Eurychaeta* Brauer & Bergenstamm (= *Helicobosca* Bezzi). Verves (1986) listed it as belonging in the Sarcophagidae. Rognes (1986c) discussed its systematic position and transferred it to the Calliphoridae. Pape (1987b) agreed with this.

Genus *Eurychaeta* Brauer & Bergenstamm, 1891

Theria Robineau-Desvoidy, 1830: 337.
Type-species: *Theria palpalis* Robineau-Desvoidy, 1830: 337, by monotypy. Junior homonym of *Theria* Hübner, 1816 (Lepidoptera).

Eurychaeta Brauer & Bergenstamm, 1891: 367 [63]; 1893: 230. Manuscript name published as a junior synonym of *Theria* Robineau-Desvoidy, and made available through adoption prior to 1961 by Townsend (1916b, 1935b, 1937) and Lopes (1955) [ICZN Art. 11(e)].
Type-species: *Sarcophaga muscaria* Meigen, 1826, by monotypy (see discussion under Notes on synonymy, below).

Helicobosca Bezzi, 1906: 49. Replacement name for *Theria* Robineau-Desvoidy, 1830: 337.
Type-species: *Theria palpalis* Robineau-Desvoidy, 1830: 337, automatic.

According to Schmitz (1910, 1917) who gives an account of the biology of a European species (either *muscaria* or *palpalis*), the female gives birth to a single very large (7mm) first instar larva at about two week intervals during the summer months. It is deposited on dying or dead pulmonate snails of the family Helicidae. The snail is completely devoured and pupariation occurs in the soil or within the shell after 4-5 weeks. The adults emerge the following year. Rognes (1986c) lists the species that are known in nature to have served as the substrate for larval development. Widespread in Western Palaearctic Region. 3 species are known, 1 occurs in Fennoscandia and Denmark. All 3 species are keyed below.

Notes on synonymy. I have previously used the name *Helicobosca* for this genus. It was also the basis for the family-group name Helicoboscina introduced by Verves (1980), given subfamily rank by Rognes (1986c). This name must stand even though *Helicobosca* itself is replaced by the older name *Eurychaeta* (cf. ICZN Art. 40 (a)). Brauer & Bergenstamm (1891: 367) mentioned this name in a remark under *Theria*: "Zu *Theria* gehört die Gattung *Eurychaeta* v.d. Wp. [van der Wulp] B.C.Am.". Bezzi (1906: 49-50) refers to this but points out that the name *Eurychaeta* Wulp cannot be found in the source cited ["B. C.-Am" = Godman, F.D. & Salvin, O (eds.), Biologia Centrali-Americana, or contributions to the knowledge of the fauna and flora of Mexico and Central America. 57 vols., London]. He also suggested that Brauer & Bergenstamm may possibly have meant *Microchaetina* Wulp, a name which is found in the publication cited, and "welcher in der Tat für *Theria* zu stehen scheint...". Bezzi & Stein (1907: 496) list *Eurychaeta* as a nomen nudum and *Microchaetina* Wulp as a questionable synonym of *Helicobosca*. *Microchaetina* Wulp, 1891 is actually a genus in the Tachinidae (Guimarães, 1971: 104.28; Stone, Sabrosky, Wirth, Foote & Coulson, 1965: 984) and need not occupy us further. Townsend (1916b: 10) maintained that the publication (by Brauer & Bergenstamm) of "*Eurychaeta* BB." "as equal to *Theria* RD" not only "gave it nomenclatorial standing, ..., but also validated it for use in place of the preoccupied *Theria* RD, whose genotype it takes.". It was also used later by Townsend (1935b, 1937). Lopes (1955: 266), in a treatment of Robineau-Desvoidy's genus group names in Sarcophagidae, notes that *Theria* Robineau-Desvoidy is preoccupied by *Theria* Hübner, and recommends its replacement by *Eurychaeta*. As to the name *Eurychaeta* van der Wulp, I cannot find it in the North American Catalogue (Stone, Sabrosky, Wirth, Foote & Coulson, 1965), in the Muscidae (Pont, 1972), Anthomyiidae (Pont, 1974), Calliphoridae (James, 1970), Tachinidae (Guimarães, 1971) or Sarcophagidae (Lopes,

1969) parts of Papavero's catalogue of the Americas south of the United States, in the Afrotropical Catalogue (Crosskey, 1980) or in the Oriental Catalogue (Delfinado & Hardy, 1977). It is therefore most likely a manuscript name, as maintained by Townsend (1916b). However, it is published by Brauer & Bergenstamm as a junior synonym of *Theria* Robineau-Desvoidy, and having been adopted by Townsend and Lopes prior to 1961 it is clearly available under the ICZN Art. 11(e). No species was actually mentioned as belonging in *Eurychaeta* (as *Theria*) by Brauer & Bergenstamm (1891), but their reference to p. 122 of their earlier work (Brauer & Bergenstamm, 1889), where *muscaria* Meigen only is included in *Theria*, makes that nominal species the type species of *Eurychaeta* by monotypy. This designation precedes the one of *Theria palpalis* by Townsend (1916b: 10, not actually named but cited as the "genotype" of *Theria*) or the one of *Sarcophaga muscaria* by Townsend (1937: 191). Townsend (1937) maintains that *Theria palpalis* was listed by Brauer & Bergenstamm (1891: 63) but this is an error.

Key to Palaearctic species of *Eurychaeta*

1 T5 shining black all over, practically without dusting; parafacialia at narrowest point about 2-3 times wider than palp at broadest point; in both sexes frons broader than an eye as seen from above and fronto-orbital plates with 2 proclinate orbital setae (North Africa)

nigrapex (Villeneuve,1910)

– T5 conspicuously dusted, with black hind margin and median stripe

2

2(1) Parafacialia at narrowest point about as wide as palp at broadest point and half as wide as first flagellomere; gena in profile about as high as width of first flagellomere; ♂: frons at narrowest point much narrower than an eye when seen from above; fronto-orbital plates without proclinate orbital setae; tarsal segment 5 of front leg 2-2.5 times longer than segment 4, and 1.5 times longer than segment 3; claws long; ♀: first flagellomere rather broad, 1.8-2 times as long as broad; frons about as wide as an eye when seen from above

26. *palpalis* (Robineau-Desvoidy, 1830)

– Parafacialia at narrowest point about 2-3 times wider than palp at broadest point and as wide as first flagellomere; gena in profile 1.5-2 times higher than width of first flagellomere; ♂: frons at narrowest point about as broad as an eye when seen from above; fronto-orbital plates with 2 proclinate orbital setae; tarsal segment 5 of front leg 1.2 times longer than segment 4, and as long as segment 3; claws short; ♀: first flagellomere narrower than in *palpalis*, about 2.5 times as long as broad; frons wider than an eye when seen from above (Central and South Europe)

muscaria (Meigen, 1826)

26. *Eurychaeta palpalis* (Robineau-Desvoidy, 1830)

Figs 22, 29, 32, 34, 37, 40, 350-365; Plate - fig 3.

Theria palpalis Robineau-Desvoidy, 1830: 337.
Helicobosca distinguenda Villeneuve, 1924: 35.

♂♀. Easily recognised by the characters given in the key.

♂. Frons 0.253-0.263 times head-width (n=5). Frontal vitta broader than fronto-orbital plate. Cerci long and narrow, wide basally in profile, apically with a small hook. Surstyli short, rounded apically. Bacilliform sclerites curved towards midline, separated, a narrow accessory linear sclerotisation present in the membrane between them.

♀. Frons 0.348-0.385 times head-width (n=5). T7, T8 and epiproct without microtrichiae. Cerci with a few microtrichiae at apex, a small group of sensillae present centrally on upper surface. [2 ovipositor slides examined.]

Length. 9-12 mm.

Distribution. Not common. In Zealand and Bornholm in Denmark. Widespread in Norway and Sweden north to NSI and Ång. In Finland north to Sa, Ta (Tiensuu, 1964). - Palaearctic, east to Altai (Verves, 1982a, 1982b).

Biology. Immature stages described by Nielsen (1917) and Rognes (1986c). Note that Thompson (1921a: plate XVII, figs. 124, 133) and Séguy (1941: 8) have described the immature stages of *muscaria*. In Denmark, puparia have been found in empty shells of *Cepaea* sp. and *Cepaea hortensis* (Müller) in early spring (April 12, May 4), and emergence took place 3-15 days afterwards (Lundbeck, 1927). No further breeding records are known from Fennoscandia and Denmark. The third stage larva described by Rognes (1986c) was reared on a cold, freshly killed (in boiling water) specimen of *Cepaea hortensis*, offered to a first-stage larva born July 27, 1984, after its mother had

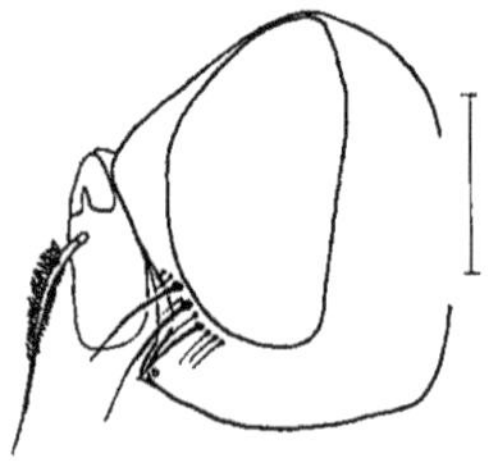

Fig. 350. *Eurychaeta palpalis* (Robineau-Desvoidy), male. Head profile. Scale = 1.0 mm.

Fig. 351. *Eurychaeta palpalis* (Robineau-Desvoidy), male. Area behind wing base, dorsoposterior view.

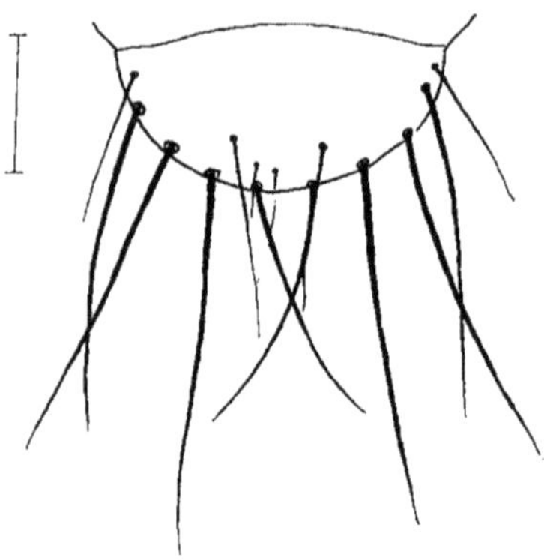

Fig. 352. *Eurychaeta palpalis* (Robineau-Desvoidy). Scutellum. Scale = 1.0 mm.

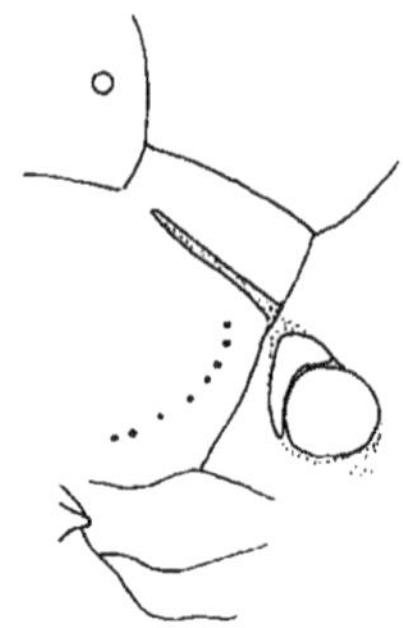

Fig. 353. *Eurychaeta palpalis* (Robineau-Desvoidy). Area around coxopleural streak and posterior spiracle.

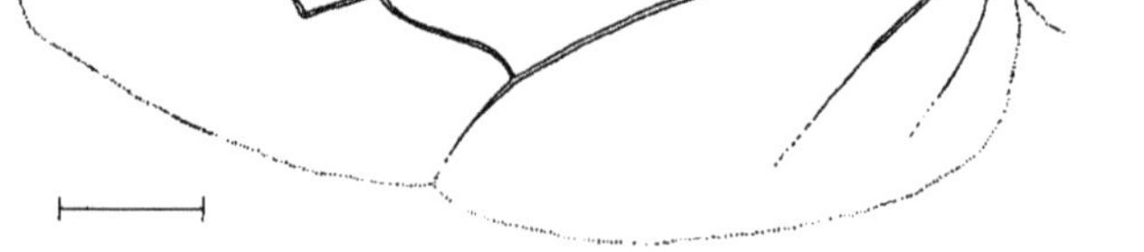

Fig. 354. *Eurychaeta palpalis* (Robineau-Desvoidy). Left wing from above. Inset: costa from below showing distribution of small hairs on its underside. Scale = 1.0 mm.

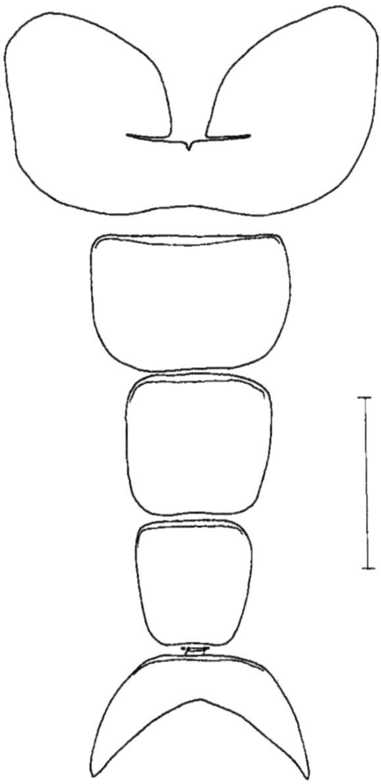

Fig. 355. *Eurychaeta palpalis* (Robineau-Desvoidy), male. ST1-5. Scale = 1.0 mm.

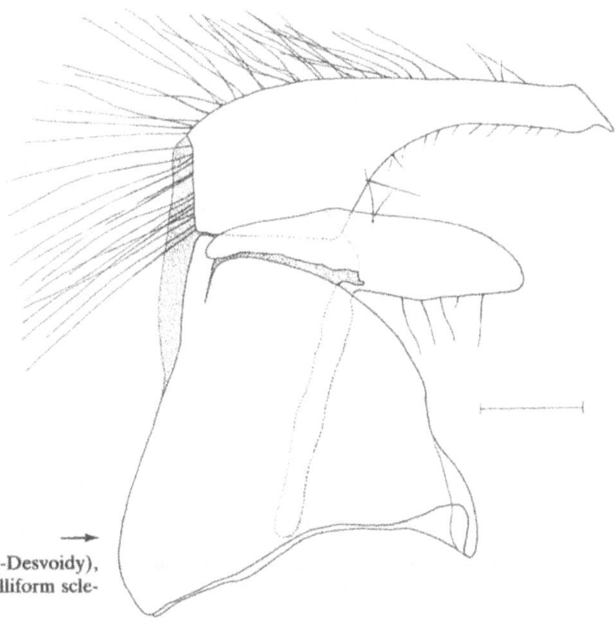

Fig. 356. *Eurychaeta palpalis* (Robineau-Desvoidy), male. Epandrium, cerci, surstylus and bacilliform sclerite, left lateral view. Scale = 0.2 mm.

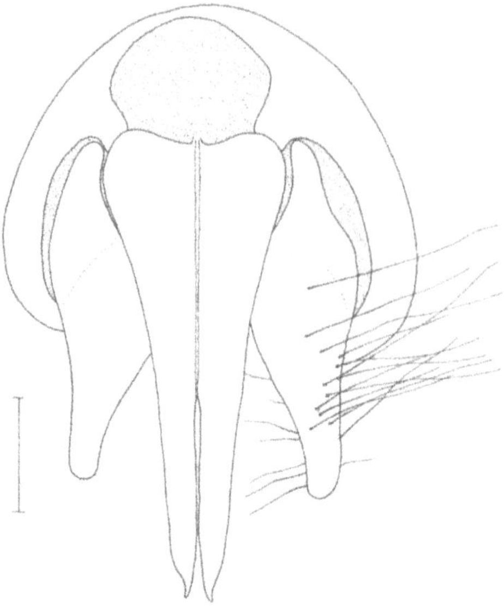

Fig. 357. *Eurychaeta palpalis* (Robineau-Desvoidy), male. Epandrium, cerci and surstyli, posterior view. Scale = 0.2 mm.

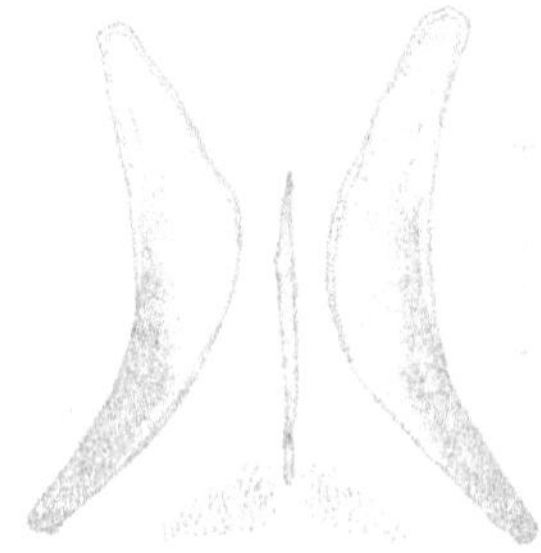

Fig. 358. *Eurychaeta palpalis* (Robineau-Desvoidy), male. Bacilliform sclerites and membrane between them. Scale = 0.2 mm.

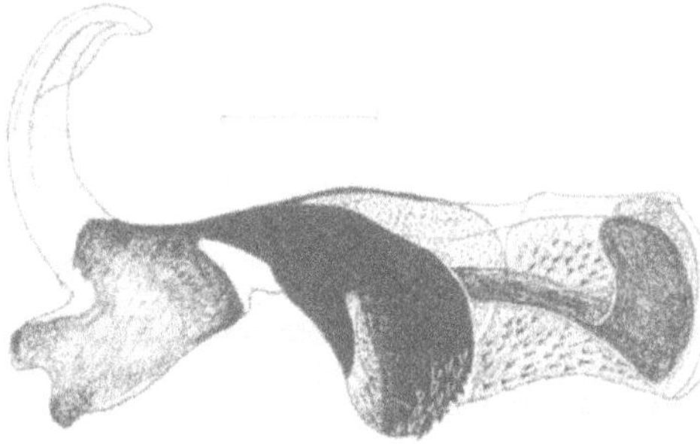

Fig. 359. *Eurychaeta palpalis* (Robineau-Desvoidy), male. Aedeagus, left lateral (and very slightly dorsal) view. Scale = 0.2 mm.

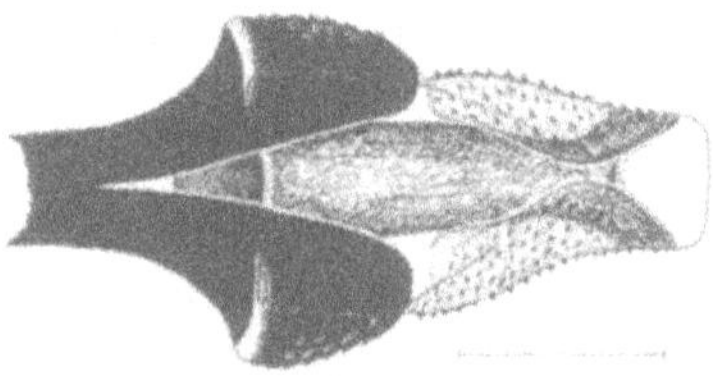

Fig. 360. *Eurychaeta palpalis* (Robineau-Desvoidy), male. Distiphallus, dorsal view. Scale = 0.2 mm.

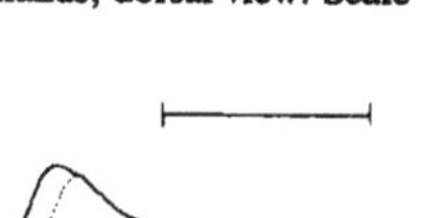

Fig. 361. *Eurychaeta palpalis* (Robineau-Desvoidy), male. Ejaculatory sclerite. Scale = 0.2 mm.

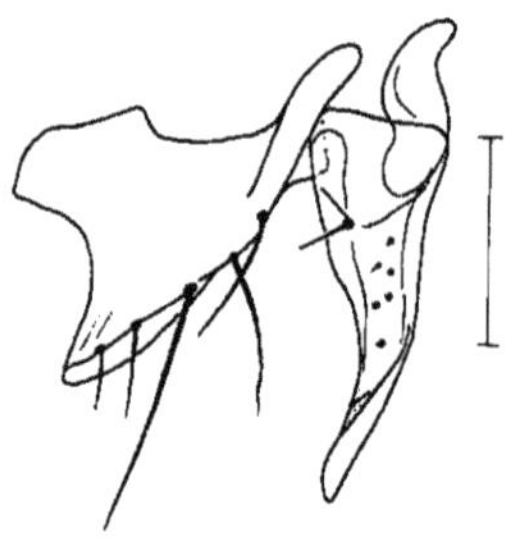

Fig. 362. *Eurychaeta palpalis* (Robineau-Desvoidy), male. Pre- and postgonites. Scale = 0.2 mm.

←
Fig. 363. *Eurychaeta palpalis* (Robineau-Desvoidy), female. Spermathecae. Scale = 0.2 mm. G.pr. 85.

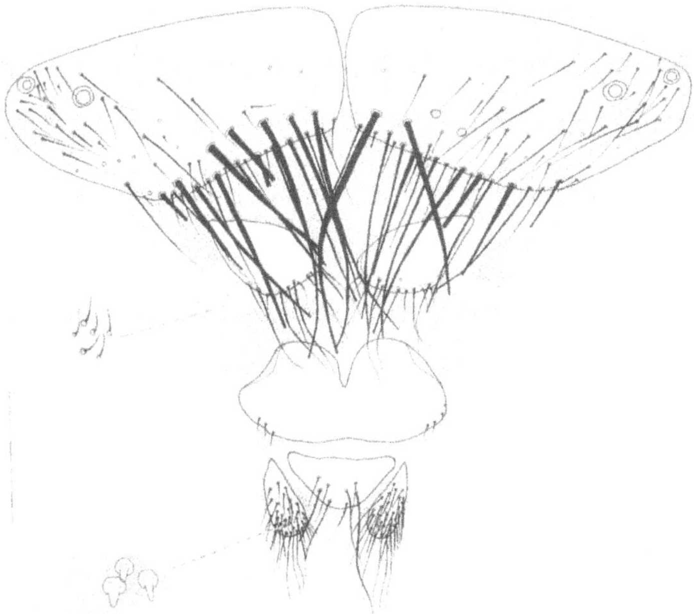

Fig. 364. *Eurychaeta palpalis* (Robineau-Desvoidy), female. Ovipositor, dorsal sclerites. Stipple indicates extent of microtrichiae. Insets: shape of individual microtrichiae, group of sensillae dorsally on cerci (both enlarged). Scale = 0.5 mm. G.pr. 85.

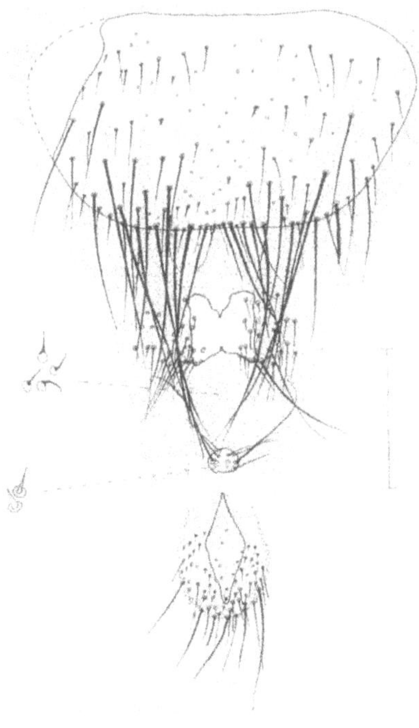

←

Fig. 365. *Eurychaeta palpalis* (Robineau-Desvoidy), female. Ovipositor, ventral sclerites. Stipple indicates extent of microtrichiae. Insets: shape of individual microtrichiae, group of sensillae on ST8 (both enlarged). Scale = 0.5 mm. G.pr. 85.

been killed with ethylacetate vapour. The fully-grown third instar larva, which stayed within the shell, was killed on August 6. The whole snail had been devoured. Adults have been collected from May to August, with maximum numbers in July.

Nomenclature. Séguy (1941: 172, 177) asserts that he had examined a ♂ type of *Theria palpalis* in MNHN.

Subfamily Luciliinae

The subfamily Luciliinae as understood here is probably a monophyletic group defined by the following ground-plan apomorphies:

(1) Parafacialia bare.
(2) Suprasquamal ridge with a posterior green metallic sclerite with numerous black setae (Fig. 397).
(3) Metakatepisternum hairy in lowermost part.
(4) Subcostal sclerite with black setulae (Fig. 9) (not all members of *Lucilia* s.lat.).

The following genera probably belong in this group: *Dyscritomyia* Grimshaw, *Hemipyrellia* Townsend, *Hypopygiopsis* Townsend, *Lucilia* Robineau-Desvoidy. It corresponds to the tribe Luciliini of Kurahashi (1977) and Shewell (1987), and its members are commonly known as greenbottle flies.

Third instar larva with or without accessory oral sclerites (Ishijima, 1967).

Widespread in all faunal regions. A single genus in Fennoscandia and Denmark.

Genus *Lucilia* Robineau-Desvoidy, 1830

Lucilia Robineau-Desvoidy, 1830: 452.
Type-species: *Musca caesar* Linnaeus, 1758: 595, by designation of Macquart, 1834: 162.
Phaenicia Robineau-Desvoidy, 1863: 750.
Type-species: *Phaenicia concinna* Robineau-Desvoidy, 1863: 778, by designation of Townsend, 1916b: 8; = *Musca sericata* Meigen, 1826.
Phumonesia Villeneuve, 1914: 307.
Type-species: *Phumonesia infernalis* Villeneuve, 1914: 307, by monotypy.
Bufolucilia Townsend, 1919: 542.
Type-species: *Lucilia bufonivora* Moniez, 1876: 25, by original designation.
Francilia Shannon, 1924: 74.
Type-species: *Francilia alaskensis* Shannon, 1924: 74, by monotypy; = *Sarcophaga magnicornis* Siebke, 1863.
Roubaudiella Séguy, 1925a: 735.
Type-species: *Roubaudiella caerulea* Séguy, 1925a: 735, by monotypy; = *Phumonesia infernalis* Villeneuve, 1914.
Viridinsula Shannon, 1926: 31.
Type-species: *Musca pionia* Walker, 1849: 880.
Dasylucilia Rodendorf, 1926: 92.
Type-species: *Lucilia pilosiventris* Kramer, 1910: 35, by monotypy.
Caesariceps Rodendorf, 1926: 93.
Type-species: *Lucilia flavipennis* Kramer, 1917: 283, by monotypy; = *Lucilia ampullacea* Villeneuve, 1922.
Luciliella Malloch, 1926: 507.
Type-species: *Lucilia fumicosta* Malloch, 1926: 507, by original designation.
Chaetophaenicia Enderlein, 1936: 211.
Type-species: *Musca silvarum* Meigen, 1826: 53, by original designation.
Acrophagella Ringdahl, 1942: 64.
Type-species: *Sarcophaga fuscipalpis* Zetterstedt, 1845: 1306, by original designation; = *Sarcophaga magnicornis* Siebke, 1863.
Sinolucilia Fan, 1965: 173.
Type-species: *Lucilia appendicifera* Fan, 1965: 173, by monotypy.

In Fennoscandia and Denmark large to medium sized flies with black ground-colour and bright green metallic colour, usually almost without dusting on thorax and abdomen. Head profiles as in Figs. 366, 367. Eyes bare. Male frons narrow without lateroclinate seta. Female frons with a lateroclinate and two proclinate orbital setae. Fronto-orbital plates dusted white in anterior half, posteriorly darkened. In female green metallic colour often shining through at vertex. Arista long plumose to pubescent (Figs. 366, 367, 411, 466). Facial carina absent. Parafacialia bare, densely silvery dusted. Proepisternal depression, prosternum, metasternal area and postalar wall hairy. Hindmost *prst acr* well in front of suture. 3+1 *h*. 2-3 *ph*, outer outside a line through *prs*. 2+2-3 *acr*. 3+3 *dc*. 0-1+2-3 *ia*. 4-6 scutellar marginals. 2 *npl*. 2+1 *kepst* setae. Lower half of anterior part of anepimeron bare (Fig. 6). Katatergite pubescent, anatergite with short hairs below lower calypter. Coxopleural streak present, except in *Lucilia ampullacea* (Figs. 368-376). Posterior spiracle with subequal lappets or anterior larger than posterior. No setae along lower edge, except in *L. ampullacea* which has a small inconspicuous bundle of small hairs on extreme ventral edge (Fig. 368). Area above hind coxa (metakatepisternum) hairy (Figs. 370, 372, 374-376). Stem-vein bare on upper surface of wing. Costa hairy below to far beyond junction with R_1 (Figs. 398). Subcostal sclerite bare or setulose (Fig. 9). R_{4+5} hairy below almost to r-m (Fig. 398). Cell r_{4+5} opens well in front of wing. Lower calypter white, bare above, broad, inner edge converging backwards with longitudinal axis of fly. Upper calypter bare on upper side. t_1 most often with 1 *pv* seta. t_2 with 1-3 *ad* setae. f_2 with or without *a* preapical seta. t_3 with strong *ad* and *d* preapical setae, *pd* preapical setae much weaker. Hind coxa bare behind. Abdominal ground-vestiture erect in males on all tergites and decumbent in females except on T5. Lateral margins of ST2 overlapping ventral margins of T2. Basal part of ST5 in males short and polished black and without setae in middle. Aedeagus variable, acrophallus always armed with denticles, five types discernible among North European species:

(1) *ampullacea* type (Figs. 381, 382). Paraphallus long and slender, gently curved anteroventrally in apical half, tip expanded, serrate dorsally. Ventral plates almost absent. Hypophallus posteriorly a dentate ventral bulge and anteriorly a cornu forming a shelf and lateral wall supporting the paraphallic tip. Acrophallus moderately long with lateral duct. Mesohypophallus narrow, tapering gradually. (*ampullacea*.)
(2) *bufonivora* type (Figs. 391, 392). Ventral plate large. Paraphallus slender, sigmoid, apically dentate on dorsal side and directed apicad. Apical part passing through a canal in the cornu of hypophallus which is heavily sclerotised and provided with an apical opening. Posterior part of hypophallus dentate. Acrophallus moderately long with lateral ducts. Mesohypophallus a long rod with parallel sides. (*bufonivora* and *silvarum*.)
(3) *caesar* type (Figs. 403, 404). Paraphallus long and slender, apically curved downwards and weakly dentate. Hypophallus shaped as low dentate ridges, no cornu differentiated. Acrophallus long. Lateral ducts not apparent. Mesohypophallus a posteriorly broad plate, tapering strongly apically. (*caesar* and *illustris*.)
(4) *magnicornis* type (Figs. 416-418). Generally very similar to the *sericata* type, but hypophallus shorter and acrophallus longer. (*magnicornis*.)
(5) *sericata* type (Figs. 459-461). Ventral plates large. Paraphallus hook-shaped, abruptly turned downwards apically, serrate here on anterior side. Hypophallus shaped as low dentate ridges. Apically the hypophallus forms a peculiar vertical passage with a posterior and lateral wall guiding the movement of the tip of the paraphallus. Acrophallus short with lateral duct. (*regalis*, *richardsi* and *sericata*.)

Pregonite narrow, the most distal seta apical in position. Postgonite with or without a basal seta. Ejaculatory sclerite usually large and fan-shaped.

Ovipositor often with shortened or reduced marginal setae on T7. Peculiar sclerotisations often present mid-ventrally in the intersegmental membranes behind ST6 and ST7. ST8 usually of full length, apically undivided. Epiproct and cerci usually shining black, without microtrichiae.

Uterovaginal tubes in some species with shallow lateral sacs (cf. Hori, 1961), these are most often unsclerotised. In *Lucilia magnicornis* and *Lucilia richardsi* (but not in other *Lucilia* species examined during this study) a pair of brown scle-

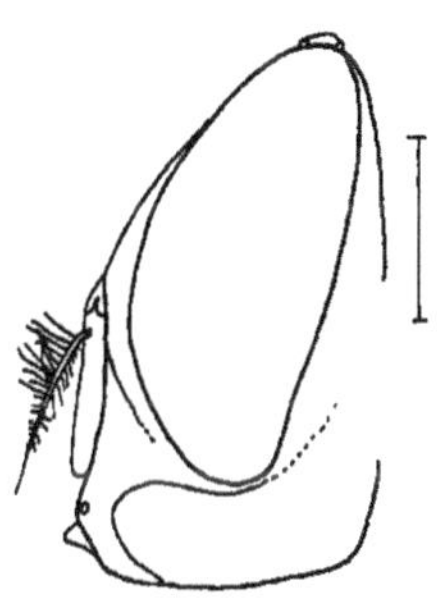

Fig. 366. *Lucilia caesar* (Linnaeus), male. Head profile. Scale = 1.0 mm.

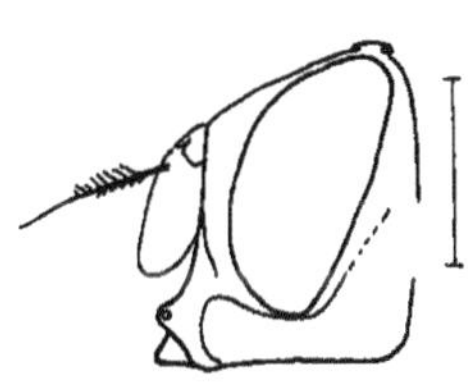

Fig. 367. *Lucilia magnicornis* (Siebke), male. Head profile. Scale = 1.0 mm.

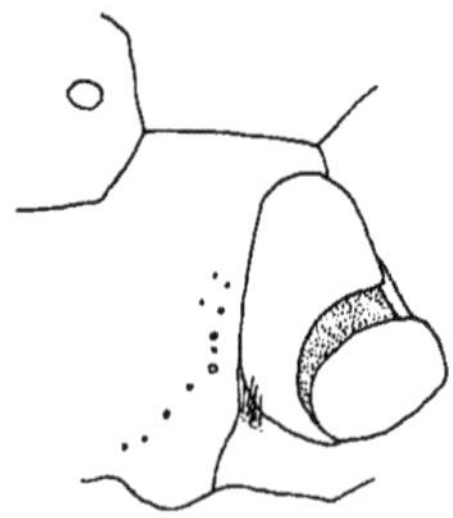

Fig. 368. *Lucilia ampullacea* Villeneuve, posterior spiracle.

Fig. 369. *Lucilia bufonivora* Moniez, posterior spiracle.

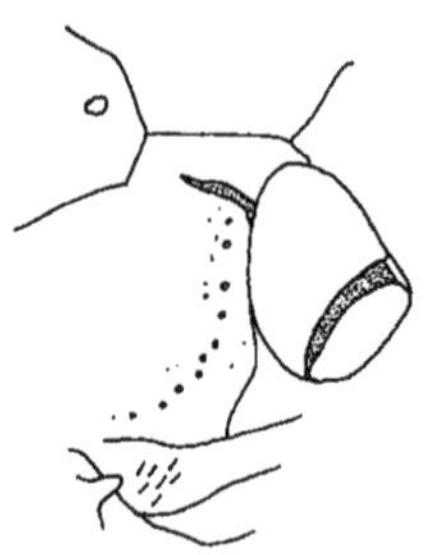

Fig. 370. *Lucilia caesar* (Linnaeus), posterior spiracle.

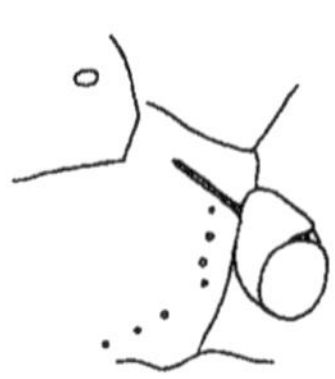

Fig. 371. *Lucilia magnicornis* (Siebke), posterior spiracle.

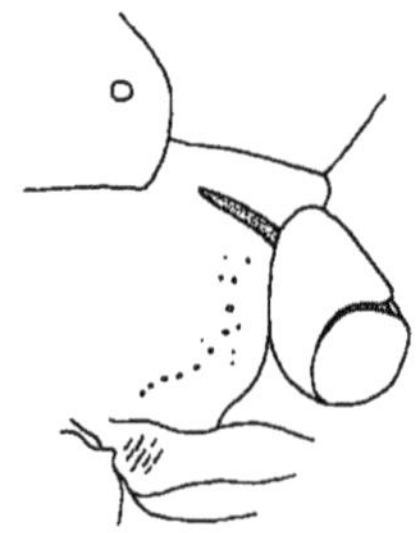

Fig. 372. *Lucilia illustris* (Meigen), posterior spiracle.

Fig. 373. *Lucilia regalis* (Meigen), posterior spiracle.

Fig. 374. *Lucilia richardsi* Collin, posterior spiracle.

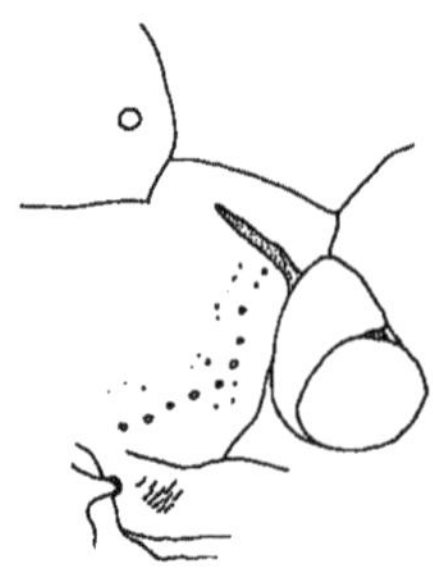

Fig. 375. *Lucilia sericata* (Meigen), posterior spiracle.

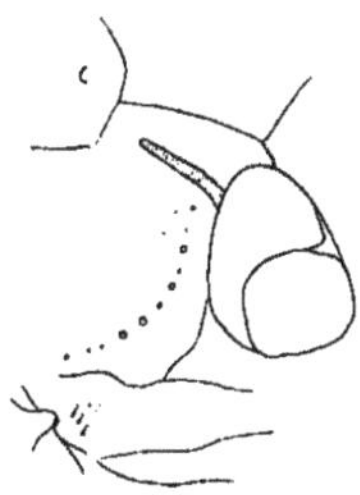

Fig. 376. *Lucilia silvarum* (Meigen), posterior spiracle.

rotised spots were found in the dorsal wall of the uterus. These may possibly be sclerotisations affecting parts of the lateral sacs. Hori (1961) found the orifices of the lateral sacs to be brown-pigmented in *L. sericata*, *L. cuprina* and *L. papuensis*. This may be the same as what Lewis & Pollock (1975) interpret in *Lucilia sericata* as "scar tissue", assumed to have been produced by the paraphallic processes piercing or abrading the wall of the uterus during copulation. I believe this sclerotisation to be the result not of any injuries to the wall of the female genital tract but of a normal maturation process. Completely sclerotised lateral sacs occur regularly in some species-groups of *Pollenia* (see below).

All species are oviparous. Larvae are primarily saprophagous in decaying animal matter. Adults visit flowers, faeces and dead animals. Many species are involved in human or animal myiasis, and readily infect wounds. The genus is of great medical, hygienic and forensic importance (Schumann, 1971; Smith, 1986). *Lucilia* species have been suspected of transmitting poliomyelitis virus to humans (Nuorteva, 1958, 1959b, 1959c, 1959d, 1959e, 1960c, 1960d, 1961; but see review by Greenberg, 1973). Descriptions of the immature stages of most species occurring in Fennoscandia and Denmark can be found in Hall (1948), Schumann (1954, 1971), Kano & Sato (1952), Zumpt (1965), Ishijima (1967) and Smith (1986).

The genus contains numerous species and occurs in all faunal regions.

9 species in Fennoscandia and Denmark.

Note on synonymy. American authors (Hall, 1948, 1965; Shewell, 1987) split *Lucilia* as understood here (s.lat.) into several genera. These are listed as subgenera by Schumann (1986). Even though several of them may represent monophyletic assemblages (cf. discussion of aedeagus-types above), no overall phylogenetic analysis of the subfamily on the basis of a broad range of characters has been performed, and no consistent classification has therefore been reached (cf. also discussion by James, 1971). For the time being I prefer to retain *Lucilia* in a traditional broad sense. The genus as here understood is possibly not monophyletic, as it is differentiated from *Hemipyrellia* and *Hypopygiopsis* only by an obvious plesiomorphy: absence of long fine setae on the katatergite (Zumpt, 1956a, 1956b; Aubertin, 1931, 1933; Senior-White, Aubertin & Smart, 1940). *Hypopygiopsis* has a quite peculiar aedeagus (cf. figures in Kurahashi, 1977). Both *Hemipyrellia* and *Hypopygiopsis* have the hind coxa hairy behind.

Key to species of *Lucilia*

1 Coxopleural streak absent (Fig. 368); a few inconspicuous hairs present at extreme lower corner of posterior spiracle (requires quite careful observation) (Fig. 368); subcostal sclerite with black setulae; occiput mostly clothed with pale hairs, only 1 irregular row of black setulae behind postocular row of setae; ♂: frons narrower than parafacialia, 0.035-0.040 times head-width; epandrium normal, narrower than length of T5 at midline; ♀: T6 with long marginal setae along whole hind margin, with no bare sections; mid-dorsal line in dried specimens straight when seen in profile

27. *ampullacea* Villeneuve

– Coxopleural streak present (Figs. 369-376); lower margin of posterior thoracic spiracle quite bare; subcostal sclerite bare or setulose

2

2(1) Basicosta white or yellow; subcostal sclerite with microscopic pubescence only, without black setulae; 3 *post acr*; ♂: frons broader than parafacialia; ♀: basal dorsal excavation of abdomen broadly separated from hind margin of T1+2

3

– Basicosta dark brown or black; t_2 with 1 *ad*

5

3(2) t_2 with 1 *ad*; palpi yellow or brown; median marginal setae of T3 weak, sometimes hair-like, adpressed, about as strong as paramedian ones, equal to or shorter than half the length of T4; ♂: frons 0.122-0.137 times head-width; ♀: abdominal dusting strong, divided along mid-line according to angle of view; fronto-orbital plates half as wide as frontal vitta

34. *sericata* (Meigen)

– t_2 with 2-3 *ad*; palpi darker

4

4(3) T5 with strong discal setae intermingled with short hairs which are less than half as long as the discal setae, shortest hairs about as long as or shorter than hairs covering disc of T4; 2-4 median marginal setae on T3 strong, erect, longer than half the length of T4; katepimeron (barette) bare, occasionally with 1-2 short fine hairs; palpi yellowish to black; ♂: frons 0.165-0.188 times head-width; ♀: abdominal dusting rather strong, divided along mid-line according to angle of view; fronto-orbital plates about half as wide as frontal vitta

32. *regalis* (Meigen)

– T5 without strong discal setae, with long hairs of rather uniform length only, shorter hairs longer than hairs covering disc of T4; median marginal setae on T3 weak, more or less adpressed, shorter than half the length of T4; katepimeron with 3-6 short fine hairs; palpi greyish-brown; ♂: frons 0.077-0.098 times head-width; ♀: abdominal dusting very weak, abdomen shining; fronto-orbital plates much less than half as wide as frontal vitta

33. *richardsi* Collin

5(2) Subcostal sclerite with microscopic pubescence only, without black setulae (Fig. 9, upper); T5 with discal setae; T1+2 black, contrasting strongly with green colour of T3-T5; palpi greyish-brown to black; ♂: frons broader than parafacialia; ♀: basal dorsal excavation of abdomen broadly separated from hind margin of T1+2

6

– Subcostal sclerite with small black setulae near apex in addition to microscopic pubescence (Fig. 9, lower); T5 discal setae hardly differentiated (♂) or absent (♀); T1+2 shining green, sometimes dark but not contrasting strongly with colour of T3-5; palpi yellow; 2 *post acr*; ♂: median marginal setae of T3 generally somewhat more erect than paramedian ones, though similar in size; ♀: median marginal setae of T3 rather weak, not different from the paramedian ones in size or direction, adpressed, shorter than half the length of T4; basal dorsal excavation of abdomen narrowly but distinctly separated from hind margin of T1+2

8

6(5) Arista with short hairs, bare on outer third or half, somewhat swollen at base (Fig. 411); 0+2(3) *ia*; 3 *post acr*; t_1 with 2 *pv*; palpi greyish-brown; first flagellomere more than half as long as greatest length of eye viewed in profile; katepimeron bare; marginal setae on T3 variable; ♂: 0.136-0.159 times head-width; face strongly protruding; ♀: first antennal flagellomere very large

30. *magnicornis* (Siebke)

– Arista normal, with long hairs (Fig. 466); 1+2 *ia*; t_1 with 1 *pv*; 2-4 median marginal setae on T3 strong, strikingly different from the paramedian ones, erect, as long as or longer than half the length of T4; first flagellomere half as long as greatest length of eye viewed in profile or less

7

7(6) 3 *post acr*; palpi brown, black on outer third or half; katepimeron usually bare, occasionally with a single hair; distance in front of suture between *acr* rows of setae equal to distance between *acr* and *dc* rows; ♂: frons 0.076-0.107 times head-width; head at lunula not protruding; most anterior frontal seta as distant from eye margin as from ptilinal suture, or slightly closer to the latter; ♀: frons about as wide as an eye, when seen from above, 0.327-0.363 times head-width; epiproct without microtrichiae, cerci normally shaped (Fig. 476)

35. *silvarum* (Meigen)

– 2 *post acr*; palpi brown; katepimeron with 0-5 short fine hairs; distance in front of suture between *acr* rows of setae distinctly less than distance between *acr* and *dc* rows; ♂: frons

0.094-0.109 times head-width; head at lunula somewhat protruding, most anterior frontal seta twice as distant from eye margin as from ptilinal suture; ♀: frons narrower than an eye, when seen from above, 0.258-0.299 times head-width; epiproct densely microtrichiose, cerci narrow (Fig. 396)

28. *bufonivora* Moniez

8(5) ♂: frons distinctly narrower than parafacialia, 0.025-0.051 times head-width; epandrium large, swollen, shining, broader than length of T5 at mid-line; ♀: T6 with short marginal setae at mid-dorsal line and posteroventrally on each side, middle section of hind margin appearing bare on each side (Fig. 407, 410); mid-dorsal edge in dried specimens more or less convex when seen in profile (Fig. 407)

29. *caesar* (Linnaeus)

– ♂: frons about as wide as parafacialia, 0.054-0.093 times head-width; epandrium of normal shape and size, narrower than length of T5 at mid-line; ♀: T6 with long marginal setae along the whole hind margin, no bare sections (Fig. 408, Fig. 432); mid-dorsal edge in dried specimens more or less straight when seen in profile (Fig. 408)

31. *illustris* (Meigen)

27. **Lucilia ampullacea** Villeneuve, 1922
Figs 35, 368, 377-386.

Lucilia flavipennis Kramer, 1917: 283. Junior primary homonym of *Lucilia flavipennis* Macquart, 1843: 296.
Lucilia ampullacea Villeneuve, 1922: 515.
Lucilia krameri Séguy, 1925b: 94. Replacement name for *Lucilia flavipennis* Kramer, 1917.
Lucilia laoshanensis Quo, 1952: 116, 118.

♂♀. Palpi yellow. First flagellomere 4-5 times as long as wide. 2 *post acr*. Occiput concave (especially in male). 3 *h* in a gently curved line. 2+1 *ph* setae. 1+2 *ia*. Basicosta black. Subcostal sclerite yellow with black setulae. t_1 with 1 *pv* seta. t_2 with 1 *ad* seta. t_3 with 0-3 *av* setae. Marginal setae on T3 weak and decumbent, those on T4 strong and erect. T5 shorter than T4.

♂. Frons 0.035-0.040 times head-width (n=5). f_2 without *a* preapical seta. Excavation in anterior end of abdomen reaching margin of T1+2. Postabdomen and ST5 of normal size. Cerci short and straight. Surstyli black but basally with yellow colour and somewhat swollen, short and blunt apically. Long and dense vestiture on surstylus and anterior surface of epandrium (Figs. 378, 379). Bacilliform sclerites elongate, separated. Membrane between them with short hairs ventrally and long setae dorsally. Aedeagus of *ampullacea* type. Pregonite with 3 setae. Postgonite with basal seta.

♀. Frons 0.284-0.319 times head-width (n=5). Frontal vitta 2 times as wide as fronto-orbital plates. f_2 with *a* preapical seta. Excavation in anterior end of abdomen narrowly separated from hind-margin of T1+2. In ovipositor T6 and T7 with a full row of rather long marginal setae. ST6 short. ST7 broad. Intersegmental membrane 7-8 with long microtrichiae. Pleural membrane 8 without microtrichiae in basal half. [3 ovipositor slides examined.]

Length. 5-11 mm.

Distribution. Widespread and apparently common in Denmark and southernmost Sweden. Absent from Norway and Finland. - Throughout Palaearctic, also in Oriental and Australian Regions.

Biology. Attracted to the stinkhorn fungus (*Phallus impudicus* Pers.) in Denmark (Nielsen, 1963, 1968). In Denmark involved in a case of myiasis in the hedgehog (*Erinaceus europeus* L.) (Nielsen, Nielsen & Walhovd, 1978). Adults have

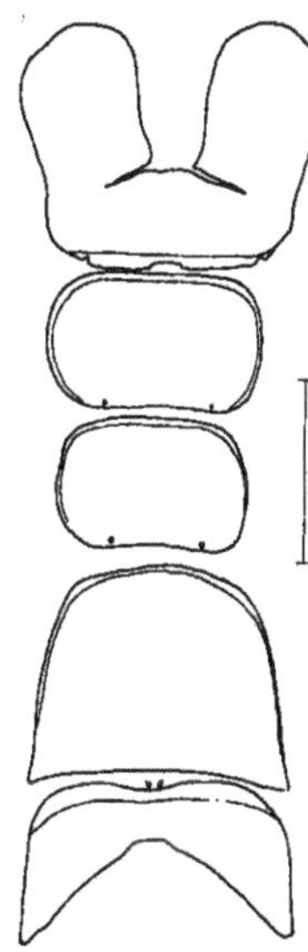

Fig. 377. *Lucilia ampullacea* Villeneuve, male. ST1-5. Scale = 0.8 mm.

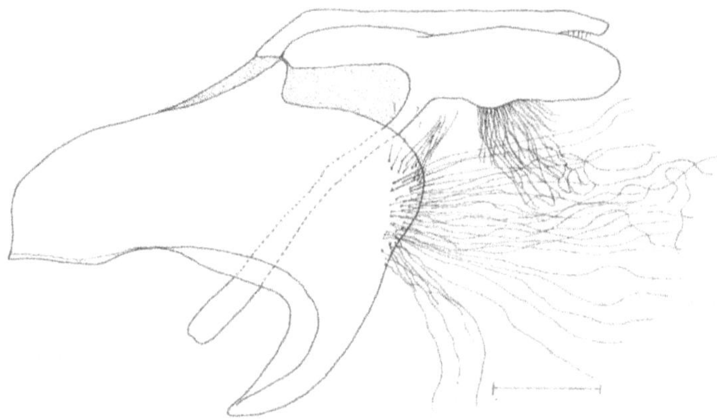

Fig. 378. *Lucilia ampullacea* Villeneuve, male. Epandrium, cerci, surstylus and bacilliform sclerite, left lateral view. Scale = 0.2 mm.

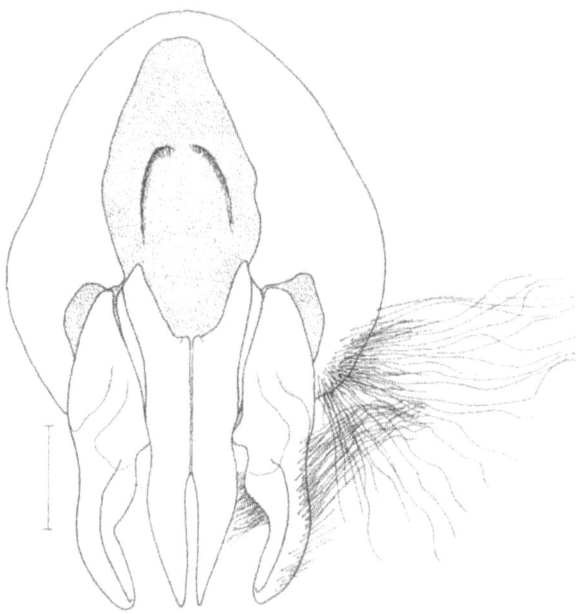

Fig. 379. *Lucilia ampullacea* Villeneuve, male. Epandrium, cerci and surstyli, posterior view. Scale = 0.2 mm.

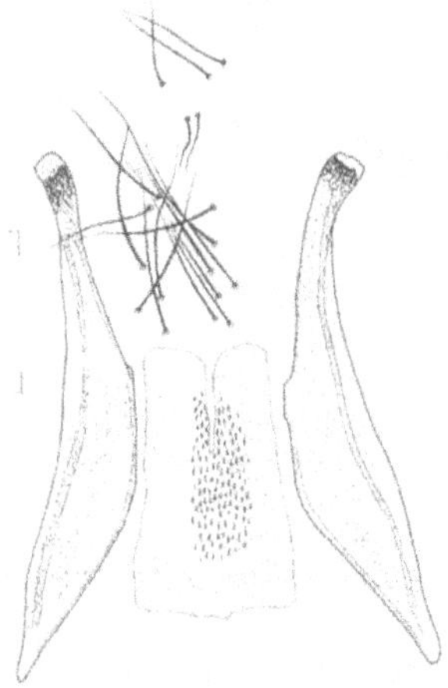

Fig. 380. *Lucilia ampullacea* Villeneuve, male. Bacilliform sclerites and vestiture on membrane between them. Scale = 0.2 mm.

Fig. 381. *Lucilia ampullacea* Villeneuve, male. Aedeagus, left lateral view. Inset: tip of paraphallus (enlarged). Scale = 0.2 mm.

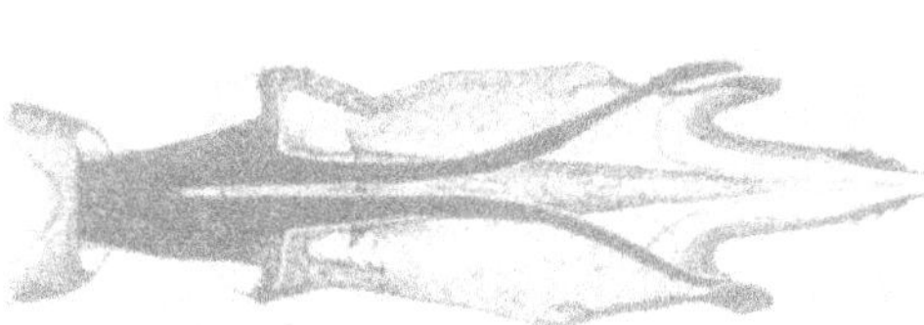

Fig. 382. *Lucilia ampullacea* Villeneuve, male. Distiphallus, dorsal view. (Left paraphallic tip has slipped out of its groove.) Scale = 0.2 mm.

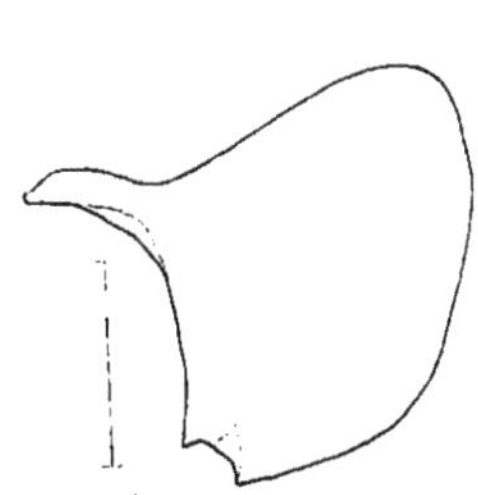

Fig. 383. *Lucilia ampullacea* Villeneuve, male. Ejaculatory sclerite. Scale = 0.2 mm.

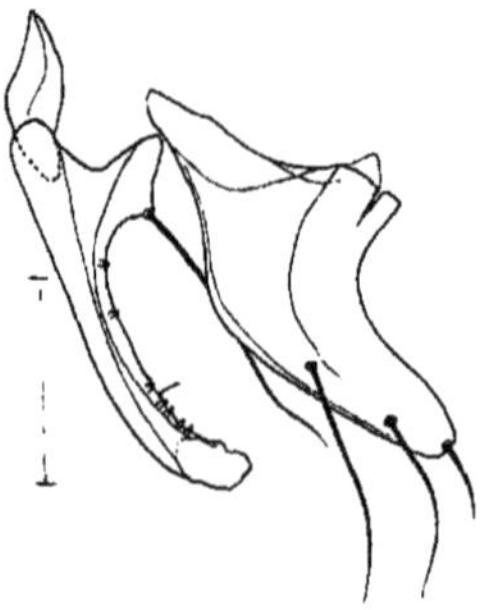

Fig. 384. *Lucilia ampullacea* Villeneuve, male. Post- and pregonites. Scale = 0.2 mm.

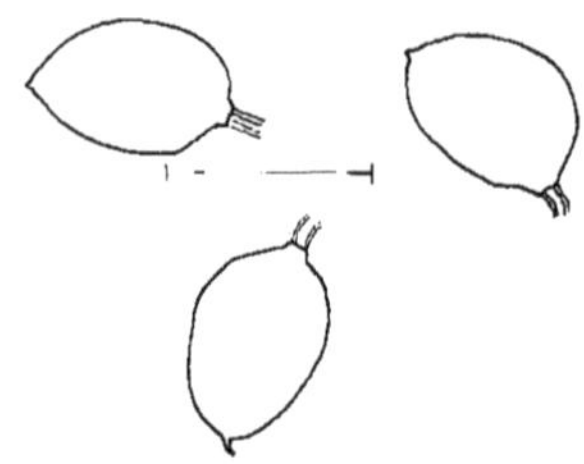

Fig. 385. *Lucilia ampullacea* Villeneuve, female. Spermathecae. Scale = 0.2 mm. G.pr. 188.

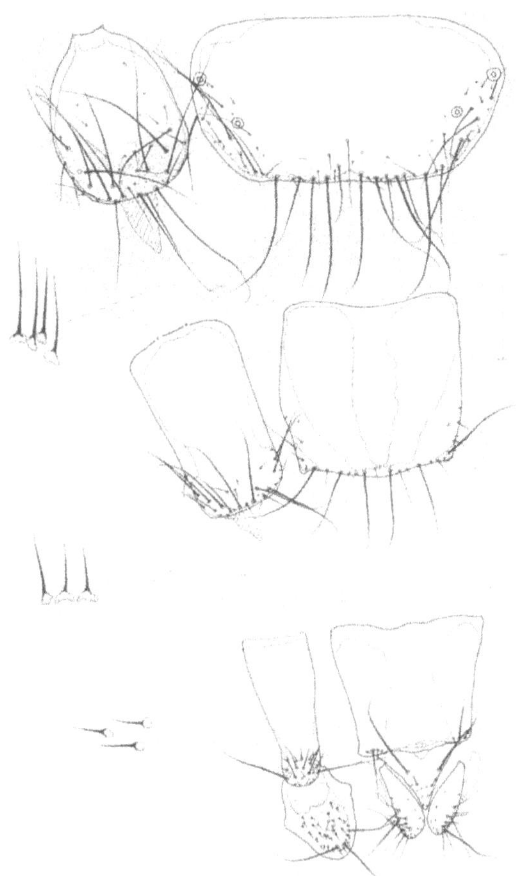

Fig. 386. *Lucilia ampullacea* Villeneuve, female. Ovipositor. Stipple indicates extent of microtrichiae. Insets: shape of individual microtrichiae in areas indicated (enlarged). Scale = 0.4 mm. G.pr. 187.

been collected from May to October, with maximum numbers in August.

28. ***Lucilia bufonivora*** Moniez, 1876

Figs 369, 387-396.

Lucilia bufonivora Moniez, 1876: 25.

♂♀. Lower squama somewhat infuscated. Palpi dark brown, yellowish proximally. First flagellomere 3 times as long as broad. Occiput flat in male, convex in female, mostly with black setae. 3+1(2) *h* setae, basals usually in a rather broken line. 2+1 *ph* setae. Occasional specimens may have 3 *post acr*, thus resembling *silvarum* in this respect. t_3 with 2 *av* setae.

♂. Frons 0.094-0.109 times head-width (n=5). Frontal vitta about as broad as or slightly broader than each fronto-orbital plate. f_2 most often without *a* preapical seta. ST5 normal. Cerci straight, slightly dilated apically in profile. Surstylus parallel-sided in profile, apically blunt. Bacilliform sclerites separated, weakly sclerotised compared with *silvarum*, with median rounded projection. Membrane between them with ventral area of hairs, setae absent. Aedeagus of *bufonivora* type.

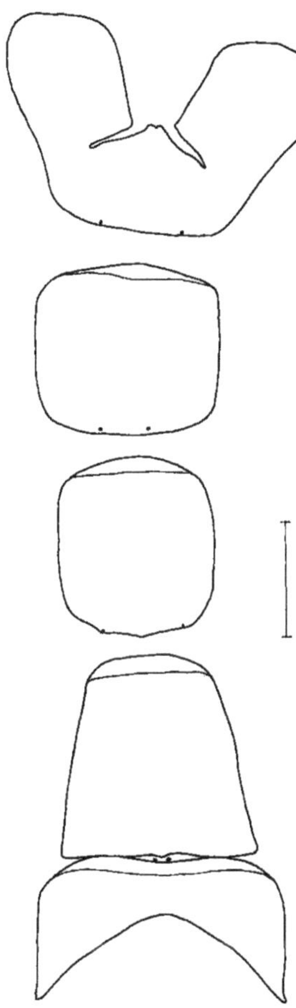

Fig. 387. *Lucilia bufonivora* Moniez, male. ST1-5. Scale = 0.4 mm.

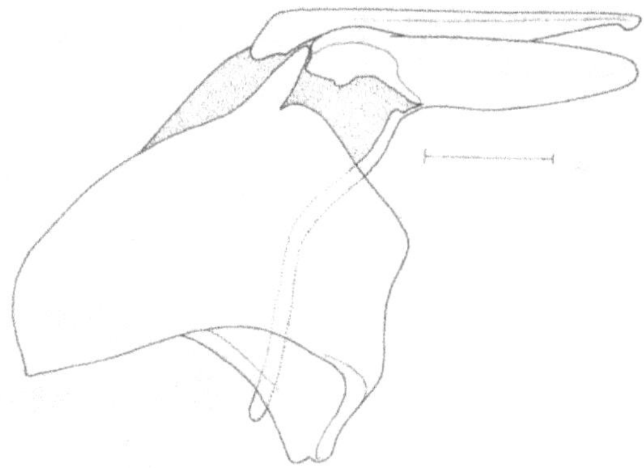

Fig. 388. *Lucilia bufonivora* Moniez, male. Epandrium, cerci, surstylus and bacilliform sclerite, left lateral view. Scale = 0.2 mm.

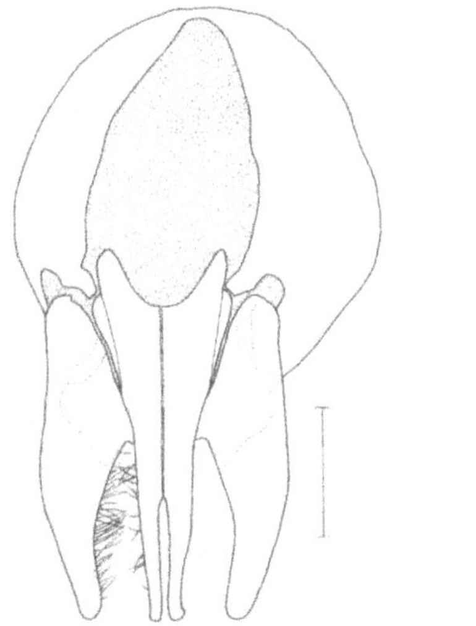

Fig. 389. *Lucilia bufonivora* Moniez, male. Epandrium, cerci and surstyli, posterior view. Scale = 0.2 mm.

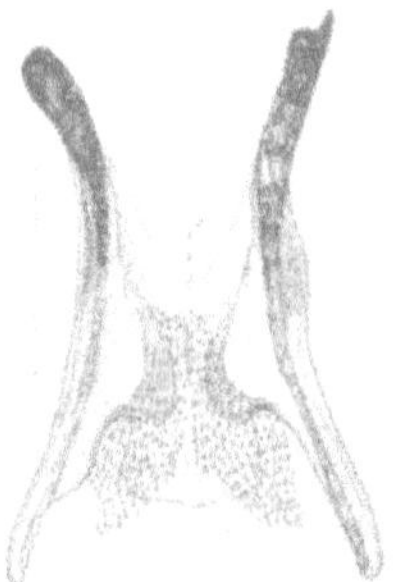

Fig. 390. *Lucilia bufonivora* Moniez, male. Bacilliform sclerites and vestiture on membrane between them. Scale = 0.2 mm.

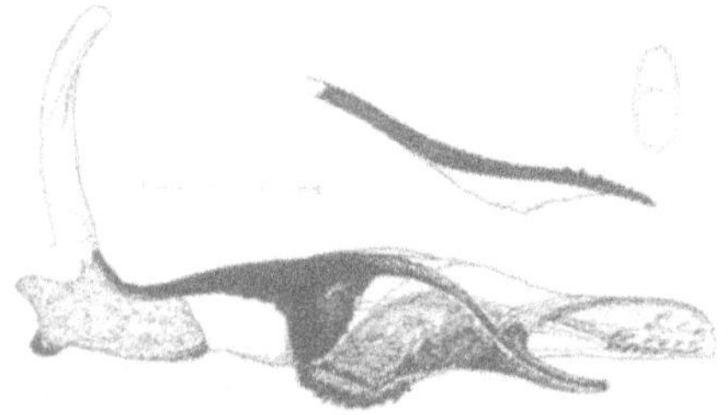

Fig. 391. *Lucilia bufonivora* Moniez, male. Aedeagus, left lateral view. Insets: tip of paraphallus (enlarged), opening of sperm duct with opening of lateral ducts above it. Scale = 0.2 mm.

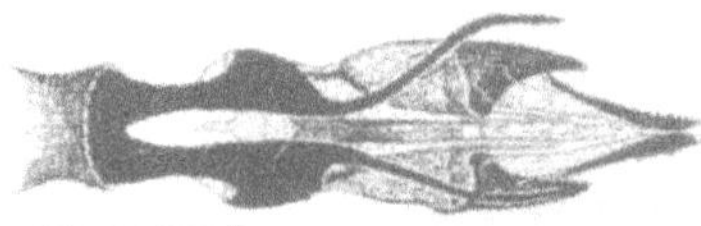

Fig. 392. *Lucilia bufonivora* Moniez, male. Distiphallus, dorsal view. (Left paraphallic tip has slipped out of its canal). Scale = 0.2 mm.

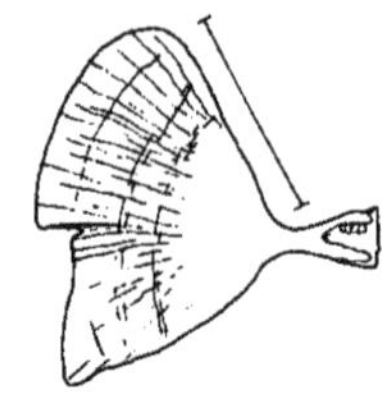

Fig. 393. *Lucilia bufonivora* Moniez, male. Ejaculatory sclerite. Scale= 0.2 mm.

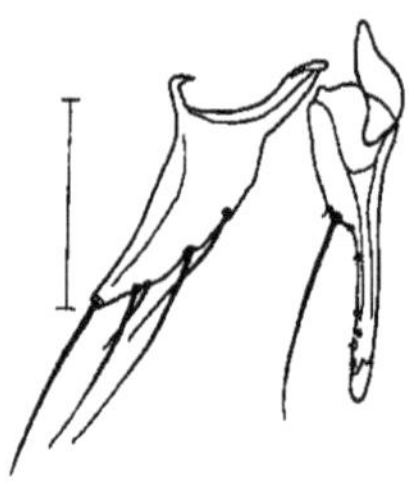

Fig. 394. *Lucilia bufonivora* Moniez, male. Pre- and postgonites. Scale = 0.2 mm.

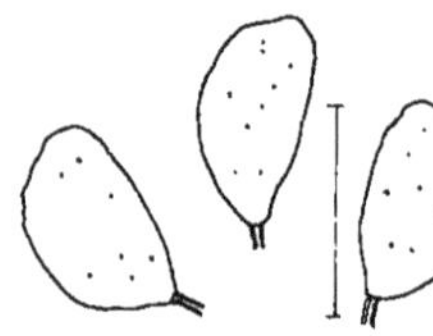

Fig. 395. *Lucilia bufonivora* Moniez, female. Spermathecae. Scale = 0.2 mm. G.pr. 196.

Pregonite with 5 setae. Postgonites with basal seta.

♀. Frons 0.258-0.299 times head-width (n=5). f_2 with *a* preapical seta. All tergites and sternites of ovipositor except T8 at least half covered by microtrichiae. T7 divided completely, halves broad. Epiproct microtrichiose. Cerci strikingly elongate with an irregular outline, without microtrichiae. ST6 and ST7 broad, the latter shaped distally like a broad semi-circle and with numerous setae. ST8 proximally weakly sclerotised. No sclerotisations in the intersegmental membranes. Pleural membrane 8 with a narrow anterior zone without microtrichiae. [3 ovipositor slides examined.]

Length. 5.5-9.5 mm.

Distribution. Rather rare. Scattered occurrences in Denmark, and southern Norway, Sweden and Finland. - Throughout Palaearctic Region, including North Africa and China.

Biology. Obligatory parasite of amphibians (Brumpt, 1934). Eggs are laid on the living host which is subsequently completely destroyed. Numerous specimens from Denmark in ZMC have been bred from the Common Toad (*Bufo bufo* L.) (see also Lundbeck, 1927, who gives further references to Danish reports). A case of parasitism involving the Common Frog (*Rana temporaria* L.) has been reported from Finland (Om: near Haapavesi) by Koskela, Itämies & Pasanen (1974). Adults are on the wing from June to August.

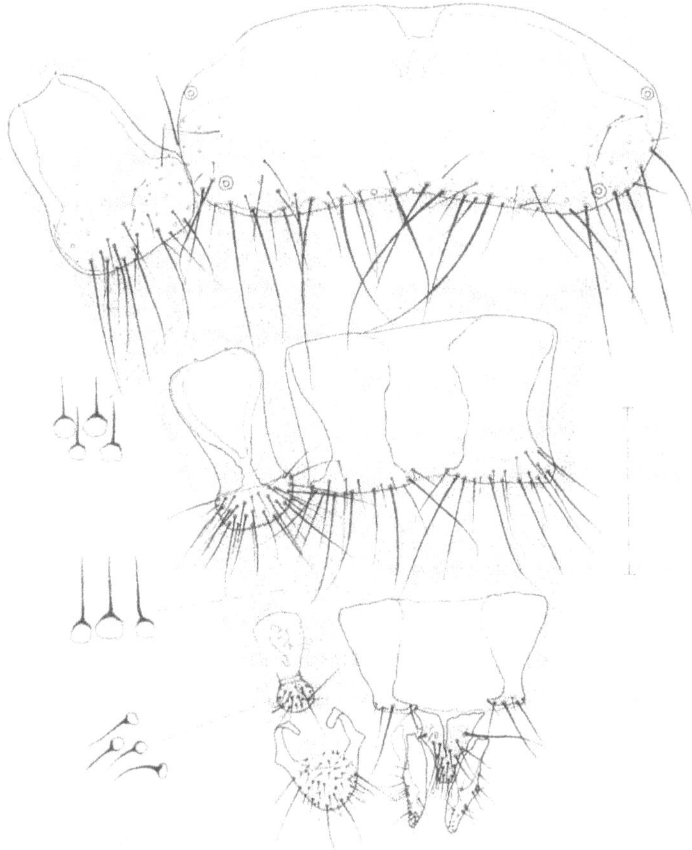

Fig. 396. *Lucilia bufonivora* Moniez, female. Ovipositor. Stipple indicates extent of microtrichiae. Insets: shape of individual microtrichiae in areas indicated (enlarged). Scale = 0.5 mm. G.pr. 196.

Note on synonymy. In the past it has been suggested that *Lucilia elongata* Shannon, 1924, is a synonym of *L. bufonivora* Moniez (Aubertin, 1933; cf. also Zumpt, 1956a; 1965). I have seen the holotype ♀ of *elongata* (in USNM) together with other ♂♀ material, and I find that it is definitely a separate species on several characters, including the shape of the ♂ surstylus and the non-microtrichiose ♀ epiproct.

29. ***Lucilia caesar*** (Linnaeus, 1758)
Figs 6, 9, 366, 370, 397-407, 409, 410; Plate - fig 4.

Musca caesar Linnaeus, 1758: 595.
Musca cadaverina Linnaeus, 1758: 595.
Musca viridicaerulea De Geer, 1776: 62. Unjustified replacement name for *Musca cadaverina* Linnaeus, 1758.
Musca splendida Meigen, 1826: 56.
Lucilia lepida Robineau-Desvoidy, 1830: 453.
Lucilia azurea Robineau-Desvoidy, 1830: 455.
Lucilia albiceps Meigen, 1838: 292.
Somomya jeddensis Bigot, 1877: 255.
Lucilia angustifrons Townsend, 1908: 120.

♂♀. Palpi yellow. First flagellomere about 3 times as long as wide. Occiput concave (especially in male), with about 2-3 irregular rows of black setulae behind postocular row, otherwise with pale hairs. 3 *h* in a weakly curved line. 2+1 *ph*. t_3 with 2 *av*.

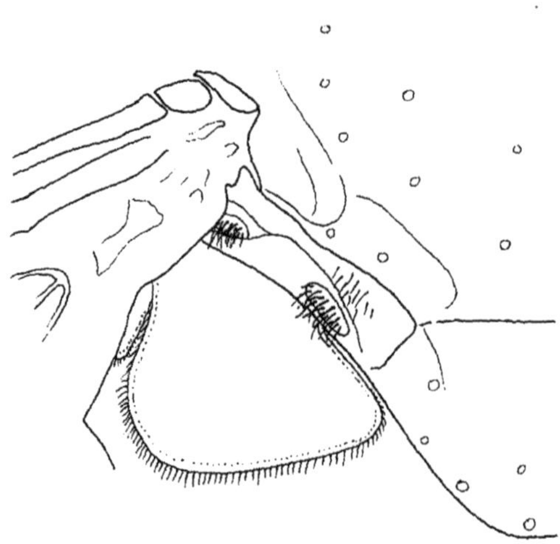

Fig. 397. *Lucilia caesar* (Linnaeus), male. Areas behind wing base, dorsoposterior view.

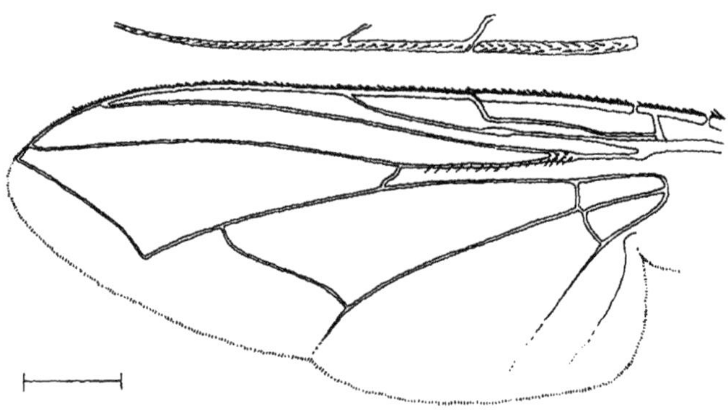

Fig. 398. *Lucilia caesar* (Linnaeus), male. Left wing from above. Inset: costa from below showing distribution of small hairs on its underside. Scale = 1.0 mm.

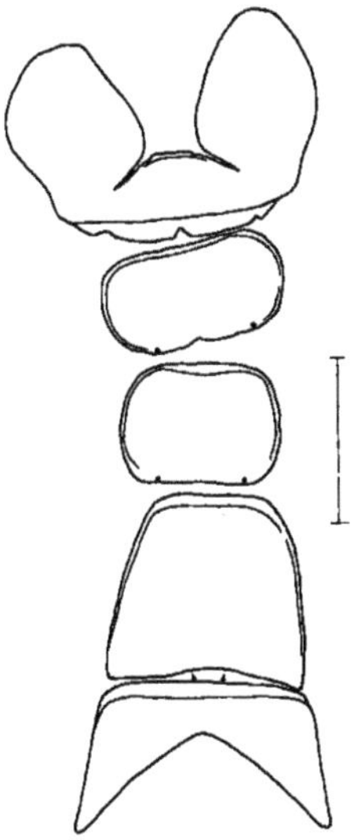

Fig. 399. *Lucilia caesar* (Linnaeus), male. ST1-5. Scale = 0.8 mm.

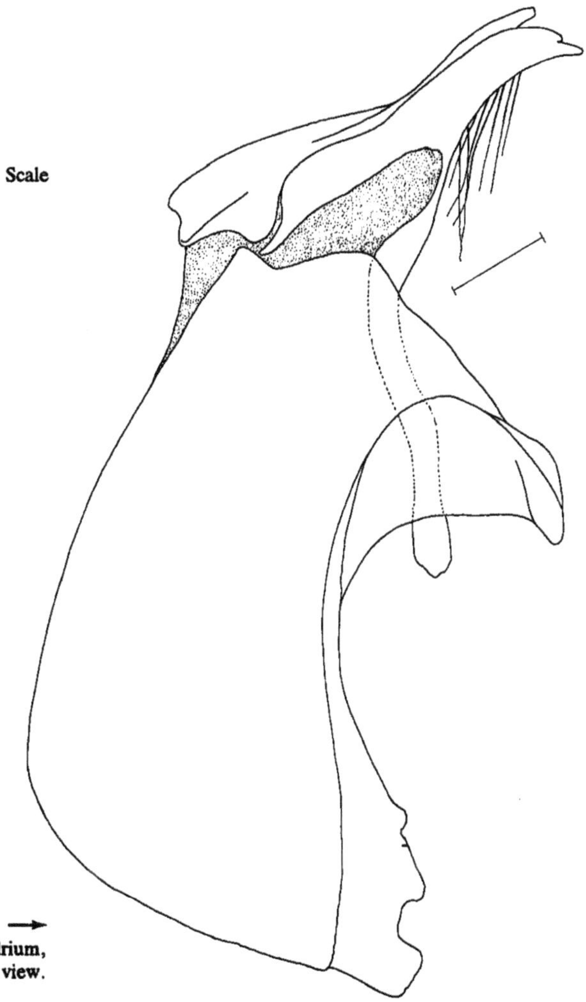

Fig. 400. *Lucilia caesar* (Linnaeus), male. Epandrium, cerci, surstylus and bacilliform sclerite, left lateral view. Scale = 0.2 mm.

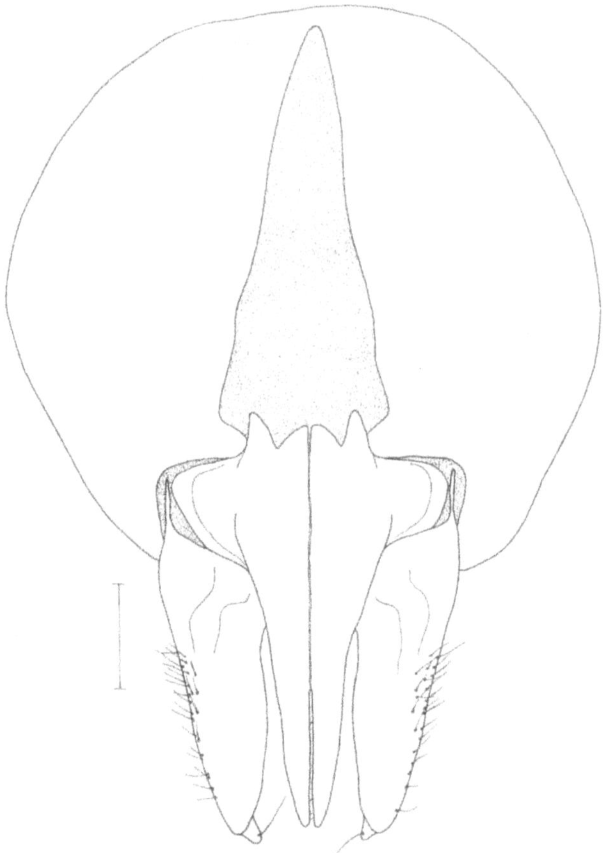

Fig. 401. *Lucilia caesar* (Linnaeus), male. Epandrium, cerci and surstyli, posterior view. Scale = 0.2 mm.

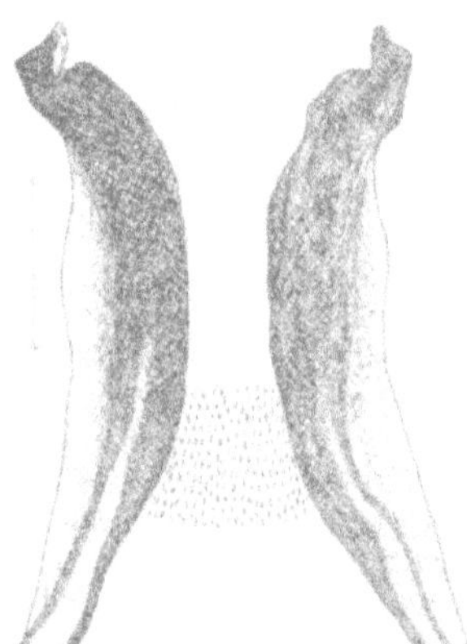

Fig. 402. *Lucilia caesar* (Linnaeus), male. Bacilliform sclerites and vestiture on membrane between them. Scale = 0.2 mm.

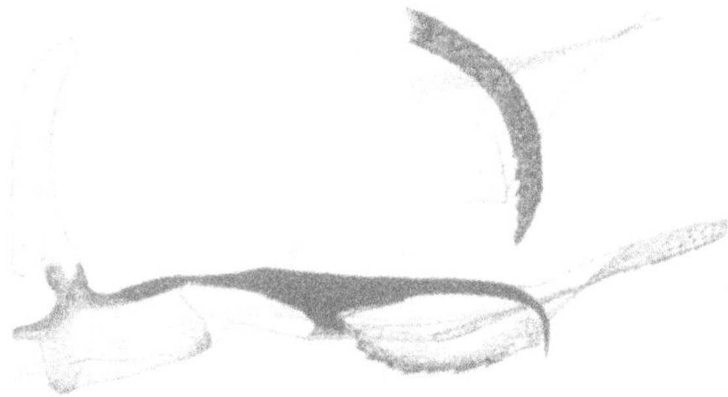

Fig. 403. *Lucilia caesar* (Linnaeus), male. Aedeagus, left lateral view. Inset: tip of paraphallus and neighbouring area (enlarged). Scale = 0.2 mm.

Fig. 404. *Lucilia caesar* (Linnaeus), male. Distiphallus, dorsal view. Scale = 0.2 mm.

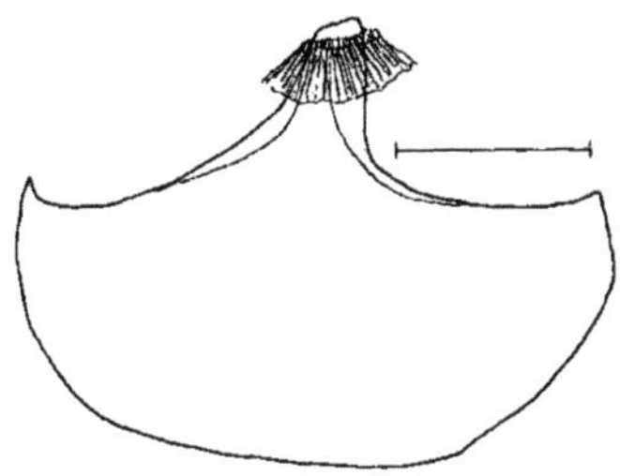

Fig. 405. *Lucilia caesar* (Linnaeus), male. Ejaculatory sclerite. Scale = 0.2 mm.

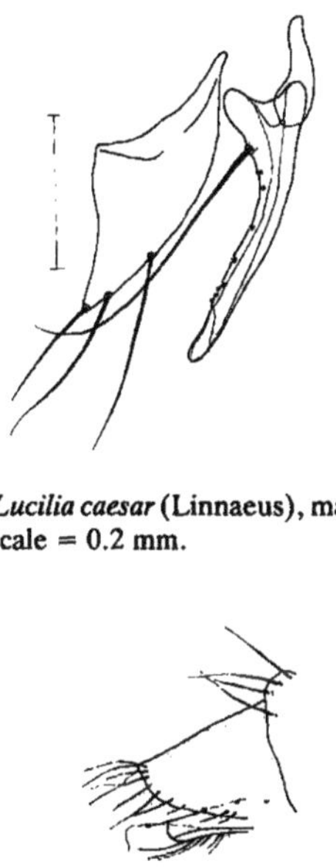

Fig. 406. *Lucilia caesar* (Linnaeus), male. Pre- and postgonites. Scale = 0.2 mm.

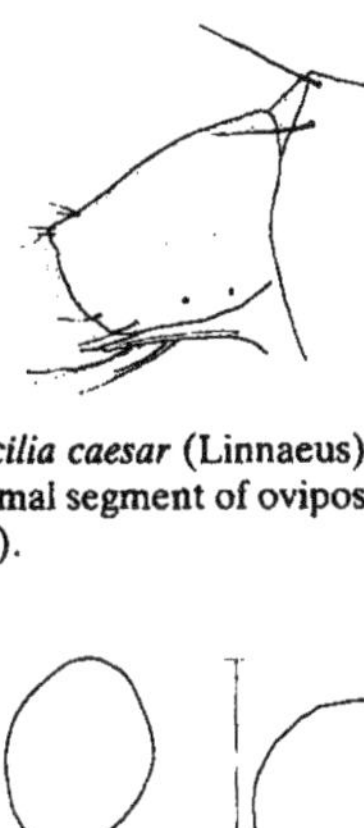

Fig. 407. *Lucilia caesar* (Linnaeus), female, left lateral view of proximal segment of ovipositor. (From Rognes, 1980: fig. 6D).

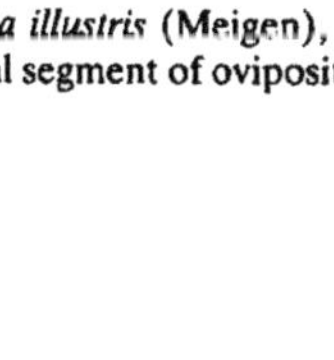

Fig. 408. *Lucilia illustris* (Meigen), female, left lateral view of proximal segment of ovipositor. (From Rognes, 1980: fig. 6H).

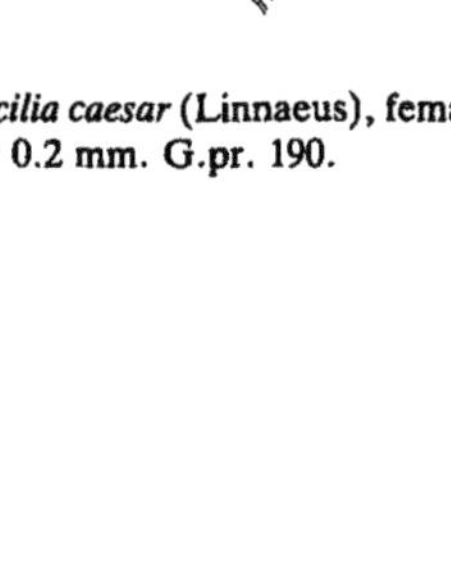

Fig. 409. *Lucilia caesar* (Linnaeus), female. Spermathecae. Scale = 0.2 mm. G.pr. 190.

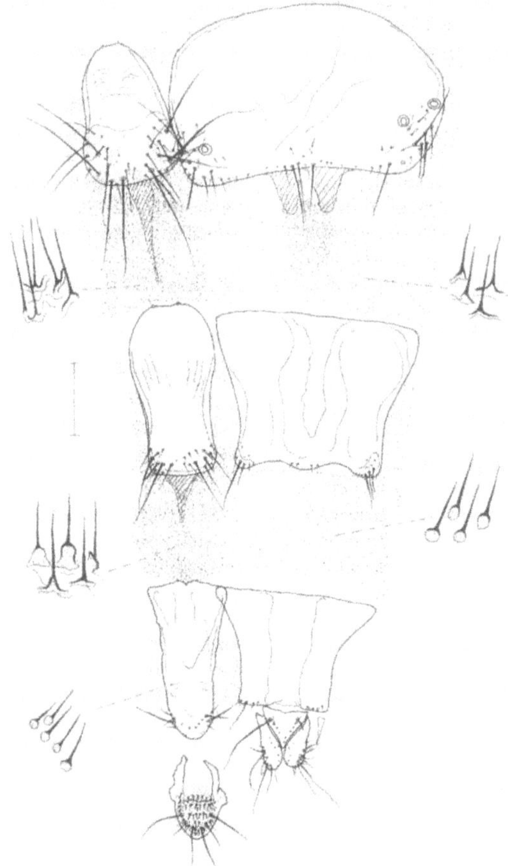

Fig. 410. *Lucilia caesar* (Linnaeus), female. Ovipositor. Stipple indicates extent of microtrichiae. Insets: shape of individual microtrichiae in areas indicated (enlarged). Scale = 0.4 mm. G.pr. 191.

♂. Frons 0.025-0.051 times head-width (n=47). Fronto-orbital plates touching. f_2 without *a* preapical. Epandrium very large. Cerci broad basally in dorsal view, apical parts slightly upturned in profile. Surstylus broad in dorsal view, slightly curved downwards in lateral view, apex bifid. Bacilliform sclerites heavily sclerotised, separated. No setae, only hairs on membrane between them. Aedeagus of *caesar* type. Pregonite with 3 setae. Postgonite with basal seta.

♀. Frons 0.296-0.333 times head-width (n=5). f_2 with *a* preapical. T6 in dried specimens convex in profile. In microscopical preparations T6 long with strong marginal setae near mid-line and near posterolateral corners, setae absent or only very short setulae present in between. T7 with similarly arranged marginals, 2-4 near mid-line and a lateral group on each side, setae absent in middle on each side. Posterior part of ST6 microtrichiose, otherwise ovipositor tergites and sternites without microtrichiae. ST7 broad. ST8 very long, almost 1.5 times as long as T8, apically rounded with a subapical transverse row of rather strong setae. Intersegmental membrane 6-7 with a double sclerotisation just behind margin of T6, a long narrow one behind ST6. A short similar prolongation behind ST7. [3 ovipositor slides examined.]

Length. 5.5-11.5 mm.

Distribution. Very common in Fennoscandia and Denmark. All over Denmark. In Norway north to NTI, in Sweden north to Dlr. and in Finland north to Sb, Kb, Om. In Kr (U.S.S.R.) north to Ahvenjärvi (formerly in Karelia pomoriensis occidentalis) (material in ZMH). Ascends to subalpine region in southern Norway. - Throughout Palaearctic Region (Schumann, 1986).

Biology. Less heliophilic than *L. illustris* (Nuorteva, 1964b). Moderately synanthropic in south Finland and more so than *L. illustris*, presumably because of its closer proximity to the northern limit of its range (Nuorteva, 1963). Attracted to the stinkhorn fungus (*Phallus impudicus* Pers.) in Denmark (Nielsen, 1963, 1968). Taken in copula in Sweden (Sk.) on May 28 and June 10 (O.Ringdahl leg.). Involved in cases of sheepstrike in Norway and Denmark (Brinkmann, 1976; Nielsen, 1984a, 1984b). In Norway probably acting as primary striker and the most important fly involved in sheep-strike. Involved in a case of myiasis in the hedgehog (*Erinaceus europaeus* L.) in Denmark (Nielsen, Nielsen & Walhovd, 1978). In Norway a larva was recovered from a cancerous wound in the external auditory meatus of a human and was reared to the adult stage (R.Mehl, pers. comm.; K.R. det.). Adults are on the wing from May to October, but single specimens have been collected in Denmark in March and April. Maximum numbers occur in June and July.

Note. The number of hairs on the underside of the arista cannot be used to distinguish this species with confidence from *L. illustris* (see Rognes, 1980).

Notes on synonymy. *Musca splendida* was described by Meigen from ♂♀ syntypes, with neither collector nor type locality given [probably West Germany: Stolberg]. Meigen Plate 120, fig. 3 (Morge, 1976a) figures both sexes. In Meigen's collection in MNHN are 1♂1♀ under this name, both labelled "Meigen" and "2111/40". The male is in good condition and is also labelled "splendida/♂ [Meigen's hand]". It is definitely the present species, and I have labelled it and here designate it as lectotype of *Musca splendida* Meigen, to conform with the synonymy suggested by Collin (1926) and established by Aubertin (1933). The male figured on Plate 120 has a rather broad frons for a *Lucilia caesar* and is possibly based on another specimen. The female is mounted very high on the pin and has lost the front and mid left legs. It bears a label in Meigen's hand reading "splendida/♀/Löfmann [?, difficult to read]". I have labelled it as paralectotype. It probably belongs to *silvarum* Meigen. *Lucilia albiceps* was described by Meigen on the basis of female(s) from "Hiesige Gegend" [West Germany, Stolberg]; neither collector nor number of specimens was indicated. Meigen Plate 299, fig. 13 (Morge, 1976b) illustrates a ♀ named "Musca albiceps" in the legends (Morge, 1975: 459, 497). In Meigen's collection in MNHN is a female labelled (1) "Meigen" and "2107/40", (2) "Lucilia / albiceps / ♀ / ... [illegible name] [in Meigen's hand]". It has the T6 of the ovipositor fully exposed. It belongs to the present species. I have attached a holotype label.

30. ***Lucilia magnicornis*** (Siebke, 1863), COMB.N.

Figs 12-17, 20, 21, 367, 371, 411-422.

Sarcophaga fuscipalpis Zetterstedt, 1845: 1306. Junior secondary homonym of *Lucilia fuscipalpis* Macquart, 1834: 28 and 1835: 253.
Sarcophaga magnicornis Siebke, 1863: 160.
Francilia alaskensis Shannon, 1924: 74.

♂♀. Palpi greyish-brown. First flagellomere broad, about 2.5 times as long as wide. Occiput strongly convex, with black setulae only behind postocular row. 3+0-1 *h*, basal humeral setae in a strongly broken line, as if at the corners of a right-angled triangle. 2+1 *ph*. f_2 with *a* preapical seta in both sexes. t_3 with 2 *av*.

♂. Frons 0.136-0.159 times head-width (n=5). Frontal vitta as wide as each fronto-orbital plate. T3 usually with a full row of rather strong marginal setae, at least half as long as T4 or longer. Abdomen undusted, with sparse erect ground-vestiture. Discal setae present on T5 and T4. ST5 with large projecting lobes. Cerci very long and narrow. Bacilliform sclerites adjoining, long, well sclerotised. Aedeagus of *magnicornis* type. Pregonite with 5 setae. Postgonite without basal seta.

♀. Frons 0.362-0.395 times head-width (n=5). T3 marginal setae variable, sometimes strong medially and laterally, sometimes almost undifferentiated. Abdominal dusting conspicuous. Ground-vestiture decumbent on T3 and T4, erect on T5. T6 with a posterior incision of weak sclerotisation which sometimes almost "divides" the sclerite. T7 and T8 halves weakly sclerotised posteriorly, both with short marginal setae. ST7 rather narrow. Hypoproct microtrichiose in apical half only, basal half quite bare, lacking both setae and microtrichiae, which is unique. [2 ovipositor slides examined.]

Length. 5-8.5 mm.

Distribution. Subalpine and low alpine regions. Widespread in Norway, also collected in the lowlands in southeastern Norway (100 m a.s.l. TEY, Drangedal). Northernmost Finland (Li and Le).

Fig. 411. *Lucilia magnicornis* (Siebke). Arista, male (left) and female (right). (From Rognes, 1980: figs. 5A,B).

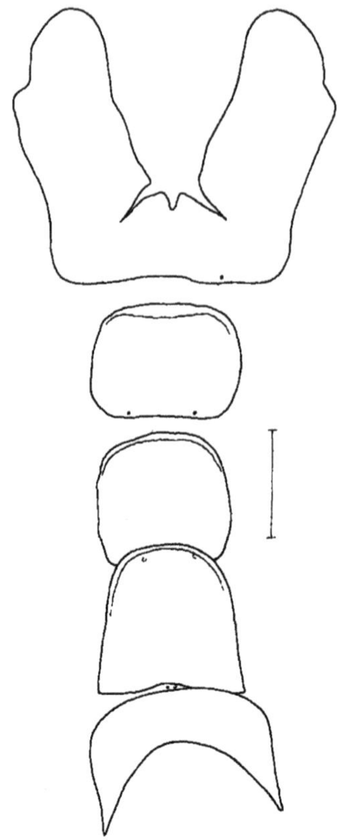

Fig. 412. *Lucilia magnicornis* (Siebke), male. ST1-5. Scale = 0.4 mm.

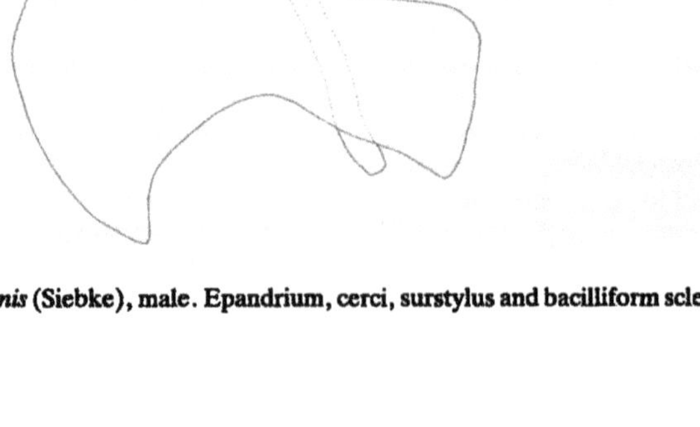

Fig. 413. *Lucilia magnicornis* (Siebke), male. Epandrium, cerci, surstylus and bacilliform sclerite, left lateral view. Scale = 0.2 mm.

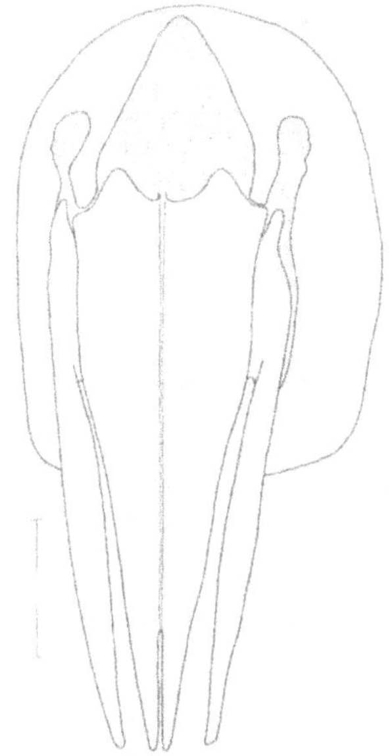

Fig. 414. *Lucilia magnicornis* (Siebke), male. Epandrium, cerci and surstyli, posterior view. Scale = 0.2 mm.

Fig. 415. *Lucilia magnicornis* (Siebke), male. Bacilliform sclerites. Scale = 0.2 mm.

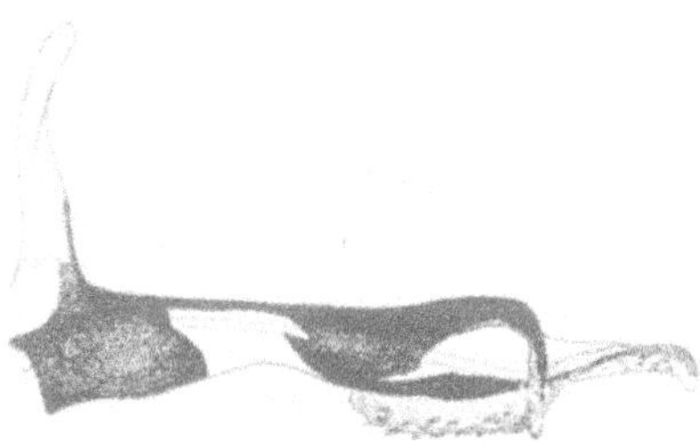

Fig. 416. *Lucilia magnicornis* (Siebke), male. Aedeagus. left lateral view. Scale = 0.2 mm.

Fig. 417. *Lucilia magnicornis* (Siebke), male. Aedeagus, tip of paraphallus and neighbouring area, left lateral view (enlarged).

Fig. 418. *Lucilia magnicornis* (Siebke), male. Distiphallus, dorsal view. Scale = 0.2 mm.

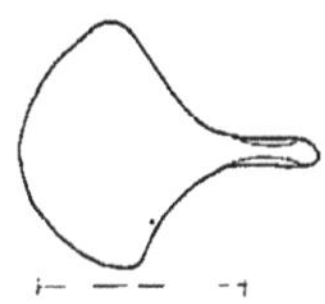

Fig. 419. *Lucilia magnicornis* (Siebke), male. Ejaculatory sclerite. Scale = 0.2 mm.

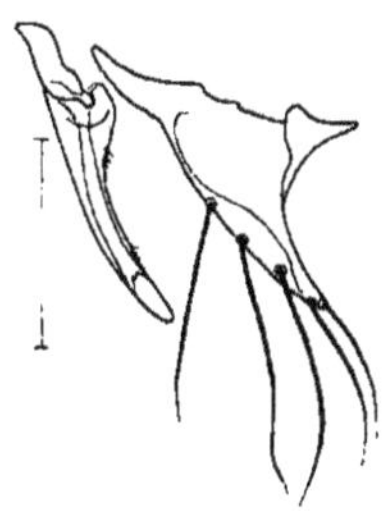

Fig. 420. *Lucilia magnicornis* (Siebke), male. Post- and pregonites. Scale = 0.2 mm.

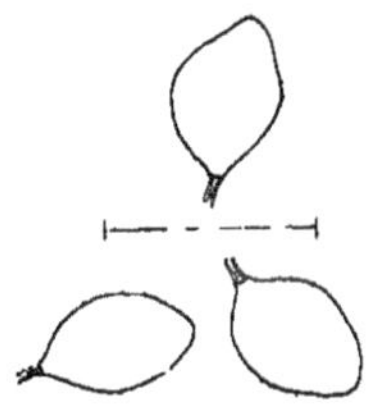

Fig. 421. *Lucilia magnicornis* (Siebke), female. Spermathecae. Scale = 0.2 mm. G.pr. 192.

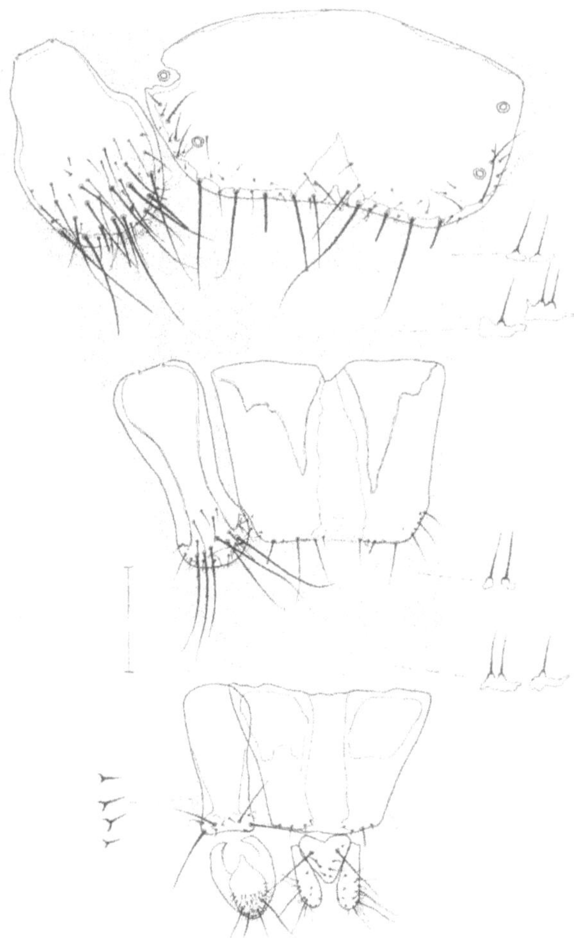

Fig. 422. *Lucilia magnicornis* (Siebke), female. Ovipositor. Stipple indicates extent of microtrichiae. Insets: shape of individual microtrichiae in areas indicated (enlarged). Scale = 0.4 mm. G.pr. 192.

Mountainous districts in Sweden from Dlr. north to T.Lpm. Absent from Denmark. - Palaearctic: apart from Fennoscandia also in Poland (Pienin) (Draber-Mońko, 1978), arctic and subarctic parts of U.S.S.R. from Kola peninsula to Taymyr, also Western Sayan mountains (Buyba) (Chernov, 1959, 1965; Grunin, 1970a). Nearctic: Alaska to Labrador (Hall, 1965; Danks, 1981).

Biology. Immature stages have not been described. Nothing is known about its life-cycle in Fennoscandia. Bred from carcasses of Middendorff's vole in the virgin tundra of East Siberia (Chernov, 1965). According to Nuorteva (1963) it is a purely asynanthropic species in Finland. Adults are on the wing from June to August, with a maximum in July.

Nomenclature. *Sarcophaga fuscipalpis* was described from 3♂ 3♀ from "... Ristansfors ... Mullfjellen ... Skalstugan ... [Jmt.] ... Suul [NTI] ...", collected on Zetterstedt's third Lapland journey in 1840. 2♂2♀ syntypes have been located in the Diptera Scandinavica collection (drawer 24) in ZML, and 1♀ in the Wallengren collection (drawer IV. 13). I have labelled and here designate as lectotype a ♂ labelled "Mullfjell" in Zetterstedt's hand and already dissected by Zumpt. 1♂3♀

which all bear a square green tag indicating capture on the 1840 journey to Lapland have been labelled as paralectotypes. The identity of the holotype ♀ of *Sarcophaga magnicornis* Siebke has been reported by Rognes (1983c).

31. ***Lucilia illustris*** (Meigen, 1826)
Figs 372, 408, 423-432.

Musca illustris Meigen, 1826: 54.
Musca parvula Meigen, 1826: 55.
Musca equestris Meigen, 1826: 57.
Phaenicia azurea Robineau-Desvoidy, 1863: 788.
Calliphora simulatrix Pandellé, 1896: 218.

♂♀. First flagellomere about 3 times as long as wide. Occiput concave (especialy in male) with mostly black setulae. 3+0-1 *h*, basal humerals in a weakly curved line. 2+1 *ph*. t_3 with 2 *av* setae.

♂. Frons 0.054-0.093 times head-width (n=39). Fronto-orbital plates touching or almost so. f_2 without *a* preapical seta. Abdominal dusting almost absent. Lobes of ST5 and epandrium of normal size. Cerci narrow, in profile bent backwards in distal half, hook-shaped apically. In dorsal view apical branches diverging, in dried specimens branches more diverging than shown in Fig. 425. Surstylus curved, distally hook-shaped in profile. Bacilliform sclerites separated, membrane between them only with hairs in ventral part, setae absent. Aedeagus of *caesar* type. Postgonite with 3 setae. Pregonite with basal seta.

♀. Frons 0.335-0.386 times head-width (n=5). Frontal vitta more than 2 times as broad as each fronto-orbital plate. f_2 with *a* preapical seta. Abdominal dusting very weak. Mid-dorsal edge of T6 in dried specimens straight when seen in profile. T6 with full row of marginal setae. T7 divided or almost so, with a full row of marginal setae, but these are short. ST6 and ST7 with small posterior prolongations into the intersegmental membranes behind them. [3 ovipositor slides examined.]

Length. 5.5-11 mm.

Distribution. Very common. All over Fennoscandia and Denmark. Ascends to subalpine region. - Palaearctic, Nearctic, Oriental and Australian Regions.

Biology. More heliophilic than *L. caesar* (Nuorteva, 1964b). In the northern parts of Finland it is strongly synanthropic, whereas it is almost independent of the conditions created by man in the south (Nuorteva, 1963). Reported from a case of myiasis in the hedgehog (*Erinaceus europaeus* L.) (Nielsen, Nielsen & Walhovd, 1978). Involved in sheep-strike in Denmark and Norway (Brinkmann, 1976; Nielsen, 1984a, 1984b), possibly as a primary striker. In Finland reported from a case of wound myiasis in an old woman (Laitinen, Nuorteva & Renkonen, 1970). Adults have been collected from April to October.

Note. The number of hairs on the underside of the arista cannot be used to distinguish this species with confidence from *L. caesar* (see Rognes, 1980).

Notes on synonymy. *Musca illustris*, *Musca parvula* and *Musca equestris* were all described by Meigen on the basis of specimens "...Von Herrn v. Winthem...", now in Vienna. All species are illustrated in Meigen's Plates (Morge, 1975, 1976a, 1976b). The types were examined by Collin (1926), and I consider him as the first reviser in establishing *illustris* as the valid name for this species. A specimen of *parvula* is also in Meigen's collection in MNHN. It is labelled (1) "Meigen" and "2110/40", (2) "7472", (3) "Lucilia / parvula / ♂ / Löfman [Meigen's hand]". It is obviously a later addition and not a syntype, as correctly pointed out by Collin (1926). It belongs to *sericata* Meigen. In Pandellé's collection in MNHN are 5♂ under *Calliphora simulatrix*. All are labelled "♂/6033" in Pandellé's hand, have the genitalia visible and belong to the present species. I have labelled and here designate one specimen as lectotype; the other 4 have been labelled as paralectotypes.

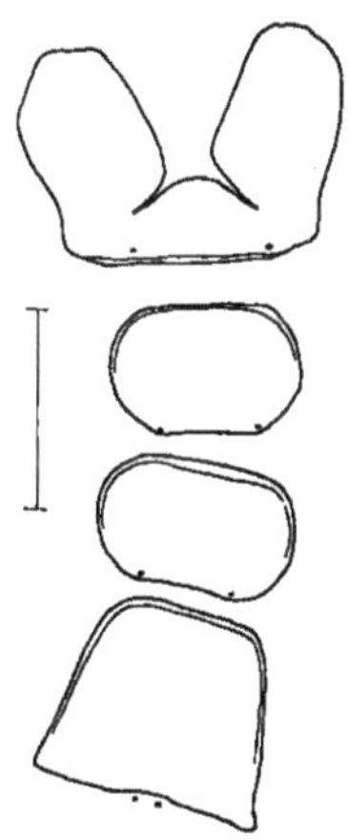

Fig. 423. *Lucilia illustris* (Meigen), male. ST2-5. Scale = 0.8 mm.

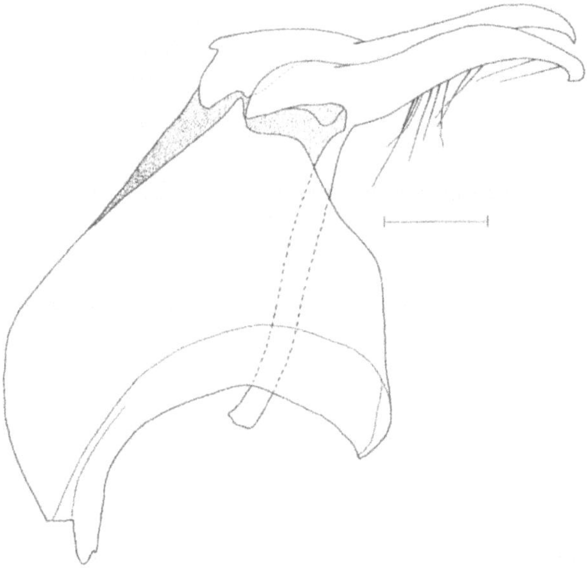

Fig. 424. *Lucilia illustris* (Meigen), male. Epandrium, cerci, surstylus and bacilliform sclerite, left lateral view. Scale = 0.2 mm.

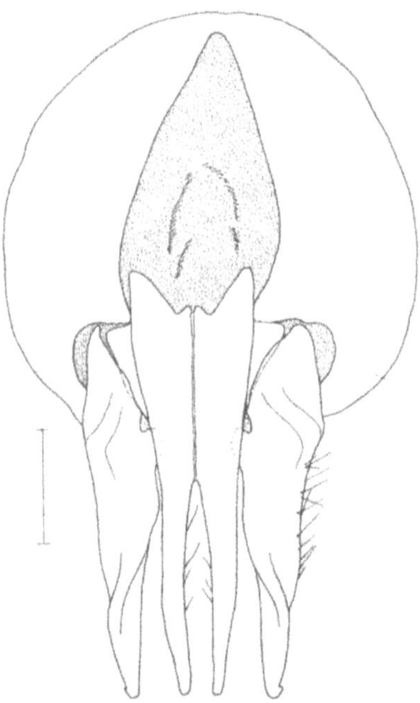

Fig. 425. *Lucilia illustris* (Meigen), male. Epandrium, cerci and surstyli, posterior view. Scale = 0.2 mm.

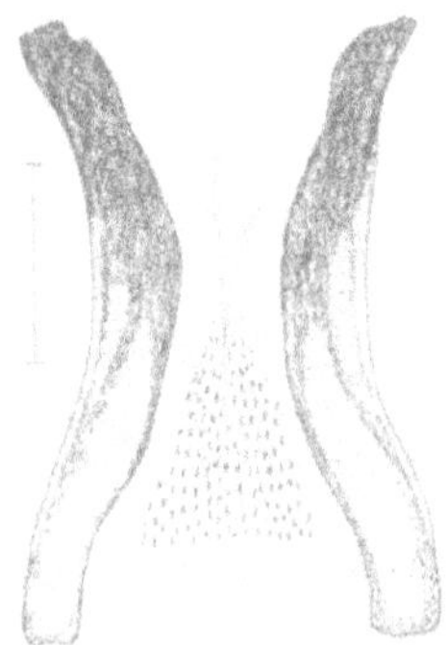

Fig. 426. *Lucilia illustris* (Meigen), male. Bacilliform sclerites and vestiture on membrane between them. Scale = 0.2 mm.

Fig. 427. *Lucilia illustris* (Meigen), male. Aedeagus, left lateral view. Inset: tip of paraphallus and neighbouring area (enlarged). Scale = 0.2 mm.

Fig. 428. *Lucilia illustris* (Meigen), male. Distiphallus, dorsal view. Scale = 0.2 mm.

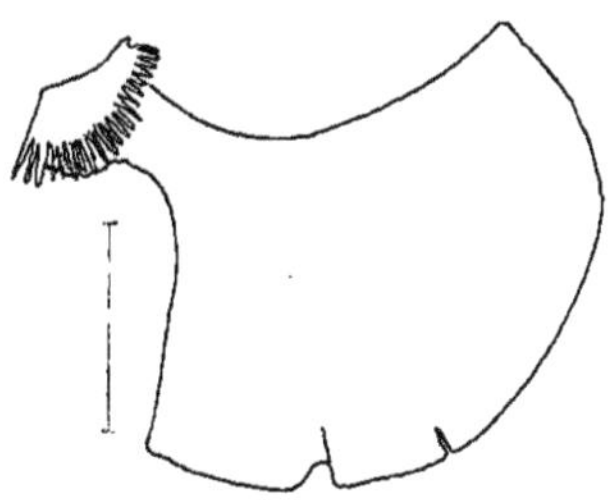

Fig. 429. *Lucilia illustris* (Meigen), male. Ejaculatory sclerite. Scale = 0.2 mm.

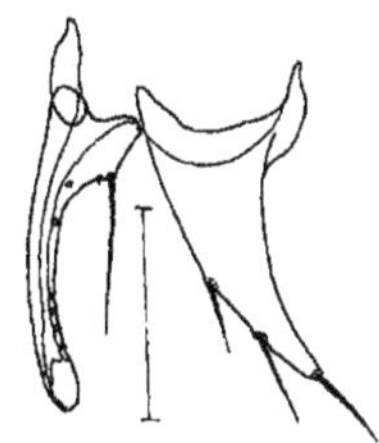

Fig. 430. *Lucilia illustris* (Meigen), male. Post- and pregonites. Scale = 0.2 mm.

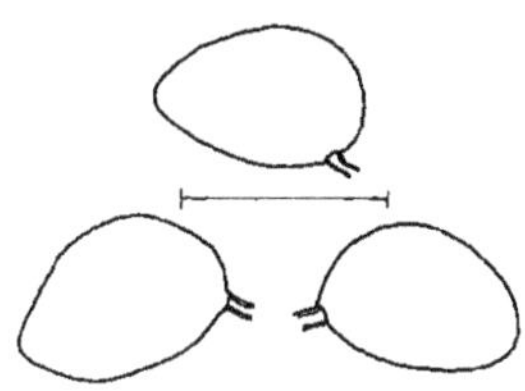

Fig. 431. *Lucilia illustris* (Meigen), female. Spermathecae. Scale = 0.2 mm. G.pr. 105.

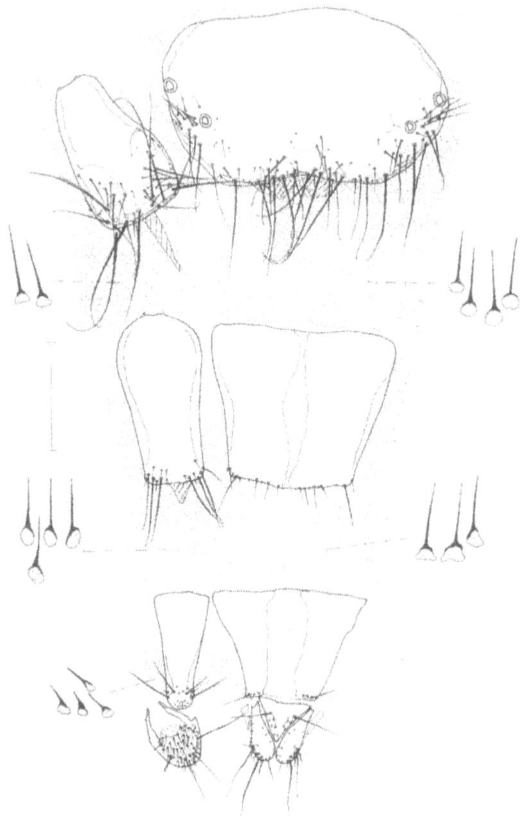

Fig. 432. *Lucilia illustris* (Meigen), female. Ovipositor. Stipple indicates extent of microtrichiae. Insets: shape of individual microtrichiae in areas indicated (enlarged). Scale = 0.6 mm. G.pr. 106.

32. ***Lucilia regalis*** (Meigen, 1826)
Figs 373, 433-443.

Musca regalis Meigen, 1826: 54.
Calliphora longilobata Pandellé, 1896: 219.

♂♀. Palpi blackish-brown. First flagellomere about 3 times as long as wide. Occiput flat in male, slightly convex in female, with black setulae only. 3+0-1 *h*, basal humerals in a weakly broken line. 2+1 *ph*. f_2 with *a* preapical seta in both sexes. t_3 with 0-2 *av* setae. Strong marginal setae on T3 may occasionally be undeveloped in both sexes. Freak specimens may also have only a single *ad* on t_2. These may be separated from *sericata* by the presence of an *a* preapical seta on f_2 (♂) and strong discal setae on T5.

♂. Frons 0.165-0.188 times head-width (n=5). Frontal vitta 2 times as wide as each fronto-orbital plate. Abdominal dusting almost absent. Lobes of ST5 large. Cerci and surstyli very long and narrow. Bacilliform sclerites long and adjoined for a long distance. No setae in membrane between them. Aedeagus of *sericata* type. Acrophallus with no ventral projection in middle in profile. Pregonite with 7 setae. Postgonite without basal seta.

♀. Frons 0.402-0.410 times head-width (n=5). Frontal vitta less than 2 times as wide as fronto-orbital plates, which are rather broad. Abdominal dusting conspicuous, shifting along mid-line according to light. T6 with a full row of marginals, posteriorly a broad band of microtrichiae. T7 almost fully divided. ST6 and ST7 with small posterior prolongations into intersegmental membrane

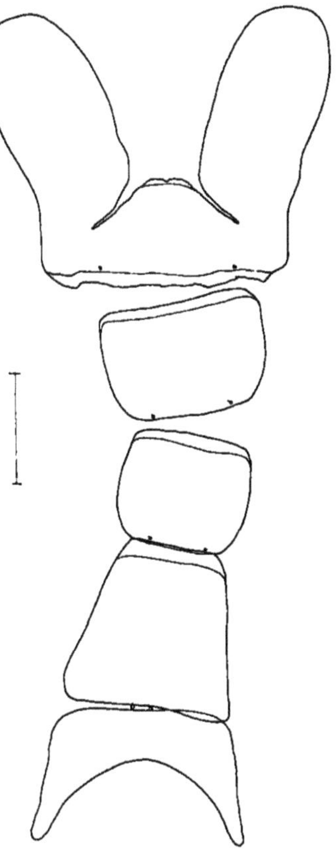

Fig. 433. *Lucilia regalis* (Meigen), male. ST1-5. Scale = 0.4 mm.

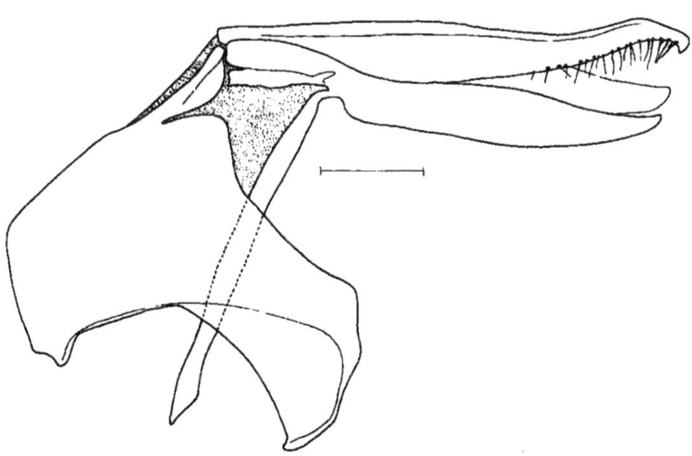

Fig. 434. *Lucilia regalis* (Meigen), male. Epandrium, cerci, surstylus and bacilliform sclerite, left lateral view. Scale = 0.2 mm.

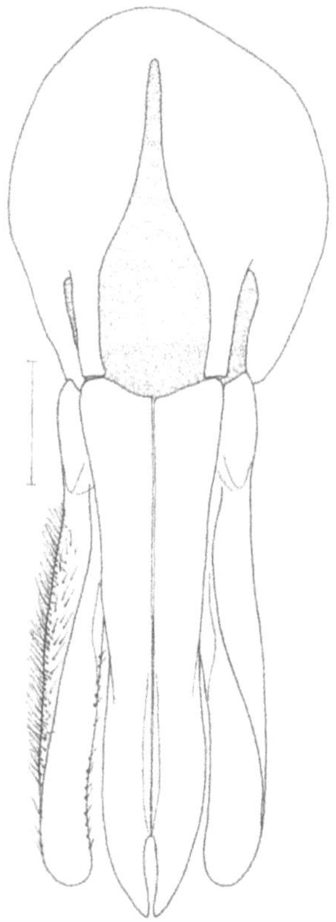

Fig. 435. *Lucilia regalis* (Meigen), male. Epandrium, cerci and surstyli, posterior view. Scale = 0.2 mm.

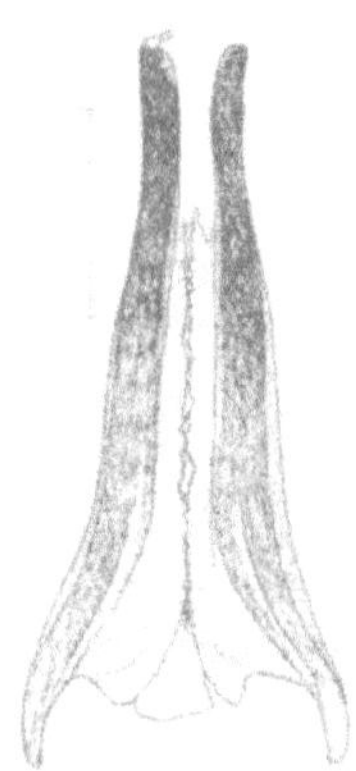

Fig. 436. *Lucilia regalis* (Meigen), male. Bacilliform sclerites. Scale = 0.2 mm.

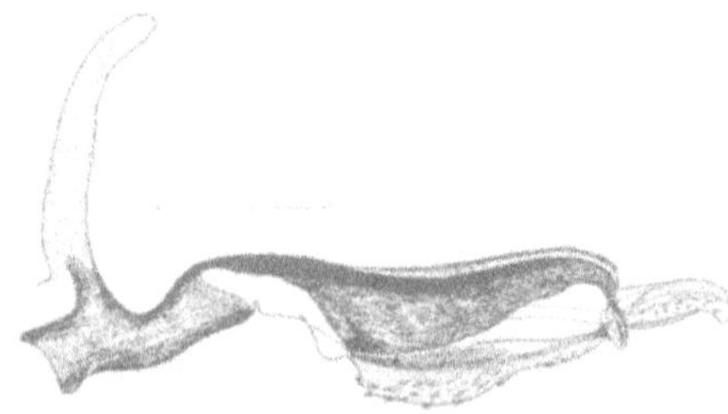

Fig. 437. *Lucilia regalis* (Meigen), male. Aedeagus, left lateral view. Scale = 0.2 mm.

Fig. 438. *Lucilia regalis* (Meigen), male. Acrophallus and tip of paraphallus, left lateral view.

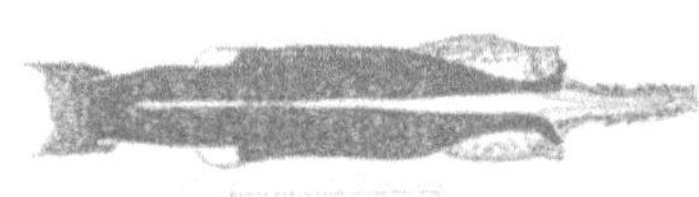

Fig. 439. *Lucilia regalis* (Meigen), male. Distiphallus, dorsal view. Scale = 0.2 mm.

Fig. 440. *Lucilia regalis* (Meigen), male. Ejaculatory sclerite. Scale = 0.2 mm.

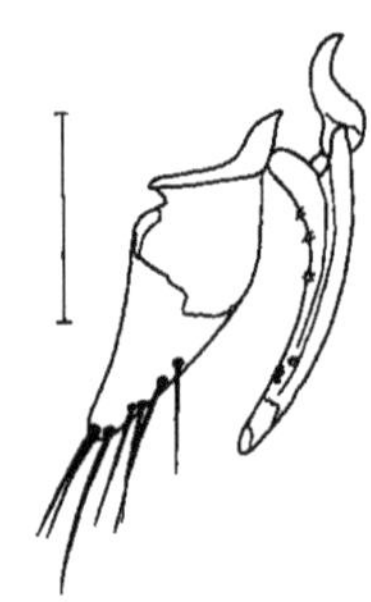

Fig. 441. *Lucilia regalis* (Meigen), male. Pre- and postgonites. Scale = 0.2 mm.

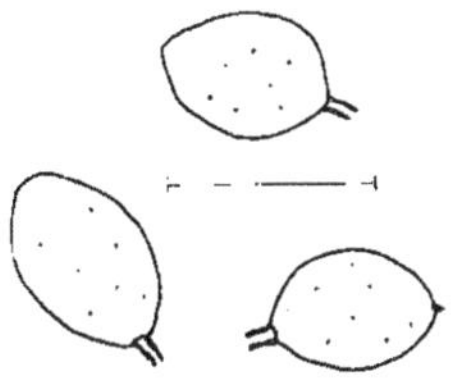

Fig. 442. *Lucilia regalis* (Meigen), female. Spermathecae. Scale = 0.2 mm. G.pr. 208.

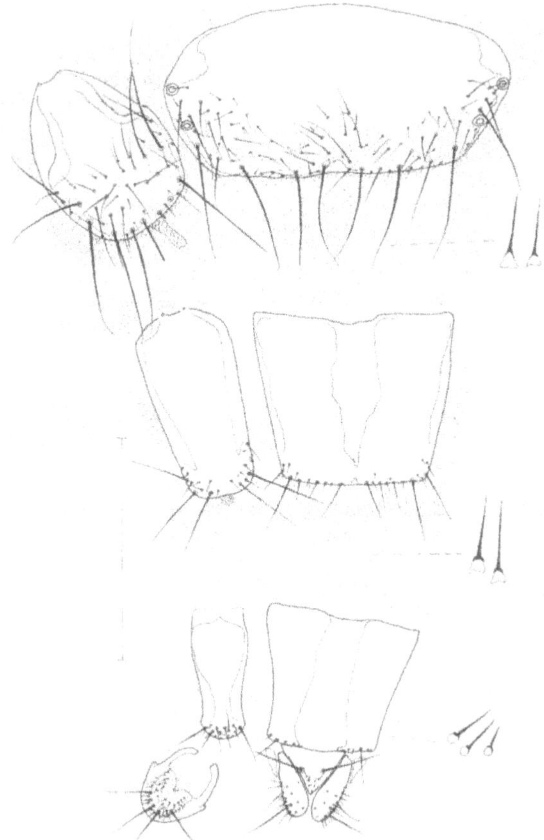

Fig. 443. *Lucilia regalis* (Meigen), female. Ovipositor. Stipple indicates extent of microtrichiae. Insets: shape of individual microtrichiae in areas indicated (enlarged). Scale = 1.0 mm. G.pr. 208.

behind them. Pleural membrane 8 with long and fine microtrichiae in posterior part.

Length. 4.5-9.0 mm.

Distribution. Rare. Widespread in Denmark. Only Sk. and Öl. in Sweden. In Norway a single record from Ø. Absent from Finland. - Widespread in the Palaearctic Region from Western Europe to Amur Oblast (U.S.S.R.), China and Mongolia. Absent from the British Isles and Japan (Ozerov, 1986; Schumann, 1986).

Biology. Imature stages unknown. Nothing is known of its life-cycle. The specimen from Norway was collected close to a refuse dump (Øra). Adults have been collected in Fennoscandia and Denmark from May to September.

Nomenclature. Only the male of *Musca regalis* was described by Meigen, on the basis of an unspecified number of specimens "von Herrn v. Winthem", now in Vienna. I have not been able to find an illustration of it in Meigen's Plates (Morge, 1975, 1976a, 1976b). Collin (1926) examined 1♂1♀ under this name in the Vienna collection. No characters of the genitalia were mentioned by Collin, but on the basis of the other characters described it appears that the name has been correctly interpreted by later authors (Zumpt, 1956a; Mihályi, 1977). *Calliphora longilobata* was described on the basis of male(s) from "Prusse Orientale". In Pandellé's collection in Paris is a single male labelled "♂/5619" in Pandellé's hand. It has the genitalia exposed and conforms to the traditional concept of *regalis* Meigen. I have labelled it as holotype.

33. ***Lucilia richardsi*** Collin, 1926
Figs 374, 444-454.

Lucilia richardsi Collin, 1926: 259.

♂♀. Palpi infuscated brown. First flagellomere about 3 times longer than wide. Occiput weakly concave in male, convex in female, with black setulae only. 3+1 *h*, basals in a weakly curved line. 2+1 *ph*. t_2 in extremely rare cases with only 1 *ad* seta. t_3 with 1-3 *av* setae.

♂. Frons 0.077-0.098 times head-width (n=5). Frontal vitta as wide as or a little wider than each fronto-orbital plate. f_2 without an *a* preapical seta. Lobes of ST5 rather large. Cerci narrow, straight and short. Surstylus parallel-sided, short, apically blunt, with very long, dense and often curly vestiture on outer and inner sides. Such vestiture is also present on ventral parts of epandrium. Bacilliform sclerites separated, on inner side with weak medially directed sclerotisation. Membrane between them with narrow stripes of short hairs ventrally and numerous long setae dorsally. Aedeagus of *sericata* type. Acrophallus relatively shorter than in *sericata*. Mesohypophallus unsclerotised. In dorsal view paraphallus divided into two branches at level of hind end of hypophallus. Pregonite with 8 setae. Postgonite without basal seta.

♀. Frons 0.352-0.375 times head-width (n=5). Frontal vitta more than 2 times as wide as fronto-orbital plates, which are relatively narrow. f_2 with an *a* preapical seta. Abdominal dusting very weak. T6 only with a very narrow zone of microtrichiae along hind edge (*sericata* has almost the posterior half microtrichiose). T7 almost completely divided. Prolongation of ST6 into the intersegmental membrane somewhat more weakly developed than in *sericata*. ST7 prolongation also present but very weak. Pleural membrane 8 with long and fine microtrichiae in posterior half, much longer and finer than those in *sericata*. [4 ovipositor slides examined.]

Length. 6.0-10.0 mm.

Distribution. Rather rare in Fennoscandia and

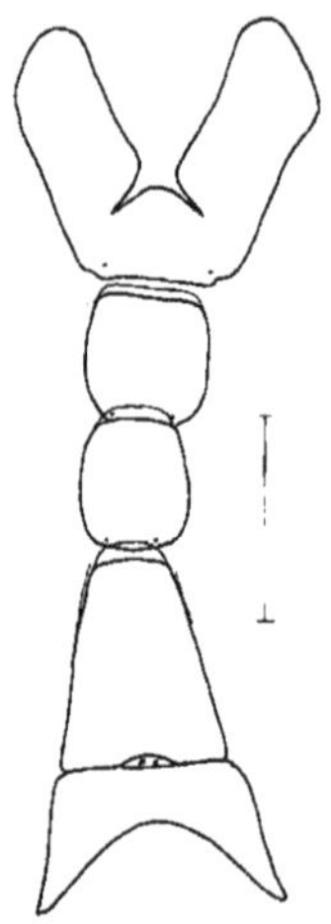

Fig. 444. *Lucilia richardsi* Collin, male. ST1-5. Scale = 0.8 mm.

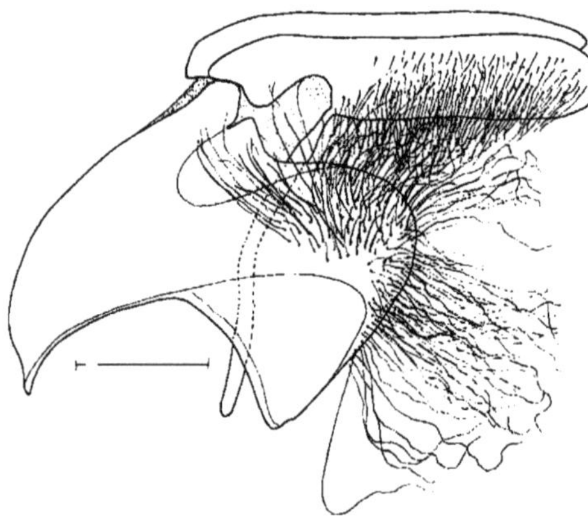

Fig. 445. *Lucilia richardsi* Collin, male. Epandrium, cerci, surstylus and bacilliform sclerite, left lateral view. Scale = 0.2 mm.

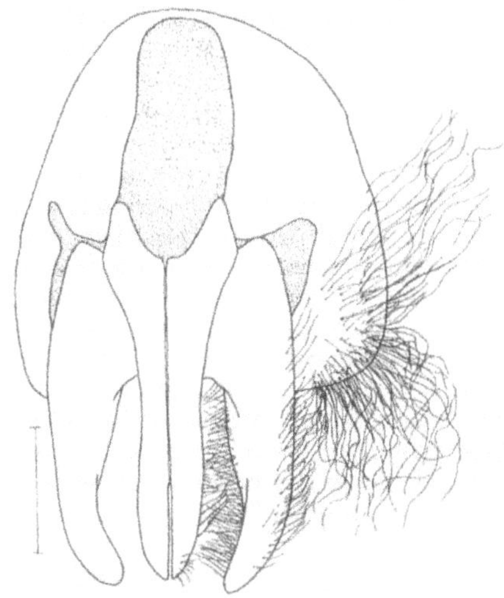

Fig. 446. *Lucilia richardsi* Collin, male. Epandrium, cerci and surstyli, posterior view. Scale = 0.2 mm.

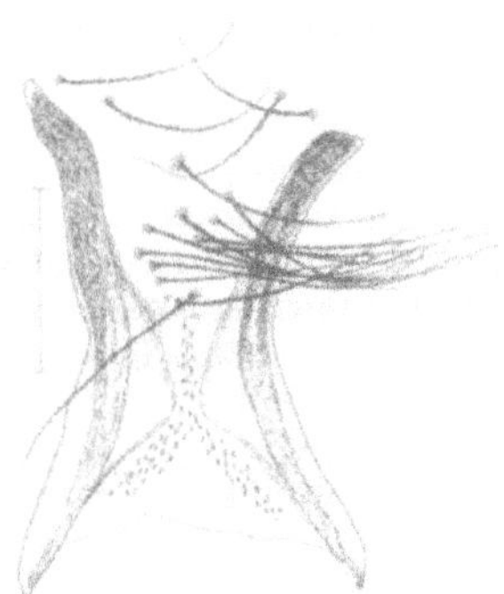

Fig. 447. *Lucilia richardsi* Collin, male. Bacilliform sclerites and vestiture on membrane between them. Scale = 0.2 mm.

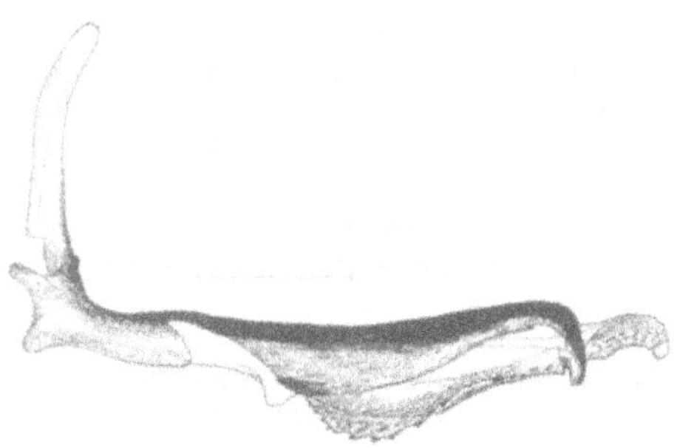

Fig. 448. *Lucilia richardsi* Collin, male. Aedeagus, left lateral view. Scale = 0.2 mm.

Fig. 449. *Lucilia richardsi* Collin, male. Acrophallus and tip of paraphallus, left lateral view.

Fig. 450. *Lucilia richardsi* Collin, male. Distiphallus, dorsal view. Scale = 0.2 mm.

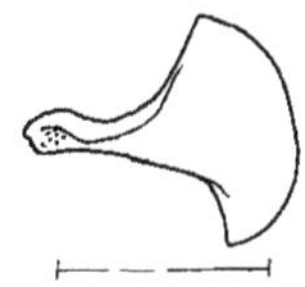

Fig. 451. *Lucilia richardsi* Collin, male. Ejaculatory sclerite. Scale = 0.2 mm.

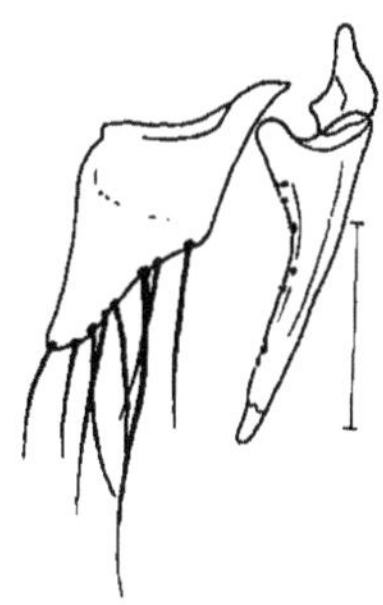

Fig. 452. *Lucilia richardsi* Collin, male. Pre- and postgonites. Scale = 0.2 mm.

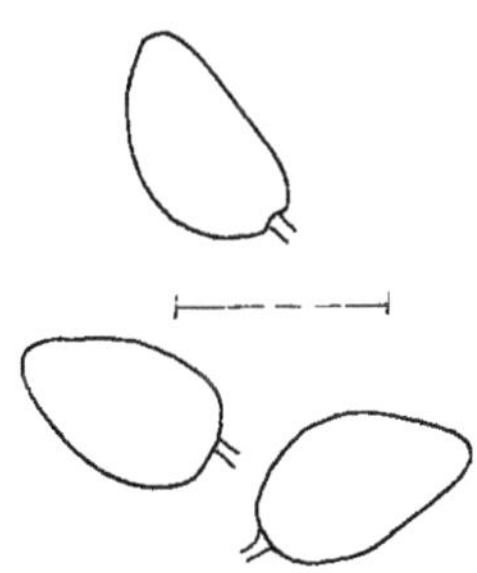

Fig. 453. *Lucilia richardsi* Collin, female. Spermathecae. Scale = 0.2 mm. G.pr. 203.

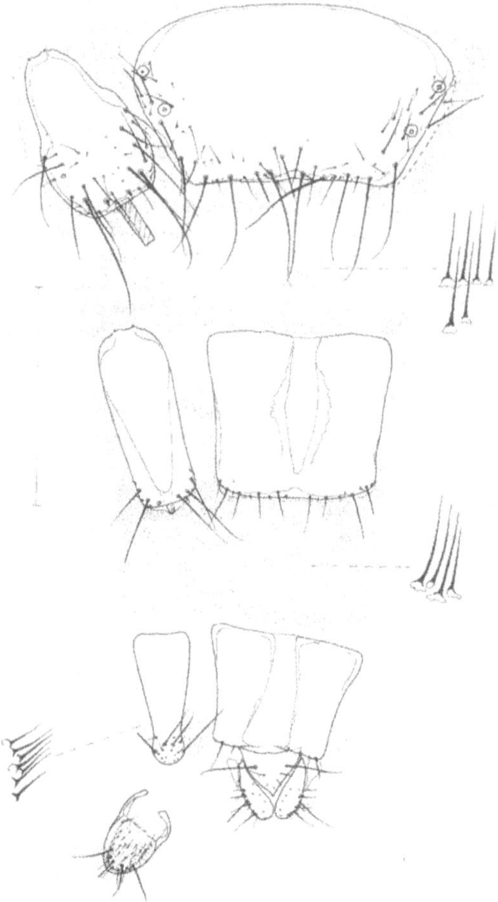

Fig. 454. *Lucilia richardsi* Collin, female. Ovipositor. Stipple indicates extent of microtrichiae. Insets: shape of individual microtrichiae in areas indicated (enlarged). Scale = 0.5 mm. G.pr. 203.

Denmark. Absent from Denmark. The entry for this country in the Palaearctic Catalog (Schumann, 1986) may be based on my own record from Denmark in a previous paper (Rognes, 1980). However, that particular record has turned out to be based on a misidentified *regalis*. South Sweden. Extreme southeastern parts of Norway from Ø to VAY. In Finland north to Ok. - Europe, including European parts of U.S.S.R., also Georgian S.S.R. and Kazakh S.S.R. (Chimkentskaya oblast) (Grunin, 1970a; Schumann, 1986).

Biology. Immature stages not known. In Finland *richardsi* has been bred form a case of wound myiasis in a nightjar (*Caprimulgus europaeus* L.) (Nuorteva, 1959a). According to Nuorteva & Skarén (1960) it shows a strong attraction to carcasses of homoiothermic animals such as birds and small mammals. Its apparent absence from Denmark is very peculiar and suggests that the species has a special biology. Adults on the wing in Fennoscandia from June to August.

34. ***Lucilia sericata*** (Meigen, 1826)
Figs 9, 23, 24, 375, 455-465.

Musca sericata Meigen, 1826: 53.
Musca nobilis Meigen, 1826: 56.
Chrysomya capensis Robineau-Desvoidy, 1830: 451.
Lucilia modesta Robineau-Desvoidy, 1830: 454.
Lucilia pubescens Robineau-Desvoidy, 1830: 454.
Musca tegularia Wiedemann, 1830: 655.
Musca chloris Haliday, 1833: 165.
Lucilia pruinosa Meigen, 1838: 294.
Lucilia flavipennis Macquart, 1843: 296.
Lucilia fulgida Zetterstedt, 1845: 1315.
Lucilia lagyra Walker, 1849: 885.
Lucilia latifrons Schiner, 1862: 590.
Phaenicia concinna Robineau-Desvoidy, 1863: 778.
Lucilia sayi Jaennicke, 1867: 375.
Lucilia barberi Townsend, 1908: 121.
Lucilia giraulti Townsend, 1908: 121.

♂♀. Palpi yellow. First flagellomere 3-4 times longer than wide. Occiput weakly concave in male, weakly convex in female, with black setulae. 3+1 *h*, basal setae in a weakly curved line. 2+1 *ph*. t_3 with 2-3 *av* setae.

♂. Frons 0.122-0.137 times head-width (n=5). Frontal vitta about 2.0-2.5 times as wide as fronto-orbital plates. f_2 without *a* preapical seta. Abdominal dusting weak. Lobes of ST5 normal. Cerci short. Surstylus short, apex blunt. Vestiture of surstylus and epandrium quite normal, quite different from *richardsi*. Bacilliform sclerites separated. Membrane between them weakly sclerotised, wholly covered with short hairs in ventral half, in dorsal half with a number of setae. Aedeagus of *sericata* type. Acrophallus with ventral projection when seen in profile. Mesohypophallus sclerotised. In dorsal view paraphallus divided into two branches far behind level of hind end of hypophallus. Pregonite with 3 setae. Postgonite without basal seta.

♀. Frons 0.356-0.392 times head-width (n=5). Frontal vitta 2 times as wide as fronto-orbital plates. f_2 with an *a* preapical seta. Abdominal dusting strong, divided along mid-line according to angle of view. T6 with almost the posterior half covered with microtrichiae. T7 divided in anterior half. Prolongation of ST6 into intersegmental membrane rather long, ST7 prolongation much weaker. Pleural membrane 8 with rather short microtrichiae in posterior half. [4 ovipositor slides examined.]

Length. 5-10 mm.

Distribution. Common in Denmark and southern Sweden. Extreme southeastern parts of Norway (also RY). Southern Finland. - Almost cosmopolitan (Zumpt, 1965; James, 1970; Dear, 1986).

Biology. Details of the life cycle of *L. sericata* are given by Zumpt (1965), Povolný & Rozsypal (1968), Nuorteva (1977), Eichler (1980) and Smith (1986). The species is of outstanding hygienic and epidemiological importance (Povolný & Rozsypal, 1968). Sterile larvae have been used to heal

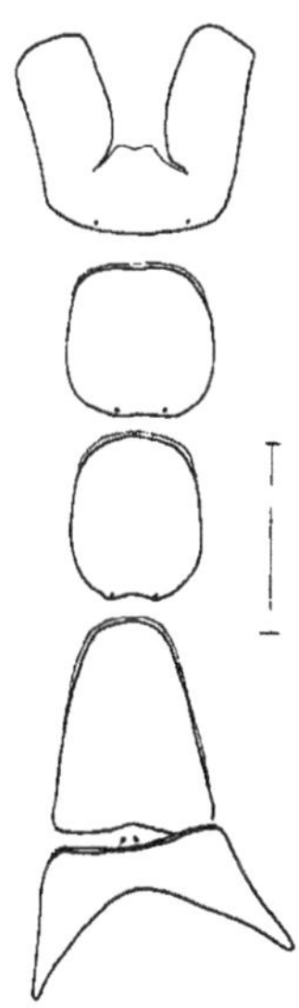

Fig. 455. *Lucilia sericata* (Meigen), male. ST1-5. Scale = 0.8 mm.

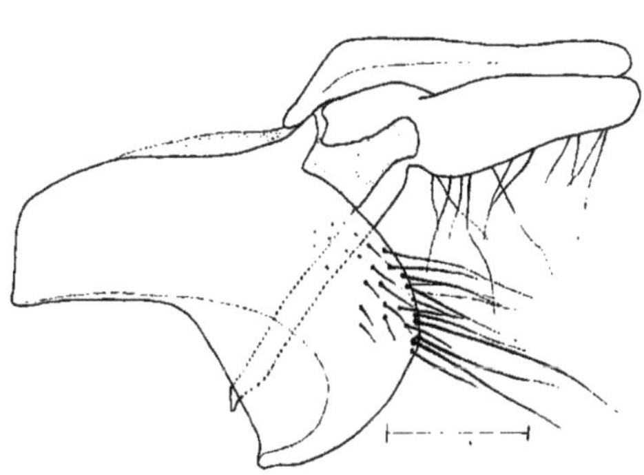

Fig. 456. *Lucilia sericata* (Meigen), male. Epandrium, cerci, surstylus and bacilliform sclerite, left lateral view. Scale = 0.2 mm.

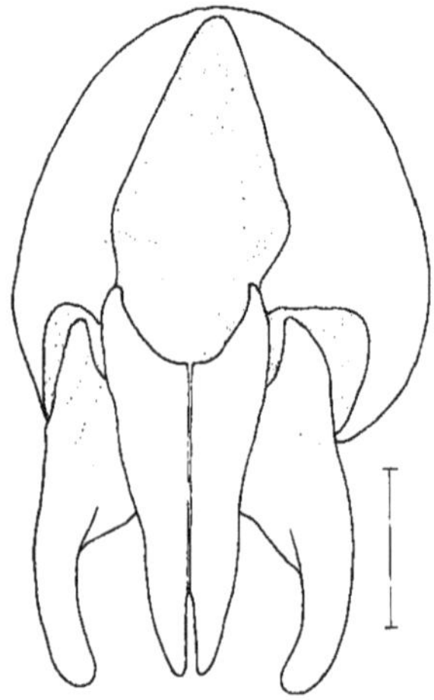

Fig. 457. *Lucilia sericata* (Meigen), male. Epandrium, cerci and surstyli, posterior view. Scale = 0.2 mm.

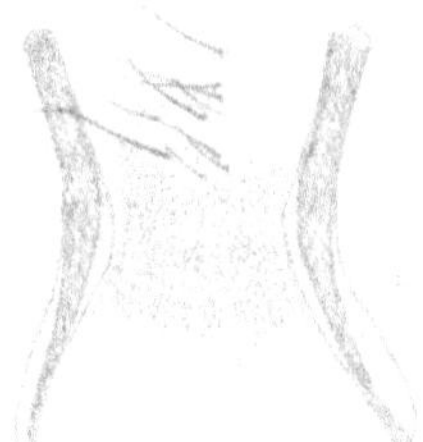

Fig. 458. *Lucilia sericata* (Meigen), male. Bacilliform sclerites and vestiture on membrane between them. Scale = 0.2 mm.

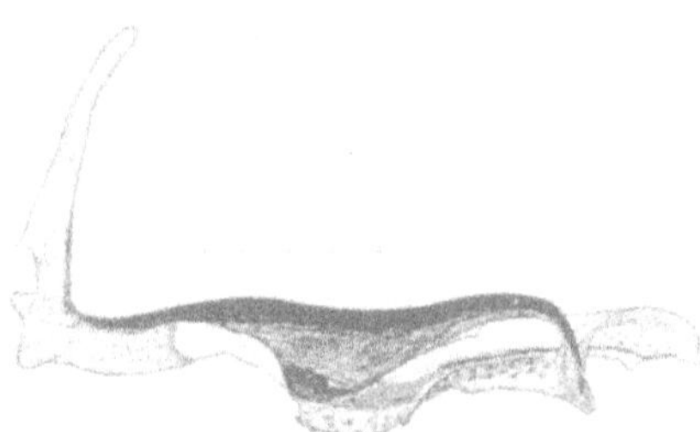

Fig. 459. *Lucilia sericata* (Meigen), male. Aedeagus, left lateral view. Scale = 0.2 mm.

Fig. 460. *Lucilia sericata* (Meigen), male. Acrophallus and tip of paraphallus, left lateral view. Scale = 0.1 mm.

Fig. 461. *Lucilia sericata* (Meigen), male. Distiphallus, dorsal view. Left paraphallic tip broken. Scale = 0.2 mm.

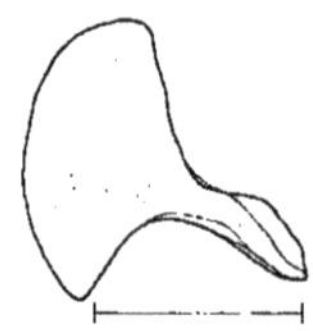

Fig. 462. *Lucilia sericata* (Meigen), male. Ejaculatory sclerite. Scale = 0.2 mm.

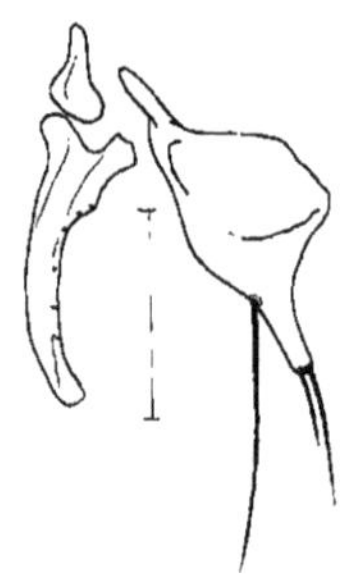

Fig. 463. *Lucilia sericata* (Meigen), male. Post- and pregonites. Scale = 0.2 mm.

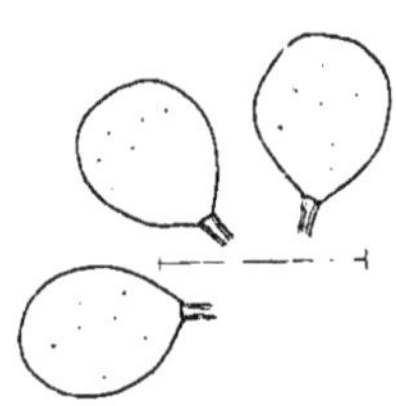

Fig. 464. *Lucilia sericata* (Meigen), female. Spermathecae. Scale = 0.2 mm. G.pr. 201.

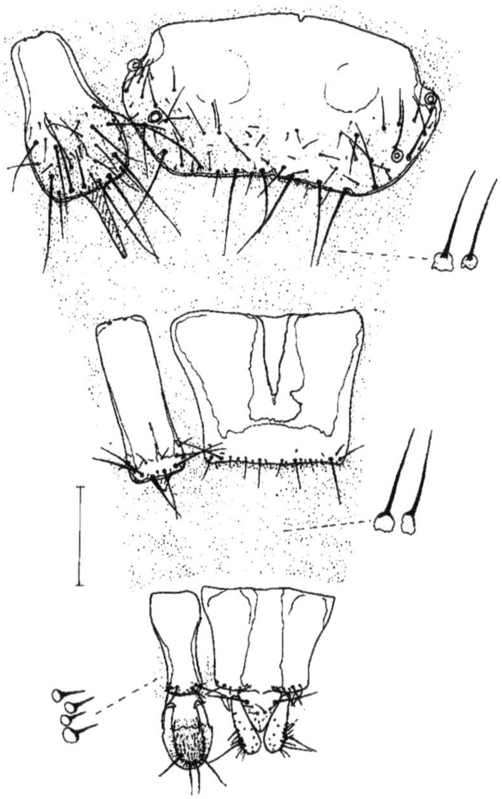

Fig. 465. *Lucilia sericata* (Meigen), female. Ovipositor. Stipple indicates extent of microtrichiae. Insets: shape of individual microtrichiae in areas indicated (enlarged). Scale = 0.5 mm. G.pr. 200.

chronically purulent wounds (Brumpt, 1933; Zumpt, 1956a; Povolný & Rozsypal, 1968; Schumann, 1971). According to Norris (1965) *sericata* larvae may develop in garbage of exclusively vegetable origin. Stewart & Foote (1974) report a case from North America of a live female toad (*Bufo* sp.) being infested with 23 full-grown larvae of *sericata*. It is involved in sheep myiasis in Great Britain (MacLeod, 1937, 1943a, 1943b) and other countries as a primary infective agent, but not in Fennoscandia and Denmark (Brinkmann, 1976; Nielsen, 1984a, 1984b). It is rarely involved in human myiasis and then usually under circumstances of personal neglect and high fly density (cf. Greenberg, 1984). In ZMC there is a ♀ bred from a full-grown larva recovered from the external auditory meatus of a human being, apparently the specimen reported by Meinert (1888, as *nobilis*). The person was infected while sleeping in the open (Meinert, 1888). Cases from Denmark of wound myiasis (ulcus cruris) in humans have been reported by Tikjøb & Haarløv (1985). It has been involved in a case of intestinal myiasis in a two-year-old child in Denmark (Haarløv, 1961). In Fennoscandia and Denmark strongly synanthropic and dependent on conditions created by man (Nuorteva, 1963). Taken in copula on June 7 and on July 1 in Sweden (Sk.) (O.Ringdahl leg.). Suspected of transmitting poliomyelitis virus to humans in Finland (Nuorteva, 1958, 1959b, 1960c, 1960d; but see review by Greenberg, 1973). In Fennoscandia and Denmark the adults are on the wing from April to October.

Nomenclature. *Musca sericata* was described by Meigen on the basis of specimens "Aus Oesterreich, von Herrn Megerle v. Mühlfeld". This collection is considered lost (Pont, 1986). Meigen Plate 277, fig. 11 shows both sexes (Morge,

1976b). I have not examined the syntypes of *Musca nobilis* described from "Beide Geschlechter von Herrn v. Winthem aus Hamburg" in Vienna, and illustrated on Meigen Plate 120, fig. 10 (Morge, 1976a). The latter shows a ♂ with a rather narrow frons for a *sericata*. The types were seen by Collin (1926), who also noted the narrow frons, and Aubertin (1933). *Lucilia pruinosa* was described from the ♀ sex and an unspecified number of specimens; no collector or locality was mentioned. I have not been able to locate a figure of it in Morge (1975, 1976a, 1976b). In Meigen's collection in MNHN is a ♀ labelled (1) "Meigen" and "2116/40", (2) "pruinosa / ♂ [sic] [in Meigen's hand]". It has lost both front legs and the left mid leg. It belongs to the present species. I have given it a holotype label. Ringdahl (1945: 32) first established the synonymy of *Lucilia fulgida* Zetterstedt with *sericata*, but this has been overlooked by subsequent authors, e.g. by Schumann (1986) who lists *fulgida* as a nomen dubium. It was described by Zetterstedt from an unspecified number of males and females (also "...in copula capta...") from "...Scania ad Lund, Lackalänga..., in Smolandia ...", all in south Sweden. 2♂2♀ from the syntypic series have been located in the Diptera Scandinavica collection in ZML (drawer 24). 1♂1♀ are on the same pin, and all pins are labelled with a small purple-red square of paper indicating that they were collected in Lund or its vicinity. I have labelled and here designate a ♂ as lectotype; the three other specimens have been labelled as paralectotypes.

35. ***Lucilia silvarum*** (Meigen, 1826)
Figs 376, 466-476.

Musca silvarum Meigen, 1826: 53.
Lucilia brunnicosa Robineau-Desvoidy, 1830: 459.
Lucilia nigripalpis Townsend, 1908: 120.
Onesia lucilioides Wulp, 1896: 288.

♂♀. Palpi brownish-black. First flagellomere 3-4 times longer than wide. Occiput weakly concave in male, weakly convex in female, with black setulae. 3+1 *h*, basals in a weakly curved line. 2+1 *ph*. Occasional specimens with only 2 *post acr* setae, thus resembling *bufonivora* in this respect. t_3 with 2-3 *av* setae. Abdominal dusting almost absent.

♂. Frons 0.076-0.107 times head-width (n=14). Frontal vitta 2 times as wide as each fronto-orbital plate. f2 without *a* preapical seta. Lower squama infuscated. Lobes of ST5 normal. Cerci long and narrow with apical hook in profile. Surstyli much longer and narrower than in *bufonivora*, apical half evenly tapering, rather pointed. Bacilliform sclerites more heavily sclerotised than in *bufonivora*. Membrane between them with numerous hairs in ventral half. Aedeagus of *bufonivora* type. Pregonite with 5 setae. Postgonite with basal seta.

♀. Frons 0.327-0.363 times head-width (n=5). Frontal vitta definitely less than 2 times as wide as each fronto-orbital plate. f_2 with *a* preapical seta. Sclerites of ovipositor without microtrichiae, except for posterior fifth of T6, posterior half of ST6 and hypoproct. T7 almost completely divided. Epiproct and cerci normal in shape and without microtrichiae. ST6 narrow. ST7 narrow apically. ST8 fully sclerotised. No sclerotisations in the intersegmental membranes. Microtrichiae on intersegmental membrane 7-8 and posterior parts of pleural membrane 8 rather short. [3 ovipositor slides examined.]

Fig. 466. *Lucilia silvarum* (Meigen), male. Arista. (From Rognes, 1980: fig. 5C).

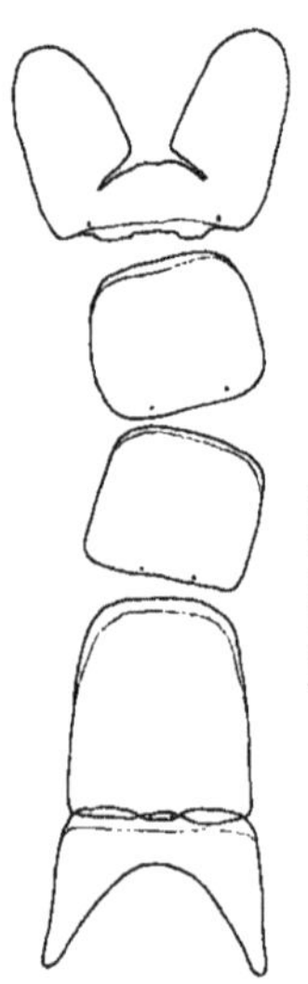

Fig. 467. *Lucilia silvarum* (Meigen), male. ST1-5. Scale = 0.8 mm.

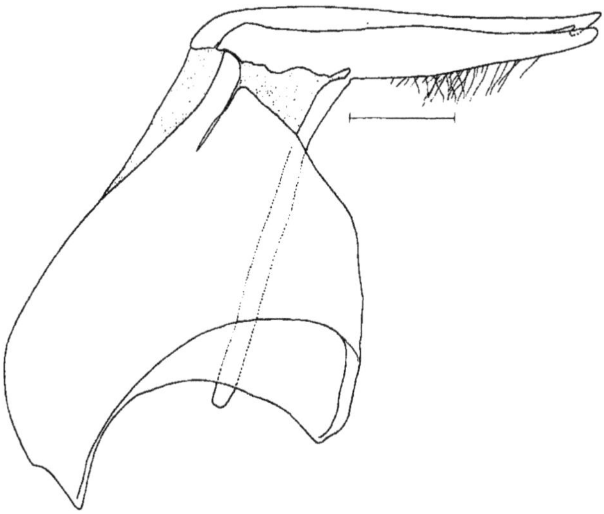

Fig. 468. *Lucilia silvarum* (Meigen), male. Epandrium, cerci, surstylus and bacilliform sclerite, left lateral view. Scale = 0.2 mm.

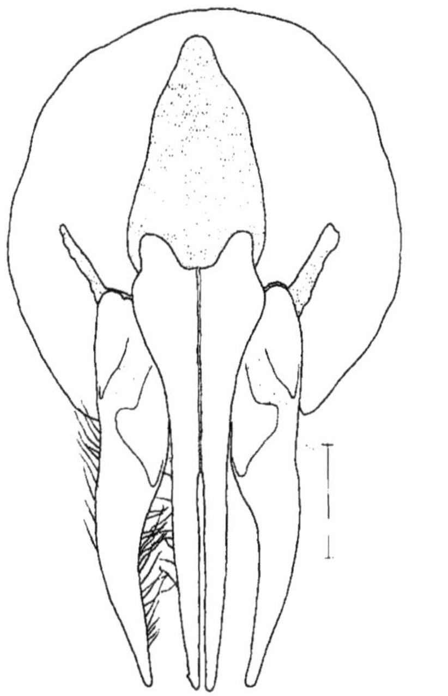

Fig. 469. *Lucilia silvarum* (Meigen), male. Epandrium, cerci and surstyli, posterior view. Scale = 0.2 mm.

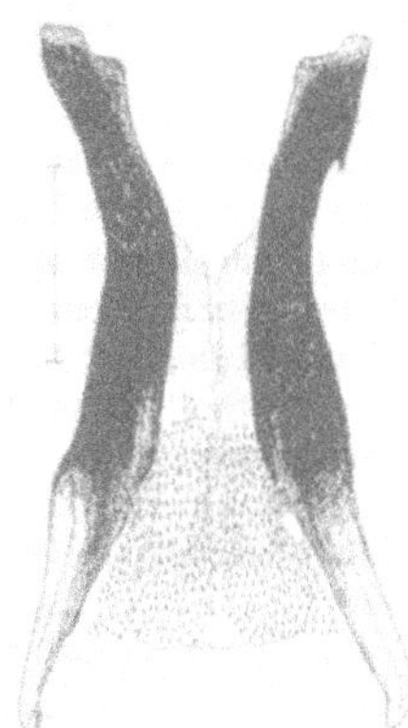

Fig. 470. *Lucilia silvarum* (Meigen), male. Bacilliform sclerites and vestiture on membrane between them. Scale = 0.2 mm.

Fig. 471. *Lucilia silvarum* (Meigen), male. Aedeagus, left lateral view. Inset: tip of paraphallus (enlarged). Scale = 0.2 mm.

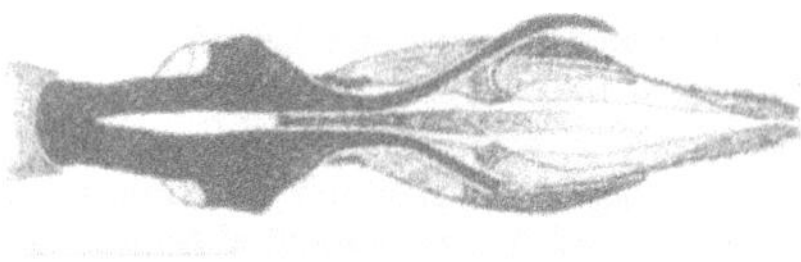

Fig. 472. *Lucilia silvarum* (Meigen), male. Distiphallus, dorsal view. Left paraphallic tip dislocated. Scale = 0.2 mm.

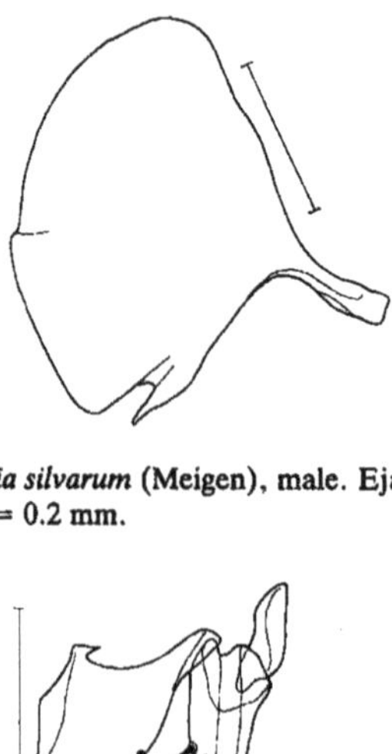

Fig. 473. *Lucilia silvarum* (Meigen), male. Ejaculatory sclerite. Scale = 0.2 mm.

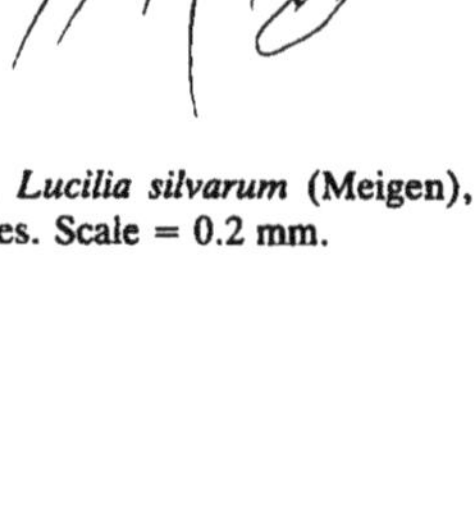

Fig. 474. *Lucilia silvarum* (Meigen), male. Pre- and postgonites. Scale = 0.2 mm.

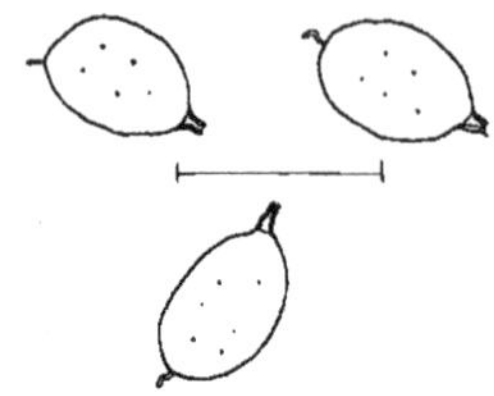

Fig. 475. *Lucilia silvarum* (Meigen), female. Spermathecae. Scale = 0.2 mm. G.pr. 197.

Length. 6-10 mm.

Distribution. Widespread in Denmark. Found only in southernmost parts of Norway. North to Nb. in Sweden and ObS in Finland. - Palaearctic: Europe, North Africa, U.S.S.R., China (Schumann, 1986). Listed from Japan by Zumpt (1956a), but not by Kano & Shinonaga (1968). Also Nearctic (Hall, 1965), and Neotropical (Mexico City) (James, 1970).

Biology. A case of several living bull-frogs (*Rana catesbeiana*) infested with numerous larvae of *silvarum* in North America has been reported by Hall (1948). James (1955) regards it as a facultative rather than an obligatory parasite of amphibians in North America. Zumpt (1956a) assumes that it lives mainly as a carrion breeder in the Palaearctic. Collected on the stinkhorn fungus (*Phallus impudicus* Pers.) in Denmark. In Fennoscandia and Denmark adults are on the wing from May to October.

Nomenclature. *Musca silvarum* was described by Meigen on the basis of "Ein Weibchen aus Oesterreich, von Herrn Megerle v. Mühlfeld". This collection is considered lost (Pont, 1986). In Meigen Plate 277, fig. 12 (Morge, 1976b) the female is illustrated.

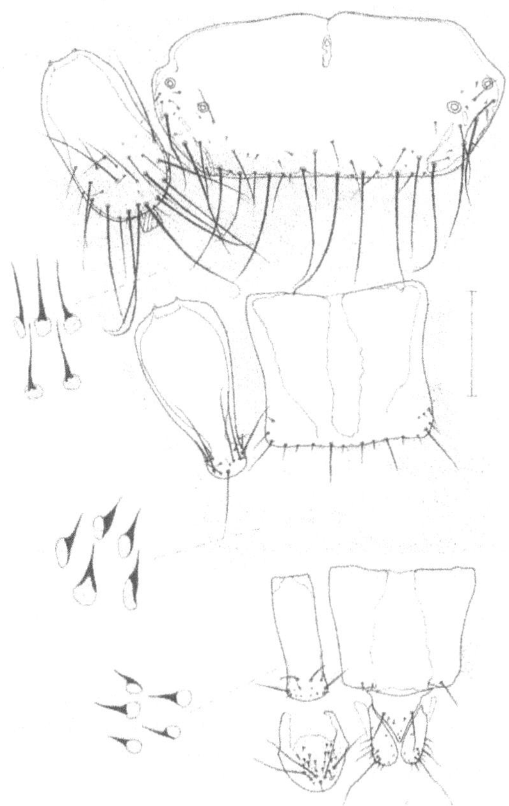

Fig. 476. *Lucilia silvarum* (Meigen), female. Ovipositor. Stipple indicates extent of microtrichiae. Insets: shape of individual microtrichiae in areas indicated (enlarged). Scale = 0.4 mm. G.pr. 197.

Subfamily Melanomyinae

The subfamily Melanomyinae is probably a monophyletic group supported by the following groundplan apomorphies:

(1) Epiproct in ovipositor fused with T8, the combined T8+epiproct often appearing shovel-shaped posteriorly.
(2) Cerci in ovipositor short and broad, wart-like, with hairs.
(3) Ovipositor sclerites (except hypoproct) without microtrichiae, thus shining black without dust (except posterior edge of T6 in *Melinda gentilis*, and T6 of *Opsodexia*).
(4) ST8 in ovipositor completely absent or very strongly reduced.
(5) Sclerites of segment 7 and 8 in ovipositor strongly elongated.
(6) Sclerites of at least segments 7 and 8 in ovipositor devoid of setae or setulae, these replaced by numerous small sensillae.
(7) Intersegmental membranes in ovipositor without microtrichiae of usual form; these entirely absent or replaced with armature of a peculiar form.
(8) Pleural membranes of ovipositor entirely without microtrichiae.
(9) Acrophallus and other parts of aedeagal walls bare, without denticles.
(10) First instar larva with labrum dorsally concave (i.e. upturned) and firmly fused with tentoropharyngeal sclerites.

There is further a tendency within the group towards

(11) Loss of metallic colour in some genera.
(12) Loss of fixed calliphoroid ground-plan pattern of wing veins, especially as regards the course of vein M in apical parts of wing.
(13) Loss of vestiture on one or more of the thoracic surfaces proepisternal depression, prosternum, metasternal area and postalar wall in several genera.

The subfamily is rather heterogenous and tentatively includes the following genera: *Angioneura* Brauer & Bergenstamm, *Eggisops* Rondani, *Glutoxys* Aldrich, *Melanomya* Rondani, *Melinda* Robineau-Desvoidy, *Opsodexia* Townsend, *Paradichosia* Senior-White, *Pseudopsodexia* Townsend and *Tricycleopsis* Villeneuve. Females of *Glutoxys* are not known (Downes, 1986). I have not seen the ovipositor of *Opsodexia*, *Paradichosia*, *Pseudopsodexia* or *Tricycleopsis*. That of *Opsodexia* is figured in Shewell (1987: Fig. 106.36-37) and its T6 described by Downes (1986: 11). The synapomorphies listed above seem to apply, except that it is impossible to state with certainty from Shewell's figure whether T8 is fused with the epiproct or not. The ovipositor of *Paradichosia* is figured by Kano (1962), Kurahashi (1970) and Lobanov (1976). The ovipositor of the single known female of the genus *Pseudopsodexia* is described briefly by Downes (1986). Nothing appears to contradict the suite of characters listed above. *Tricycleopsis* has a few inconspicuous easily rubbed abraded hairs on the lower calypter (Senior-White, Aubertin & Smart, 1940; Kurahashi, 1970, 1972a), and might be a member of Calliphorinae. However, Kurahashi (1970) treated it as closely related to *Melinda* Robineau-Desvoidy and included it within his "*Melinda*-group". I follow Kurahashi and regard it as a member of the Melanomyinae, even though its ovipositor and biology are not known.

Herting (1961), Stackelberg (1970), Crosskey (1977) and Tschorsnig (1985a) regard *Angioneura* and *Melanomya* as belonging to the Rhinophoridae. Downes (1965, 1986) considered all the genera (except *Melinda*, *Tricycleopsis* and *Paradichosia*, which were not discussed) as forming a primitive group of Calliphoridae, but he did not examine the ovipositor. I agree with him as to their status as calliphorids, but the ovipositor structure and other features suggest that they are a specialised group which has diverged considerably from the calliphorid ground-plan. He also synonymized *Angioneura*, *Eggisops*, *Melanomya* and *Opsodexia* under *Melanomya*, but retained these groups as subgenera. *Melinda* and *Paradichosia* have traditionally been treated as members of the Calliphorinae (sensu Zumpt) (Kurahashi, 1970; Schumann, 1986). Shewell (1987) suggested that *Eggisops*, *Angioneura* and *Melinda* form a tribal group, Angioneurini, within his Calliphorinae on the basis of the habit of snail parasitism and features of the first instar larva. I agree with this, but include a few more genera. Townsend's (1919b: 548) name Melanomyinae (resurrected by Downes, 1965, as tribe Melanomyini of the Polleniinae) is the oldest available name for this taxon.

The included species are oviparous (*Melinda*) or viviparous (*Eggisops*, *Paradichosia*). Uterus cylindrical, sometimes strongly elongated (*Eggisops*, *Paradichosia*) (pers.observ.; Kurahashi, 1970). Viviparity is thus not associated with shortened ovipositor sclerites as in the Calliphorinae (*Bellardia*, *Onesia*) or Sarcophagidae.

In Fennoscandia and Denmark recognisable by the following additional characters: Ground-colour black. Thorax and abdomen non-metallic or blue metallic, more or less dusted. Non-metallic members muscid-like and often small, the blue-metallic ones larger with compact build. Head profiles as in Figs. 477-480. Parafacialia without strong setae, but often with small setulae in upper part. Lower parts of anterior half of anepimeron usually bare. Coxopleural streak absent. Posterior spiracles as in Figs. 481-486. Stem-vein bare above. Costa hairy below far beyond junction with R_1 (Figs. 491, 514, 525, 536). Lower calypter bare above, narrow and subcircular with inner edge diverging from longitudinal axis of fly, or posteriorly broad with inner edge converging with body-axis. Hind coxa bare behind.

So far as their biology is known, all members are parasites of snails (Schmitz, 1917; Keilin, 1919, 1921; Emden, 1954; Čepelák & Rozkošný, 1968; Kurahashi, 1970; Downes, 1986; Shewell, 1987).

Widespread in Palaearctic, Nearctic and Oriental Regions. 4 genera in Fennoscandia and Denmark.

Key to genera of Melanomyinae

1 Thorax and abdomen metallic dark blue; M with sharp bend (Fig. 536); 3 *post ia*; ♀: T6 and ST6 without setae, only small sensillae present (Figs. 545, 554)

Melinda Robineau-Desvoidy (p. 199)

– Thorax and abdomen non-metallic; M with shallow bend (Figs. 491, 503, 514, 525); 1-3

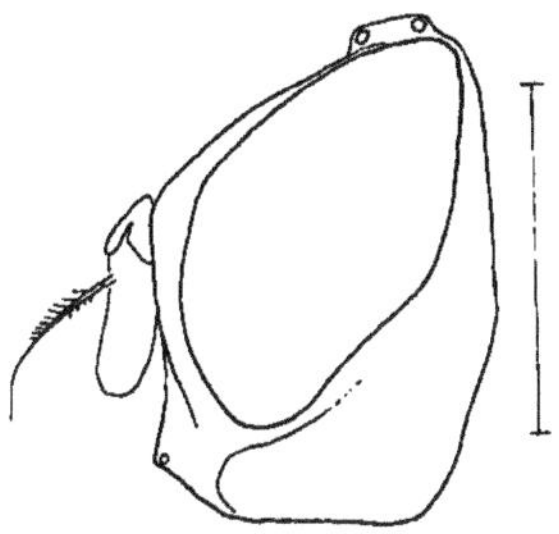

Fig. 477. *Angioneura acerba* (Meigen), head profile, male. Scale = 1.0 mm.

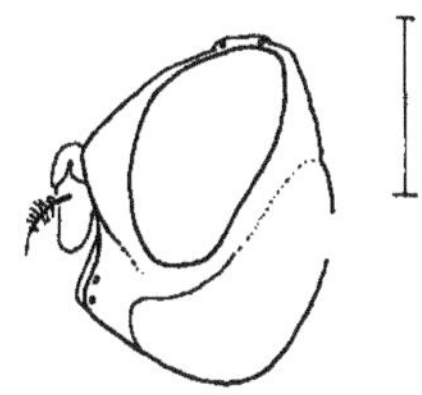

Fig. 478. *Eggisops pecchiolii* Rondani, head profile, male. Scale = 1.0 mm.

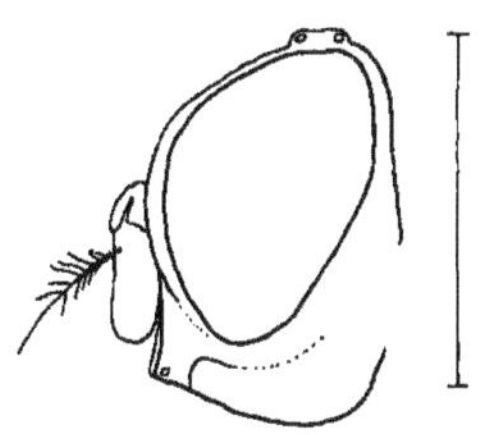

Fig. 479. *Melanomya nana* (Meigen), head profile, male. Scale = 1.0 mm.

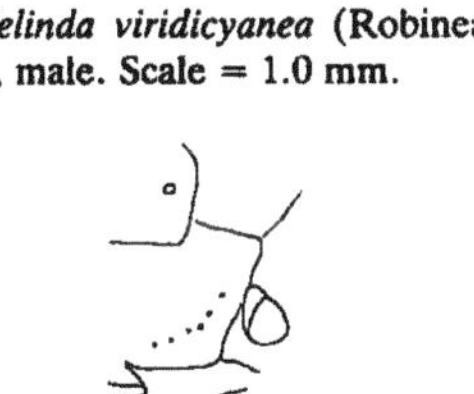

Fig. 480. *Melinda viridicyanea* (Robineau-Desvoidy), head profile, male. Scale = 1.0 mm.

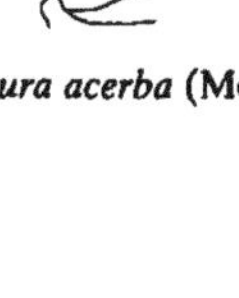

Fig. 481. *Angioneura acerba* (Meigen), posterior spiracle.

Fig. 482. *Angioneura cyrtoneurina* (Zetterstedt), posterior spiracle.

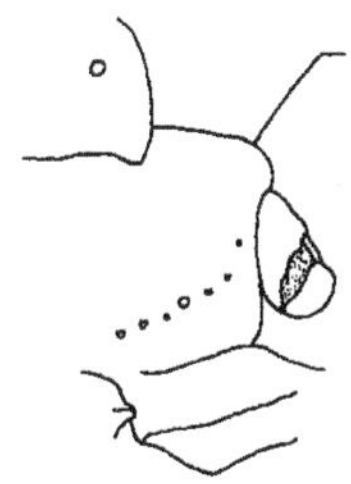

Fig. 483. *Eggisops pecchiolii* Rondani, posterior spiracle.

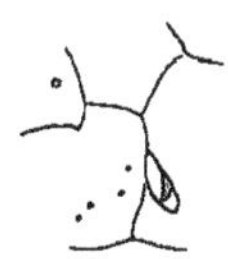

Fig. 484. *Melanomya nana* (Meigen), posterior spiracle.

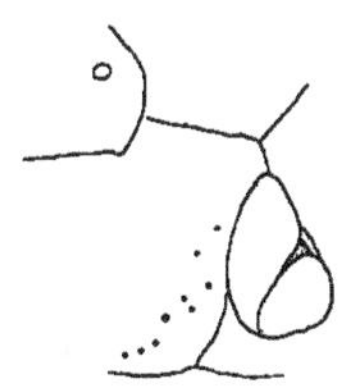

Fig. 485. *Melinda viridicyanea* (Robineau-Desvoidy), posterior spiracle.

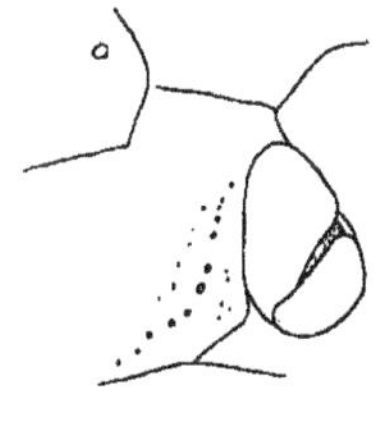

Fig. 486. *Melinda gentilis* (Robineau-Desvoidy), posterior spiracle.

post ia; ♀: T6 and ST6 with at least some normal setae (Figs. 501, 513, 524, 535)

2

2(1) Proepisternal depression setulose; lunula setulose; notopleuron hairy, especially around base of hind *npl*; scutellum with 3-5 strong marginal pairs, apicals strong (Fig. 489); *pra* strong, as strong as or stronger than first *post dc*; 1+2-3 *ia*; when only 2 *post ia*, first *post ia* closer to suture than to posterior *post ia*; first *post ia* on a line connecting *pra* and first *post dc*; lower calypter broad, inner edge converging wing longitudinal axis of fly; cell r_{4+5} open or closed with a long stalk, opening or stalk in front of apex of wing; t_1 with 2-5 *ad* setae

Eggisops Rondani (p. 193)

– Proepisternal depression bare; lunula bare; notopleuron bare or with 1-2 odd hairs; scutellum with rather weak apicals and usually only a single additional very strong pair, in subapical or lateral position (Figs. 487, 488, 490)

3

3(2) *pra* strong, much stronger than hind *npl*; arista long plumose, its width (hairing included) 1.5-2 times that of first antennal flagellomere; 1-2 *prst acr* conspicuous; 2 *post ia*; anterior strong *post ia* on a line connecting *pra* and first *post dc*; lower calypter narrow, subcircular, inner margin diverging from scutellum; posterior spiracle very narrow, compressed dorsoventrally (Fig. 484); t_1 with 1 *ad*

Melanomya Rondani (p. 197)

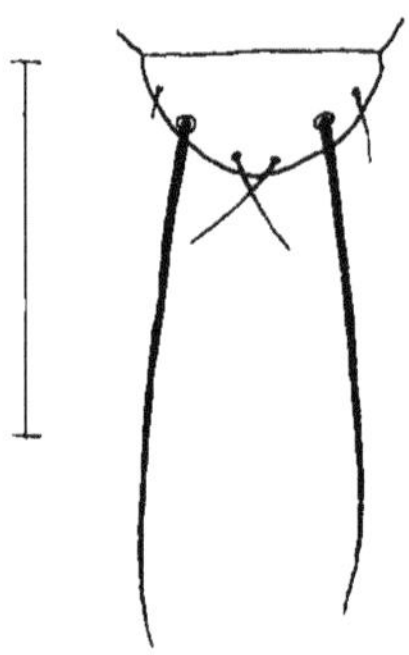

Fig. 487. *Angioneura acerba* (Meigen), scutellum. Scale = 1.0 mm.

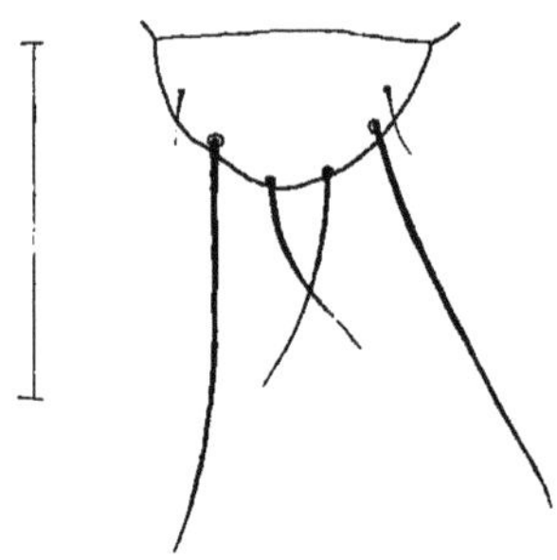

Fig. 488. *Angioneura cyrtoneurina* (Zetterstedt), scutellum. Scale = 1.0 mm.

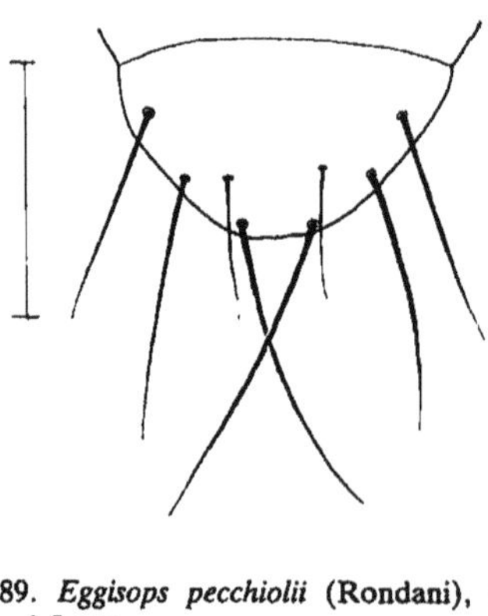

Fig. 489. *Eggisops pecchiolii* (Rondani), scutellum. Scale = 1.0 mm.

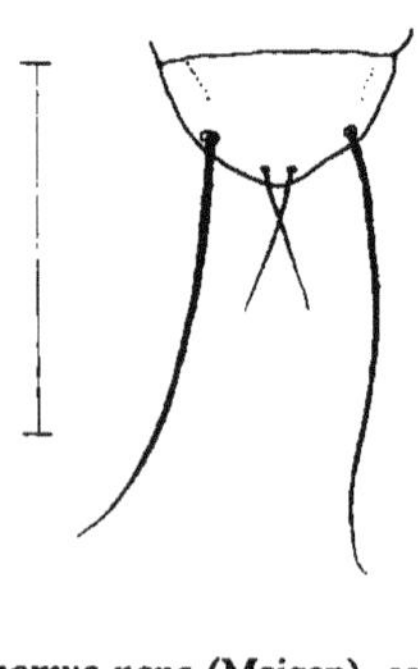

Fig. 490. *Melanomya nana* (Meigen), scutellum. Scale = 1.0 mm.

– *pra* short and weak, at most as strong as hind *npl*; arista almost bare or short plumose; 0-1 *prst acr* conspicuous; 1-2 *post ia*; anterior strong *post ia* well behind a line connecting *pra* and first *post dc*; lower calypter narrow and subcircular or broad; posterior spiracle broader (Figs. 481, 482); t_1 with 1-4 *ad*

Angioneura Brauer & Bergenstamm (p. 188)

Genus *Angioneura* Brauer & Bergenstamm, 1893

Angioneura Brauer & Bergenstamm, 1893: 161, 187.
Type-species: *Calobataemyia vetusta* Brauer & Bergenstamm, 1891: 369, by monotypy; = *Medoria acerba* Meigen, 1838.
Opelousia Townsend, 1919: 547.
Type-species: *Opelousia obscura* Townsend, 1919: 547, by original designation.
Angioneurilla Villeneuve, 1924: 31.
Type-species: *Tachina cyrtoneurina* Zetterstedt, 1859: 6135, by monotypy.
Opsodexiopsis Townsend, 1935a: 69.
Type-species: *Opsodexia abdominalis* Reinhard, 1929: 6, 9, by original designation.

Body dusted or almost shining black. Parafacialia bare or setulose in upper part. Arista almost bare or short plumose. Palpi short, brown. 0-1+3 *acr*. 2+3 *dc*. 0+1-2 *ia*. Anterior *post ia* closer to posterior *post ia* than to suture. 2-3 *h*, inner one most often rather weak. 1 *ph*, the outer absent. Prealar (first *sa*) weak or undifferentiated, at most as strong as hind *npl*. Second *sa* very strong. Posterior *sa* weak or undifferentiated. Proepisternal depression bare. Prosternum, metasternal area and postalar wall bare or hairy. Lower calypter small and subcircular, or posteriorly broad. Wing-veins pale. t_1 with 1 *pv* seta. f_2 with *a* preapical seta. t_2 with 1 *ad* seta. t_3 with 2 *av* setae. Apex of posterior arms of hypandrium close together. Bacilliform sclerites with median extensions which join each other more or less completely. Aedeagus long and of very characteristic shape. Basiphallus very long, strongly upturned posteriorly. Aedeagal membranes bare or with weak denticles ventrally along hypophallic ridges (especially in *Angioneura fimbriata* (Meigen), not treated here). Paraphallus hook-shaped. Acrophallus short with large apical opening. Pregonite long and strongly curved, with 2-3 apical setae. Postgonite without a basal seta, sensillae present apically as usual. Basal piece of postgonite ("Gelenkfortsatz") fused with main part (Tschorsnig, 1985a). Ovipositor with divided T7, T7 halves rather broad. ST7 of full length, rather broad. ST8 absent. Intersegmental membrane 6-7 quite bare. Intersegmental membrane 7-8 with or without microtrichiae. 3 spermathecae, circular, most proximal part of sperm duct sclerotised so that spermathecae appear to be equipped with a kind of narrow stalk or handle.

The first instar larva has the labrum upturned and fused to the tentoropharyngeal sclerite (Shewell, 1987). The larvae are snail parasites so far as their biology is known.

The genus is Holarctic, and 8 species are recognised at present. In the Palaearctic Region 3 species, of which 2 occur in Fennoscandia and Denmark. The previous record of *Angioneura fimbriata* (Meigen) from Finland (Hackman, 1980) was based on a misidentified ♀ of *Morinia melanoptera* (Fallén).

Notes on synonymy. I find the ♂ and ♀ genitalia too different from those of *Eggisops* and *Melanomya* to follow Downes (1986) in uniting these genera under *Melanomya* (s.lat.).

Key to species of *Angioneura*

1 Prosternum hairy; lower calypter posteriorly broad with inner edge converging with longitudinal axis of fly; parafacialia with a few black setulae in upper part; costa bare on dorsal side of wing (Fig. 491); vein A_1+CuA_2 reaching halfway to wing-margin (Fig. 491); t_3 with strong *ad* and *d* preapical seta, *pd* preapical seta weak and hardly differentiated; ♂: frons very narrow, narrower than distance between posterior ocelli inclusive, lateroclinate and proclinate setae absent; t_2 without *v* seta; ♀: abdomen brownish-grey dusted

36. *acerba* (Meigen)

– Prosternum bare; lower calypter small, with inner edge diverging from longitudinal axis of fly (Fig. 502); parafacialia bare, without setulae in upper part; costa hairy dorsally far beyond junction with R_1 (Fig. 503); vein A_1+CuA_2 almost reaching wing margin (Fig. 503); t_3 with equally strong *ad*, *d* and *pd* preapical setae; ♂: frons broad, about 3 times as wide as distance between posterior ocelli inclusive, single lateroclinate and proclinate orbitals present; t_2 with a *v* seta; ♀: abdomen shining black

37. *cyrtoneurina* (Zetterstedt)

36. ***Angioneura acerba*** (Meigen, 1838)
Figs 477, 481, 487, 491-501.

Medoria acerba Meigen, 1838: 204.
Dexia pygmaea Zetterstedt, 1844: 1274.
Calobataemyia vetusta Brauer & Bergenstamm, 1891: 369.

♂♀. Occiput convex with black setulae. Arista with short hairs on the upper side, about as long as arista itself at base, practically bare on lower side. 0-1 *prst acr*. Prosternum setulose. Metasternal area bare or setulose. Postalar wall bare or with a few setulae. Apical scutellars very weak, about 0.2-0.5 times as long as the lateral pair (Fig. 487). Posterior thoracic spiracle almost circular, lappets as in Fig. 481. Tegula black, basicosta brownish-yellow, costagium pale, not contrasting with pale wing-veins. Halteres yellow. Thorax in dorsal view with dull brown dusting.

♂. Frons 0.071-0.080 times head-width (n=3). Fronto-orbital plates narrow, about as wide as arista, silvery dusted, somewhat brownish posteriorly. Parafacialia silvery dusted with shifting spots in upper parts. t_2 without *v* seta. Abdomen densely grey dusted with brownish-black mid-dorsal stripe of weak dusting and narrow hindmarginal brownish dusted bands. ST5 as in Fig. 492. Strong marginals on T3, T4 and T5. T5 also with discal setae. Ground-vestiture of T3 and T4 erect and rather setose. Apices of cerci separated to middle, slightly bent downwards apically in lateral view. Surstylus elongate-oval. Bacilliform sclerites almost joined at mid-line. Aedeagus with meso-hypophallus differentiated. Acrophallus with sclerotisations in lateral walls. Ventral plates separated mid-ventrally. Paraphallic processes separated to level of proximal edge of ventral plates (Fig. 497).

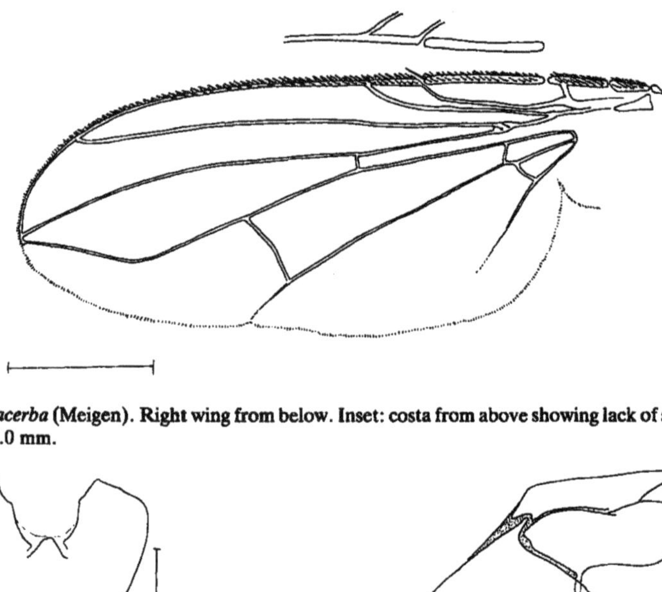

Fig. 491. *Angioneura acerba* (Meigen). Right wing from below. Inset: costa from above showing lack of small hairs on its upper side. Scale = 1.0 mm.

Fig. 492. *Angioneura acerba* (Meigen). ST5. Scale = 0.2 mm.

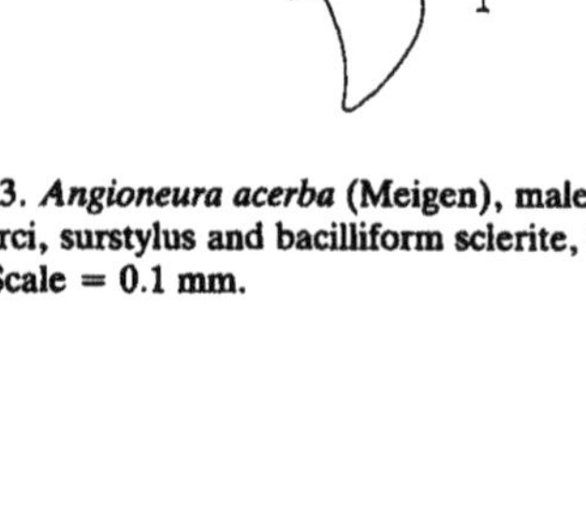

Fig. 493. *Angioneura acerba* (Meigen), male. Epandrium, cerci, surstylus and bacilliform sclerite, left lateral view. Scale = 0.1 mm.

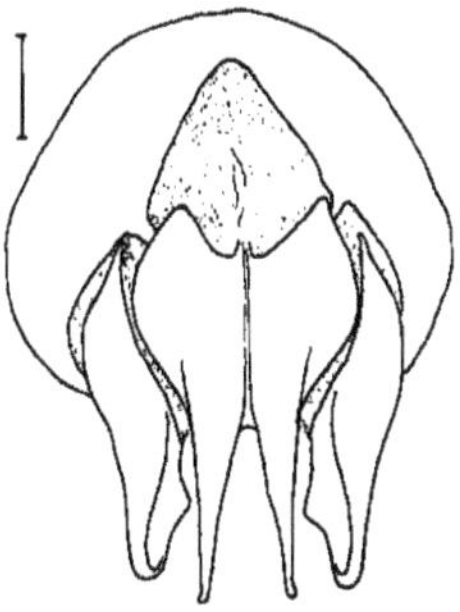

Fig. 494. *Angioneura acerba* (Meigen), male. Epandrium, cerci and surstyli, posterior view. Scale = 0.1 mm.

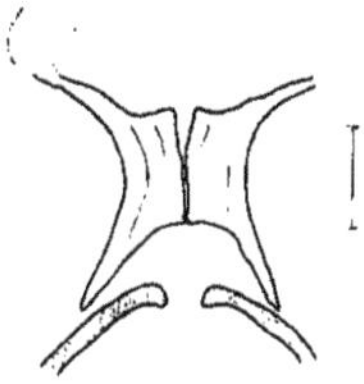

Fig. 495. *Angioneura acerba* (Meigen), male. Bacilliform sclerites. Scale = 0.1 mm.

Fig. 496. *Angioneura acerba* (Meigen), male. Aedeagus, left lateral view. Scale = 0.1 mm.

Fig. 497. *Angioneura acerba* (Meigen), male. Distiphallus, dorsal view.

Fig. 498. *Angioneura acerba* (Meigen), male. Ejaculatory sclerite. Scale = 0.1 mm.

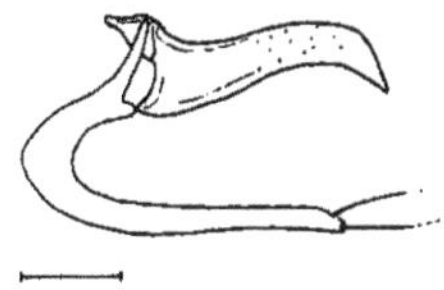

Fig. 499. *Angioneura acerba* (Meigen), male. Pre- (below) and postgonites. Scale = 0.1 mm.

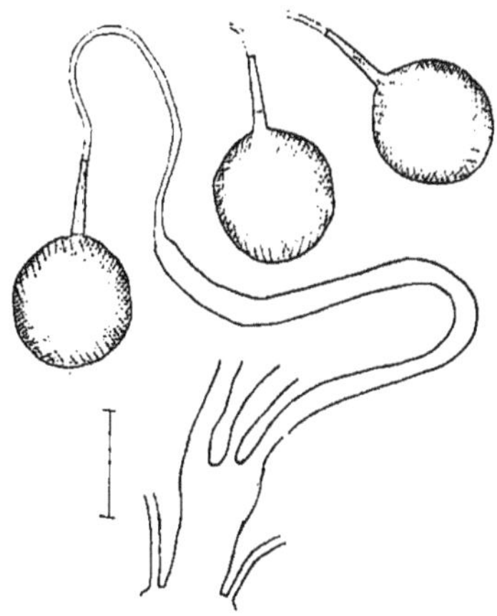

Fig. 500. *Angioneura acerba* (Meigen), female. Spermathecae, junction of spermathecal ducts. Scale = 0.1 mm. G.pr. 228.

♀. Frons 0.367-0.400 times head-width (n=4). Parafacialia and fronto-orbital plates greyish-brown dusted, former with weakly shifting spots in upper part. t_2 with a *v* seta. Abdomen greyish dusted with indistinct mid-dorsal stripe and hind-marginal bands of a different shade. Ground-vestiture weaker and more decumbent than in male. In ovipositor sensillae widespread apically on T7 halves and ST7, and subapically in a band across T8+epiproct. T8+epiproct incised in anterior third and pointed apically. Hypoproct distinct. Intersegmental membrane 7-8 armed with strong tooth-like microtrichiae. Sclerotised stalk of spermathecae more than half as long as circular main part. [3 ovipositor slides examined.]

Length. 3-5 mm.

Distribution. Rare. NEZ in Denmark, scattered in Sweden north to Dlr. A few records from AK in Norway and from southwestern Finland. - Europe, including the British Isles, south to the Pyrenees and Tyrol, east to Kr and Ukraine in the U.S.S.R. (Herting, 1961; Stackelberg, 1962, 1970).

Biology. Immature stages and life-cycle unknown. In Fennoscandia and Denmark adults

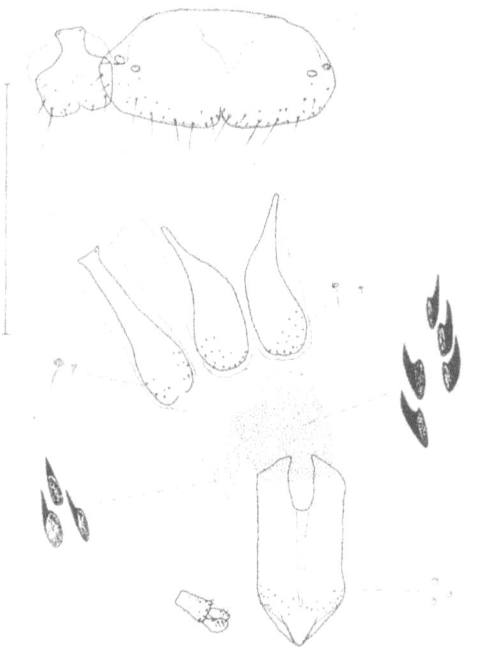

Fig. 501. *Angioneura acerba* (Meigen), female. Ovipositor. Stipple indicates extent of microtrichiae. Insets: shape of microtrichiae in indicated areas (enlarged), sensillae on sclerites (high magnification). Scale = 1.0 mm. G.pr. 228.

have been collected from May to October.

Nomenclature. *Medoria acerba* was described by Meigen from the ♂ sex and an unspecified number of specimens; no collector or locality was mentioned. A ♂ is illustrated on Meigen Plate 281, fig. 14. In the legends (Morge, 1975: 454, 489) it is named "Tachina acerba". In Meigen's collection in MNHN is a ♂ labelled (1) "Meigen" and "2090/40", (2) "Medoria / acerba [Meigen's hand]". It conforms to the current interpretation of *acerba* (cf. Herting, 1961) and is clearly the species shown on the plate. I have given it a holotype label. *Dexia pygmaea* was described by Zetterstedt from a single ♀ captured by Dahlbom in "... Gottlandia ... ad Gothem d. 17 Jul. 1841 ...". It is present in ZML and in good condition. I have given it a holotype label.

37. ***Angioneura cyrtoneurina*** (Zetterstedt, 1859)
Figs 482, 488, 502-513.

Tachina cyrtoneurina Zetterstedt, 1859: 6135.
Angioneura enigmatica Villeneuve, 1919: 307.

♂♀. Occiput convex with black hairs. Arista almost bare, hairs shorter than width of arista itself. 0 *prst acr*. Prosternum, metasternal area and postalar wall bare. Apical scutellars relatively long, at least half as long as the laterals, usually longer (Fig. 488). Posterior thoracic spiracle as in Fig. 482. Tegula black, basicosta brown to almost black, costagium infuscated, contrasting strongly with pale wing-veins. Halteres yellow. t_2 with *v* seta. Thorax in dorsal view appearing shining black, silvery dusted bands present over the notopleuron and humeral calli. A row of strong marginals on each tergite, some discals on T3-T5.

♂. Frons 0.255-0.289 times head-width (n=4). Fronto-orbital plates broad, at least as broad as frontal vitta, silvery dusted, especially in front, with 1 lateroclinate and 1 proclinate orbital setae. Parafacialia broad, densely silvery dusted without shifting spots, bare. Abdomen shining black with bands of silvery dusting on tergites. The bands occupy the anterior 3/5 of tergites laterally but, from some angles, are interrupted along the midline anteriorly so that triangular dark markings are formed. ST5 as in Fig. 504. Apices of cerci separated for less than apical half, straight in lateral view. Surstylus short, almost circular. Bacilliform sclerites joined along mid-line. Mesohypophallus not differentiated. Paraphallic processes separated to a level far distad of ventral plates (Fig. 509).

♀. Frons 0.341-0.362 times head-width (n=2). Fronto-orbital plates broad, at level of most anterior proclinate orbital seta 1.5 times as broad as frontal vitta, densely silvery dusted anteriorly, dusting thinner posteriorly and at vertex fronto-orbital plates shining black. Parafacialia broad, densely silvery dusted. Abdomen shining black without dusting. In ovipositor sensillae present apically on T7 halves, ST7 and T8+epiproct. T8+epiproct incised anteriorly almost to posterior end, posteriorly transverse. Hypoproct absent (lost during dissection?). Intersegmental membrane 7-8 without microtrichiae. Sclerotised stalk of spermathecae much less than half as long as main part. [1 ovipositor slide examined, from British material.]

Length. 4-5 mm.

Distribution. Very rare. 4 records from NWZ and NEZ in Denmark. A single record from Sweden (Upl., type-locality). Absent from Norway and Finland. - British Isles, France, Germany, Austria and Czechoslovakia, European part of the U.S.S.R. (Villeneuve, 1919; Wainwright, 1932; Herting, 1961; Stackelberg, 1962, 1970; Čepelák & Rozkošný, 1968).

Biology. Third instar larva and puparium de-

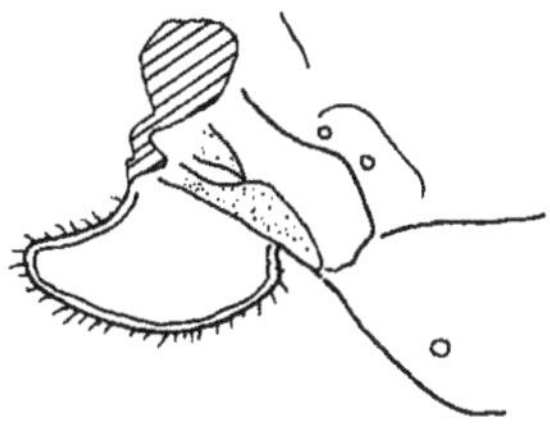

Fig. 502. *Angioneura cyrtoneurina* (Zetterstedt). Area behind wing base, dorsoposterior view.

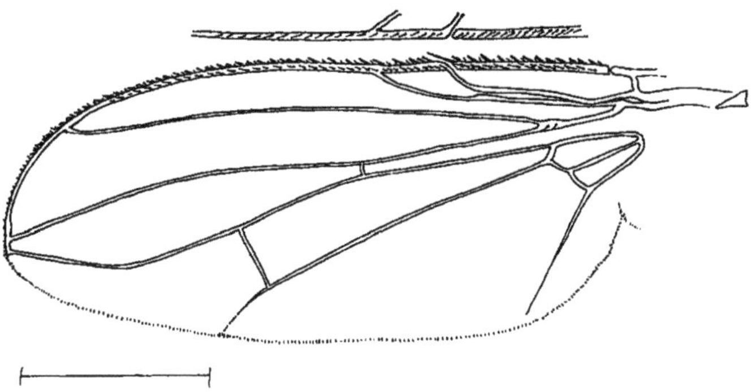

Fig. 503. *Angioneura cyrtoneurina* (Zetterstedt), male. Right wing from below. Inset: costa from above showing distribution of small hairs on its upper side. Scale = 1.0 mm.

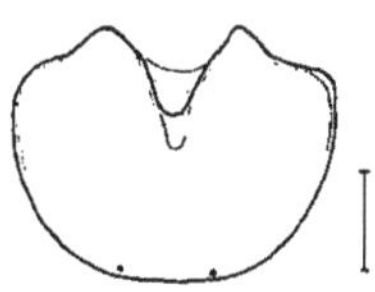

Fig. 504. *Angioneura cyrtoneurina* (Zetterstedt), male. ST5. Scale = 0.2 mm.

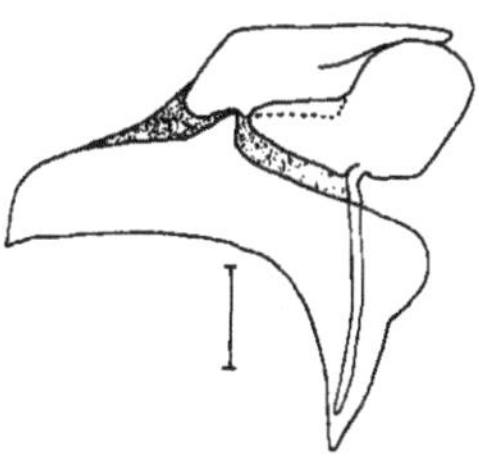

Fig. 505. *Angioneura cyrtoneurina* (Zetterstedt), male. Epandrium, cerci, surstylus and bacilliform sclerite, left lateral view. Scale = 0.1 mm.

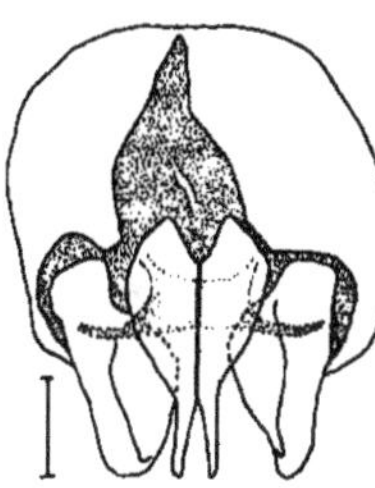

Fig. 506. *Angioneura cyrtoneurina* (Zetterstedt), male. Epandrium, cerci and surstyli, posterior view. Scale = 0.1 mm.

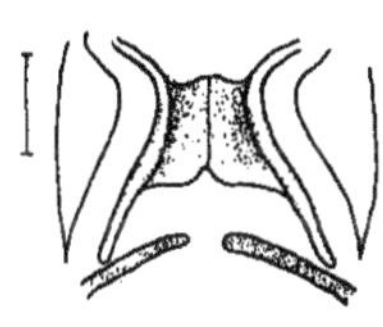

Fig. 507. *Angioneura cyrtoneurina* (Zetterstedt), male. Bacilliform sclerites. Scale = 0.1 mm.

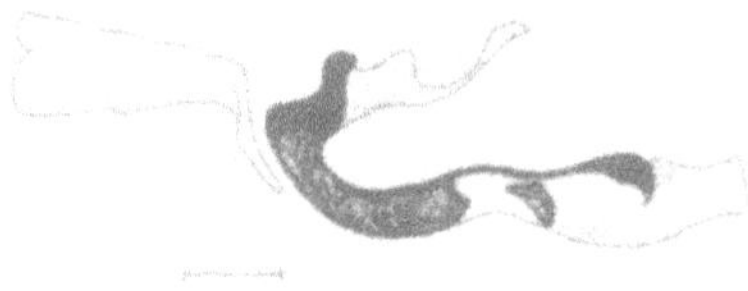

Fig. 508. *Angioneura cyrtoneurina* (Zetterstedt), male. Aedeagus and phallapodeme, left lateral view. Scale = 0.1 mm.

Fig. 509. *Angioneura cyrtoneurina* (Zetterstedt), male. Distiphallus, dorsal view. Scale = 0.1 mm.

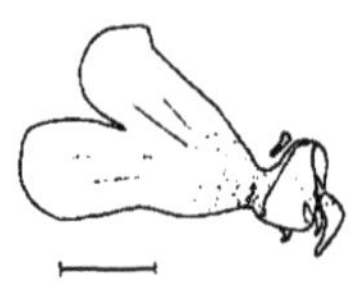

Fig. 510. *Angioneura cyrtoneurina* (Zetterstedt), male. Ejaculatory sclerite. Scale = 0.1 mm.

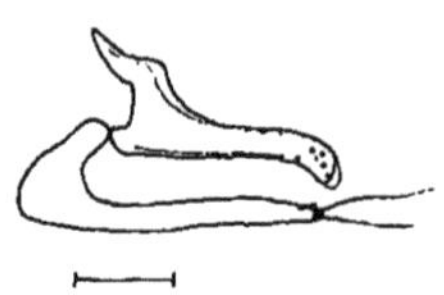

Fig. 511. *Angioneura cyrtoneurina* (Zetterstedt), male. Pre- (below) and postgonites. Scale = 0.1 mm.

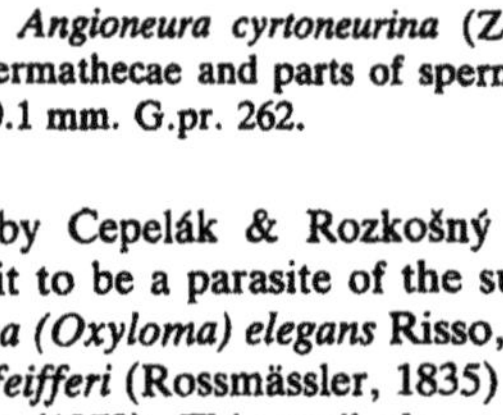

Fig. 512. *Angioneura cyrtoneurina* (Zetterstedt), female. Spermathecae and parts of spermathecal ducts. Scale = 0.1 mm. G.pr. 262.

scribed by Cepelák & Rozkošný (1968), who showed it to be a parasite of the succineid snail "*Succinea (Oxyloma) elegans* Risso, 1826" = *Oxyloma pfeifferi* (Rossmässler, 1835) of Kerney & Cameron (1979). This snail also occurs in Fennoscandia and Denmark and may well act as host there. In Fennoscandia and Denmark *cyrtoneurina* has been collected in June and August.

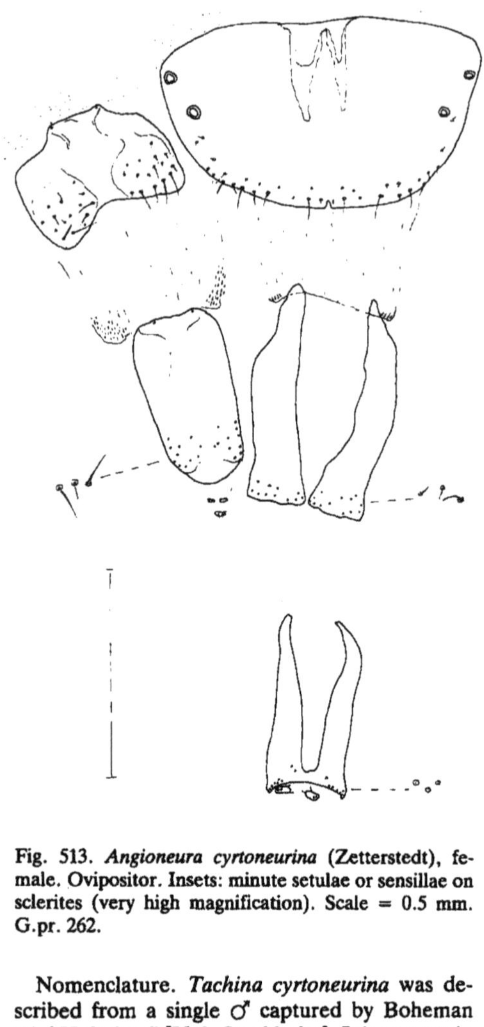

Fig. 513. *Angioneura cyrtoneurina* (Zetterstedt), female. Ovipositor. Insets: minute setulae or sensillae on sclerites (very high magnification). Scale = 0.5 mm. G.pr. 262.

Nomenclature. *Tachina cyrtoneurina* was described from a single ♂ captured by Boheman "Ad Holmiam" [Upl.:Stockholm]. It is present in NHRM and is in very good condition. It conforms to the current interpretation of *cyrtoneurina* (cf. Herting, 1961). I have given it a holotype label.

Genus *Eggisops* Rondani, 1862

Eggisops Rondani, 1862: 152, 177.
Type-species: *Eggisops pecchiolii* Rondani, 1862: 178, by monotypy.
Engyzops, *Engyops*: incorrect subsequent spellings.

Ground-colour black. Body dusted brownish-grey. Head profile as in Fig. 478, lower posterior part of head swollen. Occiput with black setulae. Parafacialia setulose for their whole length. Lunula setulose or with weak hairs. Arista with very short hairs, about as long as diameter of arista at base, or plumose with hairs almost as long as width of first antennal flagellomere. First flagellomere short, hardly longer than pedicel. Distance from tip of antenna to lower facial margin at least as long as width of first flagellomere. Palpi about as long as antenna, narrow or clavate. Chaetotaxy very variable. 1-3+2-3 *acr*. 2-2+3-4 *dc*. 1+2-4 *ia*, if only 2, middle one absent. Anterior *post ia*, if only 2 *post ia* are present, closer to suture than to posterior *post ia*. 2-7 *h*. 1-7 *ph* setae, outer present or absent. *Pra* close to suture, very strong, 2 times as long as hind *npl*, also longer than front *npl*. The first *sa* seta behind the *pra* very strong, the ones behind it weak. Prosternum bare. Proepisternal depression hairy. Metasternal area bare or setulose. Postalar wall usually with a few short setae. Cell r_{4+5} open or stalked. Lower calypter broad posteriorly. Calypters and wings more or less infuscated. f_1 with a few *av* setae apically. t_1 with 1-5 *pv* setae. t_2 with 2-3 *ad*. t_3 with 2-4 *av* and equally strong *ad*, *d*, and *pd* preapical setae. T1+2 with excavation very far from reaching hind margin, without median marginals. T3, T4 and T5 with full rows of marginal setae, with or without discal setae. Abdomen narrow. ST1-5 broad, broadly exposed, lateral margins of T2 overlapping margins of ST2. ST5 with shallow posterior incision in male. Posterior ends of hypandrial arms close together. Bacilliform sclerites with median extensions almost joining each other. Acrophallus long and thin without armature, forming an angle with the proximal part of aedeagus. Hypophallus short, strongly sclerotised with denticles externally and along ventral edge. Paraphalli short, thin, slightly curved, apically pointed, separated all the way to base of distiphallus. Mesohypophallus present. Pregonite with a few small apical setae. Postgonite without basal seta. In ovipositor T7 divided in basal half, T7 halves narrow. T8 divided in more than anterior half, halves narrow. ST7 of full length. ST8 absent. Hypoproct present. Intersegmental membranes 6-7 and 7-8 with characteristically shaped microtrichiae. 2 normal sized spermathecae, and 1 very small one. Uterus long and cylindrical, in mature females reaching anterior end of abdomen, with numerous first stage larvae.

First instar larva has teeth on the movable mandibles (lateral hooks) (Thompson, 1921b). The labrum (median hook) is upturned and fused to the tentoropharyngeal sclerite. The larvae are snail parasites so far as their biology is known (Emden, 1954; Downes, 1986).

Palaearctic, 2 species. A new species (*petiolata*) has recently been described from Japan (Pape, 1988). 1 species in Fennoscandia and Denmark.

Notes on synonymy. I find the ♂ and ♀ genitalia too different from those of *Angioneura* and *Melanomya* to follow Downes (1986) in uniting these genera under *Melanomya* (s.lat.).

38. ***Eggisops pecchiolii*** Rondani, 1862.
Figs 478, 483, 489, 514-524; Plate - fig 5.

Eggisops pecchiolii Rondani, 1862: 178.
Engyops macronyx Brauer & Bergenstamm, 1889: 168.
Morinia alpicola Strobl, 1894: 33.

♂♀. Arista plumose, aristal width (including hairing) about as broad as first flagellomere. Palpi brown. Posterior spiracle as in Fig. 483. Scutellum with 3-5 pairs of strong marginal setae (Fig. 489). Haltere yellow, head with some infuscate spots. t_1 with 1 *pv* seta. f_2 with or without *a* preapical seta. On all legs largest tibial setae at least twice as long as tibial diameter. Wings and calypters pale, more or less infuscated yellow. Costa with occasional hairs on dorsal side from base to junction with R_1. Cell r_{4+5} open. Vein A_1+CuA_2 almost reaching hind margin of wing (Fig. 514). Abdomen narrow, thinly dusted with distinct shifting pattern. Discal setae on T4 and T5.

♂. Frons 0.071-0.121 times head-width (n=5), at least as broad as ocellar triangle. Fronto-orbital plates separated. Frontal vitta 1-2 times width of fronto-orbital plates. Parafacialia 1.5 times as broad as first flagellomere. Parafacialia and fronto-orbital plates dusted dirty grey, more brownish towards vertex. Apices of cerci free and curving somewhat towards mid-line. Surstylus short, broad, in dorsal view stout and curving inwards apically.

♀. Frons 0.307-0.347 times head-width (n=5). T6 and ST6 in ovipositor with a proximal narrow "handle" and a wide setulose distal part. Spiracles in membrane outside T6 and ST6. Sensillae present on apical parts of T7, ST7 and T8+epiproct. T8+epiproct apically transverse. Hypoproct linear, elongate. [2 ovipositor slides examined.]

Length. 5-10 mm.

Distribution. 14 specimens are known from LFM and F in Denmark. Absent from Sweden, Finland and Norway. Also Austria, Czechoslovakia, East and West Germany, British Isles, Italy (Schumann, 1986) and France.

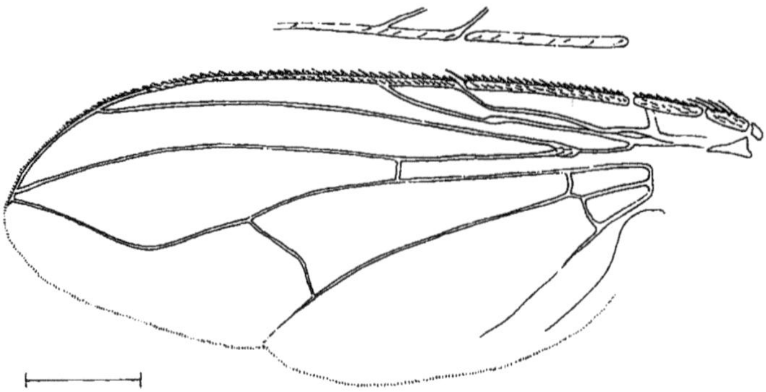

Fig. 514. *Eggisops pecchiolii* Rondani, male. Right wing from below. Inset: costa from above showing distribution of small hairs on its upper side. Scale = 1.0 mm.

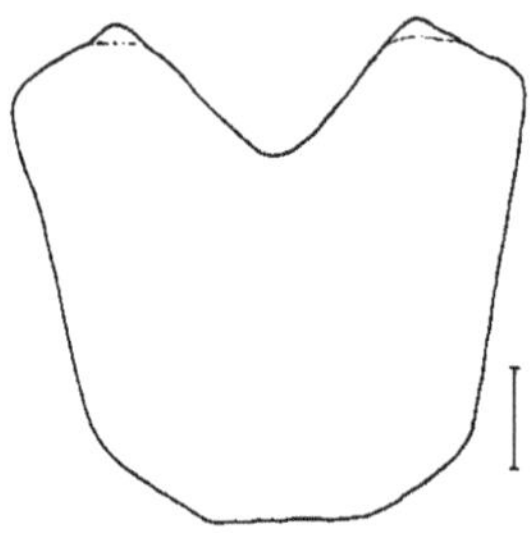

Fig. 515. *Eggisops pecchiolii* Rondani, male. ST5. Scale = 0.2 mm.

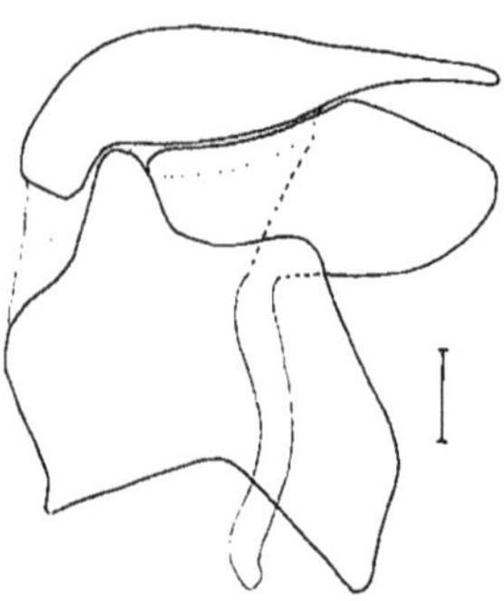

Fig. 516. *Eggisops pecchiolii* Rondani, male. Epandrium, cerci, surstylus and bacilliform sclerites, left lateral view. Scale = 0.1 mm.

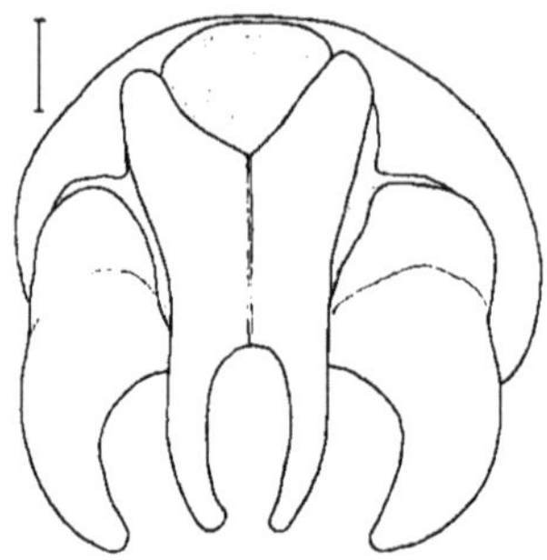

Fig. 517. *Eggisops pecchiolii* Rondani, male. Epandrium, cerci and surstyli, posterior view. Scale = 0.1 mm.

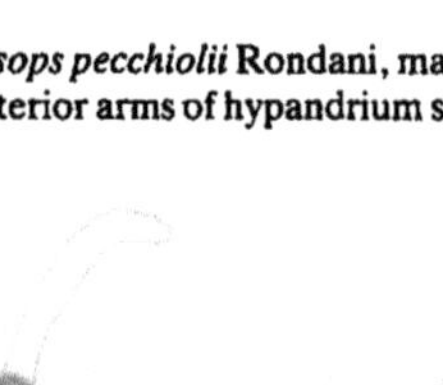

Fig. 518. *Eggisops pecchiolii* Rondani, male. Bacilliform sclerites. (Posterior arms of hypandrium stippled). Scale = 0.1 mm.

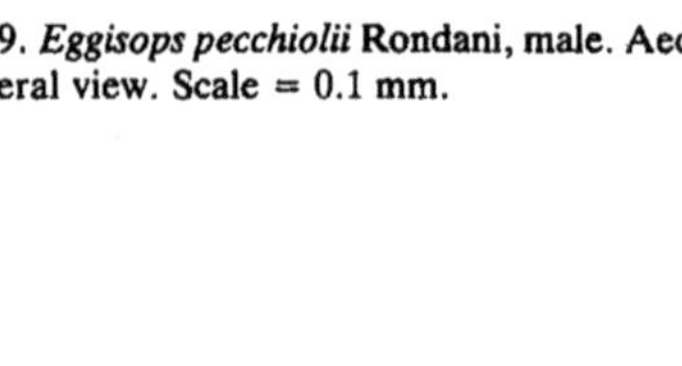

Fig. 519. *Eggisops pecchiolii* Rondani, male. Aedeagus, left lateral view. Scale = 0.1 mm.

Fig. 520. *Eggisops pecchiolii* Rondani, male. Aedeagus and right postgonite, dorsal view. Scale = 0.1 mm.

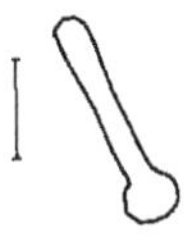

Fig. 521. *Eggisops pecchiolii* Rondani, male. Ejaculatory sclerite. Scale = 0.1 mm.

Fig. 522. *Eggisops pecchiolii* Rondani, male. Pre- and postgonites. Scale = 0.1 mm.

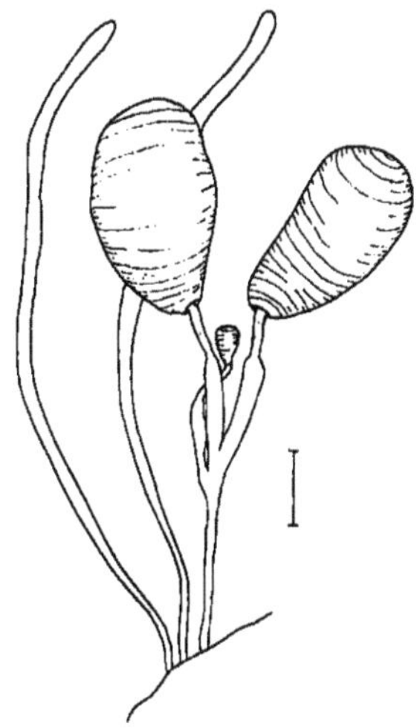

Fig. 523. *Eggisops pecchiolii* Rondani, female. Spermathecae, spermathecal ducts and accessory glands. Scale = 0.1 mm. G.pr. 227.

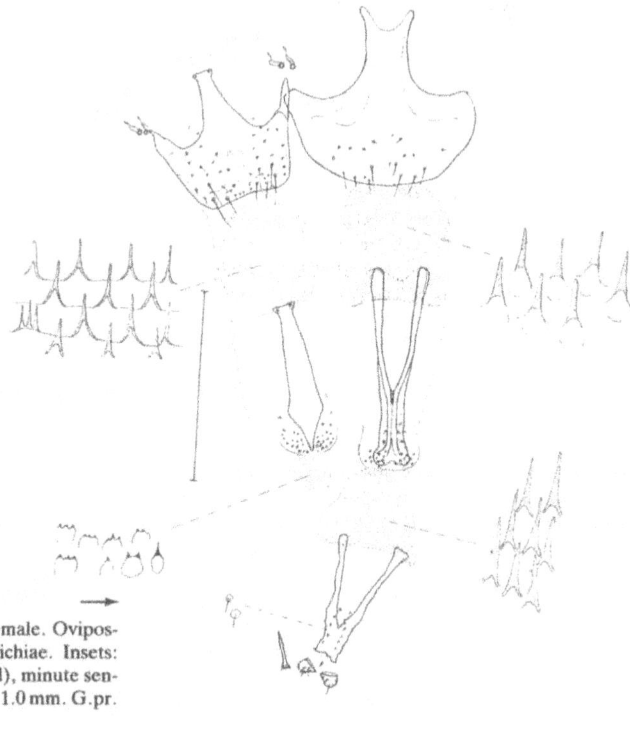

Fig. 524. *Eggisops pecchiolii* Rondani, female. Ovipositor. Stipple indicates extent of microtrichiae. Insets: microtrichiae in areas indicated (enlarged), minute sensillae on T8 (high magnification). Scale = 1.0 mm. G.pr. 227.

Biology. Viviparous. I found 63 first instar larvae in the uterus of a female from D, Bayern: Hohenschwangau (KR). First instar larvae described by Thompson (1921b). Hosts unknown. In Denmark adults have been collected in June and July.

Genus *Melanomya* Rondani, 1856

Melanomya Rondani, 1856: 88.
Type-species: *Dexia nana* Meigen, 1826: 37, by original designation.
Melanomyia, incorrect subsequent spelling.

Ground-colour black. Body thinly dusted brownish-grey. Head profile as in Fig. 479. Parafacialia hardly half as wide as first flagellomere, with a few setulae in a single row in upper half. Occiput with black setulae. Lunula bare. Arista plumose, about as wide (hairing included) as first antennal flagellomere. First flagellomere 1.5-2 times as long as pedicel. Palp about as long as first flagellomere, brownish-black. 1+2-3 *acr* (first ones behind suture weak). 2+3 *dc*. 0+2 *ia*, anterior *ia* closer to suture than to posterior *ia*. 2-3 *h*, inner ones weak. 1 *ph*, outer absent. *pra* strong, near suture, as long as anterior *npl*, more than 2 times as long as posterior *npl*. A single strong *sa* behind *pra*. Prosternum with 1-2 inconspicuous hairs on margin (most often) or bare. Proepisternal depression, metasternal area and postalar wall bare. Scutellum with a weak apical and a strong lateral pair of marginal setae (Fig. 490). Posterior spiracle as in Fig. 484, elongate and very narrow dorsoventrally. Lower calypter small and semicircular, inner edge diverging from longitudinal axis of body. Costa bare on dorsal side of wing. Cell r_{4+5} open at apex of wing. Vein A_1+CuA_2 reaching halfway to margin. Haltere black. t_3 with strong *ad* and *d* preapical setae, a *pd* preapical not differentiated. Abdomen narrow. ST2-5 broadly exposed, margins of ST2 overlapped by margins of T2. Full row of marginals on T3-T5, some discals on T4 and T5. Bacilliform sclerites with median extension, quite separated. Acrophallus broad and thick, with rod-like longitudinal sclerotisations in lateral wall. Mesohypophallus absent. Aedeagal membrane without armature. Paraphalli heavily sclerotised, beak-like, separated to base of distiphallus. Pregonite triangular with a few apical setae. Postgonite with a basal seta. According to Tschorsnig (1985a) the basal piece ("Gelenkfortsatz") is fused to the main part of the postgonite. In the specimen figured (Fig. 533) this is not the case. In ovipositor T7 shaped as long rods, partly joined in apical 1/4. T8+epiproct broad, a narrow unsclerotised zone mid-dorsally far backwards from anterior edge. ST7 sclerotised in anterior half only. ST8 absent. Intersegmental membrane 7-8 with microtrichiae shaped as in Fig. 535. Uterus resembling that of *Eurychaeta* in having a large forwardly-directed blind pouch, and in having the accessory gland ducts and spermathecal ducts entering the uterus dorsally far from tip. 3 small spherical spermathecae.

Palaearctic, 1 species.

Notes on synonymy. I find the ♂ and ♀ genitalia too different from those of *Angioneura* and *Eggisops* to follow Downes (1986) in uniting these genera under *Melanomya* (s.lat.).

39. *Melanomya nana* (Meigen, 1826)
Figs 479, 484, 490, 525-535.

Dexia nana Meigen, 1826: 37.
Morinia parva Robineau-Desvoidy, 1830: 265.
Musca minima Zetterstedt, 1838: 654.

♂♀. Parafacialia and fronto-orbital plates white, latter brownish posteriorly. t_1 with 1 *ad*, 0-1 *pv* setae. t_2 with 1 *ad* and 1 *v* setae. f_2 with *a* preapical seta in both sexes. t_3 with 2 *av* setae.

♂. Frons 0.063-0.100 times head-width (n=5). Fronto-orbital plates separated. Frontal vitta about as broad as each fronto-orbital plate. Cerci with upper edge sinuate in profile. In dorsal view divided to middle, apices pointed. Surstylus short and apically blunt in lateral view, in dorsal view curved towards mid-line.

♀. Frons 0.338-0.366 times head-width (n=3). Frontal vitta about 2 times as wide as fronto-orbital plates. These are rather shining at vertex. In ovipositor T6 and ST6 of characteristic shape, strongly setose. A narrow zone of microtrichiae behind ST6. No sensillae on T7 or ST7. T8+epiproct posteriorly rounded, with sensillae scattered over apical 2/3. ST8 absent. Hypoproct broad. [2 ovipositor slides examined.]

Length. 3-4.5 mm.

Distribution. Widespread in Denmark and South Norway. In Sweden north to Nb., in Finland north to Ks. - Widely distributed in Europe. North Kazakhstan (Stackelberg, 1970). Japan (Hokkaido) (Pape, 1988).

Biology. Immature stages and hosts unknown. In Fennoscandia and Denmark adults have been collected from June to September.

Nomenclature. *Dexia nana* was described by Meigen from several specimens, "Sehr gemein im Frühling und Sommer in Hekken, auf Schirm-

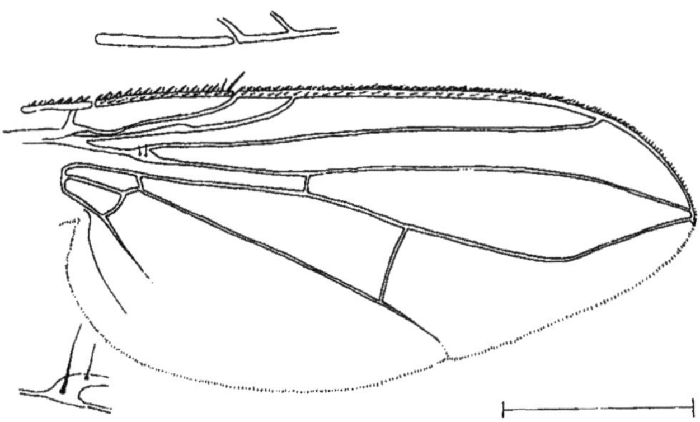

Fig. 525. *Melanomya nana* (Meigen), male. Right wing from below. Insets: parts of costa, node at junction of R2+3 and R4+5, both from above (enlarged). Scale = 1.0 mm.

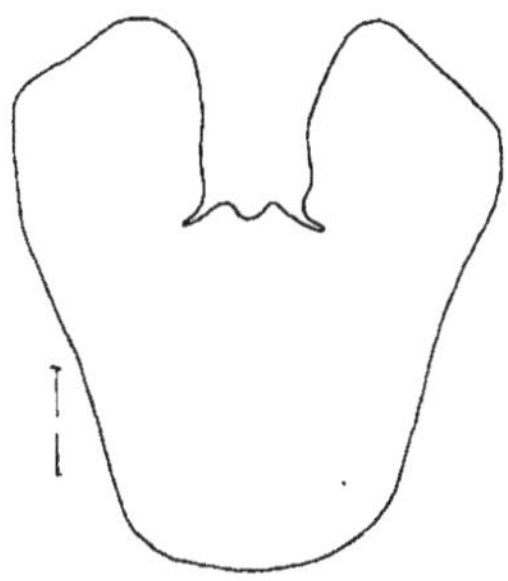

Fig. 526. *Melanomya nana* (Meigen), male. ST5. Scale = 0.1 mm.

Fig. 527. *Melanomya nana* (Meigen), male. Epandrium, cerci and surstylus, left lateral view. Scale = 0.1 mm.

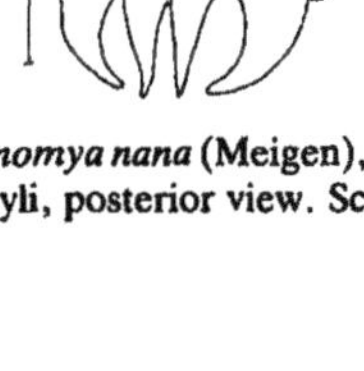

Fig. 528. *Melanomya nana* (Meigen), male. Epandrium, cerci and surstyli, posterior view. Scale = 0.1 mm.

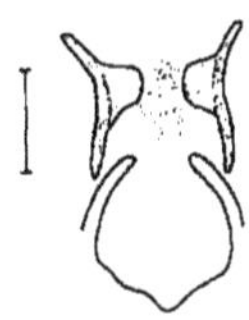

Fig. 529. *Melanomya nana* (Meigen), male. Bacilliform sclerites (posterior hypandrial arms below). Scale = 0.1 mm.

Fig. 530. *Melanomya nana* (Meigen), male. Aedeagus, right lateral view. Scale = 0.1 mm.

Fig. 531. *Melanomya nana* (Meigen), male. Aedeagus, dorsal view. Scale = 0.1 mm.

Fig. 532. *Melanomya nana* (Meigen), male. Ejaculatory sclerite. Scale = 0.1 mm.

Fig. 533. *Melanomya nana* (Meigen), male. Post- and pregonites. Scale = 0.1 mm.

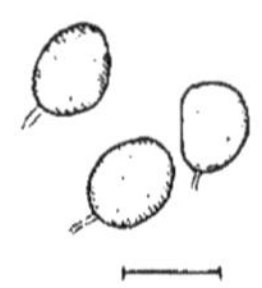

Fig. 534. *Melanomya nana* (Meigen), female. Spermathecae. Scale = 0.1 mm. G.pr. 186.

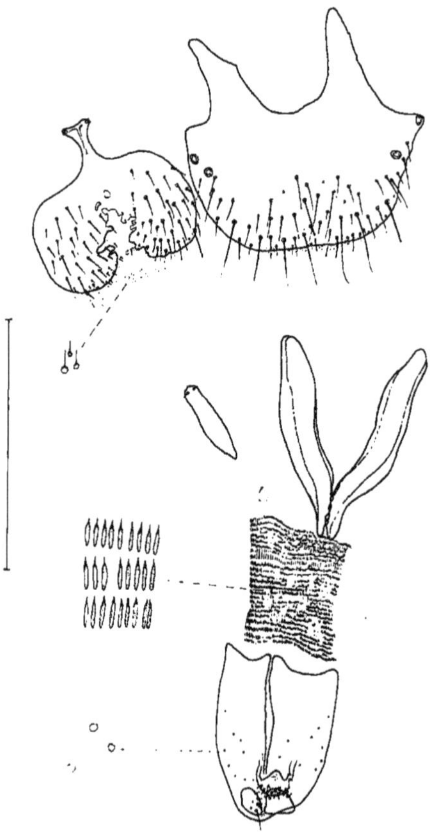

Fig. 535. *Melanomya nana* (Meigen), female. Ovipositor. Stipple indicates extent of microtrichiae. Insets: shape of microtrichiae in areas indicated (enlarged), shape of sensillae on T8 (high magnification). Scale = 0.5 mm. G.pr. 186.

gewächsen und im Grase". It was assigned to *Morinia* by Meigen (1838: 275). Meigen's Plate 114, fig. 7 shows both sexes, Plate 296, fig. 14 shows a ♂ (Morge, 1976a, 1976b). In Meigen's collection in MNHN are 3♂3♀ under *nana*, all conforming to the current concept of *nana* (cf. Herting, 1961). 5 are labelled "Meigen" and "2042/40", whilst the sixth has the number "2955/40". 1♂2♀ have additional determination labels in Meigen's hand. I have labelled and here designate a ♂ pinned from the side and in good condition as the lectotype of *Dexia nana*. The remaining 5 specimens have been labelled as paralectotypes. *Musca minima* was described from ♂♀ syntypes collected "Lapponia Norvegica; ... in Dowre ... D. BOHEMAN [ON or STI] ... Suecia MARKLIN ... Gottland., ipse ...". Only a single ♂ labelled "M. minima / ♂. Suec. Markl." in Zetterstedt's hand is present in ZML. I have labelled it and here designate it as lectotype.

Genus *Melinda* Robineau-Desvoidy, 1830

Melinda Robineau-Desvoidy, 1830: 439.
Type-species: *Musca coerulea* Meigen, 1826: 63, by designation of Hendel, 1901: 32; = *Melinda gentilis* Robineau-Desvoidy, 1830.
Xerophilophaga Enderlein, 1933: 120.
Type-species: *Melinda gentilis* Robineau-Desvoidy, 1830: 441, by original designation.

Ground-colour black. Head profiles as in Fig. 480. Lower facial margin somewhat projecting below and in front of vibrissae. Parafacialia densely silvery dusted with shifting spots, setulose in upper part. Fronto-orbital plates silvery dusted except for extreme posterior part. Occiput with short black setae. First antennal flagellomere about 2 times as long as pedicel. Arista plumose. Palpi translucent yellowish-brown to almost black. Thorax and abdomen metallic blue, with white dusting. Thorax dorsally with stripes. 2+3 *acr*, posterior *prst acr* far in front of suture. 2-3+3 *dc*. 0-1+3 *ia*. 3+1 *h*, basals in a broken line. 1-2+1 *ph*, outer present and outside a line through *prs*. Prosternum, proepisternal depression, metasternal area and postalar wall hairy. Posterior spiracle dark with large anterior lappets (Figs. 485, 486). Scutellum with 3 pairs of strong marginals. Lower calypters white, bare on dorsal surface, posteriorly broad with inner edge converging with longitudinal axis of fly. Wing-veins dark. Costa hairy below beyond junction with R_1 (Fig. 536). t_1 with 1 *pv* and full row of short *ad* setae. f_2 with *a* preapical seta in both sexes. t_2 with 3-4 *ad*, *v* seta present in both sexes. t_3 with 3-5 *av* setae, *ad* and *d* preapical setae strong and of the same size, a *pd* preapical hardly differentiated. Abdomen with at least some white dusting, basal excavation practically reaching margin of T1+2. Full marginal rows of strong setae on T3 and T4. Discal setae on T5. Bacilliform sclerites in male with median extensions. In distiphallus hypophallus shaped as laterally situated dentate ridges. All other parts of aedeagal membrane without armature. Paraphallus long and apically curved. Ventral plates fused mid-ventrally and continuing apicad to some extent. Mesohypophallus absent. Pregonite with 2-3 small apical setae. Postgonite with strong basal seta. In ovipositor T6 and ST6 without setae, only small sensillae present (*M. septentrionis* Xue, 1983 appears to have ordinary setae here). T6 with anterior areas of unsclerotised membrane. T7 with U-shaped areas of strong sclerotisation. T8+epiproct apically more or less pointed. ST7 sclerotised only at the extreme proximal end around the alphasetae (cf. also *M. septentrionis*), but outline of unsclerotised part indicated by distribution of sensillae distally. ST8 present as a small narrow rod in front of the hypoproct with a few apical sensillae. Hypoproct microtrichiose. 3 oval or almost spherical spermathecae. (*Paradichosia* ♀ appears to have a completely sclerotised ST7, cf. Kano, 1962; Kurahashi, 1970; Lobanov, 1976).

First instar larvae of *Melinda* have small teeth on the movable mandibles (lateral hooks) (Keilin, 1919). The labrum (median hook) is upturned and fused to the tentoropharyngeal sclerite.

Oviparous. Parasite of pulmonate snails. Eggs are laid in the pulmonary cavity (Schmitz, 1917; Keilin, 1919).

Palaearctic. 6 species; 5 are listed by Schumann (1986), but *Melinda septentrionis* Xue, 1983, from China (Liaoning, Benxi) was not included in the Palaearctic Catalogue. 2 species in Fennoscandia and Denmark.

Note. The nomenclature of the North European species has had a confused history, involving several names, mainly due to the fact that various authors have not been able to examine all the relevant type material (Emden, 1954; Schumann, 1973) or have misidentified it (Séguy, 1928a).

Key to species of *Melinda*

1 *prst ia* usually absent; palpi black; ♂: frons about as wide as first flagellomere, broader than posterior ocelli inclusive; T3 and T4 with dorsal erect ground-vestiture much denser and finer than the ventral one; ♀: frons narrower than each eye when seen from above; T4 very often with some discal setae; length of T5 1.5-2 times length of T4; T3 and T4 in some lights appearing dusted at antero-lateral corners only, without tessellations; T6 in ovipositor shining black all over, also hind margin

40. *viridicyanea* (Robineau-Desvoidy)

– *prst ia* usually present; palpi often more translucent brownish; ♂: frons narrower than first flagellomere, narrower than posterior ocelli inclusive; T3 and T4 with dorsal erect ground-vestiture not denser or finer than the ventral one; ♀: frons about as wide as an eye when seen from above; T4 without discal setae; length of T5 1.5 times length of T4; T3 and T4 more strongly dusted all over, with some weak tessellations; T6 in ovipositor with conspicuously dusted hind margin

41. *gentilis* Robineau-Desvoidy

40. *Melinda viridicyanea* (Robineau-Desvoidy, 1830)

Figs 480, 485, 536-546.

Onesia viridicyanea Robineau-Desvoidy, 1830: 368.

Musca anthracina Meigen, 1838: 301.

cognata: authors (e.g. Keilin, 1919; Schumann, 1973; 1986), not Meigen, 1830; misidentifications.

coerulea: authors (e.g. Wainwright, 1928; Emden, 1954; Zumpt, 1956a; Grunin, 1970a), not Meigen, 1826; misidentifications.

♂♀. Mesonotum in posterior view with 3 dark stripes before the suture, the lateral one rather broad. *prst ia* almost always absent, occasionally present.

♂. Frons 0.072-0.087 times head-width (n=5). Frontal vitta at most as wide as fronto-orbital plates, not interrupted, though narrow. Cerci evenly tapering from base and apically cleft to middle in dorsal view, hook-shaped apically when seen in profile. Surstylus almost circular in distal part, somewhat shorter than cerci. Anterior part of epandrium rounded in profile, not projecting. Paraphalli separated to the level of middle of ventral plates, evenly diverging, rather thick, apically weakly curved and pointed.

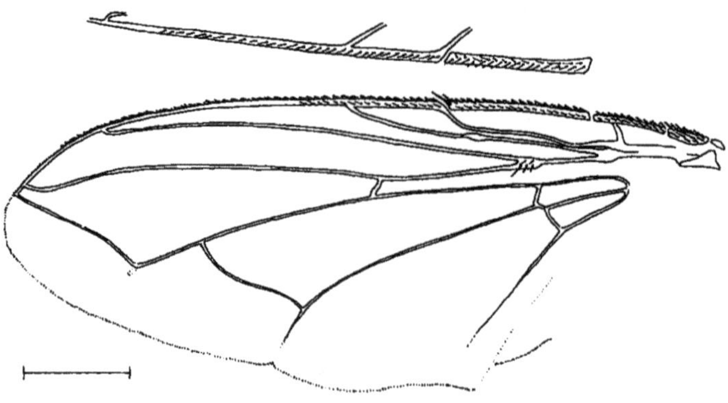

Fig. 536. *Melinda* spp., wing. Main figure: *M. gentilis* Robineau-Desvoidy, male, underside. Inset: costa from below of *M. gentilis*, female, and *M. viridicyanea* (Robineau-Desvoidy), male and female. Scale = 1.0 mm.

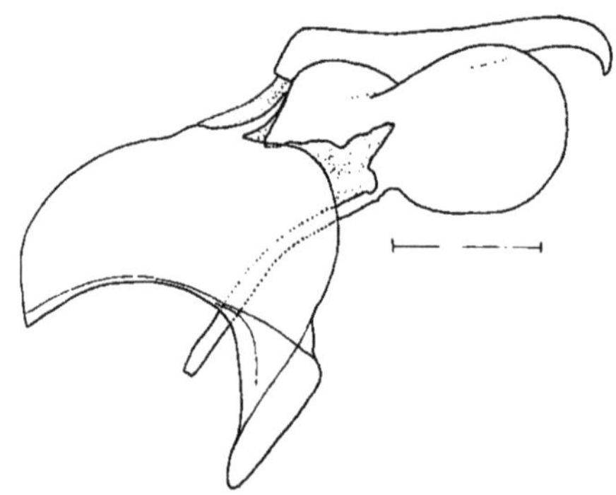

Fig. 537. ***Melinda viridicyanea*** **(Robineau-Desvoidy), male. Epandrium, cerci, surstylus and bacilliform sclerite, left lateral view. Scale = 0.2 mm.**

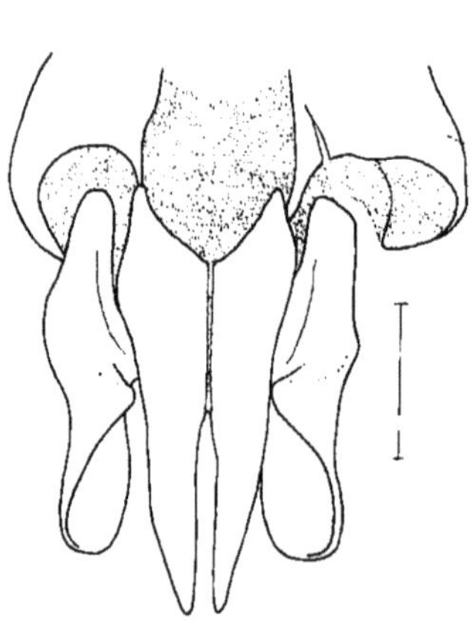

Fig. 538. ***Melinda viridicyanea*** **(Robineau-Desvoidy), male. Epandrium (parts), cerci and surstyli, posterior view. Scale = 0.2 mm.**

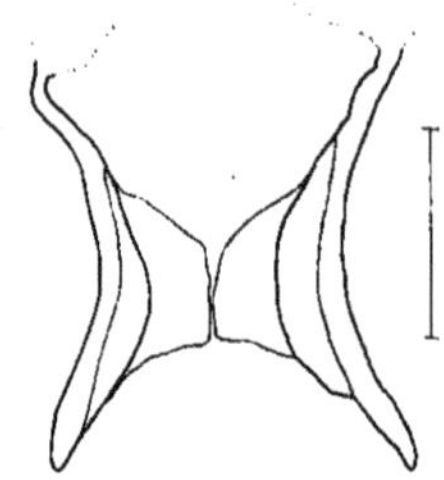

Fig. 539. ***Melinda viridicyanea*** **(Robineau-Desvoidy), male. Bacilliform sclerites. Scale = 0.2 mm.**

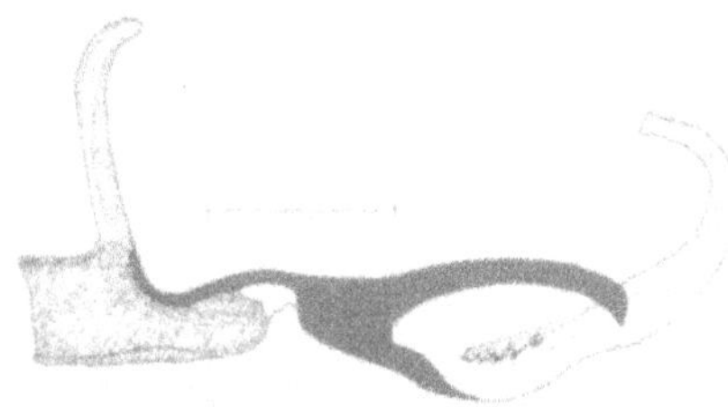

Fig. 540. ***Melinda viridicyanea*** **(Robineau-Desvoidy), male. Aedeagus, left lateral view. Scale = 0.2 mm.**

Fig. 541. ***Melinda viridicyanea*** **(Robineau-Desvoidy), male. Aedeagus, dorsal view. Scale = 0.2 mm.**

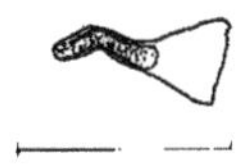

Fig. 542. *Melinda viridicyanea* (Robineau-Desvoidy), male. Ejaculatory sclerite. Scale = 0.2 mm.

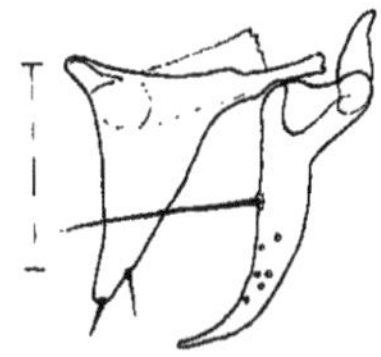

Fig. 543. *Melinda viridicyanea* (Robineau-Desvoidy), male. Pre- and postgonites. Scale 0.2 mm.

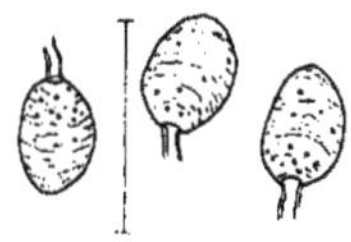

Fig. 544. *Melinda viridicyanea* (Robineau-Desvoidy), female. Spermathecae. Scale = 0.2 mm. G.pr. 209.

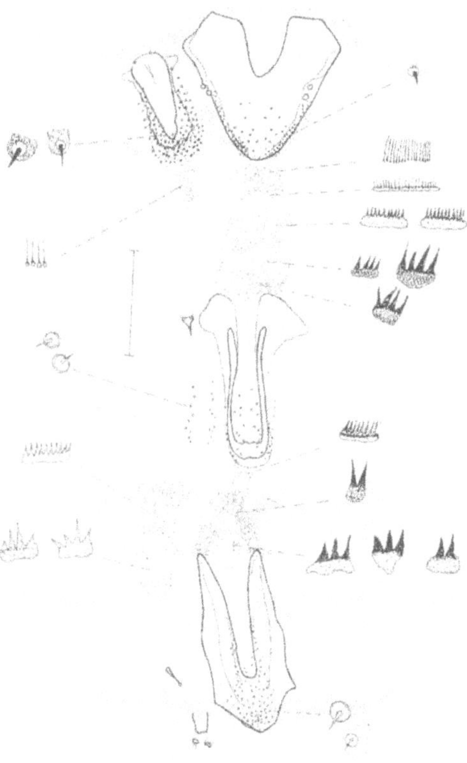

Fig. 545. *Melinda viridicyanea* (Robineau-Desvoidy), female. Ovipositor. Stipple indicates extent of microtrichiae. Insets: shape of microtrichiae in indicated areas (enlarged), minute sensillae on sclerites (high magnification). Scale = 0.8 mm. G.pr. 209.

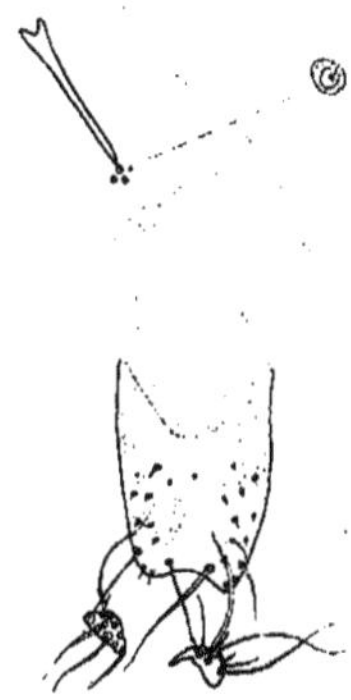

Fig. 546. *Melinda viridicyanea* (Robineau-Desvoidy), female. ST8, hypoproct and cerci of ovipositor (enlarged). Stipple indicates extent of microtrichiae. Inset: minute sensillae at tip of ST8 (very high magnification). G.pr. 209.

♀. Frons 0.262-0.292 times head-width (n=5). T6 rather broad basally. T7 with sensillae over most of distal half. T8+epiproct with sensillae mid-dorsally in distal half, also apically. Cerci with rather few pale setae. ST6 entire, sensillae along margins almost to base. ST7 with sensillae along margin of distal half of unsclerotised part. Apical border of hypoproct concave. Intersegmental membranes 6-7 and 7-8 dorsally and intersegmental membrane 7-8 ventrally with numerous, rather large multidentate sclerotisations. [3 ovipositor slides examined.]

Length. 5-9 mm.

Distribution. Common in Denmark. Widespread in southern Norway, north to NSI. In Sweden north to Jmt. In Finland only known from N. - Widely distributed in Western Palaearctic, including North Africa, east to Kazakhstan (Schumann, 1986; Grunin, 1970a).

Biology. Immature stages and biology described by Keilin (1919) and Schumann (1973). Proved hosts: *Helicella virgata* (da Costa) (Keilin, 1919), "*Patula rotundata*" = *Discus rotundatus* (Müller) (Schmitz, 1917). The map in Kerney & Cameron (1979) of the latter species fits the known distribution of *viridicyanea* in Fennoscandia and Denmark rather well. In Fennoscandia and Denmark adults have been taken from April to October.

Nomenclature. *Onesia viridicyanea*, found rarely "parmi les fleurs des prairies", was described by Robineau-Desvoidy from an unspecified number of specimens. No specimen remains in Robineau-Desvoidy's collection in MNHN. In the general collection of MNHN under *Onesia* is a ♀ under *viridicyanea*. It is labelled (1) "195", (2) "viridicyanea" in Macquart's hand. It is probably on the basis of this specimen identified by Macquart that Robineau-Desvoidy's name has been associated with a species of *Melinda* (cf. Séguy, 1928a: 126; Zumpt, 1956a; Schumann, 1973, 1986). The head and thorax is a ♀ *Melinda* with a rather narrow frons, without *prst ia* seta and with 3 broad presutural dark stripes, and thus = "*cognata*" of Schumann (1973). However, the abdomen, which is not conspecific with the rest, has been glued to the thorax and is from a *Bellardia* ♀. I do not accept this specimen as a syntype. However, in the interest of nomenclatural stability I find it desirable to use *viridicyanea* for this species, as it is not one of the names which has so far contributed to the confused nomenclature of the North European species of *Melinda*. I have therefore labelled and here designate a recent specimen in good condition as neotype of *Onesia viridicyanea* Robineau-Desvoidy, in order to fix its identity (= "*cognata*" of Schumann, 1973). It is labelled (1) "316", (2) "MUSEUM PARIS / RAMBOUILLET / (Seine-à-Oise) / E. SÉGUY 19". I have given it a neotype label and placed it under *Onesia* in box 12 of Robineau-Desvoidy's collection in Paris. *Musca anthracina* was described by Meigen from the ♀ sex and from several specimens, "...mehrmahlen in hiesiger Gegend, auch aus der Lütticher...". Meigen Plate 296, fig. 16 also shows a ♀. In Meigen's collection in MNHN are 3♀, all labelled "Meigen" and "2139/40". All have a narrow frons, lack a *prst ia* seta and have the T6 of the ovipositor, whose tip is visible, shining black and wholly without dusting marginally, thus fitting "*cognata*" of Schumann (1973). I have labelled and here designate as lectotype a specimen labelled "anthra-/cina/♀" in Meigen's hand and also with Schumann's determination label. The other two ♀ have been labelled as paralectotypes.

41. ***Melinda gentilis*** Robineau-Desvoidy, 1830
Figs 486, 536, 547-555.

Musca coerulea Meigen, 1826: 63. Junior primary homonym of *Musca coerulea* Wiedemann, 1819: 23. SYN.N.
Melinda gentilis Robineau-Desvoidy, 1830: 441.
Musca cognata Meigen, 1830: 374. Replacement name for *Musca coerulea* Meigen, 1826. SYN.N.
Onesia villeneuvi Kramer, 1917: 284.
Xerophilophaga noelleri Enderlein, 1933: 121.
anthracina: Wainwright, 1928, Emden, 1954, not Meigen, 1838; misidentifications.

♂♀. Mesonotum in posterior view with 5 dark stripes before the suture, the lateral one of *viridicyanea* being subdivided by dusted stripe. *prst ia* almost always present, very rarely absent.

♂. Frons 0.042-0.058 times head-width (n=5). Fronto-orbital plates touching. Cerci heart-shaped, broadest at middle in dorsal view, apically cleft in apical fifth. In profile slightly bent downwards apically. Surstylus triangular and pointed apically. Anterior part of epandrium strongly projecting, shaped as a triangle. Paraphalli separated only apically, in profile rather abruptly curved distally, apex blunt, almost transverse.

♀. Frons 0.311-0.333 times head-width (n=5). T6 rather narrow basally. T7 and T8+epiproct each with sensillae in a small subapical transverse zone. Cerci with numerous pale setae. ST6 divided distally in two sclerotisations, sensillae restricted to these parts. ST7 with a few sensillae only at extreme distal tip of unsclerotised part. Apical border of hypoproct convex. Intersegmental membranes 6-7 and 7-8 dorsally and ventrally only with numerous, small microtrichiae, strikingly short on ventral side of ovipositor. [3 ovipositor slides examined.]

Length. 5-10 mm.

Distribution. Common in Denmark. Widespread in southernmost part of Norway. In Sweden north to Dlr. Not known from Finland. - Widespread in the Palaearctic, south to Tunisia (Zumpt, 1956a), east to Far East and Japan (?) (Schumann, 1986).

Biology. Immature stages and biology described by Keilin (1919) and Schumann (1973). Hosts: *Helicella virgata* (da Costa), *H. candidula* (Studer), *H. ericetorum* (Müller) (Helicidae) (Keilin, 1919; Enderlein, 1933). According to maps in Kerney & Cameron (1979), *H. virgata* is absent from Fennoscandia and Denmark, *H. candidula* occurs in SZ, Sk. and Gtl., and *H. ericetorum* in Zealand and eastern Jutland in Den-

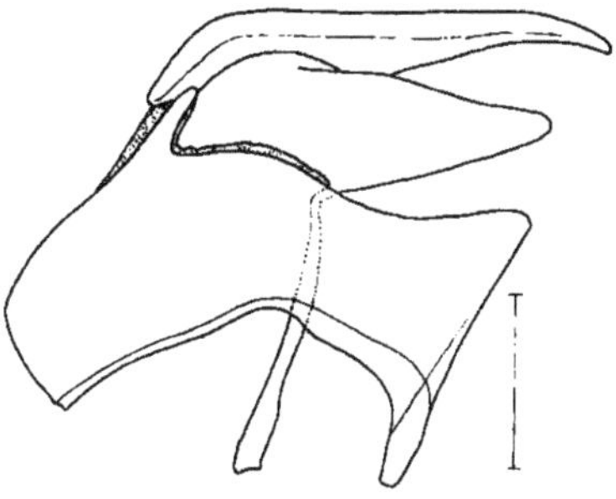

Fig. 547. *Melinda gentilis* Robineau-Desvoidy, male. Epandrium, cerci, surstylus and bacilliform sclerite, left lateral view. Scale = 0.2 mm.

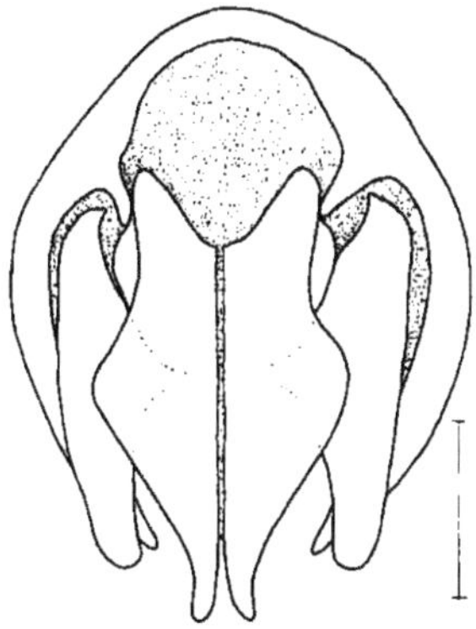

Fig. 548. *Melinda gentilis* Robineau-Desvoidy, male. Epandrium, cerci and surstyli, dorsal view. Scale = 0.2 mm.

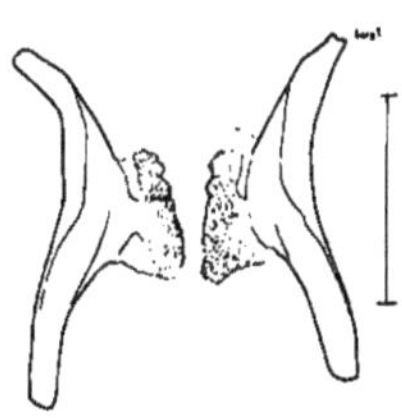

Fig. 549. *Melinda gentilis* Robineau-Desvoidy, male. Bacilliform sclerites. Scale = 0.2 mm.

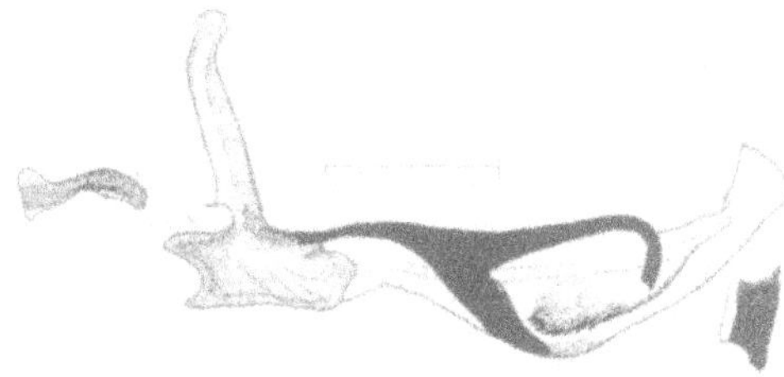

Fig. 550. *Melinda gentilis* Robineau-Desvoidy, male. Aedeagus (with ejaculatory sclerite in situ), left lateral view. Inset: tip of paraphallus (enlarged). Scale = 0.2 mm.

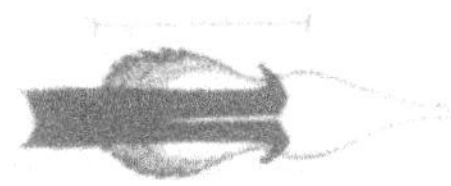

Fig. 551. *Melinda gentilis* Robineau-Desvoidy, male. Distiphallus, dorsal view. Scale = 0.2 mm.

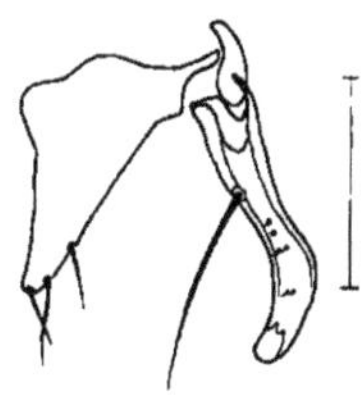

Fig. 552. *Melinda gentilis* Robineau-Desvoidy, male. Post- and pregonites. Scale = 0.2 mm.

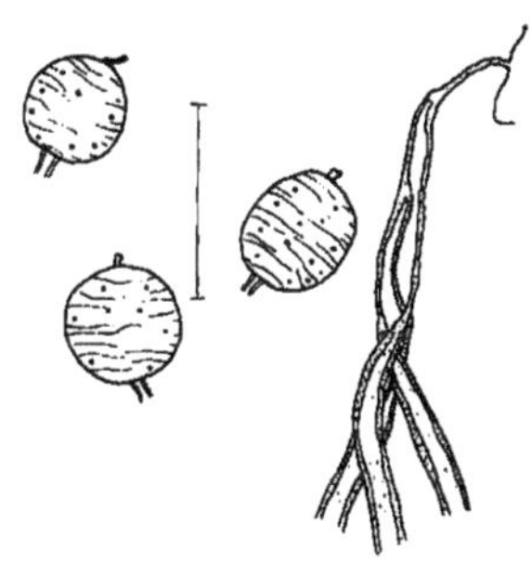

Fig. 553. *Melinda gentilis* Robineau-Desvoidy, female. Spermathecae. Inset: junction of spermathecal ducts (enlarged). Scale = 0.2 mm. G.pr. 211.

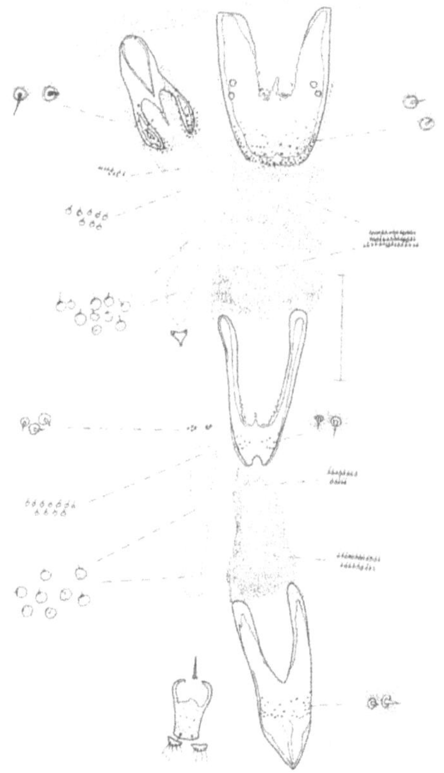

Fig. 554. *Melinda gentilis* Robineau-Desvoidy, female. Ovipositor. Stipple indicates extent of microtrichiae. Insets: shape of microtrichiae in areas indicated (enlarged), minute sensillae on sclerites (high magnification). Scale = 0.5 mm. G.pr. 212.

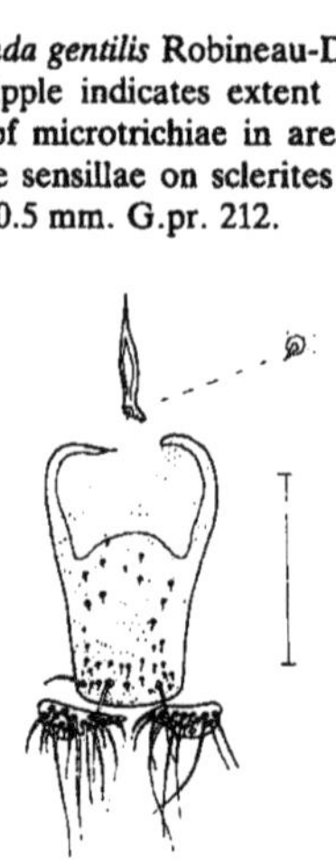

Fig. 555. *Melinda gentilis* Robineau-Desvoidy, female. ST8, hypoproct and cerci of ovipositor (enlarged). Stipple indicates extent of microtrichiae. Inset: minute sensillae at tip of ST8 (very high magnification). Scale = 0.2 mm. G.pr. 212.

mark. Apparently other species must serve as hosts for *gentilis* in Fennoscandia and Denmark. Adults have been collected from April to October.

Nomenclature. *Melinda gentilis* was described by Robineau-Desvoidy as the seventh species in his genus *Melinda*. No material of this species remains in Robineau-Desvoidy's collection in MNHN. *Musca coerulea* Meigen, "nicht gemein in Hekken", was described on the basis of several specimens; neither collector nor locality was stated. Meigen Plate 118, fig. 2, illustrates both ♂ and ♀ (Morge, 1976a), in the legends named as "cognata (coerulea)" (Morge, 1975: 418, 480). Meigen (1830) himself proposed *cognata* as a replacement name. In Meigen's collection in MNHN are 1♂1♀ under *Musca cognata*. Both are labelled "Meigen" and "2137/40" and are in good condition. The ♂, which is rather large, also has a label in Meigen's hand reading "Musca / cognata / (coerulea olim)". The ♀ has a label in Meigen's hand reading "cognata / ♀ / coerulea". Both have 1 *prst ia* seta, and in other characters agree with the current concept of *gentilis* Robineau-Desvoidy (cf. Schumann, 1973). I have labelled and here designate the ♂ as lectotype of *Musca coerulea* Meigen, and the ♀ has been labelled as paralectotype. Séguy (1928a: 125) had acess to Meigen's material of *cognata* but misidentified it. Schumann (1973) unfortunately was unable to study the Meigen types himself and apparently relied on Séguy. Hence the subsequent misapplication of the name *cognata* by Schumann (1973) and in the Palaearctic Catalogue (Schumann, 1986). Zumpt (1956a) and Schumann (1973, 1986) list "*germanorum* Kramer, 1911" as a synonym of *gentilis*. However, there is no evidence that Kramer described a new species under this name in his 1911 paper. Rather he misidentified certain specimens as belonging to *germanorum* Villeneuve (now = *Calliphora loewi*), but corrected his mistake in the 1917 paper by assigning them to a new species, i.e. *villeneuvi*: "n.n. Germanorum der T.d.O. [= "Tachiniden der Oberlausitz", i.e. Kramer, 1911]. Da meine *O.germanorum* nicht Dr. Villeneuve's Art ist muss sie neu benannt werden." The name *villeneuvi* is available, being tied bibliographically to a figure of the genitalia of the misidentified specimens in the 1911 paper.

Subfamily Polleniinae

The subfamily is probably a monophyletic group supported by the following ground-plan apomorphies:

(1) Non-metallic body-colour.
(2) Facial carina more or less developed.
(3) Prosternum, proepisternal depression and metasternal area bare, postalar wall setose.
(4) Base of surstylus with bundle of long setae internally and ventrally.
(5) Epiproct in ovipositor microtrichiose all over.
(6) Cerci in ovipositor long and narrow.
(7) Cerci in ovipositor microtrichiose in apical half.
(8) Sclerites of segment 7 and 8 in ovipositor sometimes invaded posteriorly by microtrichiae.
(9) Pleural membrane 7 in ovipositor wholly microtrichiose, no zone without microtrichiae anteriorly.
(10) Acrophallus without denticles, unarmed.
(11) First instar larva with labrum separate from tentoropharyngeal sclerites (*Pollenia*).

The following genera probably belong to this subfamily: *Dexopollenia* Townsend, *Melanodexia* Williston, *Morinia* Robineau-Desvoidy, *Nesodexia* Villeneuve, *Pollenia* Robineau-Desvoidy, *Xanthotryxus* Aldrich, *Wilhelmina* Schmitz & Villeneuve. The inclusion of several genera is tentative, as they are not sufficiently known. *Nepenthomyia* Kurahashi & Beaver was originally assigned to this group, but the ovipositor (figured by Kurahashi & Beaver, 1979) is reminiscent of that of the Melanomyinae. *Nepenthomyia* and *Nesodexia* have an armed acrophallus, the former an unsclerotised (cf. figure in Kurahashi & Beaver, 1979) and the latter an only apically sclerotised ST8 in the ovipositor (ZMC, G.pr. 282).

Until now *Morinia* has been regarded as a member of the Rhinophoridae (Herting, 1961; Crosskey, 1977; Tschorsnig, 1985a), but Pape (1986) and Rognes (1986d) considered it to be a calliphorid genus and Rognes (1986c) found its assignment to the Polleniinae the best-founded.

In Fennoscandia and Denmark the subfamily can be recognised by the following additional characters apart from those mentioned in the key: Body more or less dusted. Head profiles as in Figs. 556, 557. Parafacialia bare or with numerous rather long hairs. Lower parts of anterior half of anepimeron bare. Coxopleural streak present. Posterior spiracles with distinct anterior and posterior lappets, posterior one most often the largest (Figs. 560-569). Stem-vein bare above. Lower calypter bare above. Hind coxa bare behind.

Widely distributed in Palaearctic, Nearctic, Oriental and Australian Regions. 2 genera in Fennoscandia and Denmark.

Key to genera of Polleniinae

1 Thorax with yellow crinkly vestiture in addition to ordinary black vestiture; parafacialia hairy; outer *ph* present or absent; 1 *prst ia*; lower calypter broad, inner margin converging with the longitudinal axis of the fly; node at base of R_{4+5} hairy on both wing surfaces; costa bare below beyond junction with Sc; t_3 with *d* and *ad* preapicals, a *pd* preapical hardly differentiated; 3-6 marginal scutellars; ♀: fronto-orbital plate with lateroclinate seta

Pollenia Robineau-Desvoidy (p. 212)

\- Thorax without yellow crinkly hairs in addition to black vestiture; parafacialia bare (hairy in a species from Caucasus); outer *ph* absent; 0 *prst ia*; lower calypter narrow, inner margin diverging from the longitudinal axis of the fly; node at base of R_{4+5} bare on both

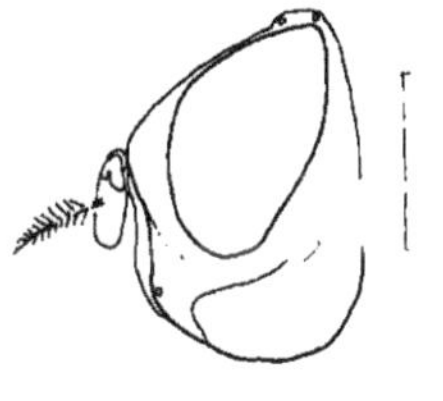

Fig. 556. *Morinia melanoptera* (Fallén), head profile, male. Scale = 1.0 mm.

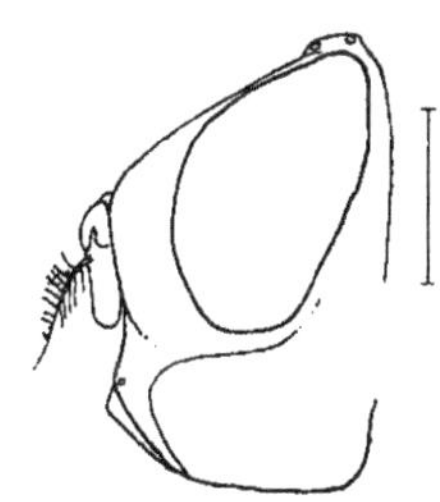

Fig. 557. *Pollenia pediculata* Macquart, head profile, male. Scale = 1.0 mm.

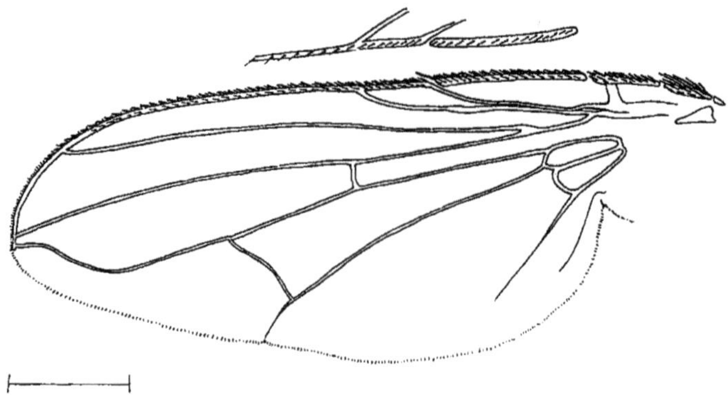

Fig. 558. *Morinia melanoptera* (Fallén), right wing from below. Inset: parts of costa from above to show distribution of small hairs on its upper side. Scale = 1.0 mm.

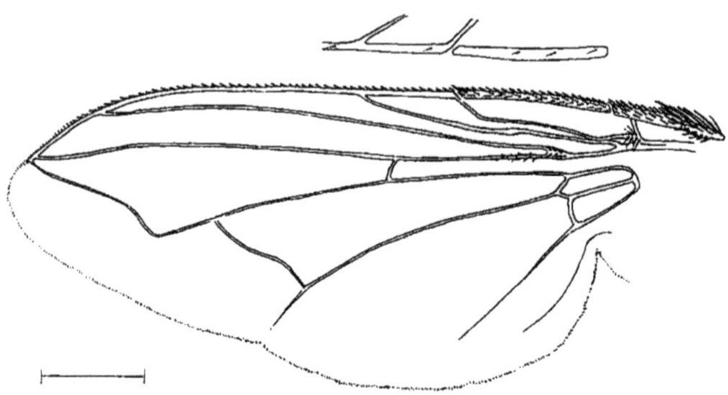

Fig. 559. *Pollenia pediculata* Macquart, right wing from below. Inset: parts of costa from above to show distribution of small hairs on its upper side. Scale = 1.0 mm.

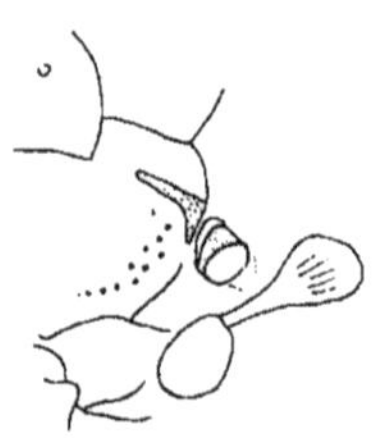

Fig. 560. *Morinia melanoptera* (Fallén), posterior spiracle.

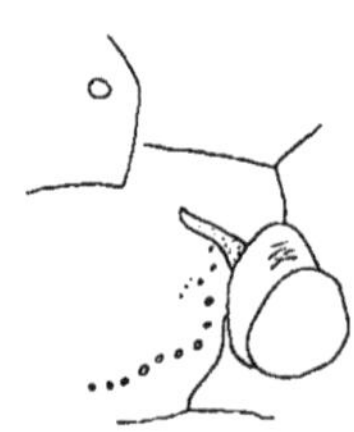

Fig. 561. *Pollenia amentaria* (Scopoli), posterior spiracle.

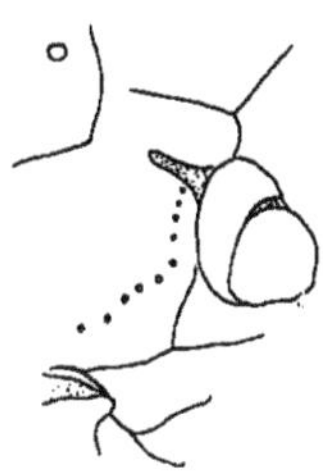

Fig. 562. *Pollenia angustigena* Wainwright, posterior spiracle.

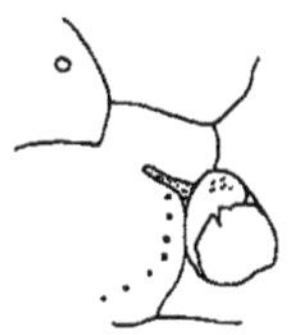

Fig. 563. *Pollenia griseotomentosa* (Jacentkovský), posterior spiracle.

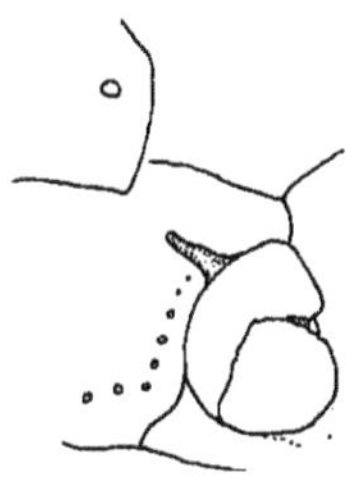

Fig. 564. *Pollenia hungarica* Rognes, posterior spiracle.

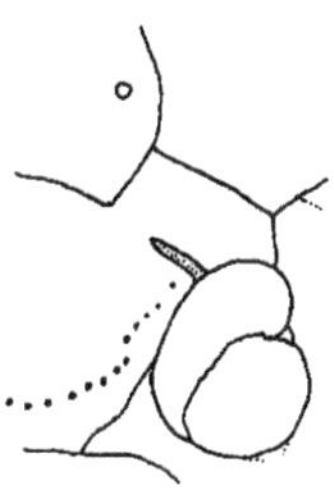

Fig. 565. *Pollenia labialis* Robineau-Desvoidy, posterior spiracle.

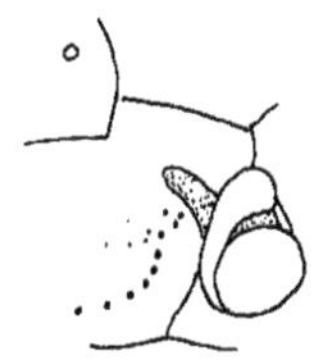

Fig. 566. *Pollenia pallida* Rodendorf, posterior spiracle.

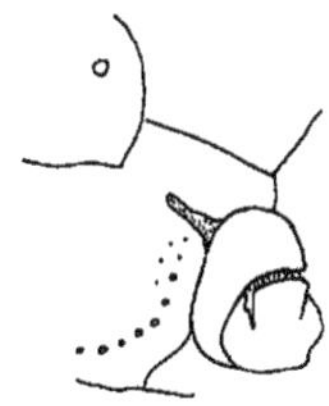

Fig. 567. *Pollenia pediculata* Macquart, posterior spiracle.

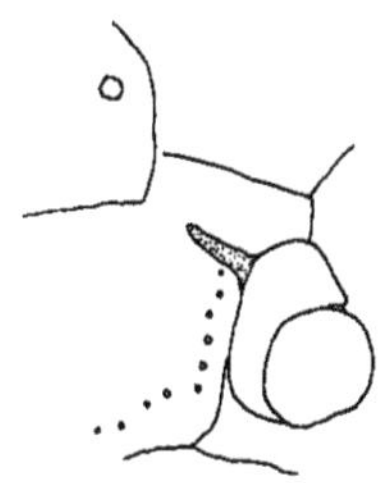

Fig. 568. *Pollenia rudis* (Fabricius), posterior spiracle.

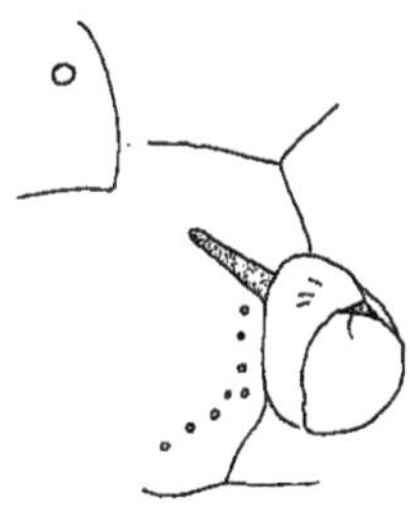

Fig. 569. *Pollenia vagabunda* (Meigen), posterior spiracle.

wing surfaces; t_3 with *ad*, *d*, and *pd* preapical seta equally well differentiated; 2 marginal scutellars; ♀: fronto-orbital plate without lateroclinate seta

Morinia Robineau-Desvoidy (p. 209)

Genus *Morinia* Robineau-Desvoidy, 1830.

Morinia Robineau-Desvoidy, 1830: 264.
Type-species: *Morinia velox* Robineau-Desvoidy, 1830: 265, by designation of Rondani, 1862: 159; = *Musca melanoptera* Fallén, 1817 (cf. Herting, 1974: 31).
Anthracomya Rondani, 1856: 87.
Type species: *Anthracomya geneji* Rondani, 1856: 87 (as *genèji*; *genei* Rondani, 1862: 147, 148, as *genéi*), by original designation; = *Musca melanoptera* Fallén, 1817 (cf. Herting, 1961: 8).
Calobatémyie Macquart, 1854: 736.
Unavailable, vernacular name.
Calobatemyia Macquart, 1855: 33.
Type-species: *Calobatemyia nigra* Macquart, 1855: 34, by monotypy; = *Musca melanoptera* Fallén, 1817 (cf. Herting, 1976: 7).
Disticheria Enderlein, 1934: 188.
Type-species: *Musca melanoptera* Fallén, 1817: 253, by monotypy.

Morinia appears more closely related to the Nearctic *Melanodexia* Williston than to any other genus. With this genus it shares the lack of crinkly yellow vestiture, lack of outer *ph* and *prst ia* setae, the narrow lower calypter, the completely bare node at base of R_{4+5} and the presence of three equally strong *ad*, *d* and *pd* preapical setae on t_3. Several of these characters are apomorphies. However, *Melanodexia* shares with *Pollenia* the hairy parafacialia and the presence of a lateroclinate seta on the female fronto-orbital plates, both of which are plesiomorphies.

Body dark with a thin layer of brownish dusting, no obvious stripes or bands. Arista plumose, total width exceeding width of first flagellomere, which is a little longer than pedicel. Antenna short, tip of first flagellomere separated from lower facial margin by about the length of the first flagellomere. Vibrissa well above lower facial margin. Occiput concave in both sexes, with black hairs only. Palpi short, black. 1+1 *acr*. 2+3 *dc*. 0+2 *ia*. 2 *h*. 1 *ph*, the inner. *pra* strong, longer than posterior *npl*. A single *sa* behind it. Notopleuron hairy. 1+1 *kepst* setae (0+1 *kepst* in *Morinia skufyini* Khitsova, 1983 from Caucasus). 2 pairs of strong scutellar marginals (Fig. 570). Posterior thoracic spiracle as in Fig. 560, anterior lappet small, both lappets dark brown. Basicosta black. Wing-veins and membrane infuscated. Vein M with a shallow bend, its extreme apical part almost parallel with R_{4+5}. Opening of cell r_{4+5} at apex of wing (Fig. 558). Haltere black. Abdomen without alphasetae on several sternites. Bacilliform sclerites present (*argenticincta*) or absent (*melanoptera*). Paraphallus long and narrow, evenly curved, more strongly bent apically. Ventral plates very strongly reduced, absent from ventral half of distiphallus. Mesohypophallus very narrow, widened apically. Dorsal wall of acrophallus with irregular sclerotisations, apically with a ring-like sclerotisation open dorsally. Hypophallus a triangular lobe, anteriorly apparently fused with paraphallus, posteriorly strongly sclerotised, denticles present along ventral edge. Postgonite without basal seta. Ovipositor with short and broad sclerites. T6-T8 and ST6-ST8 more or less microtrichiose posteriorly. Marginal setae of T8 completely marginal. ST8 of full length. Epiproct and at least distal half of cerci microtrichiose. Cerci and hypoproct about as long as ST8. All intersegmental and pleural membranes wholly microtrichiose. Hypoproct without rows of short setae in basal half. 3 spermathecae, each with a short and narrow basal part, pale and internally papillate, and a distal longer part, more strongly sclerotised and transversely striped (Fig. 577). [Description based on *Morinia melanoptera* and *Morinia argenticincta* (Senior-White, 1923) = *Morinia nigerrima* Herting, 1961. On the basis of an examination of ♂ and ♀ *nigerrima* from Japan (identified by Kurahashi as *Pollenia argenticincta*), the ♀ paratype of *Idiopsis argenticincta* Senior-White in BMNH (cf. Crosskey, 1977), and the descriptions by Senior-White (1923) and Senior-White, Aubertin & Smart (1940), I regard *argenticincta* and *nigerrima* as synonyms, even though the holotype of *argenticincta* has not been located (cf. Crosskey, 1977).]

The genus occurs in the Palaearctic and Oriental Regions; 3 species are known. Note that *argenticincta* is listed in *Pollenia* by James (1977) and Schumann (1986). A single species in Fennoscandia and Denmark.

Biology unknown (Herting, 1961).

Note. Séguy's (1941) fig. 445 is erroneously assigned to "*Calobataemyia melanoptera* Fallén". The genitalia illustrated are actually of *Nyctia halterata* (Panzer) (Sarcophagidae).

42. ***Morinia melanoptera*** (Fallén, 1817)
Figs 556, 558, 560, 570-578.

Musca melanoptera Fallén, 1817: 253.
Morinia velox Robineau-Desvoidy, 1830: 265.
Morinia fuscipennis Robineau-Desvoidy, 1830: 265.
Calobatemyia nigra Macquart, 1855: 34.
Anthracomya geneji Rondani, 1856: 87 (as *genèji*; *genei* Rondani, 1862: 147, 148, as *genéi*).

♂♀. Parafacialia broader than length of first flagellomere, densely silvery dusted all over in both sexes. A weak facial carina present dorsally. Genal dilation rather swollen. Thorax dark with a conspicuous band of white dusting across katepisternum and anepisternum towards notopleuron when viewed from certain directions. Surface below lower calypter bare or at most with 1-2 occasional hairs. Costa hairy below far beyond junction with R_1 (Fig. 558). t_1 with *ad, d, pd, p* and *pv* preapicals, and 1-2 *ad*, 1-3*p* and 1-2 *pv* setae on shaft. t_2 with 1 *ad*. t_3 with 1-2 *av*. Wings infuscated, particularly in anterior half and along veins. Full rows of marginal setae on T3-T5, very long in the male, sometimes a few discals medially on T5.

♂. Frons 0.051-0.078 times head-width (n=5), broader than anterior ocellus. Fronto-orbital plates about as broad as frontal vitta. Frontal vitta strongly widening out towards lunula, black. Lower calypter yellowish to brown. Bacilliform sclerites absent. Cerci rather broad, tapering gradually from middle to apex in dorsal view. Apical cleft very short. Apex slightly upturned in profile. Surstylus slender, narrower in distal than in proximal half.

♀. Frons 0.302-0.384 times head-width (n=5). Fronto-orbital plates just in front of anterior ocellus broader than half the width of the frontal vitta; with 1-2 proclinate orbitals (which may be directed outwards to a considerable extent) but no lat-

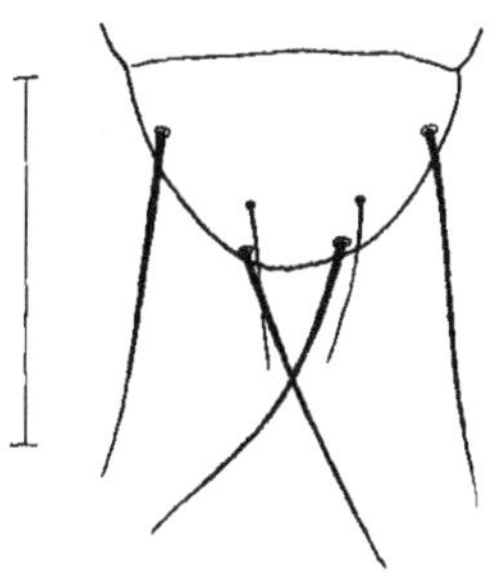

Fig. 570. *Morinia melanoptera* (Fallén). Scutellum. Scale = 1.0 mm.

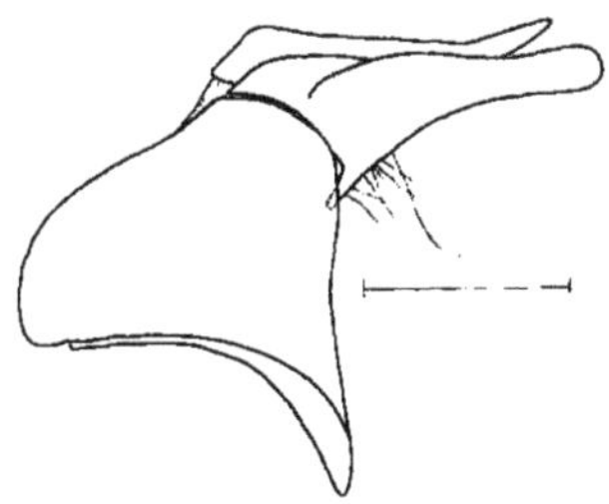

Fig. 571. *Morinia melanoptera* (Fallén), male. Epandrium, cerci and surstylus, left lateral view. Scale = 0.2 mm.

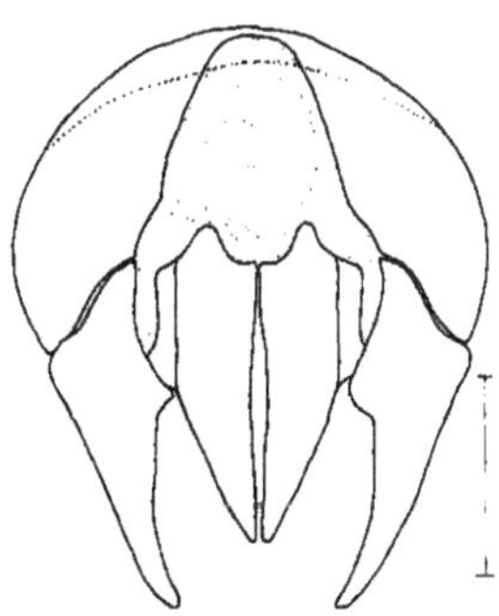

Fig. 572. *Morinia melanoptera* (Fallén), male. Epandrium, cerci and surstyli, posterior view. Scale = 0.2 mm.

Fig. 573. *Morinia melanoptera* (Fallén), male. Aedeagus, left lateral view. Scale = 0.2 mm.

Fig. 574. *Morinia melanoptera* (Fallén), male. Aedeagus, dorsal view. Scale = 0.2 mm.

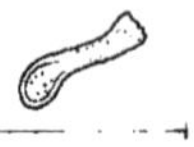

Fig. 575. *Morinia melanoptera* (Fallén), male. Ejaculatory sclerite. Scale = 0.2 mm.

Fig. 576. *Morinia melanoptera* (Fallén), male. Pre- and postgonites. Scale = 0.2 mm.

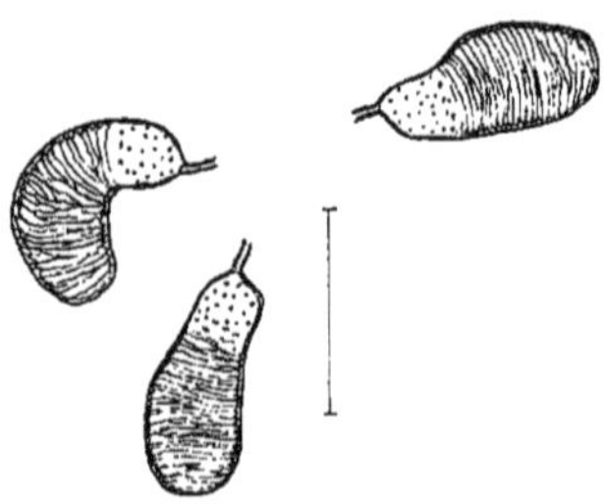

Fig. 577. *Morinia melanoptera* (Fallén), female. Spermathecae. Scale = 0.2 mm. G.pr. 179.

eroclinate seta (prevertical); polished black in posterior 2/3, otherwise silvery dusted. Ovipositor with rather narrow zones of microtrichiae along hind edge of sclerites on segments 6-8. In one of the dissected specimens T7 with a mid-dorsal stripe of microtrichiae for its whole length. Cerci microtrichiose in more than apical half. Epiproct 3 times longer than broad. [2 ovipositor slides examined.]

Length. 5-7 mm.

Distribution. Scattered in Denmark. Only known from Lårgård (ON, Sel) in Norway (Siebke leg.). Scattered in Sweden north to Ly.Lpm. In Finland north to ObN. - Europe, east

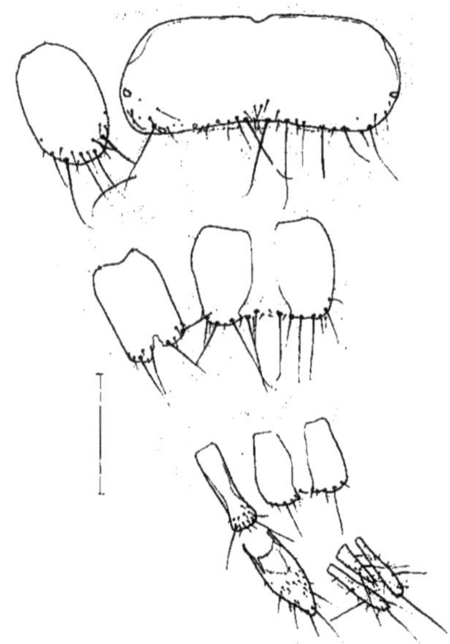

Fig. 578. *Morinia melanoptera* (Fallén), female. Ovipositor. Stipple indicates extent of microtrichiae. Scale = 0.4 mm. G.pr. 179.

to Ukraine (Herting, 1961; Stackelberg, 1970). Not in the British Isles.

Biology. A pupa was found in decaying wood, from which a specimen emerged in May ("Pupa hujus speciei a Cl. Baumhauer in ligno putrido inventa, mense Majo exclusa, teste Meigen.") (Zetterstedt, 1838, 1844, citing Meigen). In Fennoscandia and Denmark adults have been collected from May to August.

Nomenclature. Fallén (1817) described *melanoptera* from ♂♀ syntypes from "Westergöthland och Östergöthland..." (Vg., Ög.). In Fallén's collection in NHRM are 4♂1♀ under *melanoptera*. All are conspecific and mostly in good condition. They conform to the current concept of *melanoptera* (cf. Herting, 1961), fit the original description, and are recognised as syntypes. 1♂ is labelled "T.melano-/ptera ♂. / Lärket. Ostr." [=Lärketorp, Ög.] in Zetterstedt's hand (a syntype?); 1♂ is labelled "♂ var. β [underlined]" in Fallén's (?) hand; 2♂ are without original labels; the ♀ is labelled "M.mela-/noptera [underlined] ♀" in Fallén's (?) hand. I have labelled and here designate a ♂ without original labels as lectotype. It is lacking the left antenna, the left fore tarsus and the left mid leg. The other specimens have been labelled as paralectotypes.

Genus *Pollenia* Robineau-Desvoidy, 1830

Pollenia Robineau-Desvoidy, 1830: 412.
Type species: *Musca rudis* Fabricius, 1794: 314, by original designation.
Nitellia Robineau-Desvoidy, 1830: 417.
Type species: *Musca vespillo* Fabricius, 1794: 318, by designation of Coquillett (1910: 576). The type species has been misidentified. An application should be made to the Commission to fix *Musca atramentaria* Meigen, 1826: 65 as type species. For details, see section on Nomenclature below.
Cephysa Robineau-Desvoidy, 1863: 655, 677.
Type species: *Cephysa muscidea* Robineau-Desvoidy, 1863: 677, by monotypy.
Orizia Robineau-Desvoidy, 1863: 655, 678.
[For 6 nominal species; no type species designated].
Chaetopollenia Enderlein, 1936: 211.
Type species: *Musca vespillo* Fabricius, 1794: 318, by monotypy. The type species has been misidentified. An application should be made to the Commisssion to designate *Musca amentaria* Scopoli, 1763 as type species. See also section on Nomenclature below.
Micronitellia Enderlein, 1936: 211. Unavailable. [No type species designated.] Validated by Lehrer, 1967: 255.
Trichopollenia Enderlein, 1936: 211.
Type species: *Musca vagabunda* Meigen, 1826: 72, by monotypy.
Polleniella Jacentkovský, 1941a: 15, 16. Nomen nudum. [No taxonomic characters.] Validated by Jacentkovský, 1941b: 20, 22.
Buresiella Jacentkovský, 1941b: 21, 22 (as *Burešiella*).
Type species: *Pollenia pallida* Rodendorf, 1926, by monotypy.
Dasypollenia Jacentkovský, 1941b: 20, 22. Unavailable. [No type species designated.] Validated by Lehrer, 1967: 256.
Polleniella Jacentkovský, 1941b: 20, 22.
Type species: *Polleniella distincta* Jacentkovský, 1941b: 20, by monotypy; = *Pollenia mayeri* Jacentkovský, 1941a: 14,15 (cf. also Jacentkovský, 1942: 220 (17); 1944b: 118).
Polleniomyia Jacentkovský, 1941b: 20, 23. Unavailable. [No type species designated.] Validated by Jacentkovský, 1942: 200 (28).
Pseudopollenia Jacentkovský, 1941b: 21, 22.
Type species: *Pollenia vera* Jacentkovský, 1936a: 9, by monotypy (cf. also Jacentkovský, 1942: 205 (13)).

Bureschiella Jacentkovsky, 1941c: 31. Unjustified emendation of *Buresiella* Jacentkovský, 1941b: 21, 22.
Type species: *Pollenia pallida* Rodendorf, 1926, automatic.
Chaetopollenia Jacentkovský, 1941c: 31. Nomen nudum. [No taxonomic characters.]
Dasypollenia Jacentkovský, 1941c: 31. Nomen nudum. [No taxonomic characters. No type species designated.]
Polleniella Jacentkovský, 1941c: 31. Subsequent citation of *Polleniella* Jacentkovský, 1941b: 20, 22.
Polleniomyia Jacentkovský, 1941c: 31. Nomen nudum. [No taxonomic characters. No type species designated.]
Pseudopollenia Jacentkovský, 1941c: 31. Subsequent citation of *Pseudopollenia* Jacentkovský, 1941b: 21, 22.
Bureschiella Jacentkovský, 1942: 213 [21].
Subsequent citation of *Buresiella* Jacentkovský, 1941c: 31.
Dasypollenia Jacentkovský, 1942: 210 (18). Unavailable. [No type species designated.]
Polleniomyia Jacentkovský, 1942: 220 (28).
Type species: *Pollenia labialis* Robineau-Desvoidy, 1863 (2): 671, by original designation.
Polleniomyma Jacentkovský, 1944b: 119. Unnecessary replacement name for *Polleniomyia* Jacentkovský, 1942: 220 (28). [The latter name is different by one letter from *Pollenomyia* Séguy, 1935 (cfr. ICZN, Art. 56 (b)).]
Eupollenia Lehrer, 1963: 290.
Type species: *Musca rudis* Fabricius, 1794: 314, by original designation.
Jacentkovskyiomyia Lehrer, 1963: 292.
Type species: *Polleniella griseotomentosa* Jacentkovský, 1944a: 45, 48, by original designation.
Mariomyia Lehrer, 1963: 292.
Type species: *Pollenia mayeri* Jacentkovský, 1941a: 14, 15, by original designation.
Parapollenia Lehrer, 1963: 290.
Type species: *Pollenia dasypoda* Portschinsky, 1881: 143, by original designation.
Rohdendorfiomyia Lehrer, 1963: 292.
Type species: *Musca vespillo* Fabricius, 1794, by original designation. The type species has been misidentified. An application should be made to the Commission to designate *Musca amentaria* Scopoli, 1763 as type species. See also section on Nomenclature below.
Sachtlebeniola Lehrer, 1963: 291. Unavailable. [No type species designated.]

Seguyiomyia Lehrer, 1963: 293.
Type species: *Musca vagabunda* Meigen, 1826: 72, by original designation.
Zumptiomyia Lehrer, 1963: 292.
Type species: *Pollenia bisulca* Pandellé, 1896: 152, by original designation.
Dasypollenia Lehrer, 1967: 256.
Type species: *Pollenia dasypoda* Portschinsky, 1881: 142, by original designation.
Micronitellia Lehrer, 1967: 255.
Type species: *Musca varia* Meigen, 1826: 66, by original designation; = *Musca rudis* Fabricius, 1794.

In Fennoscandia with ground colour black, only the extreme anterior parts of gena, vibrissal angles, facial ridges, lunula and parts of antenna more or less reddish. Thorax usually rather weakly dusted, abdomen densely dusted with shifting tessellations or almost shining black. Head profile as in Fig. 557. Arista long plumose. Antenna moderately long, first flagellomere about 2 times as long as pedicel. Tip of first flagellomere widely separated from lower facial margin. Parafacialia with numerous rather long dark hairs. Vibrissae high above lower facial margin. Lower part of facial plate vertical in profile, not produced anteriorly, narrowed at the level of the vibrissae when seen from in front. Occiput concave in both sexes, but less so in the female. Occiput with 1-4 rows of black setulae behind postocular row, otherwise with thin pale hairs. Palpi usually very dark. 2+3 *acr*. 2+3 *dc*. 1+2 *ia*. 3 *h*, in a straight line. Usually 1+1 *ph*. Outer *ph* in line with *prs*. *pra* strong, close to suture, 2 *sa* behind it. Notopleuron hairy, 2 *npl*. 1+1 *kepst*. Postalar wall with bundle of numerous long pale hairs, sometimes black. 3-6 pairs of strong scutellar marginals. Posterior thoracic spiracle as in Figs. 561-569, lappets usually pale yellow, dark in a few species. Basicosta black or various shades between yellow and dark brown. Wing-veins pale. Stem-vein bare above. Costa hairy below to junction with Sc, bare below beyond this point (Fig. 559). Vein M with a sharp bend, opening of cell r_{4+5} well in front of wing-tip (Fig. 559). Calypters white or weakly infuscated. Haltere yellow. Abdominal ST2 much broader than sternites behind it, its lateral edges overlapping lateral margins of T2. Bacilliform sclerites present as long, widely separated, unmodified rods. Aedeagus variable, acrophallus always without denticles, 6 types discernible among material from Fennoscandia and Denmark:

(1) *amentaria*-type (Figs. 586, 587): Paraphallic processes very long and slender, distally armed with minute denticles. Sclerotisation at base of distiphallus (ventral plate) short, continuous ventrally with proximal end of mesohypophallus. Mesohypophallus long. No median hypophallic lobe. Hypophallus short, square, dentate posteriorly, its posterior end widely separated from distal edge of ventral plate. (*amentaria*).

(2) *pallida*-type (Figs. 632, 634): Paraphallic processes very long and slender, unarmed. Sclerotisation at base of distiphallus (ventral plate) long, continuous ventrally with proximal end of mesohypophallus. Mesohypophallus long. No median hypophallic lobe. Hypophallus low, elongate, square, dentate posteriorly, its posterior end widely separated from distal edge of ventral plate. (*pallida*).

(3) *vagabunda*-type (Figs. 664, 665): Paraphallic processes shorter. Sclerotisation at base of distiphallus (ventral plate) rather long, weakly joining the thickened proximal end of the mesohypophallus mid-ventrally. Mesohypophallus long. No median hypophallic lobe. Hypophallus about as long as acrophallus, posteriorly dentate, its posterior end close to distal edge of ventral plate. (*vagabunda*).

(4) *rudis*-type (Figs. 596, 597,614, 615, 643, 644, 654, 655): Paraphallic processes reaching a little further than tip of hypophallus, in dorsal view proceeding distad parallel with longitudinal axis of aedeagus and distally unsclerotised in a manner that makes them appear transversely or obliquely truncated, sometimes armed with minute denticles. Sclerotisation at base of distiphallus (ventral plate) short, variously shaped, two types discernible (Fig. 583, Type I and II). Median hypophallic lobe present, strongly sclerotised in posterior half. Mesohypophallus short. Hypophallus triangular, dentate posteriorly, pointed apically, strengthened centrally by a strong sclerotisation. (*angustigena*, *hungarica*, *pediculata*, *rudis*).

(5) *labialis*-type (Figs. 624, 625): similar to *rudis*-type, but paraphallic tips unarmed, narrow, curved inwards and not appearing transversely or obliquely truncated apically. Hypophallus with long, acute prolongation apically. (*labialis*).

(6) *griseotomentosa*-type (Figs. 606, 607): similar to *rudis*- and *labialis*-types, but tip of paraphallic process unarmed, median hypophallic lobe absent and central sclerotisation

of hypophallus present only distally. Sclerotisation at base of distiphallus (ventral plate) long. (*griseotomentosa*).

Pregonite narrow, with numerous setae. Postgonite with a basal seta.

Ovipositor differently shaped according to species and species-group. Several types discernible, all congruent with those based on the structure of the aedeagus:

(1) *amentaria*-type (Fig. 591): T8 marginal setae not quite marginal. Epiproct with or without microtrichiae. Cerci microtrichiose except in basal half. Tip of ST8 without microtrichiae. Pleural membrane 8 without microtrichiae.
(2) *pallida*-type (Fig. 637, 638, 639): T8 marginals far in front of margin. Epiproct, cerci and tip of ST8 without microtrichiae. Pleural membranes 6 without microtrichiae in about anterior half. Pleural membrane 7 without microtrichiae in at least anterior three fourths. Pleural membrane 8 wholly without microtrichiae. Epiproct, cerci, tip of ST8 and tip of hypoproct with long, slender, straight spines.
(3) *vagabunda*-type (Fig. 669): T8 marginals in front of margin. Epiproct microtrichiose. Cerci and tip of ST8 without microtrichiae. Pleural membrane 8 without microtrichiae.
(4) *rudis*-type (Fig. 603, 621, 650, 661): T8 marginals not quite marginal. Microtrichiae on epiproct, on at least apex of cerci, on tip of ST8 and on about posterior half of pleural membrane 8.
(5) *labialis*-type (Fig. 628): T8 marginals absolutely marginal. Epiproct and cerci completely covered with microtrichiae. Tip of ST8 microtrichiose. Pleural membrane 8 completely covered with microtrichiae.
(6) *griseotomentosa*-type (Fig. 611): T8 marginals absolutely marginal. Epiproct microtrichiose all over. Cerci microtrichiose except at extreme base. Posterior half of ST8 microtrichiose. Pleural membrane 8 fully microtrichiose. Mid-dorsal broad complete bands of microtrichiae on T7 and T8.

Sclerites on segments 6-8 variously shaped and invaded by microtrichiae to varying extents. All intersegmental membranes fully microtrichiose. Pleural membranes 6 and 7 usually fully microtrichiose (except *pallida*-group). Hypoproct with several rows of short setae in basal half. 3 spermathecae, spherical to elongate. Uterus in several species with sclerotised tube-shaped lateral sacs of varying shape, which receive the paraphallic tips during copulation.

Oviparous. Some species are known to be predators or parasites of earthworms in the larval stages. There is also evidence that larvae or pupa of Lepidoptera (Séguy, 1941) and also adult honeybees (Ibrahim, 1984) are attacked. Larval stages and biology have been described by Keilin (1915), Yahnke & George, 1972; Thomson & Davies (1973a, 1973b, 1974), Shewell (1987), but the exact specific identity of the described specimens is unknown (cf. Rognes, 1987b). The life-history and immature stages of *Pollenia dasypoda* Portschinsky (not in North Europe) have been described by Tawfik & El-Husseini (1971).

Several species overwinter indoors, where they have been reported as a nuisance (Séguy, 1941; Hall, 1948).

The genus is known from the Palaearctic, Nearctic, Oriental and Australian Regions and more than 100 species have been described.

Boyes & Shewell (1975) regard the genus as an immigrant into the Nearctic, possibly introduced in soil by early European settlers, and they suggest that its distribution is dependent on that of non-native earthworm species (not specified). A similar argument is used by Shewell (1987) when suggesting that *Bellardia* in North America is parasitic upon an introduced European earthworm. As long as the earthworm species are not specified, it is difficult to evaluate this argument, but possibly it is based on the prevalent opinion that all the European earthworm species in North America have been introduced by European settlers during the last two centuries (cf. discussion in Omodeo, 1963). Lindroth (1957) regards 94 % of the earthworms common to Europe and North America as introduced. Omodeo (1963) opposes this view and considers them as an integral part of an ancient Holarctic fauna, now separated because of continental drift.

On the basis of Lindroth's (1957) criteria of introduced species and the known distribution of the Nearctic species (cf. Rognes, 1987b), I find it likely that *Pollenia labialis* and *P. vagabunda* are introduced. The status of *P. angustigena*, *P. pediculata* and *P. rudis* cannot be decided, but in the absence of evidence to the contrary I regard them provisionally as belonging to the ancient fauna of North America.

9 species in Fennoscandia and Denmark.

Note. Even in the broad sense adopted here, the generic limits of *Pollenia* on a world basis are obscure and it is possible that it is not a monophyletic group. Dear (1986) treated a number of species from New Zealand under *Pollenia* even though all of them lack a facial carina and the long crinkly yellow thoracic ground-vestiture. Some of them are even metallic green. *Xanthotryxus* Aldrich, strikingly *Pollenia*-like, and *Dexopollenia* Townsend, both of which have crinkly yellow vestiture, should possibly also be included in *Pollenia*. The following monophyletic species-groups within *Pollenia* have been defined by Rognes (1987a, 1987b, 1988): *pallida*-group, *tenuiforceps*-group, *rudis*-group, *semicinerea*-group, *labialis*-group (= *intermedia*-group), *griseotomentosa*-group. Work on other species-groups is in progress.

Nomenclature. As regards the type designation for *Nitellia*, Coquillett's designation was made as follows: "*Nitellia* Desvoidy, Essai Myod., p. 417, 1830. 2 species. Type, *Musca vespillo* Fabricius, the first species, by present designation. Equals *Pollenia* Desvoidy, 1830". Robineau-Desvoidy gives the name of the first species in his list as "*Nitellia vespillo* [and on the next line:] *Musca vespillo* Fabr. Meig.". "*Musca vespillo* Fabricius" as interpreted by Meigen, 1826: 65, equals *Musca amentaria* Scopoli, 1763 (see below). This species is a *Pollenia* which almost always has an open cell r_{4+5}. However, Robineau-Desvoidy diagnoses his genus *Nitellia* as having this cell closed and with a small stalk ("un peu petiolée") and even refers *Musca atramentaria* Meigen to this genus. ("Je pense qu'il faut rapporter à ce genre le *Musca atramentaria* de Meigen."). This makes it credible that Robineau-Desvoidy's "*Nitellia vespillo*" equals *Musca atramentaria* Meigen, 1826, which usually has a stalked cell r_{4+5}, and not "*Musca vespillo* Fabricius" sensu Meigen or the true *Musca vespillo* Fabricius (a *Bellardia*, see above), both of which usually have an open cell r_{4+5}. According to the ICZN Art. 70 (b), which concerns misidentified type species, the case should be referred to the Commission. I think stability is best served if Coquillett's designation is considered to apply to *Musca atramentaria* Meigen, as currently understood (i.e. with hairy stem-vein), in accordance with the views of James (1977) and Schumann (1986). If the true *Musca vespillo* Fabricius is designated as type species, *Nitellia* becomes a senior synonym of *Bellardia*. This will result in confusion, as *Nitellia* is currently used by some authors for a group of species within *Pollenia* (e.g. Lehrer, 1972). It would also be possible to interpret the designation to apply to *Musca amentaria* Scopoli, 1763 (= "*Musca vespillo* Fabricius" sensu Meigen, 1826), which occasionally has the cell r_{4+5} stalked or closed in margin, as it is conceivable that an abnormal specimen of this species was before Robineau-Desvoidy when he erected his new genus. *Pollenomyia* Séguy, 1935: 149 (type species: *Pollenomyia sinensis* Séguy, 1935: 149, by original designation) was listed by Schumann (1986) as a synonym of *Pollenia*. I have examined the holotype of the type species (in MNHN). It lacks yellow crinkly hairs on the thorax; the proepisternal depression, prosternum and postalar wall have long hairs; the coxopleural streak is absent; and in the wing the costa is hairy below all the way to the tip of R_{2+3}. It is not a *Pollenia*, and I have therefore removed *Pollenomyia* from the synonymy of *Pollenia*.

Key to species of *Pollenia*

1 Abdomen shining black, at most with very weak dusting if seen very obliquely from behind, shifting pattern not conspicuous; scutellum with 5-6 pairs of strong marginals of uniform size; facial keel sharp-edged; *iv* hair-like, not different from neighbouring postocular setae; lappets of posterior spiracle dark brown; t_2 2-3 *ad*; ♀: f_2 with *a* preapical seta

43. *amentaria* (Scopoli)

– Abdomen with dense conspicuous dusting, most often with a shifting tessellated pattern; scutellar chaetotaxy different, only 3-4 marginal pairs of setae; ♀: f_2 without an *a* preapical seta

2

2(1) Thorax with a dark median stripe between *prst acr*, usually extending forwards to extreme anterior slope of thorax; 3-5 *h*; 2 inner *ph*; abdomen bluish-black, with a mid-dorsal stripe; lateral chequered markings on anterior part of tergites almost reaching mid-line; t_2 with 1 *ad*

51. *vagabunda* (Meigen)

– No presutural dark stripe between *prst acr*; 3 *h*; 1 inner *ph*; abdominal chequered markings reaching halfway to mid-line

3

3(2) Node at junction of humeral crossvein and vein Sc with a bundle of pale hairs below

(Fig. 640); facial keel distinct, usually rather broadened and somewhat flattened on top; palpi black; f_2 and f_3 with black *pv* hairs; ♂: ventral abdominal vestiture black, semi-decumbent and normally spaced, as dense as the dorsal vestiture, hairs similar in size to those dorsally; t_3 without erect *av* hairs, only decumbent setulae present in addition to *av* setae which are rather short (Fig. 581)

49. *pediculata* Macquart

– Node at junction of humeral crossvein and vein Sc bare below

4

4(3) Outer *ph* absent; facial keel absent or very inconspicuous; lappets of posterior thoracic spiracle yellow; t_2 with 1 *ad*; size small; abdomen in posterior view with a median dark stripe widening out posteriorly to cover whole posterior margin; f_2 and f_3 mostly with black hairs on *pv* surface; ♂: frons very narrow, about as broad as anterior ocellus; tarsi shorter than tibiae; ♀: frontal vitta black, reddish near lunula; occipital dilation black anteriorly

45. *griseotomentosa* (Jacentkovský)

– Outer *ph* usually present, if absent, then f_2 and f_3 with yellow hairs on *pv* surface; frons in ♂ usually broader; tarsi in ♂ at least as long as tibiae

5

5(4) Facial keel very low or absent; lappets of posterior thoracic spiracle very dark brownish; basicosta black; t_2 with 2-3 *ad*; ♂: no erect *av* hairs beside the 3-5 *av* setae on t_3

47. *labialis* Robineau-Desvoidy

– Facial keel usually present; lappets of posterior thoracic spiracle yellow; basicosta yellow, brown, rarely very dark

6

6(5) Facial keel very broad and low (Fig. 629); palpi mostly yellow, darkened basally; t_1 with 1 *pv*; f_2 and f_3 with black *pv* hairs; ♂: ventral abdominal vesture black, semi-decumbent, not particularly dense; t_3 with no erect *av* hairs besides *av* setae; ♀: numerous slender spines at tip of ovipositor (usually visible without dissection) (Figs. 638, 639)

48. *pallida* Rodendorf

– Facial keel usually narrow, relatively sharp-edged (as in Fig. 651); palpi dark; t_1 usually with 2 *pv*; other characters not present in combination; ♀: no stiff slender spines on ovipositor, only soft slightly wavy hairs

7

7(6) t_2 usually with 1 *ad*; f_2 and f_3 with yellow *pv* hairs; ventral abdominal vestiture predominantly black; ♂: frons narrow; ventral abdominal vestiture semi-decumbent and normally spaced, as dense as the dorsal vestiture, hairs similar in size to those dorsally; t_3 with a varying number of semi-erect to erect *av* hairs besides *av* seta, often with none at all (Fig. 579); ♀: spermathecae oval, without a narrow "handle" basally (Fig. 601)

44. *angustigena* Wainwright

– t_2 usually with 2-3 *ad*; ♂: ventral abdominal vestiture erect, dense, much denser than the dorsal vestiture, hairs finer than those dorsally

8

8(7) f_2 and f_3 usually with black *pv* hairs; ventral abdominal vestiture most often predominantly black; ♂: t_3 with numerous long, erect *av* hairs on middle third besides *av* setae (Fig. 582); ♀: spermathecae oval, without a narrow "handle" basally (Fig. 659)

50. *rudis* (Fabricius)

– f_2 and f_3 with yellow *pv* hairs; ventral abdominal vestiture predominantly yellow (especially on T3); ♂: t_3 with very few, if any, erect *av* hairs besides *av* setae, which are rather long (Fig. 580); ♀: at least numerous yellow hairs on ventral part of T3; spermathecae elongate, with a narrow "handle" proximally (Fig. 619)

46. *hungarica* Rognes

Fig. 579. *Pollenia angustigena* Wainwright, male, hind tibia in anterodorsal view. (From Rognes, 1987b: fig. 15).

Fig. 581. *Pollenia pediculata* Macquart, male, hind tibia in anterodorsal view. (From Rognes, 1987b: fig. 19).

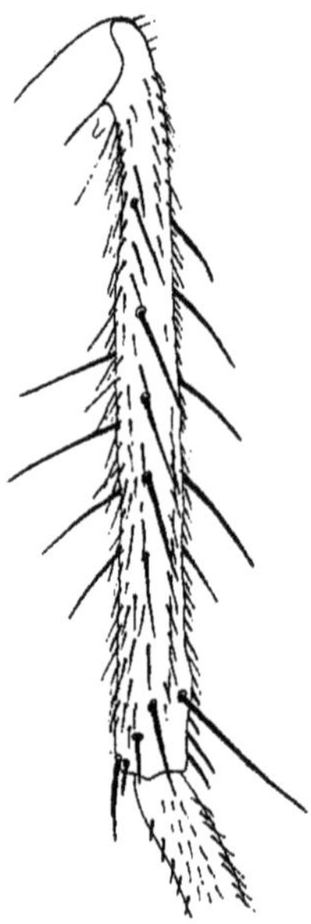

Fig. 580. *Pollenia hungarica* Rognes, male, hind tibia in anterodorsal view. (From Rognes, 1987b: fig. 16).

Fig. 582. *Pollenia rudis* (Fabricius), male, hind tibia in anterodorsal view. (From Rognes, 1987b: fig. 20).

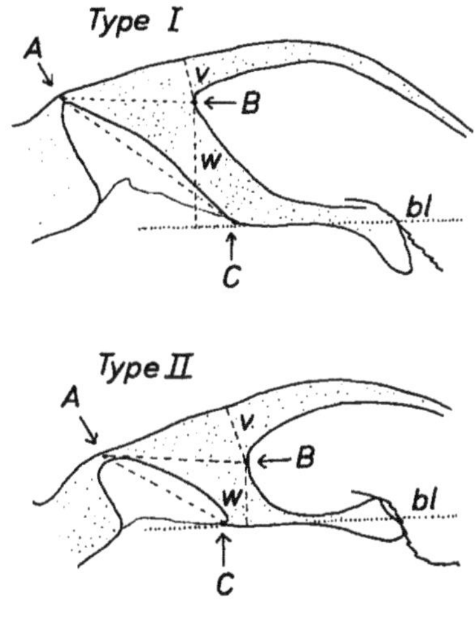

Fig. 583. *Pollenia* spp. (*rudis* species-group), male, shape of sclerotisation at base of distiphallus in profile. Type I: high, vaulted, imagined transverse section of base of distiphallus at level of point C narrow; point B close to dorsal wall, i.e. *v* much shorter than *w*; *w* meets *bl* at a right angle at or proximad of point C; AB much shorter than AC (*P. angustigena* and *P. rudis*). Type II: low, imagined transverse section at level of point C rounded; point B about midway between dorsal and ventral wall, i.e. *v* about as long as *w*; *w* meets *bl* at a right angle distad of point C; AB longer than or equal to AC (*P. hungarica* and *P. pediculata*). (From Rognes, 1987b: fig. 21).

43. ***Pollenia amentaria*** (Scopoli, 1763), COMB.N.

Figs 561, 584-591.

Musca amentaria Scopoli, 1763: 335.
Pollenia micans Robineau-Desvoidy, 1830: 416. SYN.N.
Musca nigrina Meigen, 1838: 305. Junior primary homonym of *Musca nigrina* Fallén, 1817 (Tachinidae). SYN.N.
Musca nitens Zetterstedt, 1845: 1340. Junior primary homonym of *Musca nitens* Villers, 1789: 549 (Syrphidae). SYN.N.
Chaetopollenia soudeki Jacentkovský, 1941b: 21, 22 (key). SYN.N.
? *Chaetopollenia moravica* Jacentkovský, 1941b: 21, 23 (key). SYN.N.
? *Chaetopollenia pseudobisulca* Jacentkovský, 1941b: 21, 23 (key). SYN.N.
vespillo: authors, not Fabricius, 1794; misidentifications.

♂♀. Facial keel sharp. Facial membrane black, with little dust. Gena in front of genal dilation, vibrissal corner, facial ridges, lunula, and most of antenna except arista reddish. Parafacialia with short hairs. Posterior edge of anepisternum with rows of long pale hairs behind *anepst* setae. Katepisternum with pale ground-vestiture. Scutellum with characteristic vestiture of 5-6 strong subequal marginal setae, often with a few additional smaller ones. Posterior spiracles very dark brown. t_1 normally with 2 *pv* setae. f_2 with an *a* preapical seta. t_2 with 2-3 *ad* setae. t_3 with 3-5 *av* setae. Tarsi a little shorter than tibiae. Basicosta black.

♂. Frons 0.033-0.050 times head-width (n=5), about 1.0-1.5 times as wide as anterior ocellus. Fronto-orbital plates touching. Cerci rather broad in dorsal view, cleft in apical fourth, apex bare and shining black. Surstyli dilated apically in profile, bent towards mid-line when seen in dorsal view. Aedeagus of *amentaria*-type.

♀. Frons 0.331-0.347 times head-width (n=5). Frontal vitta 2 times as wide as each fronto-orbital plate. Parafacialia and fronto-orbital plates with pale golden shimmering dusting. T7 in ovipositor of characteristic shape, invaded in front by microtrichiae, and with a mid-dorsal separate "rod" between T7 halves. T8 distally curved medially. ST7 long, broad basally, slender apically, a narrow zone of microtrichiae posteriorly. Spermathecae spherical. Lateral sacs unsclerotised. [4 ovipositor slides examined.]

Length. 7-11 mm.

Distribution. Widely distributed and common in Denmark. In Sweden north to Gstr. and Dlr. In Norway from southern and extreme southeastern parts. South Finland. - Widely distributed in Europe and European parts of U.S.S.R., including Armenia and Georgia (Schumann, 1986; Rognes, 1988).

Biology. Immature stages and biology unknown. Females have been attracted to fish-bait in Finland. In Fennoscandia and Denmark adults have been collected from March to September.

Nomenclature. Scopoli described *Musca amentaria* from an unspecified number of specimens of unspecified sex occurring copiously on *Salix* catkins in March in Yugoslavia. Scopoli's collection is known to have been destroyed (in a shipwreck on the Danube, A.C.Pont in litt.). I have labelled and here designate as neotype of *Musca amentaria* Scopoli a ♂ in BMNH labelled: (1) "Jugoslavia, road below Kr. Gora & Tolbin 13.viii.1960"; (2) "71843"; (3) "O.W.Richards Coll. Brit.Mus. 1967-510". It is in excellent condition except for the loss of the two hind tarsi. Robineau-Desvoidy de-

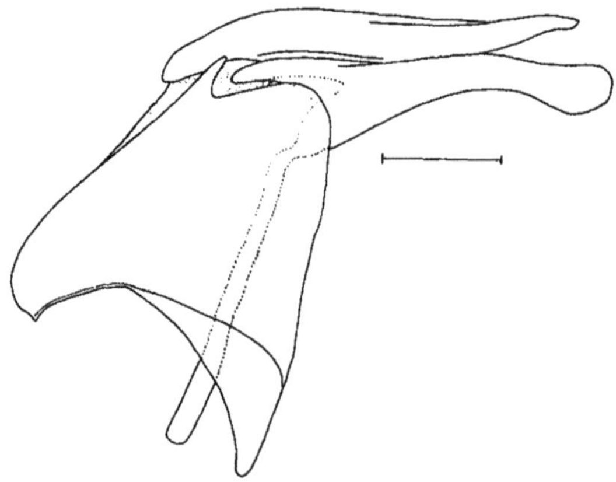

Fig. 584. *Pollenia amentaria* (Scopoli), male. Epandrium, cerci, surstylus and bacilliform sclerite, left lateral view. Scale = 0.2 mm.

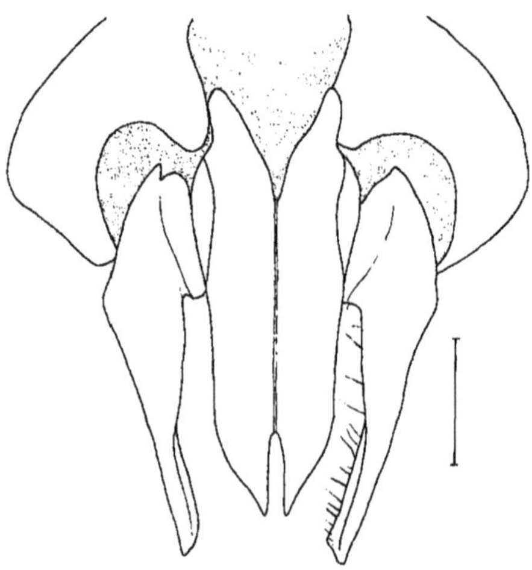

Fig. 585. *Pollenia amentaria* (Scopoli), male. Epandrium (parts), cerci and surstyli, posterior view. Scale = 0.2 mm.

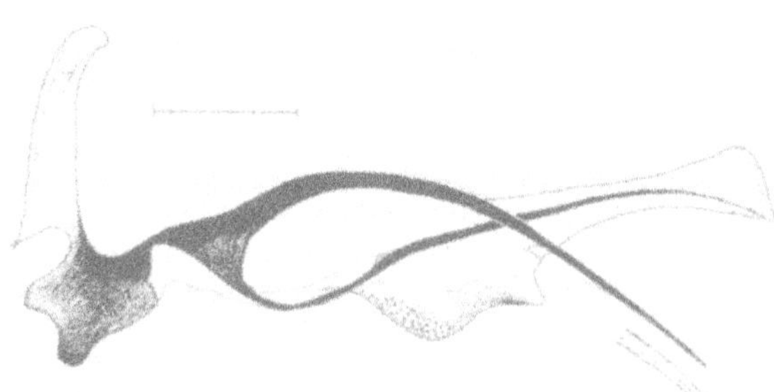

Fig. 586. *Pollenia amentaria* (Scopoli), male. Aedeagus, left lateral view. Inset: tip of paraphallus (enlarged). Scale = 0.2 mm.

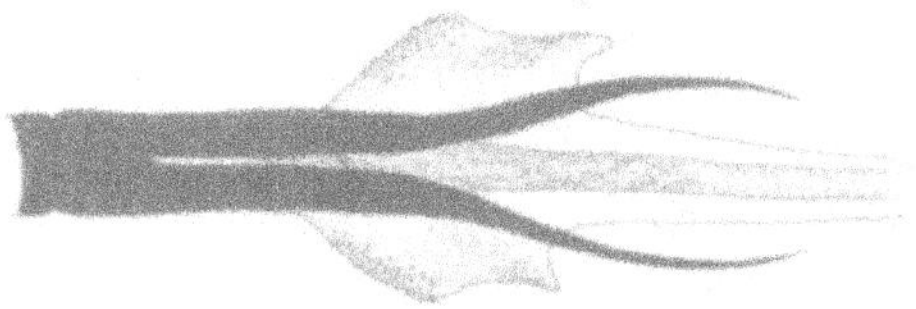

Fig. 587. *Pollenia amentaria* (Scopoli), male. Distiphallus, dorsal view. Scale = 0.2 mm.

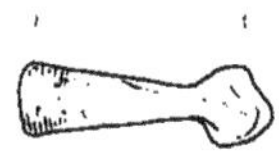

Fig. 588. *Pollenia amentaria* (Scopoli), male. Ejaculatory sclerite. Scale = 0.2 mm.

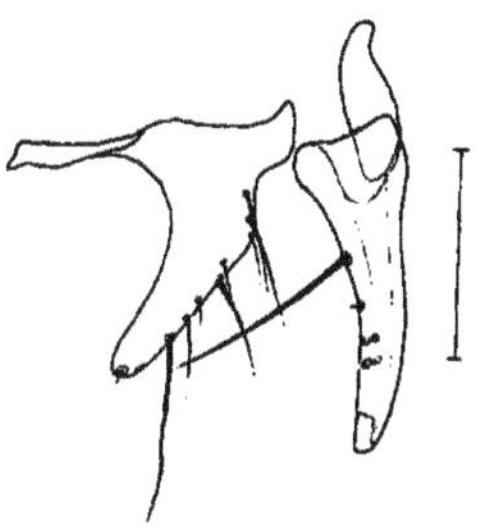

Fig. 589. *Pollenia amentaria* (Scopoli), male. Pre- and postgonites. Scale = 0.2 mm.

Fig. 590. *Pollenia amentaria* (Scopoli), female. Spermathecae. Scale = 0.2 mm. G.pr. 96.

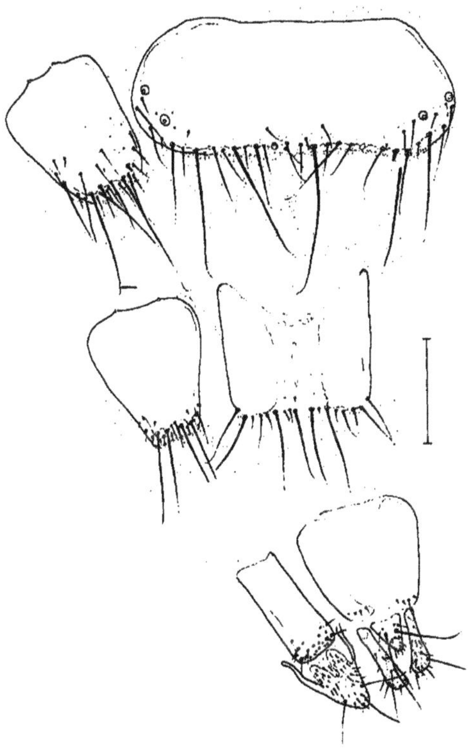

→

Fig. 591. *Pollenia amentaria* (Scopoli), female. Ovipositor. Stipple indicates extent of microtrichiae. Scale = 0.4 mm. G.pr. 96.

scribed *Pollenia micans* without mention of locality or number of specimens. The name is listed as a nomen dubium in the Palaearctic Catalogue (Schumann, 1986). No specimen remains in Robineau-Desvoidy's collection in MNHN, but in the General collection in MNHN under *Pollenia* is a specimen labelled (1)"456", (2) "Guérin / Menue- / ville", (3) "Pollenia / micans RD / (Lat.)" [old label in Robineau-Desvoidy's hand glued to new card, latter with "Det. Rob. Desvoidy" in pencil]. This specimen is referred to by Séguy (1928a: 178). It may be a syntype. *Musca nigrina* was described in the ♂ sex by Meigen on the basis of an unspecified number of specimens from "Hiesige Gegend" [West Germany, probably Stolberg]. I have not found it figured in Morge (1975, 1976a, 1976b). In Meigen's collection in MNHN is a single ♂ under *nigrina*. It is labelled (1) "Meigen" and "2190/40", (2) "nigrina / ♂ [Meigen's hand]". I have given it a holotype label. Its stem-vein lacks a row of fine setae above and otherwise agrees with North European material of the present species. I have therefore removed *nigrina* Meigen from the synonymy of *atramentaria* Meigen where it is placed in the Palaearctic Catalogue (Schumann, 1986). Zetterstedt introduced the name *Musca nitens* for a species misidentified by Meigen and Macquart as "*Musca vespillo* Fabricius". It was based on an unspecified number of ♂♀ from Sweden and Denmark. I have located 7 conspecific specimens from the syntypic series in ZML (Gothenburg collection, drawer 3: 1♂; Diptera Scandinaviae collection, drawer 24: 4♂2♀). I have labelled and here designate a specimen labelled (1) "87"; (2) "M.nitens"; (3) "M.vespillo / Stæger" (the last two in Zetterstedt's hand) as lectotype of *nitens*. The other syntypes have been labelled as paralectotypes. *Chaetopollenia soudeki*, *Chaetopollenia moravica* and *Chaetopollenia pseudobisulca* were treated as new species by Jacentkovský also in 1941c and 1942. Jacentkovský (1941b) described *soudeki* from 2♂ taken on August 20, 1936, in "Adamovské lesy" (the school forest of the Agricultural University of Brno) in Czechoslovakia. 1♂ is present in NMP (no. 33061). It is labelled (1) "Typus" (red label); (2) "Skolni statek Adamov / Kanice 2 / 20/8-1936" (all printed except for date); (3) Cat.no. 33061 (red label). It agrees with the description given in his 1942 paper. I accept this as one of the original syntypes. I have labelled and here designate it as lectotype of *soudeki*. I have seen 1♂ "cotype" of *moravica* and 2♂ "cotypes" of *pseudobisulca* in NMP (Cat. No. 33062 and 33059, 33060, respectively). *P. moravica* and *pseudobisulca* appear to be the same species, but both differ from the neotype of *amentaria* by having a much broader frons (3-4 times as wide as front ocellus, broader than distance between posterior ocelli inclusive), and a more strongly and in posterior view evenly dusted abdomen with tessellations evident only in obliquely posterior view. I have rather hesitantly listed them as questionable synonyms, as I am not quite confident that only one species is involved here. Through the neotype designation the name *amentaria* is restricted to the species with narrow frons and shining black abdomen. The lectotype of *soudeki* also has narrow frons and shining very weakly dusted abdomen. Should it later be decided that a second species is contained in *amentaria* (s.lat. = *vespillo*: authors), the name *moravica* is available for it. I have seen material corresponding to Jacentkovský's *moravica* (= *pseudobisulca*) only from central Europe (Austria, Czechoslovakia, Hungary and Yugoslavia), so the nomenclature of Scandinavian material is not affected. I have found no differences between *moravica* and *amentaria* in ♂ and ♀ genitalia, but it has not been possible to investigate the problem in sufficient detail.

44. ***Pollenia angustigena*** Wainwright, 1940
Figs 41, 562, 579, 583, 592, 594-603.

Pollenia angustigena Wainwright, 1940: 444.

♂♀. Facial carina usually conspicuous and rather sharp, but occasionally weakly developed. Palpi vary from black all over to transparent yellow; the usual condition is dark with tip yellowish. Outer *ph* seta rarely absent but apparently more often so than in other species. t_1 with (1-)2 *pv* setae. Basicosta yellow, occasionally dark.

♂. Frons 0.029-0.046 times head-width (n=5), thus rather narrow. Fronto-orbital plates touching. t_3 with or without some erect *av* hairs besides *av* setae, these hairs never conspicuous or as striking as in e.g. *rudis*. Cerci broad, straight in profile. Surstylus with parallel edges, sometimes weakly curved, a little longer than cerci, apically with stiff vestiture on the inner side. Aedeagus of *rudis*-type. Base of distiphallus of Type I (cf. Fig. 583). Tip of paraphallus armed with minute denticles. Central sclerotisation in hypophallus strikingly broad, usually broader than clear area in front of it.

♀. Frons 0.318-0.341 times head-width (n=5). Just in front of anterior ocellus frontal vitta more than 2 times as wide as each fronto-orbital plate. Eye in profile higher than gena. Ovipositor of the *rudis*-type. T6 rather short. Strongly sclerotised

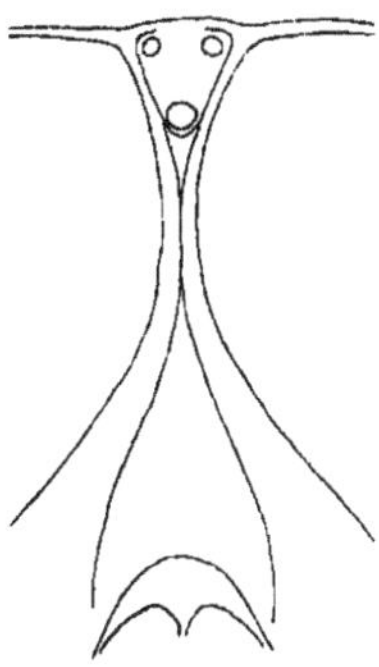

Fig. 592. *Pollenia angustigena* Wainwright, male frons. (From Rognes, 1987b: fig. 1).

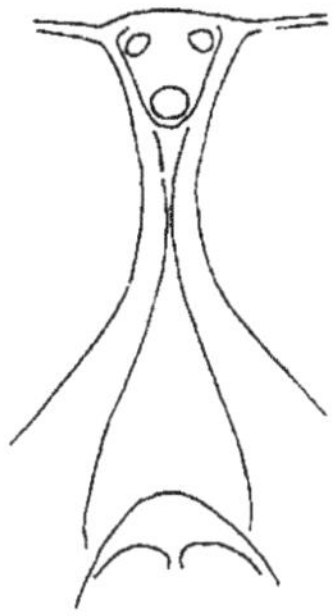

Fig. 593. *Pollenia pediculata* Macquart, male frons. (From Rognes, 1987b: fig. 5).

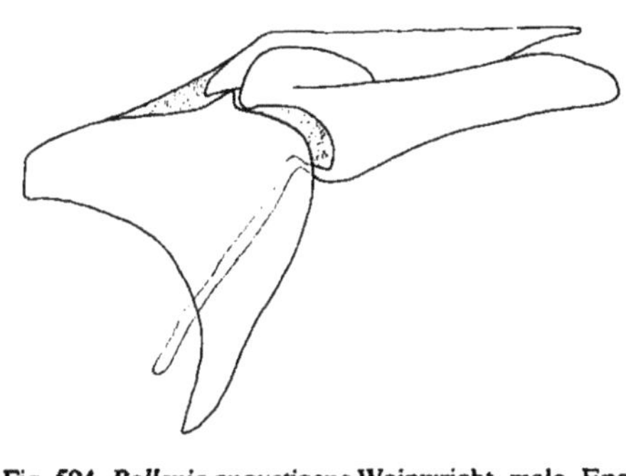

Fig. 594. *Pollenia angustigena* Wainwright, male. Epandrium, cerci, surstylus and bacilliform sclerite, left lateral view. (From Rognes, 1987b: fig. 23).

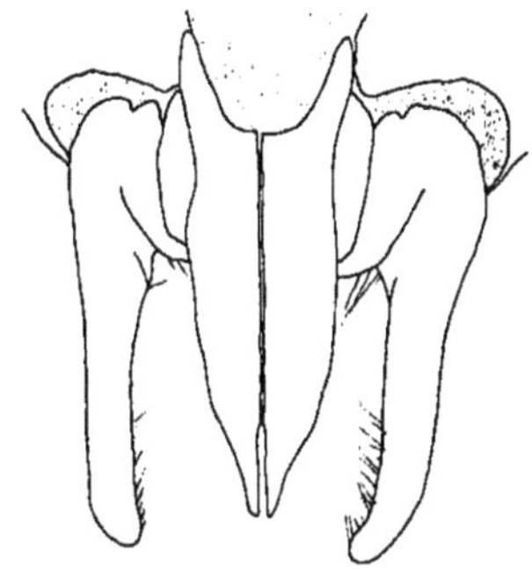

Fig. 595. *Pollenia angustigena* Wainwright, male. Epandrium (parts), cerci and surstyli, posterior view. (From Rognes, 1987b: fig. 22).

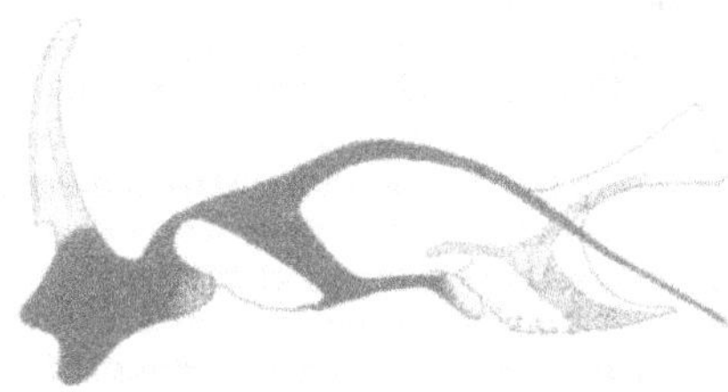

Fig. 596. *Pollenia angustigena* Wainwright, male. Aedeagus, left lateral view. (From Rognes, 1987b: fig. 24).

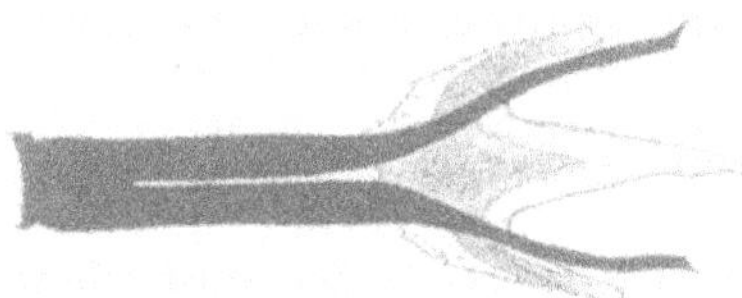

Fig. 597. *Pollenia angustigena* Wainwright, male. Distiphallus, dorsal view. (From Rognes, 1987b: fig. 25).

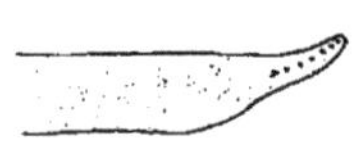

Fig. 598. *Pollenia angustigena* Wainwright, male. Tip of paraphallus. (From Rognes, 1987b: fig. 26).

Fig. 599. *Pollenia angustigena* Wainwright, male. Ejaculatory sclerite. (From Rognes, 1987b: fig. 28).

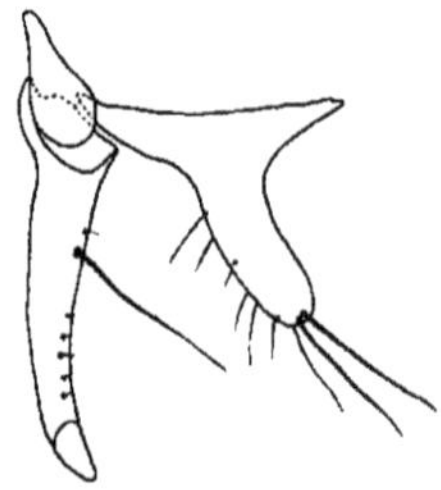

Fig. 600. *Pollenia angustigena* Wainwright, male. Post- and pregonites. (From Rognes, 1987b: figs. 27).

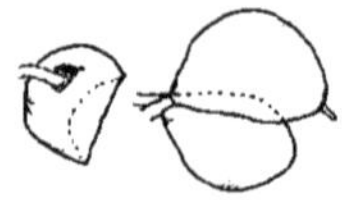

Fig. 601. *Pollenia angustigena* Wainwright, female. Spermathecae. (From Rognes, 1987b: figs. 77).

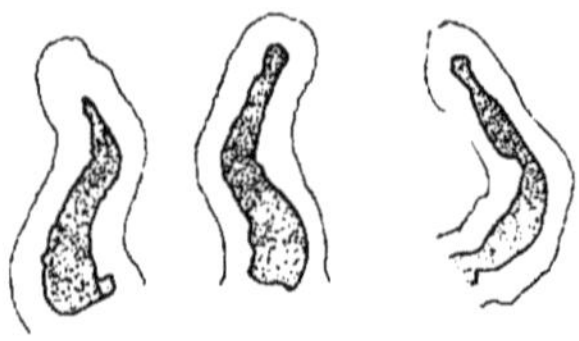

Fig. 602. *Pollenia angustigena* Wainwright, female. Lateral sacs (dorsal view to the left, lateral view to the right). (From Rognes, 1987b: figs. 78).

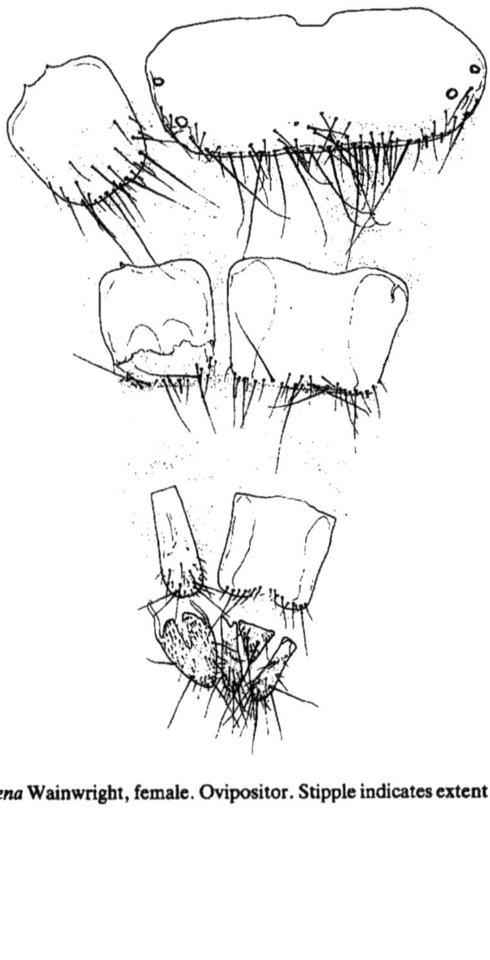

Fig. 603. *Pollenia angustigena* Wainwright, female. Ovipositor. Stipple indicates extent of microtrichiae. (From Rognes, 1987b: fig.29). G.pr. 112.

halves of T7 as wide as or slightly narrower than the almost unsclerotised area between them. Basal fifth or less of cerci devoid of microtrichiae. ST6 and ST7 short. ST7 hardly longer than broad, its posterior third microtrichiose and weakly sclerotised. Spermathecae short, oval. Lateral sacs sclerotised, strongly curved, not as long as in *rudis*. [8 ovipositor slides examined.]

Length. 5.0-9.5 mm.

Distribution. Widely distributed in Denmark. In Sweden north to Vrm. In Norway only in the very southern and extreme southeastern parts. Extreme southwestern part of Finland. - Palaearctic: also Austria, Czechoslovakia, France, Great Britain, Hungary, Italy, Portugal, Spain, Switzerland, West Germany and Yugoslavia; Nearctic: U.S.A. and Canada (Rognes, 1987b).

Biology. Earthworm host species unknown. Larval stages unknown. Puparium described by Rognes (1987b). In Fennoscandia and Denmark adults are on the wing mostly from March to October, with a pronounced peak in numbers in April.

45. ***Pollenia griseotomentosa*** (Jacentkovský, 1944a)

Figs 563, 604-611.

Polleniella griseotomentosa Jacentkovský, 1944a: 45 (Czech), 48 (German).

varia: Grunin (1970a), Emden (1954), not Meigen, 1826; misidentifications.

♂♀. Palpi dark. Scutellum with 3 marginals. Basicosta usually very dark, occasionally brown. Sometimes middle pair of *post acr* absent. Legs often rather stout. t_1 with 1(-2) *pv* setae.

♂. Frons 0.026-0.038 times head-width (n=5), thus very narrow, fronto-orbital plates touching. Cerci narrow, cleft in apical two fifths. Surstyli very narrow, strongly curved. Aedeagus of *griseotomentosa*-type. t_3 without *av* hairs besides *av* setae.

♀. Frons 0.278-0.343 times head-width (n=5). Frontal vitta just in front of anterior ocellus more than 2 times as wide as each fronto-orbital plate. Eye in profile much higher than gena. Ovipositor of *griseotomentosa*-type. T7 halves not as broad as the weakly sclerotised area between them. ST6 and ST7 about as long as broad. Posterior half of ST7 unsclerotised and microtrichiose. Spermathecae spherical. Lateral sacs unsclerotised. [2 ovipositor slides examined.]

Length. 4.5-8.0 mm.

Distribution. Not very common. Widely distributed in Denmark. In Sweden north to Ly.Lpm. Scattered captures in southern Norway. In Finland north to ObS. - Widely distributed in Europe: also East Germany, France, Great Britain, Czechoslovakia, Hungary, Poland, Spain, Switzerland, West Germany, and European parts of U.S.S.R. (Grunin, 1970a).

Biology. Unknown. Larval stages unknown. In the single puparium available (Rozkošný leg. and cult., coll. KR) the anterior spiracle has 5 branches on both sides. Puparium narrowed at level of puparial horns which are very long. Posterior spiracular slits arranged as in Fig. 41. Ratio of distance between posterior spiracles inclusive and greatest width of puparium equals 0.151 (n=1). In Fennoscandia and Denmark adults are on the wing from March to October.

Nomenclature. *Polleniella griseotomentosa* was described from an unspecified number of ♂ from "Borotína" / "Borotin" (CS), Eberswalde (DDR), "Bialowieské puszcze" / "Bjelowieska"

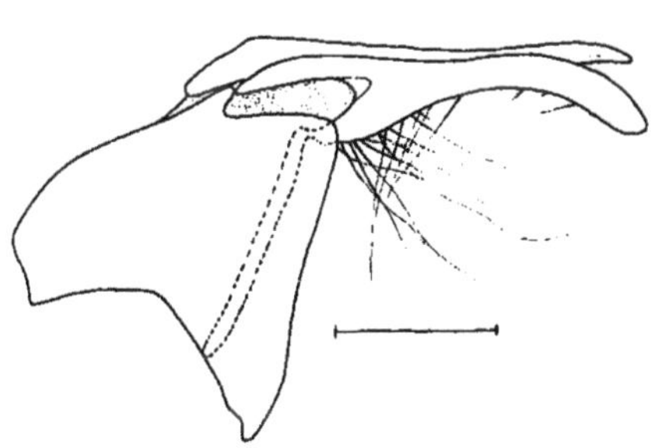

Fig. 604. *Pollenia griseotomentosa* (Jacentkovský), male. Epandrium, cerci, surstylus and bacilliform sclerites, left lateral view. Scale = 0.2 mm.

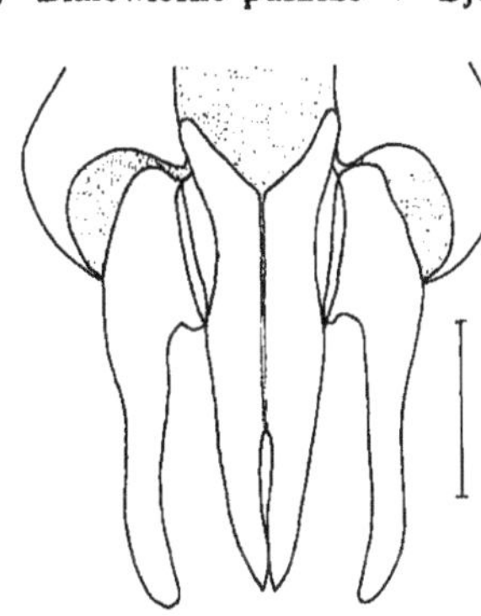

Fig. 605. *Pollenia griseotomentosa* (Jacentkovský), male. Epandrium (parts), cerci and surstyli, posterior view. Scale = 0.2 mm.

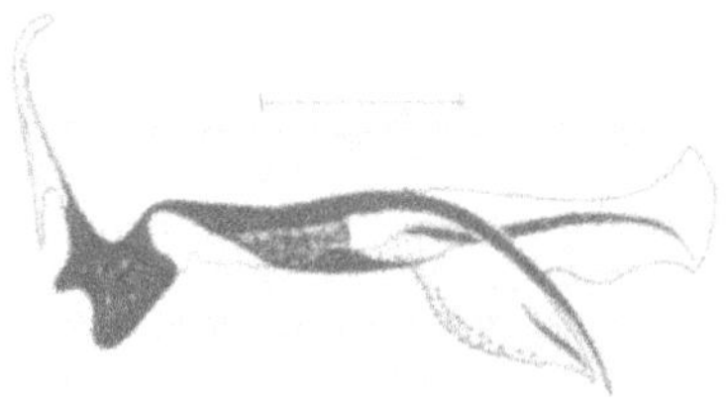

Fig. 606. ***Pollenia griseotomentosa*** (Jacentkovský), male. Aedeagus, left lateral view. Scale = 0.2 mm.

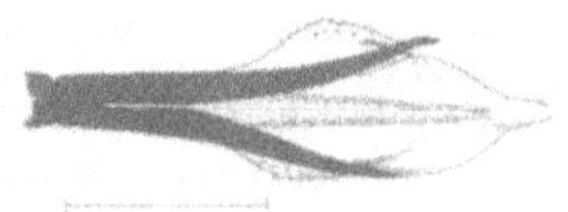

Fig. 607. ***Pollenia griseotomentosa*** (Jacentkovský), male. Distiphallus, dorsal view. Scale = 0.2 mm.

Fig. 608. ***Pollenia griseotomentosa*** (Jacentkovský), male. Ejaculatory sclerite. Scale = 0.2 mm.

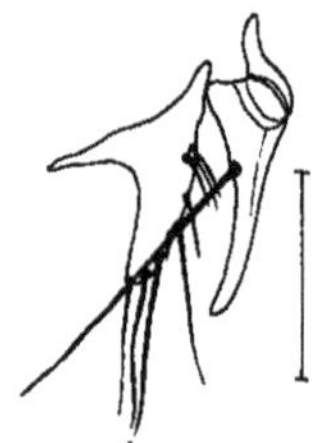

Fig. 609. ***Pollenia griseotomentosa*** (Jacentkovský), male. Pre- and postgonites. Scale = 0.2 mm.

Fig. 610. ***Pollenia griseotomentosa*** (Jacentkovský), female. Spermathecae. Scale = 0.2 mm. G.pr. 98.

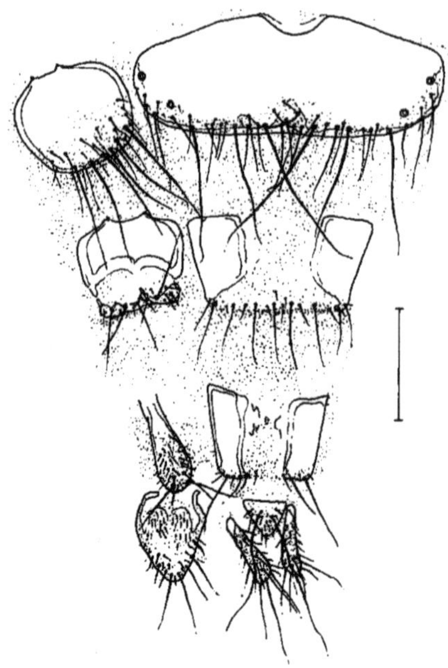

Fig. 611. ***Pollenia griseotomentosa*** (Jacentkovský), female. Ovipositor. Scale = 0.4 mm. G.pr. 99.

and "Otwocku" / "Otwock" (PL, Pawlowicz leg.). The mention of "Prague" (Cepelak leg.) is ambiguous and may refer to *mayeri* as well as *griseotomentosa*, although the former is most likely. Jacentkovský probably designated a holotype in the sense of the ICZN Art. 73 (a) (i) when he stated that "Die Typen beider Arten befinden sich in meiner Sammlung, die Cotypen in den Sammlungen ... Brünn" [in statement concerning *Polleniella griseotomentosa* and *Onesia borotinensis*]. However, he does not state how the specimen is to be recognised nor its provenance. Possibly it was labelled "Type", like some Jacentkovský material in NMP, although such material may have had curatorial labels added later. In MMBCS is a ♂ from Poland labelled "Struga / nad / wzeka / 11.4.35 / Pawlowicz leg." and with Jacentkovský's determination label from 1944. It lacks any "Type" or "Cotype" label and carries a red museum label reading "Syntype". It is in good condition except for the loss of left hind tarsal segments 2-5. The genitalia are concealed. I have not been able to find "Struga" on a map near the Polish localities cited above, and it is not listed among the Polish localities cited by Jacentkovský. The specimen may thus not belong to the type-series. However, it agrees exactly with the above concept of *griseotomentosa* first established by Zumpt (1956a). Jacentkovský also sent material to Villeneuve. No material is in Villeneuve's calliphorid collection in ZIKIEL (F. Sick in litt. March 4, 1988). No material has been traced in NMP. In the interest of stability of nomenclature, it is desirable to fix the identity of *Polleniella griseotomentosa*. The status of the MMBCS specimen is quite dubious, but as it is most likely either a paratype or not a member of the type-series at all, it is in no way eligible for lectotype selection if the holotype is lost (ICZN, Art. 72 (b) (v)). I have therefore labelled and here designate the MMBCS specimen as neotype of *Polleniella griseotomentosa*.

46. ***Pollenia hungarica*** Rognes, 1987b
Figs 564, 580, 583, 612-621.

Pollenia hungarica Rognes, 1987b: 483.

♂♀. Facial keel conspicuous and most often rather sharp. Palpi usually black, sometimes tip slightly yellowish, very rarely all yellow. Outer *ph* present. 3(-4) marginal scutellars. Basicosta yellow. Abdominal dusting in some cases rather dense, in which case tessellated pattern not conspicuous. t_1 with 2 *pv* setae.

♂. Frons 0.029-0.058 times head-width (n=5). Fronto-orbital plates contiguous. t_3 with 3-4 *av* setae which are rather long, about 2 times tibial diameter, practically no erect *av* vestiture besides *av* setae. Cerci broad, straight in profile. Surstylus with parallel edges, sometimes weakly curved, a little longer than cerci, apically with stiff vestiture on the inner side. Aedeagus of *rudis*-type. Base of distiphallus of type II (Fig. 583). Central sclerotisation of hypophallus strengthened distally. Paraphallic tips with minute denticles apically.

♀. Frons 0.319-0.346 times head-width (n=5). Frontal vitta more than 2 times as broad as each fronto-orbital plate. Ovipositor of *rudis*-type. T7 halves broader than the almost unsclerotised area between them. Basal two fifths of cerci devoid of microtrichiae. ST6 and ST7 about as long as broad. ST7 microtrichiose and weakly sclerotised posteriorly. Spermathecae elongate-oval, often with a constricted "handle"-like part proximally. Lateral sacs sclerotised and curved when seen in profile, but never to the extent shown by *angustigena* and *rudis*, sometimes very weakly curved. [2 ovipositor slides examined.]

Length. 5.5-9.5 mm.

Distribution. Very rare in Fennoscandia. Absent from Denmark. 3 records from Norway (AK, Frogn: Håøya, F. Midtgaard leg.). 8 captures in southern Sweden (Sk., Bl., Sm., Boh.). 2 specimens from southern Finland (N, Ta). - Palaearctic: also Austria, Czechoslovakia, Hungary, Italy, Switzerland and West Germany.

Biology. Reared from the lumbricid *Eisenia rosea* (Savigny). Larval stages unknown. The single puparium available described by Rognes (1987b). In Fennoscandia collected from April to July.

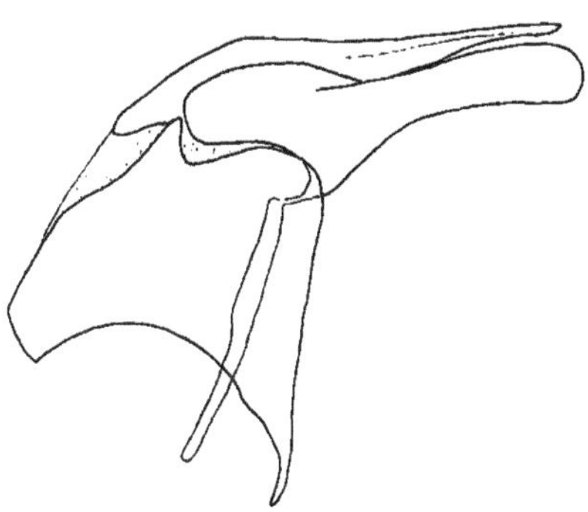

Fig. 612. *Pollenia hungarica* Rognes, male. Epandrium, cerci, surstylus and bacilliform sclerite, left lateral view. (From Rognes, 1987b: fig. 31).

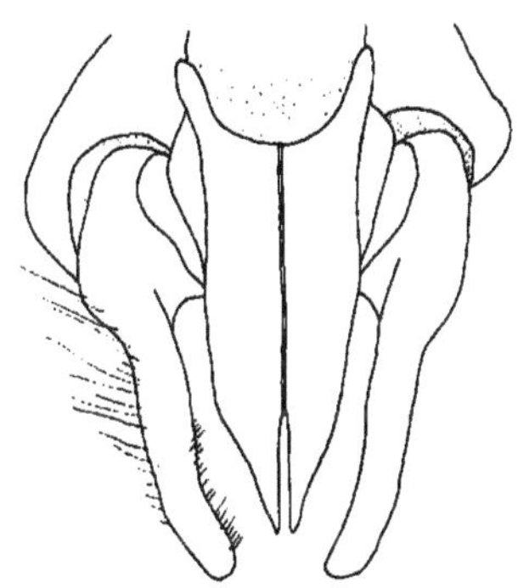

Fig. 613. *Pollenia hungarica* Rognes, male. Epandrium (parts), cerci and surstyli, posterior view. (From Rognes, 1987b: fig. 30).

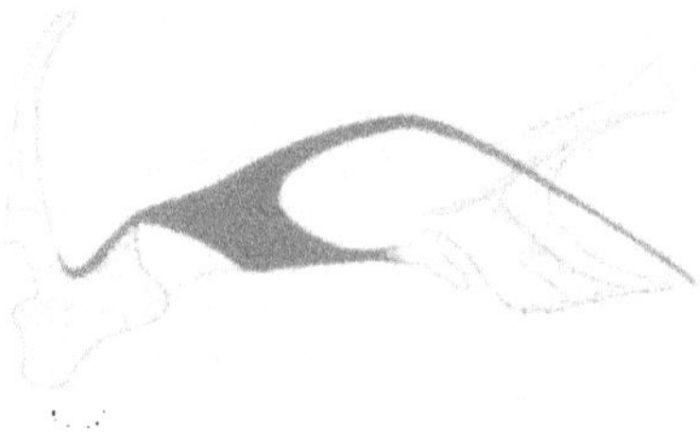

Fig. 614. *Pollenia hungarica* Rognes, male. Aedeagus, left lateral view. (From Rognes, 1987b: fig. 32).

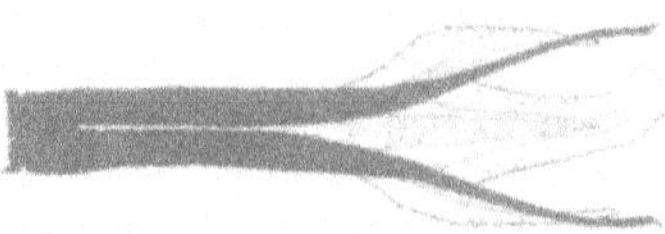

Fig. 615. *Pollenia hungarica* Rognes, male. Distiphallus, dorsal view. (From Rognes, 1987b: fig. 33).

Fig. 616. *Pollenia hungarica* Rognes, male. Tip of paraphallus. (From Rognes, 1987b: fig. 34).

Fig. 617. *Pollenia hungarica* Rognes, male. Ejaculatory sclerite. (From Rognes, 1987b: fig. 36).

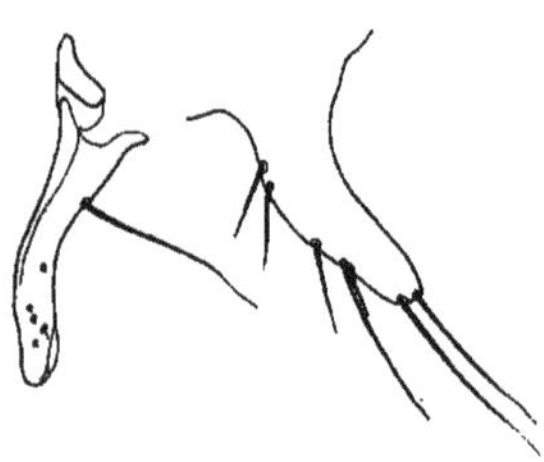

Fig. 618. *Pollenia hungarica* Rognes, male. Post- and pregonites. (From Rognes, 1987b: fig. 35).

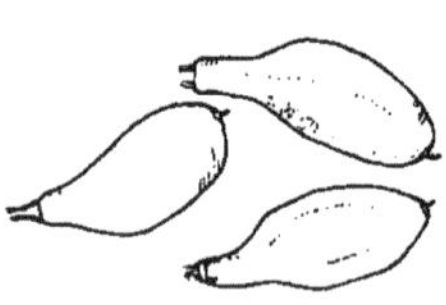

Fig. 619. *Pollenia hungarica* Rognes, female. Spermathecae. (From Rognes, 1987b: fig. 79).

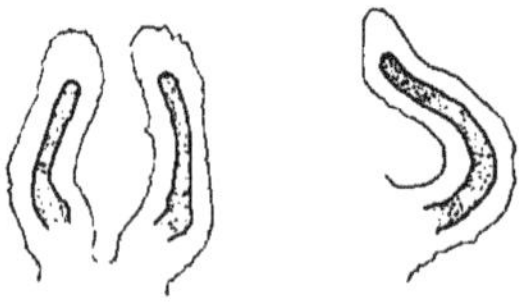

Fig. 620. *Pollenia hungarica* Rognes, female. Lateral sacs (dorsal view to the left, lateral view to the right). (From Rognes, 1987b: fig. 80).

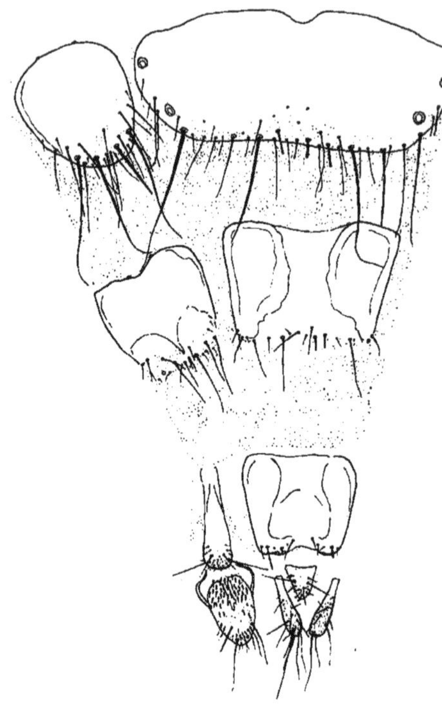

Fig. 621. *Pollenia hungarica* Rognes, female. Ovipositor. Stipple indicates extent of microtrichiae. (From Rognes, 1987b: fig.37). G.pr. 132.

47. ***Pollenia labialis*** Robineau-Desvoidy, 1863 Figs 565, 622-628.

Musca depressa Meigen, 1826: 67. Junior primary homonym of *Musca depressa* Swederus, 1787 (Syrphidae); *Musca depressa* Villers, 1789; *Musca depressa* Gmelin, 1790. SYN.N.
Pollenia labialis Robineau-Desvoidy, 1863: 671.
Pollenia excarinata Wainwright, 1940: 442.
intermedia: authors, not Macquart, 1835; misidentifications.

♂♀. Facial keel absent or indistinct. Palpi black. t_1 with 2(-1) *pv* setae. Antennae unusually dark, except for areas immediately bordering junction between pedicel and first flagellomere. Basicosta always black.

♂. Frons 0.048-0.070 times head-width (n=5). First antennal flagellomere wholly black, except for extreme base. Ventral vestiture on abdominal tergites erect and rather thin. Cerci broad, acute. Surstylus with parallel edges, weakly curved in profile, sinuous apically in dorsal view. Aedeagus of *labialis*-type. Tips of paraphallus weakly sclerotised apically and curved strongly towards midline, tips pointing towards axis of distiphallus at right angles. Most proximal point of clear area in middle of distiphallus when seen in profile closer to ventral wall or midway between ventral and dorsal wall (arrow in Fig. 624).

♀. Frons 0.297-0.327 times head-width (n=5). Frontal vitta more than 2 times as wide as each fronto-orbital plate. 2 rows of setulae outside row of frontal setae. f_2 often (about 50%) with an *a* preapical seta (on left, right or both sides). Ovipositor of *labialis*-type. T6 with a strong "incision" of weak or absent sclerotisation on posterior edge. ST6 and ST7 about as long as wide. Usually less than anterior half of ST7 strongly sclerotised, posterior weakly sclerotised part microtrichiose. T8 without microtrichiae in front of marginal setae. Spermathecae oval. Lateral sacs unsclerotised. [3 ovipositor slides examined.]

Length. 6-10 mm.

Distribution. Common and widespread in all of Denmark and south Sweden north to Jmt. Widespread in southern Norway, where it ascends to subalpine zone. In Finland north to ObS. - Palaearctic: Widely distributed all over Europe (Schumann, 1986; Rognes, 1987a). Nearctic: recently collected in Canada (Ontario): Ottawa, 1♂ July 28, 1984; Nepeau, Rideau River, 3♀ August 8, 1983; Point Pelée, 1♂1♀ August 21, 1982 (all L. Huggert leg., in ZML).

Biology. Unknown. Females have been attracted to fish bait in Finland. In Fennoscandia and Denmark adults are on the wings from March to November.

Nomenclature. Meigen described *Musca depressa* from an unspecified number of ♂♀ "...Aus Oesterreich, von Berlin, Hamburg, auch hier, aber selten...". Meigen Plate 248, fig. 1 shows both sexes, Plate 267, fig. 8 a ♀ (Morge, 1976b). The name was listed as a nomen dubium in the Palaearctic Catalogue (Schumann, 1986). In Meigen's collection in MNHN are 1♂1♀ under *depressa*. Both are labelled "Meigen" and "2148/40". The ♂ is in good condition and is labelled "depressa/♂" in Meigen's hand. It belongs to *Pollenia rudis* (Fabricius). The ♀ is small, mouldy and in bad condition. It is labelled "depressa/♀" in Meigen's hand and belongs to the present species. I have labelled and here designate the ♀ as lectotype of *depressa*. The ♂ has been labelled as paralectotype. Macquart (1835: 272) described *Pollenia intermedia* from a syntypic series of both sexes "Du Nord de la France". In MNHN is a ♀ labelled (1) "263" [old paper in Maquart's hand], (2) "586", (3) "intermedia" [old long paper label in Macquart's hand], (4) "Métatype" [green label in Séguy's hand]. I see no reason for not regarding this as one of the original syntypes, as Séguy (1928a) did too. I have labelled it and here designate it as the lectotype of *intermedia* Macquart. On account of the presence of 6 strong pairs of marginal scutellar setae, a distinct facial carina (cf. key in Séguy, 1928a: 172) and several other features, it does not fit the current concept of *intermedia* of authors (e.g. Rognes, 1987a). I believe it to be *Pollenia vera* Jacentkovský, 1936a, but will not make a firm decision as to its identity before *vera* is better known in both sexes. The name *Pollenia labialis* Robineau-Desvoidy was reintroduced by Séguy (1928a: 176) for a species distinct from *intermedia* Macquart. However, it is currently catalogued as a synonym of "*intermedia* Macquart" by Schumann (1986), following Zumpt (1956a). Robineau-Desvoidy described *labialis* from ♂ specimen(s) taken in May. No locality was stated. No original material remains in Robineau-Desvoidy's collection in MNHN. In the interests of stability of nomenclature, however, and to fix its identity I here designate as neotype of *Pollenia labialis* Robineau-Desvoidy a ♀ in MNHN in good condition which has already been selected by Séguy for this purpose. It is labelled (1) "Rambouillet / 2.5.17", (2) "Museum Paris / Coll. E. Séguy", (3) "Néotype" [green label in Séguy's hand]. It fits the current concept of *intermedia* of authors, not Macquart. I have placed it in the General collection in MNHN in one of the three boxes for *Pollenia*.

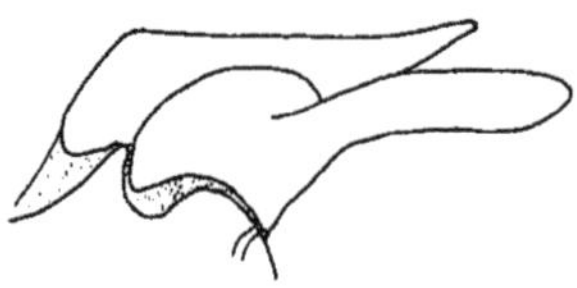

Fig. 622. *Pollenia labialis* Robineau-Desvoidy, male. Cerci and surstylus, left lateral view. (From Rognes, 1987a: fig. 6).

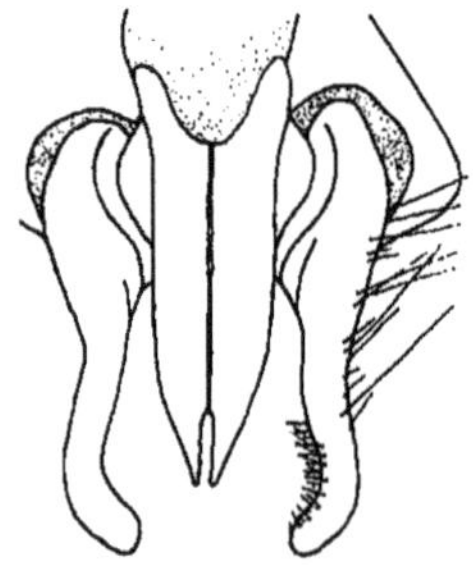

Fig. 623. *Pollenia labialis* Robineau-Desvoidy, male. Cerci and surstyli, posterior view. (From Rognes, 1987a: fig. 4).

Fig. 624. *Pollenia labialis* Robineau-Desvoidy, male. Aedeagus, left lateral view. (From Rognes, 1987a: fig. 9).

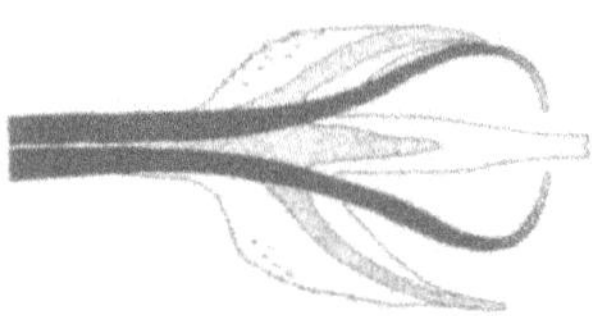

Fig. 625. *Pollenia labialis* Robineau-Desvoidy, male. Distiphallus, dorsal view. (From Rognes, 1987a: fig. 12)

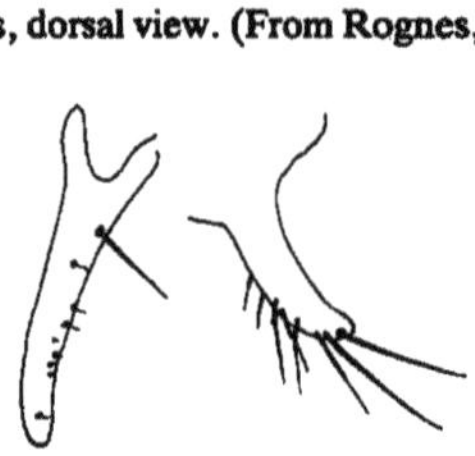

Fig. 626. *Pollenia labialis* Robineau-Desvoidy, male. Post- and pregonites. (From Rognes, 1987a: fig. 10).

Fig. 627. *Pollenia labialis* Robineau-Desvoidy, female. Spermathecae. (From Rognes, 1987a: fig. 16). G.pr. 83.

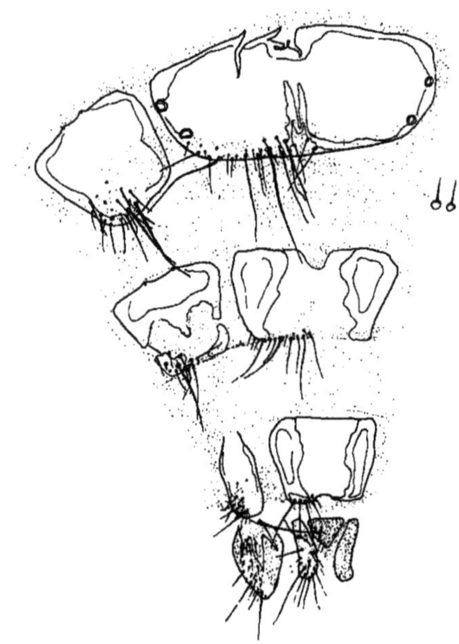

Fig. 628. *Pollenia labialis* Robineau-Desvoidy, female. Ovipositor. Stipple indicates extent of microtrichiae. Inset: shape of individual microtrichiae. (From Rognes, 1987a: fig. 14). G.pr. 83.

48. ***Pollenia pallida*** Rodendorf, 1926
Figs 566, 629-639.

Musca fulvicornis Zetterstedt, 1859: 6186. Junior secondary homonym of *Pollenia fulvicornis* Robineau-Desvoidy, 1830: 413. SYN.N.
Pollenia pallida Rodendorf, 1926: 103; 1928: 338.
Pollenia rudis luciensis Mercier, 1930: 320, figs. 3, 5. SYN.N.
Pollenia carinata Wainwright, 1940: 442.

♂♀. Facial keel usually low, broad and rounded, sometimes sharpened. Palpi yellowish, at least apically. 4 marginal scutellar setae. t_1 with 1 *pv* seta. t_2 with 2-3 *ad* setae. t_3 with (1-)2 short *av* setae.

♂. Frons 0.066-0.082 times head-width (n=5) (Danish material). Cerci broad in dorsal view, cleft in apical third. Surstylus rather broad, slightly curved in profile, slightly bent towards mid-line in dorsal view. Aedegus of *pallida*-type.

♀. Frons 0.348-0.372 times head-width (n=5). Frontal vitta just in front of anterior ocellus about 2 times as broad as each fronto-orbital plate. Outer *ph* sometimes absent. Ovipositor of *pallida*-type. Sclerites elongate. T6 not or very narrowly, ST6 rather broadly microtrichiose posteriorly. Ovipositor sclerites of segments 7 and 8 without microtrichiae. ST6 and ST7 2 times as long as wide. Spermathecae spherical. Lateral sacs unsclerotised. [13 ovipositor slides examined.]

Length. 6.5-9.5 mm.

Distribution. Rare. Several records from Denmark. Apart from the holotype of *fulvicornis* only a single record from Sweden (Sk.: Reslöv, H. Rosén leg.). Absent from Norway and Finland. - Widely distributed in Western Palearctic: Albania, Czechoslovakia, Bulgaria, France, Great Britain, Greece, Hungary, Israel, Lebanon, Turkey, Poland, Romania, U.S.S.R. (Georgia, Armenia, Azerbaijan, Kazakhstan, Kirgizia, Uzbekistan) (partly from Schumann, 1986).

Biology. Unknown. All captures from Fennoscandia and Denmark are from July, August and September, with a maximum in August. *Pollenia pallida* is apparently a late summer or autumn species at these latitudes, thus differing sharply from other species of *Pollenia*.

Nomenclature. *Musca fulvicornis* was described from a single ♂ "d. 31 Jul captus, Museum Lundense", i.e. not in Zetterstedt's own collection. In the Wallengren collection in ZML is a male labelled (1) "Lk 31/7" and (2) "M. fulvi-/cornis n./sp. Z. ♂" in Zetterstedt's hand. The locality label is typical for specimens collected by J. B. v. Borck (1821-1855) in Lackalänga, NW of Lund (Sk.). It fits the description ("...Palpi ferrugineo-flavi ..."). I consider this to be the holotype and have labelled it as such. It is *pallida* Rodendorf. In the Gothenburg collection is a second specimen labelled "M.fulvi-/cornis Zett./n.sp. ♂" in Zetterstedt's hand. It also carries a red "Type" label. It does not fit the description (palpi dark). It is *Pollenia rudis*. I believe that this specimen is a later addition, thought to be conspecific with the holotype and put there by Zetterstedt for inclusion in his own collection. Rodendorf (1926) described *Pollenia pallida* only on the basis of the structure of the aedeagus. From a later paper (Rodendorf, 1928), it is evident that it was based on several ♂ syntypes from "Ak-Tash-Gebirge, Turkestan (50 km nordöstlich von Tashkent)" [U.S.S.R., Uzbek S.S.R.] collected by E. S. Smirnov and B. B. Rodendorf. I have located 8 ♂ syntypes collected on August 21 and 25, 1922: 7 in ZMMSU, 1 in ZILEN. I have labelled and here designate as lectotype a ♂ in ZMMSU dissected by me; the others have been labelled as paralectotypes. *Pollenia luciensis* was described (as a subspecies of *rudis*) from an unspecified number of ♂ from Luc, France. I have not been able to locate the type-material (not in MNHN). I have examined the holotype of *carinata* in BMNH.

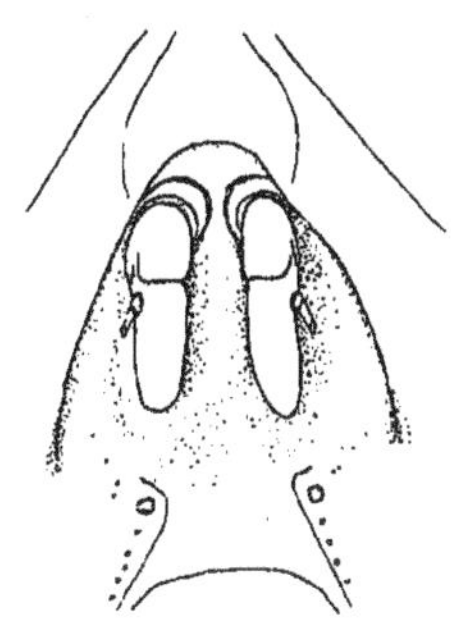

Fig. 629. *Pollenia pallida* Rodendorf, male. Facial area to show facial carina.

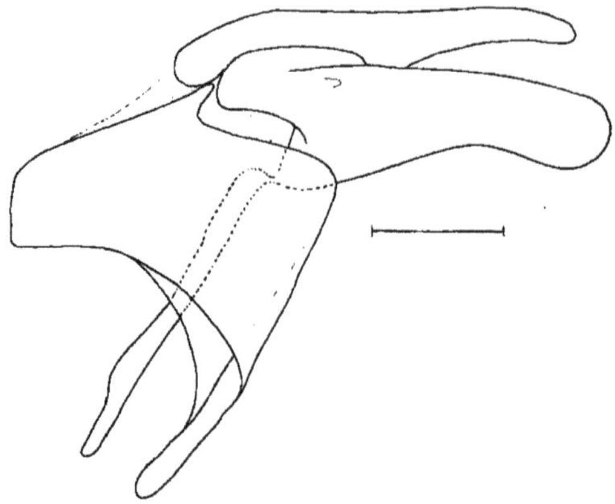

Fig. 630. ***Pollenia pallida*** Rodendorf, male. Epandrium, cerci, surstylus and bacilliform sclerite, left lateral view. Scale = 0.2 mm.

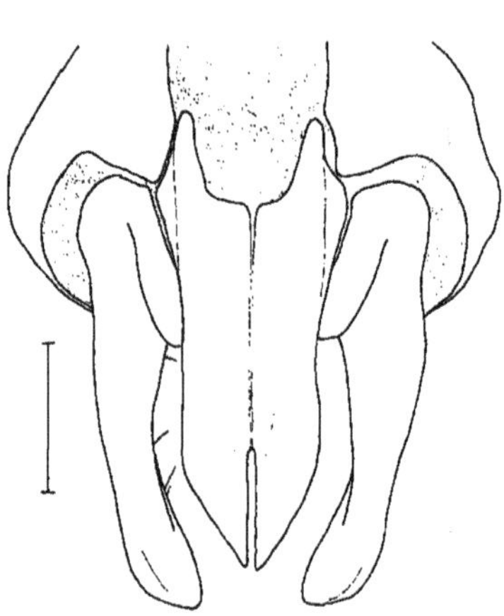

Fig. 631. ***Pollenia pallida*** Rodendorf, male. Epandrium (parts), cerci and surstyli, posterior view. Scale = 0.2 mm.

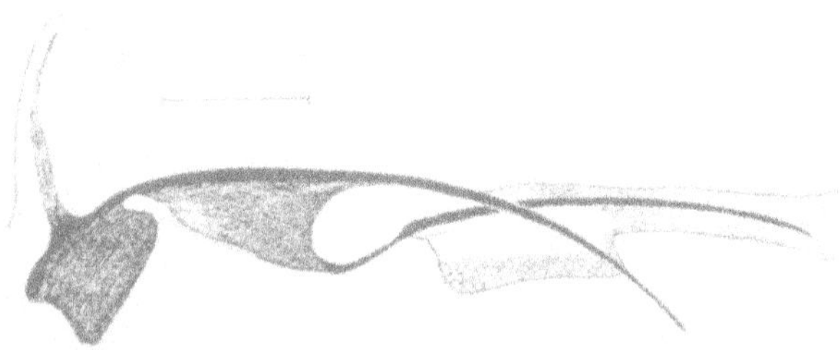

Fig. 632. ***Pollenia pallida*** Rodendorf, male. Aedeagus, left lateral view. Scale = 0.2 mm.

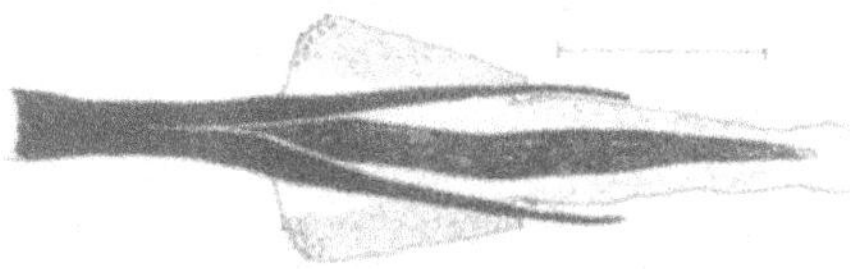

Fig. 633. *Pollenia pallida* Rodendorf, male. Distiphallus, dorsal view. Scale = 0.2 mm.

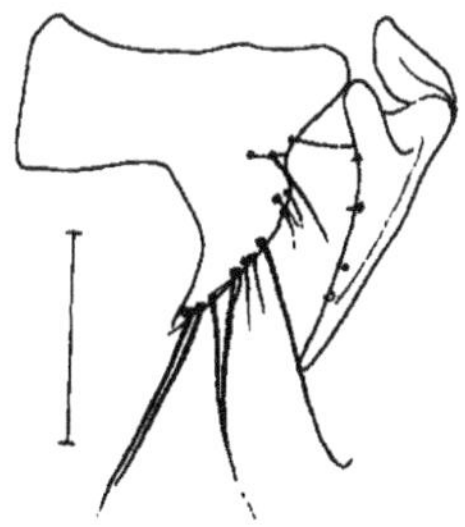

Fig. 635. *Pollenia pallida* Rodendorf, male. Pre- and postgonites. Scale = 0.2 mm.

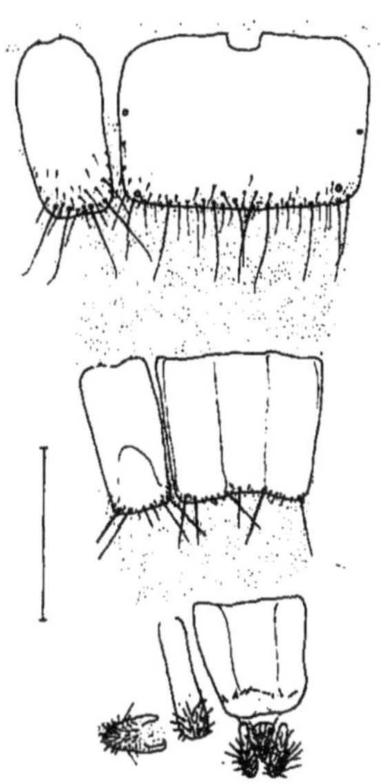

Fig. 637. *Pollenia pallida* Rodendorf, female. Ovipositor. Stipple indicates extent of microtrichiae. Scale = 1.0 mm. G.pr. 231.

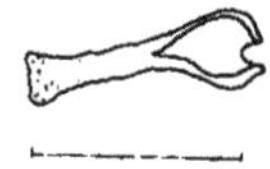

Fig. 634. *Pollenia pallida* Rodendorf, male. Ejaculatory sclerite. Scale = 0.2 mm.

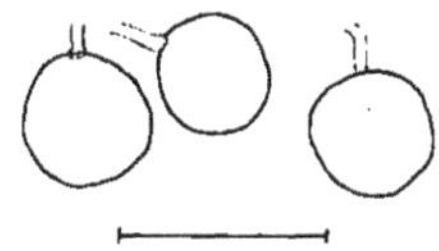

Fig. 636. *Pollenia pallida* Rodendorf, female. Spermathecae. Scale = 0.2 mm. G.pr. 231.

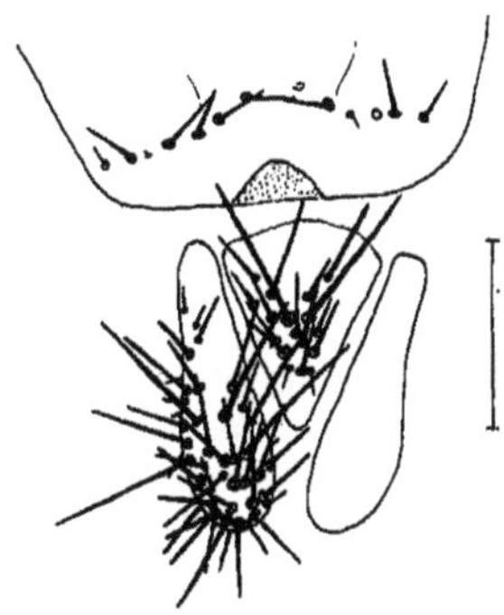

Fig. 638. *Pollenia pallida* Rodendorf, female. Tip of ovipositor: T8, epiproct and cerci. Stipple indicates extent of microtrichiae. Scale = 0.2 mm. G.pr. 231.

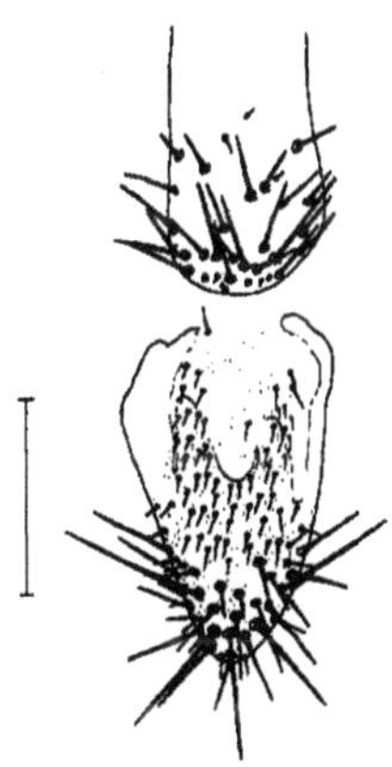

Fig. 639. *Pollenia pallida* Rodendorf, female. Tip of ovipositor: parts of ST8, hypoproct. Stipple indicates extent of microtrichiae. Scale = 0.2 mm. G.pr. 231.

49. ***Pollenia pediculata*** Macquart, 1834
Figs 36, 557, 559, 567, 581, 583, 593, 640-650.

Pollenia pediculata Macquart, 1834: 155.
Pollenia coerulescens Macquart, 1834: 153. SYN.N.
Pollenia obscura Bigot, 1888a: 597. Junior secondary homonym of *Musca obscura* Fabricius, 1794: 315 (= *Musca rudis* Fabricius, 1794). SYN.N.
Pollenia pseudorudis Rognes, 1985b: 90. Replacement name for *Pollenia obscura* Bigot, 1888a. SYN.N.

♂♀. A rather dark species easily recognised by the presence of a bundle of pale hairs on the node at junction of humeral crossvein and vein Sc on the underside of the wing. Antennae mostly very dark. Facial keel variable, from rather narrow and sharp to rather broad, occasionally even with a weakly indicated furrow along summit. Palpi black, even at tip. Outer *ph* present. 3(-4) marginal scutellars. Basicosta yellow. t_1 with 2 *pv* setae. t_2 with 2-4 *ad* setae. t_3 2-6 *av* setae.

♂. Frons 0.048-0.064 times head-width (n=5). t_3 with no erect *av* hairs in addition to the *av* setae. Cerci broad, straight in profile. Surstylus with parallel edges, sometimes weakly curved, a little longer than cerci, apically with stiff vestiture on the inner side. Aedeagus of *rudis*-type. Base of distiphallus of type II (Fig. 583). Hypophallus with rather oblique central sclerotisation. Tip of paraphallus apparently unarmed, in frontal view rather broad distally, often appearing more transversely than obliquely cut off.

♀. Frons 0.331-0.351 times head-width (n=5). Frontal vitta more than 2 times as wide as each fronto-orbital plate. In ovipositor T6 rather short. Strongly sclerotised halves of T7 wider than pale area between them. ST6 and ST7 about as long as broad. ST7 weakly sclerotised and microtrichiose posteriorly. Spermathecae oval. Lateral sacs sclerotised, almost straight. [4 ovipositor slides examined.]

Length. 6.0-9.5 mm.

Distribution. Very common. Widely distributed in Denmark. North to Jmt. in Sweden and NSY in Norway. In southern Norway it ascends to upper parts of subalpine region. In Finland north to Oa. - Palaearctic: From Madeira, Portugal, Spain and British Isles, east to Northern India and Pakistan, south to Corsica, Cyprus, Greece, Israel, Jordan, Iran, Yemen; Nearctic: Canada, U.S.A., widespread from Pacific to Atlantic coast and south to New Mexico; Australian: New Zealand (cf. Rognes, 1987b).

Biology. Females have been attracted to fish bait in Finland. The lumbricid *Eisenia rosea* (Savigny) is known as the host and substrate for larval development. Possibly the description of larval stages by Yahnke & George (1972) refers to this species (cf. Rognes, 1987b). Puparium described by Rognes (1987b). Adults collected all through the year, except in February, with maximum in April and July.

Nomenclature. *Pollenia pediculata* was described from an unspecified number of female specimens from the "Environs de Lille". In Macquart's collection in MHNL (Box G 22) is a single female under this name. It fits the original description. The pin carries no label but the drawer label reads "L." [= Lille]. I accept this specimen as the holotype of *Pollenia pediculata* Macquart and I have labelled it as such. It is very dirty and the cell r_{4+5} has a short stalk (hence the name). The presence of a bundle of pale hairs on the node at junction of humeral crossvein and vein Sc on un-

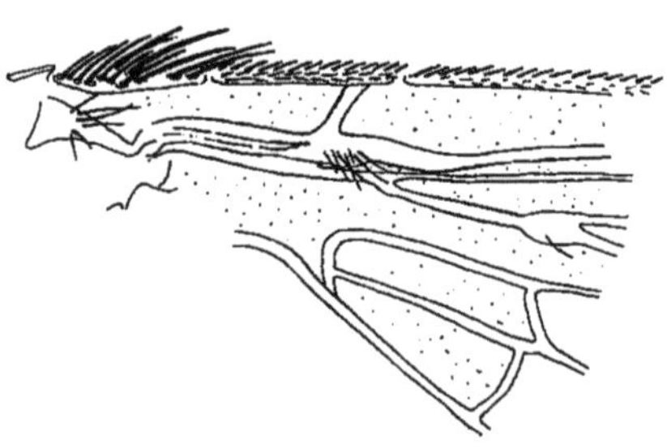

Fig. 640. *Pollenia pediculata* Macquart. Anterior part of wing base from below. (From Rognes, 1987b: fig. 14).

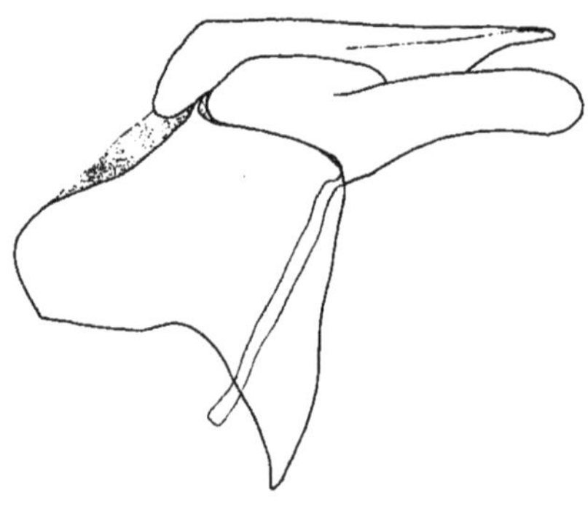

Fig. 641. *Pollenia pediculata* Macquart, male. Epandrium, cerci, surstylus and bacilliform sclerite, left lateral view. (From Rognes, 1987b: fig. 55).

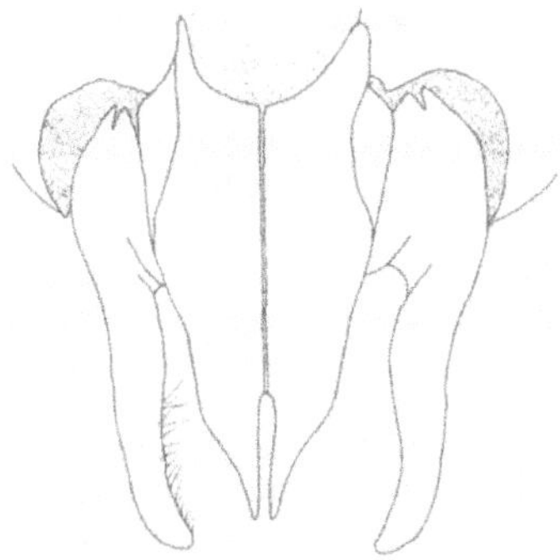

Fig. 642. *Pollenia pediculata* Macquart, male. Epandrium (parts), cerci and surstyli, posterior view. (From Rognes, 1987b: fig. 54).

Fig. 643. *Pollenia pediculata* Macquart, male. Aedeagus, left lateral view. (From Rognes, 1987b: fig. 56).

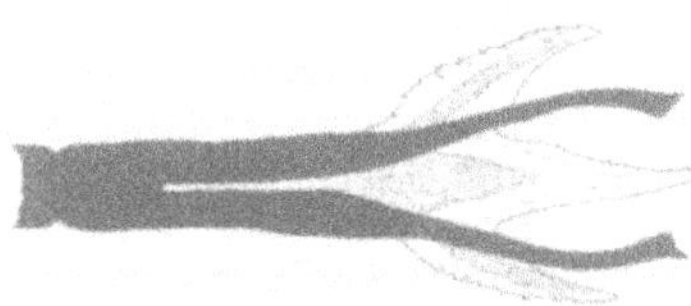

Fig. 644. *Pollenia pediculata* Macquart, male. Distiphallus, dorsal view. (From Rognes, 1987b: fig. 57).

Fig. 645. *Pollenia pediculata* Macquart, male. Tip of paraphallus. (From Rognes, 1987b: fig. 58).

Fig. 646. *Pollenia pediculata* Macquart, male. Ejaculatory sclerite. (From Rognes, 1987b: fig. 60).

Fig. 647. *Pollenia pediculata* Macquart, male. Post- and pregonites. (From Rognes, 1987b: fig. 59).

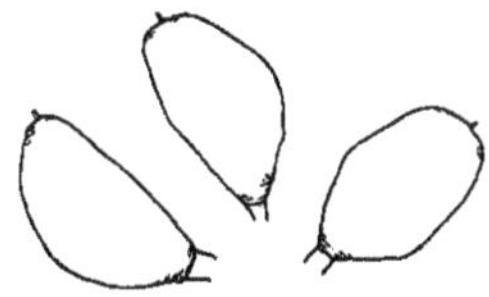

Fig. 648. *Pollenia pediculata* Macquart, female. Spermathecae. (From Rognes, 1987b: fig. 85).

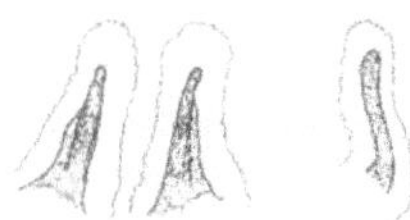

Fig. 649. *Pollenia pediculata* Macquart, female. Lateral sacs (dorsal view to the left, lateral view to the right). (From Rognes, 1987b: fig. 86).

derside of wing and other features clearly identifies the specimen as = *pseudorudis* Rognes. *Pollenia coerulescens* was described from an unspecified number of specimens of unspecified sex from the "Environs de Lille", although the male sex was mentioned explicitly. In Macquart's collection in MHNL (Box G 22) are 1♂ 1♀ under this name. Both fit the original description. The pins carry no labels but the drawer label reads "L." [= Lille]. I accept these specimens as syntypes of *Pollenia coerulescens* Macquart. The male is clearly identifiable as the present species on the presence of a few pale hairs on the node at junction of humeral crossvein and vein Sc on underside of wing. The female belongs to a not readily identifiable species of *Pollenia*. I have labelled and here designate the male as lectotype and the female as paralectotype of *Pollenia coerulescens* Macquart. As first reviser, I have selected the hitherto unused name *pediculata* as the valid name for the present species. The name *coerulescens* has previously been misassociated with other species of *Pollenia* (e.g. by Strobl, 1894; Lundbeck, 1927) and its reintroduction is therefore likely to cause confusion.

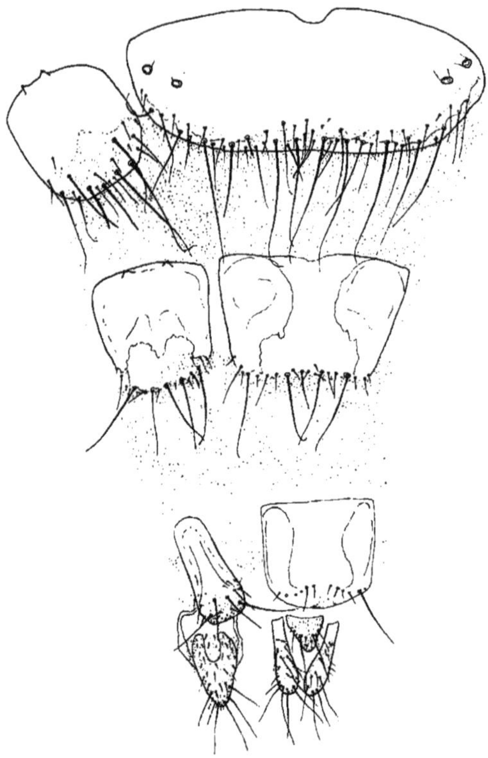

Fig. 650. *Pollenia pediculata* Macquart, female. Ovipositor. (From Rognes, 1987b: fig. 61). G.pr. 110.

50. ***Pollenia rudis*** (Fabricius, 1794)
Figs 10, 11, 568, 582, 583, 651-661; Plate - fig 6.

Musca rudis Fabricius, 1794: 314.
Musca obscura Fabricius, 1794: 315.
Musca varia Meigen, 1826: 66. Junior primary homonym of *Musca varia* Gmelin, 1790. SYN.N.
Musca familiaris Harris, 1869: 336. Junior primary homonym of *Musca familiaris* Panzer, 1804.

♂♀. Facial carina conspicuous and usually rather sharp. Palpi black, tip usually pale. Usually 4 marginal scutellar setae, the lateral weak. t_1 with 1-2 *pv* setae. t_2 with 2-4 *ad* setae. f_2 and f_3 with black vestiture on *p* and *pv* surface (specimens from central Europe may have many yellow hairs in this position not unlike *angustigena* and *hungarica*). Basicosta yellow to brown. Node at junction of humeral crossvein and vein Sc bare on underside of wing.

♂. Frons 0.054-0.066 times head-width (n=5). Fronto-orbital plates touching or almost so. t_3 always with numerous erect *av* hairs in addition to *av* setae. Cerci broad, straight in profile. Surstylus with parallel edges, sometimes weakly curved, a little longer than cerci, apically with stiff vestiture on the inner side. Aedeagus of *rudis*-type. Base of distiphallus of type I (cf. Fig. 583). Central sclerotisation of hypophallus rather narrow, narrower than the clear area in front of it. Tip of paraphallus with a row of minute denticles. Paraphallus strongly curved, highest point much closer to base of distiphallus than to tip of paraphallus when seen in profile.

♀. Frons 0.307-0.358 times head-width (n=5). Frontal vitta just in front of anterior ocellus a little more than 2 times as wide as each fronto-orbital plate. Eye in profile about as high as gena. In ovipositor T6 strikingly long, a line connecting the two anterior spiracles across T6 divides the sclerite into anterior and posterior halves, of which the anterior one is practically as long as posterior one. T7 rather long, its strongly sclerotised halves narrow, usually much narrower than transparent

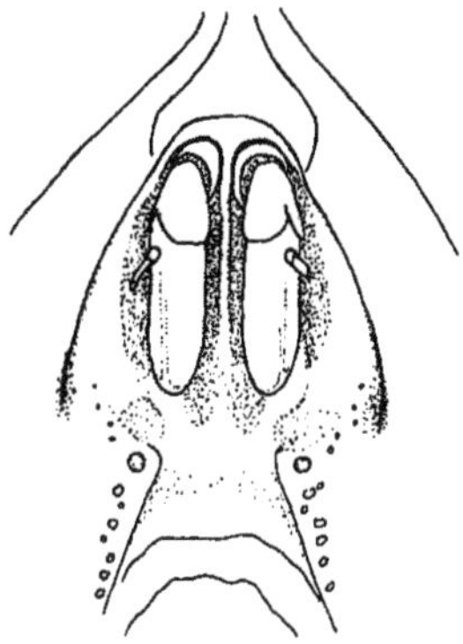

Fig. 651. *Pollenia rudis* (Fabricius), male. Facial area to show facial keel. (From Rognes, 1987b: fig. 13).

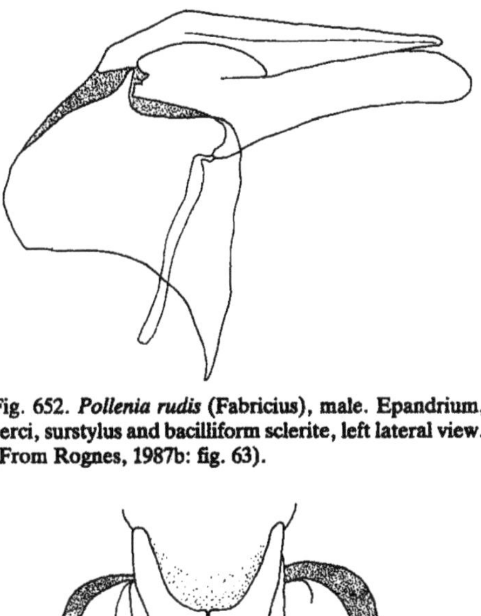

Fig. 652. *Pollenia rudis* (Fabricius), male. Epandrium, cerci, surstylus and bacilliform sclerite, left lateral view. (From Rognes, 1987b: fig. 63).

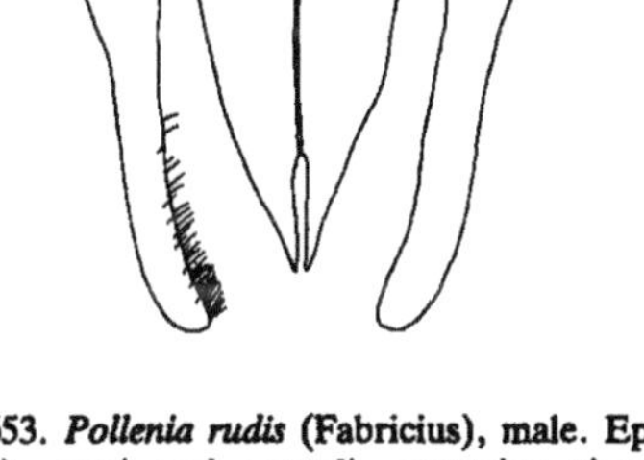

Fig. 653. *Pollenia rudis* (Fabricius), male. Epandrium (parts), cerci and surstyli, posterior view. (From Rognes, 1987b: fig. 62).

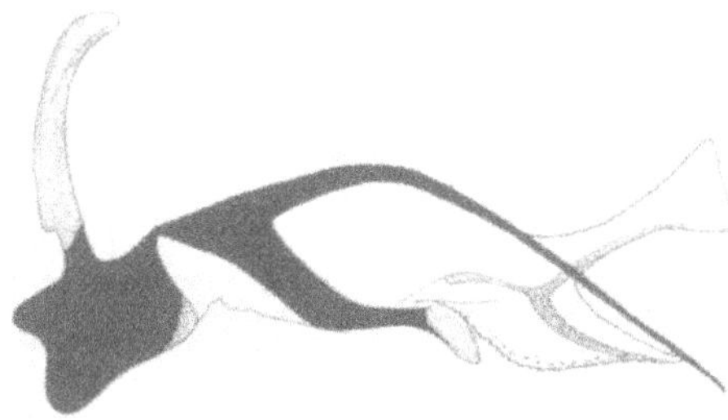

Fig. 654. *Pollenia rudis* (Fabricius), male. Aedeagus, left lateral view. (From Rognes, 1987b: fig. 64).

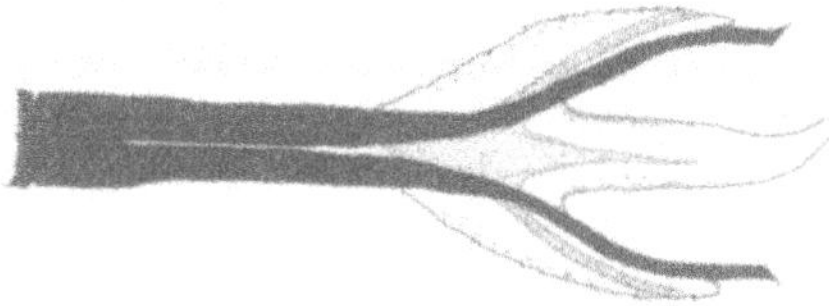

Fig. 655. *Pollenia rudis* (Fabricius), male. Distiphallus, dorsal view. (From Rognes, 1987b: fig. 65).

Fig. 656. *Pollenia rudis* (Fabricius), male. Tip of paraphallus. (From Rognes, 1987b: fig. 66).

Fig. 657. *Pollenia rudis* (Fabricius), male. Ejaculatory sclerite. (From Rognes, 1987b: fig. 68).

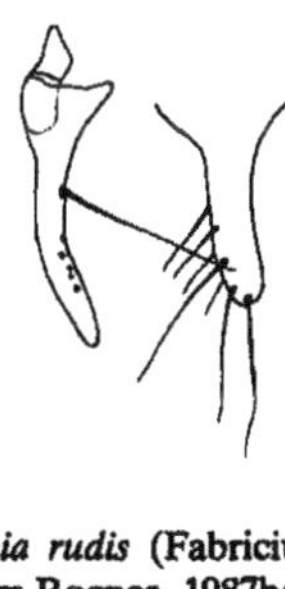

Fig. 658. *Pollenia rudis* (Fabricius), male. Post- and pregonites. (From Rognes, 1987b: fig. 67).

Fig. 659. *Pollenia rudis* (Fabricius), female. Spermathecae. (From Rognes, 1987b: fig. 87).

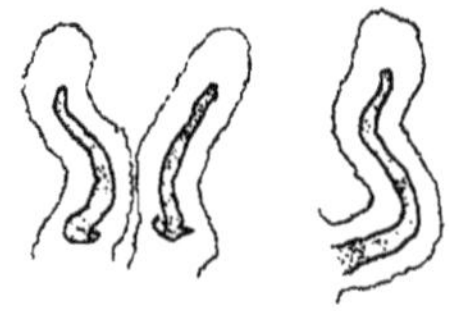

Fig. 660. *Pollenia rudis* (Fabricius), female. Lateral sacs (dorsal view to the left, lateral view to the right). (From Rognes, 1987b: fig. 88).

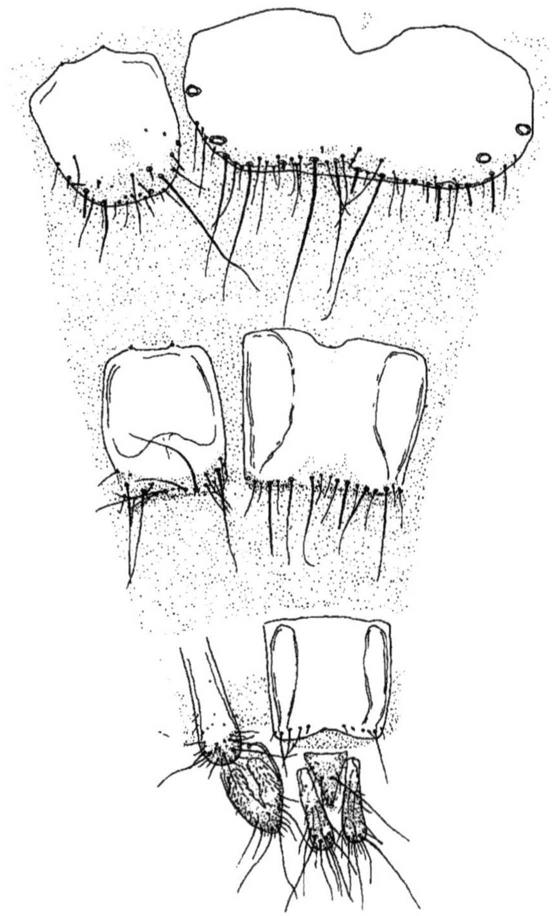

Fig. 661. *Pollenia rudis* (Fabricius), female. Ovipositor. Stipple indicates extent of microtrichiae. (From Rognes, 1987b: fig. 69) G.pr. 108.

area between them. Cerci rather long, truncate at tip, broadest point close to tip. Spermathecae oval, shorter than in *hungarica* and without "handle". Lateral sacs very long and spirally coiled tubes, longer than in *angustigena*. [8 ovipositor slides examined.]

Length. 5.0-10.5 mm.

Distribution. Very common. Widely distributed in Denmark. North to Vb. in Sweden. North to TRY in Norway. In South Norway it ascends to subalpine region. In Finland north to Sb. - Palaearctic: West to Spain and Portugal (including Madeira), east to Northern India (Jammu & Kashmir) south to Morocco, Algeria, Tunisia, Cyprus, Israel; Nearctic: Canada, U.S.A., widespread from Atlantic to Pacific coasts, also Hawaii; Australian: New Zealand (cf. Rognes, 1987b).

Biology. Bred from the lumbricid *Eisenia rosea* (Savigny). In MNHN there is a ♂ reared from the chrysalid of *Chondrostegia maghrebica* Joan. (Lepidoptera) taken in Marrakech (Morocco). Larval stages unknown. Previous reports of the eggs and larvae of a species named "*rudis*" are unreliable as to exact identity of the species studied (cf. Rognes, 1987b). Puparium described by Rognes (1987b). Adults have been collected all through the year, but maximum occurrences in Fennoscandia and Denmark are in April and July.

Nomenclature. *Musca varia* Meigen was described from an unspecified number of specimens of unspecified sex, "In unserer Gegend selten" [West Germany, Stolberg]. The cell r_{4+5} in the wing was described as closed at margin. Meigen Plate 119, fig. 10 (Morge, 1976a) and Plate 298, fig. 8 (Morge, 1976b) both show a ♀ specimen, and the latter figure also illustrates the detail of the wing. In Meigen's collection in MNHN is a single ♂ under *varia*. It is labelled (1) "Meigen" and "2147/40", (2) "varia/♂" in Meigen's hand, and is possibly a syntypic specimen. It belongs to *rudis* Fabricius. I have unfortunately given it a holotype label. I here designate it as lectotype of *varia*.

51. ***Pollenia vagabunda*** (Meigen, 1826)
Figs 569, 662-669.

Musca vagabunda Meigen, 1826: 72.
Pollenia hasei Séguy, 1928b: 370.

♂♀. Facial plate short, deepened at level of first flagellomere. Facial ridges in profile appearing short and strikingly concave. Facial keel low but sharp. Palpi black. Antennae rather dark. 3 marginal scutellar setae. t_1 with 2 *pv* setae. t_3 with 2-4 *av* setae. Presutural area of thorax with 3 distinct stripes, the median one is diagnostic. Basicosta black. Abdomen often with bluish sheen.

♂. Frons 0.047-0.064 times head-width (n=5). Fronto-orbital plates touching. Cerci bent backwards in apical half when seen in profile, widened apically when seen dorsally, and cleft in apical two fifths. Surstylus with parallel edges, slightly bent ventrally in profile, and inwards in dorsal view. Aedeagus of *vagabunda*-type. Apical part of paraphallus broad and dentate, saw-like. In profile a distinct gap between anterior edge of ventral plate and proximal end of mesohypophallus. Hypophallus elongate without projections.

♀. Frons 0.316-0.333 times head-width (n=5). Frontal vitta a little more than 2 times as wide as each fronto-orbital plate. Junction of fronto-orbital plate and parafacial with a broad dark zone. Ovipositor of *vagabunda*-type. T6 rather long. ST6 and ST7 each about 2 times as long as wide. Spermathecae spherical. Lateral sacs unsclerotised. [3 ovipositor slides examined.]

Length. 6-11 mm.

Distribution. Not common. Only NWZ and NEZ in Denmark. In Sweden north to Jmt. Widespread in southern Norway. Southernmost Finland. - Palaearctic: Europe, including European parts of the U.S.S.R.; Nearctic: Canada (British Columbia, Nova Scotia and Prince Edward Island) (Shewell, 1961; Schumann, 1986).

Biology. Unknown. Females have been attracted to fish bait in Finland. Adults have been collected all the year round in Fennoscandia and Denmark, with maximum numbers in April and in the autumn. Apparently it may overwinter indoors. According to Séguy (1928b, 1941) puparia and adults have been found in a stem of *Zea mais* infested by *Sesamia nonagrioides* (Lepidoptera).

Nomenclature. Meigen described only the ♀ sex of *Musca vagabunda*, which was "... Nur zweimal gefangen." No localities were given. Meigen Plate 118 fig. 7 (Morge, 1976a) also shows a ♀. In Meigen's collection in MNHN are 1♂1♀ under *vagabunda*, labelled (1) "Meigen and "2149/40" and (2) "vagabunda/♂ [♀]" in Meigen's hand. Both fit the description (except for the sex as regards the ♂) and conform to the current interpretation of the name. The ♂ is in relatively good condition, but somewhat eaten and lacking both front legs. The ♀ has lost the head and abdomen, but is clearly recognisable. I have labelled and here designate the ♀ as lectotype. I have labelled the ♂ as paralectotype, but one has to assume a lapsus on the part of Meigen to regard it as a syntype at all. *Pollenia hasei* was accurately

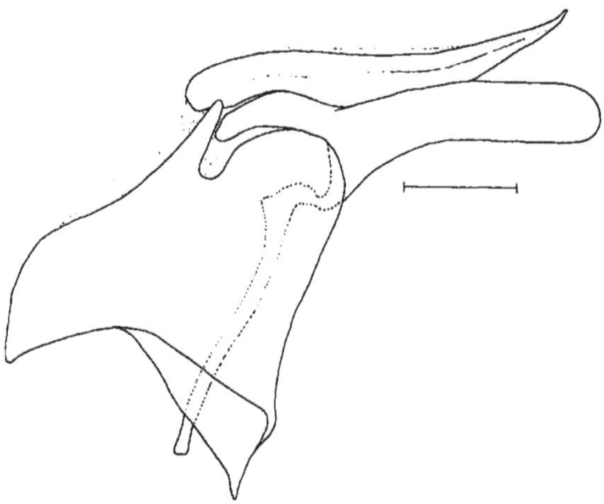

Fig. 662. ***Pollenia vagabunda*** (Meigen), male. Epandrium, cerci, surstylus and bacilliform sclerite, left lateral view. Scale = 0.2 mm.

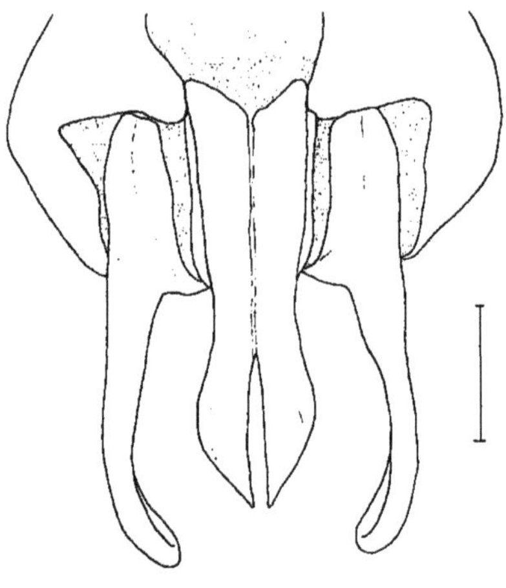

Fig. 663. ***Pollenia vagabunda*** (Meigen), male. Epandrium (parts), cerci and surstyli, posterior view. Scale = 0.2 mm.

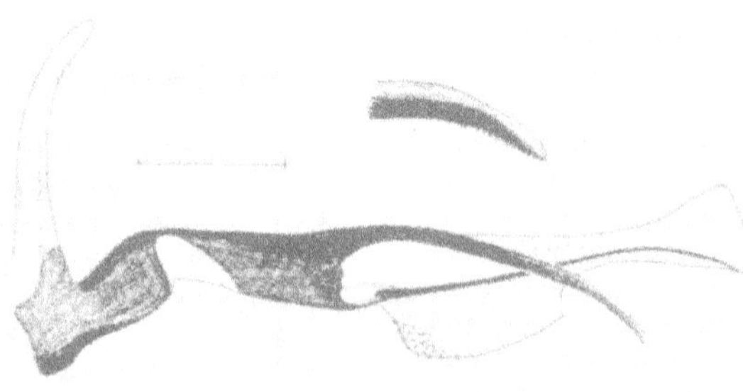

Fig. 664. ***Pollenia vagabunda*** (Meigen), male. Aedeagus, left lateral view. Inset: tip of paraphallus. Scale = 0.2 mm.

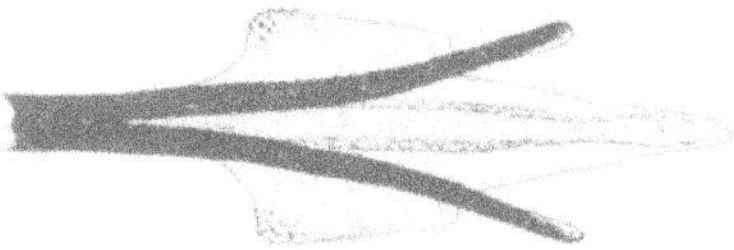

Fig. 665. *Pollenia vagabunda* (Meigen), male. Distiphallus, dorsal view. Scale = 0.2 mm.

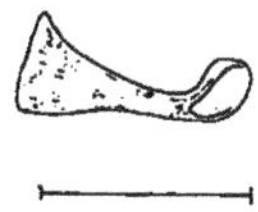

Fig. 666. *Pollenia vagabunda* (Meigen), male. Ejaculatory sclerite. Scale = 0.2 mm.

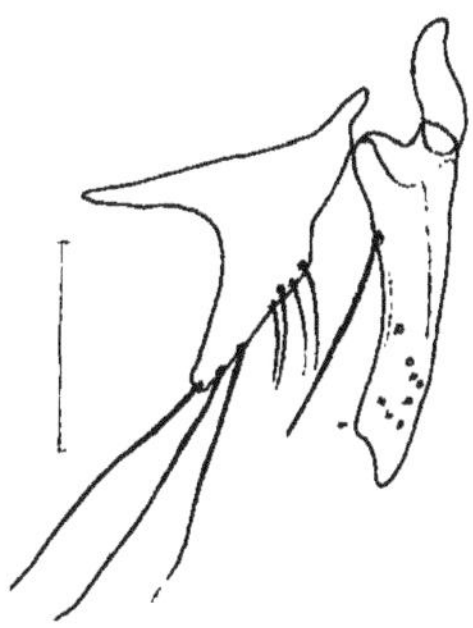

Fig. 667. *Pollenia vagabunda* (Meigen), male. Pre- and postgonites. Scale = 0.2 mm.

Fig. 668. *Pollenia vagabunda* (Meigen), female. Spermathecae. Scale = 0.2 mm. G.pr. 94.

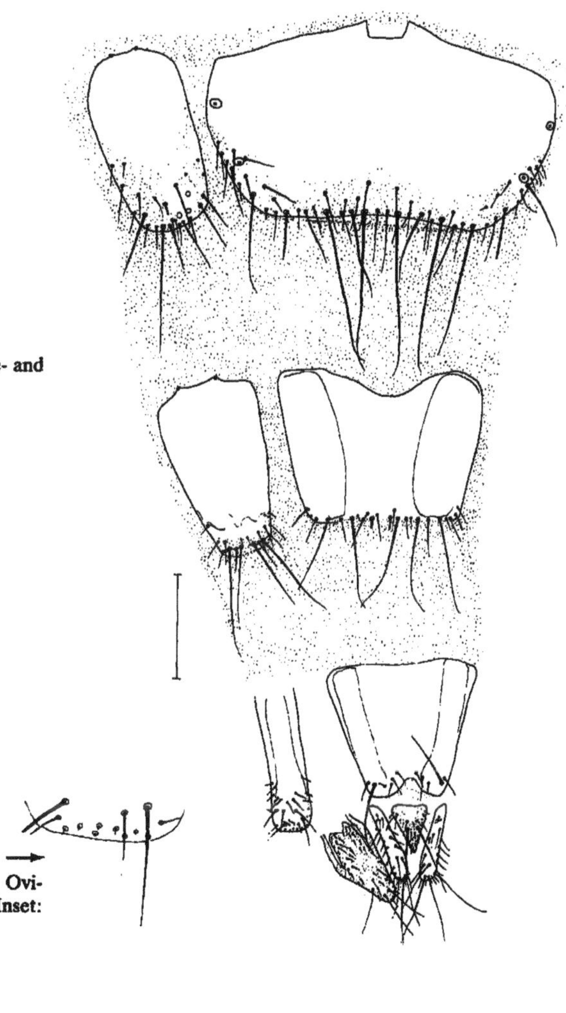

Fig. 669. *Pollenia vagabunda* (Meigen), female. Ovipositor. Stipple indicates extent of microtrichiae. Inset: tip of ST8 (enlarged). Scale = 0.4 mm. G.pr. 94.

described by Séguy (1928b) and there can be no doubt about the synonymy. However, in the original publication no holotype or paratypes were designated. In MNHN are several specimens labelled "cotype", including slide preparations of the ♂ and ♀ genitalia. I assume that a specimen labelled "Type" is present in ZMBERL.

Subfamily Rhiniinae

The subfamily is probably a monophyletic group with the following ground-plan apomophies:

(1) Lower part of face protruding strongly in front of vibrissae, these high above lower facial margin (Fig. 670).
(2) Upper half of occiput with a bare glossy submarginal band across almost its entire width.
(3) Stem-vein with row of setae on dorsal surface of wing (Fig. 673).
(4) Lower calypter narrow, inner edge diverging from the longitudinal axis of the fly (Fig. 671).
(5) Aedeagus of very characteristic structure, ventral plates fused into a ring around base of distiphallus, paraphalli fused, distally broadened into a kind of shield on dorsal side, acrophallus short.

It contains numerous genera, mostly in the Afrotropical, Oriental and Australian Regions but also in the southern part of the Palaearctic Region. A single species in Nearctic (only Bermuda Is).

Oviparous.

A single genus in Fennoscandia and Denmark with a single immigrant species captured once in 1844. For the sake of consistency the subfamily is treated in the same manner as the other subfamilies in this work.

Genus *Stomorhina* Rondani, 1861

Idia Wiedemann, 1820: 21. Junior homonym of *Idia* Hübner, 1809 (Lepidoptera).
Type-species: *Musca lunata* Fabricius, 1805: 292, by designation of Townsend, 1916b: 7.
Stomorhina Rondani, 1861: 9. Replacement name for *Idia* Wiedemann, 1820.
Type-species: *Musca lunata* Fabricius, 1805: 292, automatic.
Euidiella Townsend, 1917: 192.
Type-species: *Musca discolor* Fabricius, 1794: 320, by original designation.

Arista pectinate, with hairs on dorsal side only (Fig. 670). Outer *ph* present. 2 *prepst* setae, *prepm* setae absent. Vein dm-cu weakly sigmoid (Fig. 673). Cell r_{4+5} open. t_3 with a dense row of short *ad* setae of uniform strength, sometimes 2 or 3 a little stronger and longer than the others. Lower calypter bare above. Abdomen black with more or less developed yellow pattern (Fig. 674) or dusted with distinct black undusted spots around setal bases.

Several species in Nearctic, Palaearctic, Oriental and Australian Regions.

1 immigrant species in Fennoscandia and Denmark.

52. ***Stomorhina lunata*** (Fabricius, 1805)
Figs 670-685.

Musca lunata Fabricius, 1805: 292.
Idia rostrata Wiedemann, 1820: 22.
Idia fasciata Meigen, 1826: 9.
Idia syrphoidea Robineau-Desvoidy, 1830: 421.
Idia cinerea Robineau-Desvoidy, 1830: 422.
Idia myoidea Bigot, 1860: 538.
Stomorhyna maculata Rondani, 1865: 228.
Stomorhina melanorhina Bigot, 1888a: 592.
Stomorhina muscoidea Brauer, 1899: 516.
Stomorhina selgae Lehrer, 1979: 82, 89.

♂♀. Parafacialia hairy, dusted in upper half, shining black in lower half. Facial plate dusted except for a broad glossy black band across lower projecting part. Upper part of facial carina, lunula, facial ridges and anterior parts of gena glossy black and quite bare. A sharp line running from the lower extremity of the eye to the extreme anterior point of the horizontal lower border of the gena marks the anterior border of the densely dusted and yellow pilose part of gena. Occiput strongly convex with pale hairs only. Palpi long, black and apically wide and flattened. Dorsal surface of thorax black or olive, thinly white dusted with three broad stripes, small black undusted spots around base of hairs or setae. Prosternum long yellow pilose. Proepisternal depression bare. Suprasquamal ridge and postalar wall bare. Anepisternum and katepisternum densely yellow or white pollinose, with numerous pale crinkly hairs. Lower parts of anterior half of anepimeron bare. Fringes of anterior spiracle pale white or yellow, posterior spiracle with dark brown lappets, posterior lappet larger than anterior one. No row of setae or hairs present along its lower edge. Coxopleural streak present. Posterior margin of anepisternum with

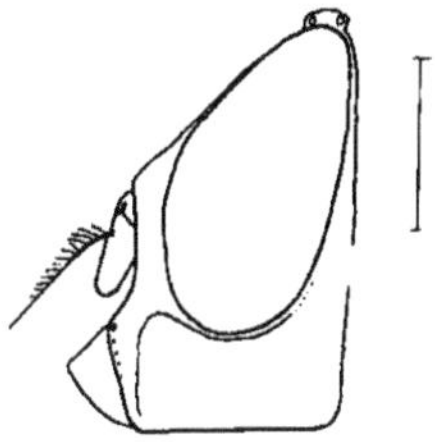

Fig. 670. *Stomorhina lunata* (Fabricius), male. Head, left lateral view. Scale = 1.0 mm.

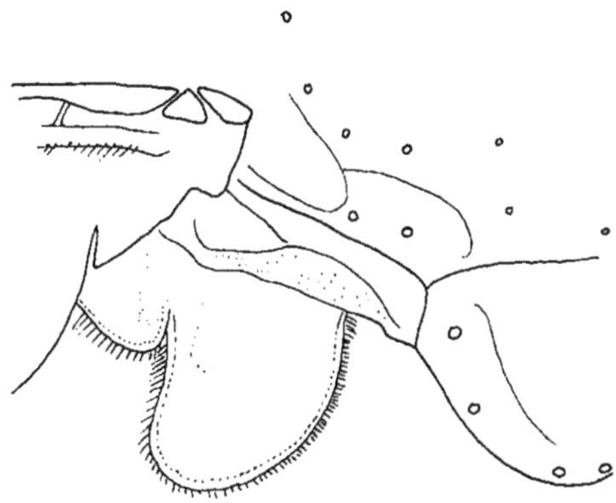

Fig. 671. *Stomorhina lunata* (Fabricius), female. Area behind wing base, dorsoposterior view.

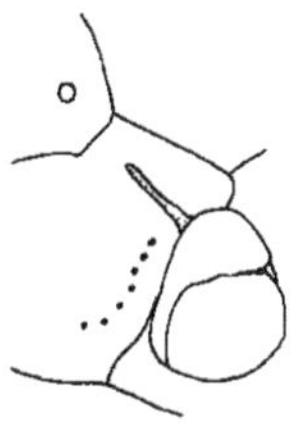

Fig. 672. *Stomorhina lunata* (Fabricius). Posterior spiracle.

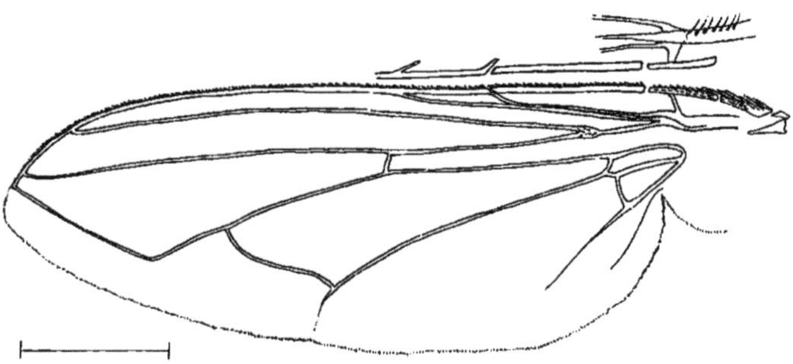

Fig. 673. *Stomorhina lunata* (Fabricius), female. Right wing from below. Inset: parts of upper wing surface. Scale = 1.0 mm.

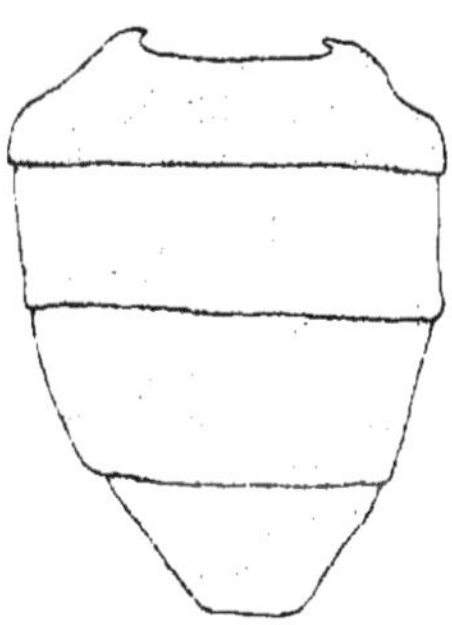

Fig. 674. *Stomorhina lunata* (Fabricius), male. Abdomen, dorsal view.

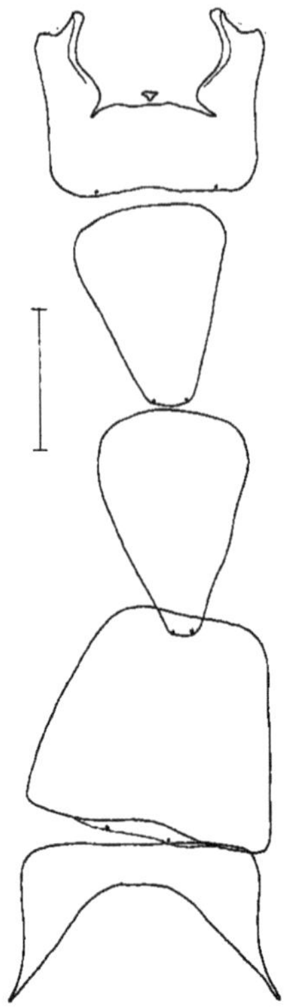

Fig. 675. *Stomorhina lunata* (Fabricius), male. ST1-5. Scale = 0.4 mm.

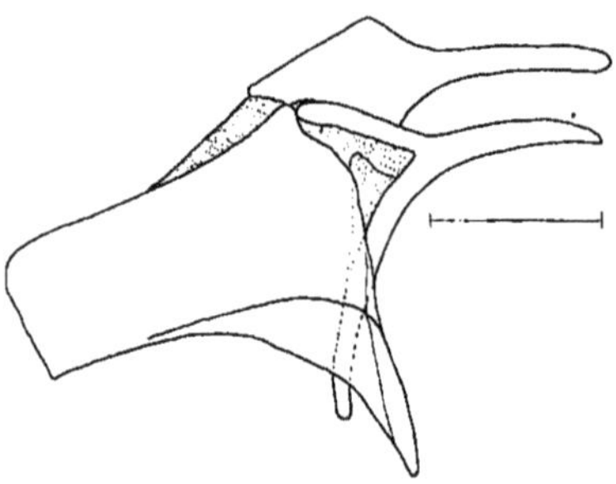

Fig. 676. *Stomorhina lunata* (Fabricius), male. Epandrium, cerci, surstylus and bacilliform sclerite, left lateral view. Scale = 0.2 mm.

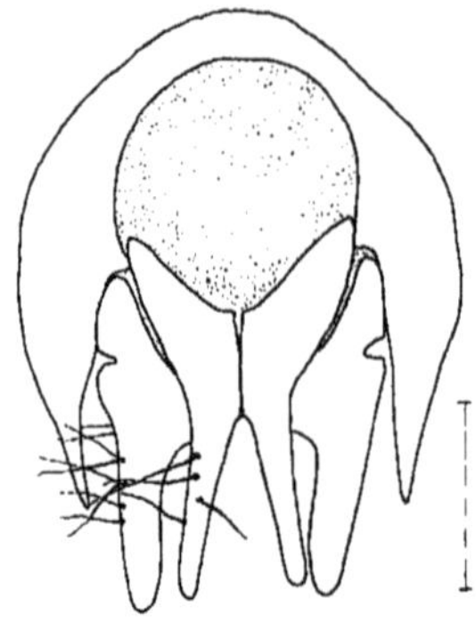

Fig. 677. *Stomorhina lunata* (Fabricius), male. Epandrium, cerci and surstyli, posterior view. Scale = 0.2 mm.

Fig. 678. *Stomorhina lunata* (Fabricius), male. Bacilliform sclerite, frontal and lateral view. Scale = 0.2 mm.

Fig. 679. *Stomorhina lunata* (Fabricius), male. Aedeagus, dorsal view.

Fig. 680. *Stomorhina lunata* (Fabricius), male. Aedeagus (incl. ejaculatory sclerite), left lateral view. Scale = 0.2 mm.

Fig. 681. *Stomorhina lunata* (Fabricius), male. Aedeagus, ventral view.

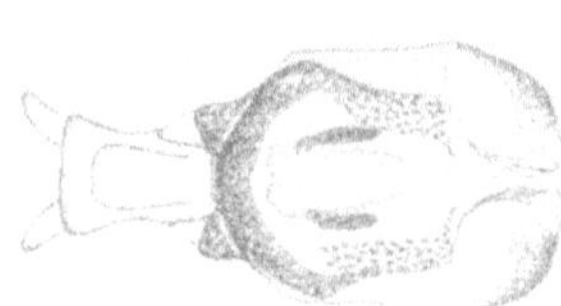

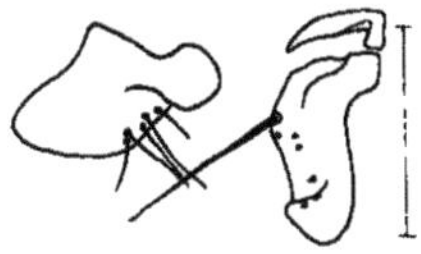

Fig. 682. *Stomorhina lunata* (Fabricius), male. Pre- and postgonites. Scale = 0.2 mm.

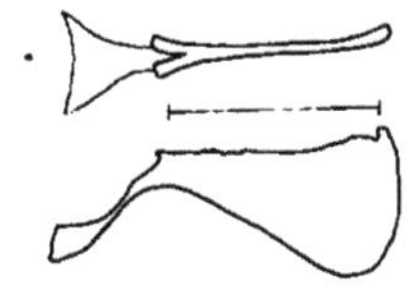

Fig. 683. *Stomorhina lunata* (Fabricius), male. Phallapodeme, frontal and lateral view. Scale = 0.2 mm.

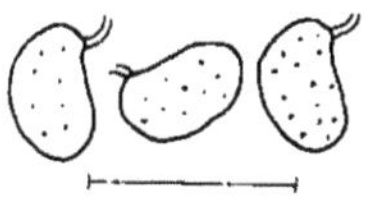

Fig. 684. *Stomorhina lunata* (Fabricius), female. Spermathecae. Scale = 0.2 mm. G.pr. 254.

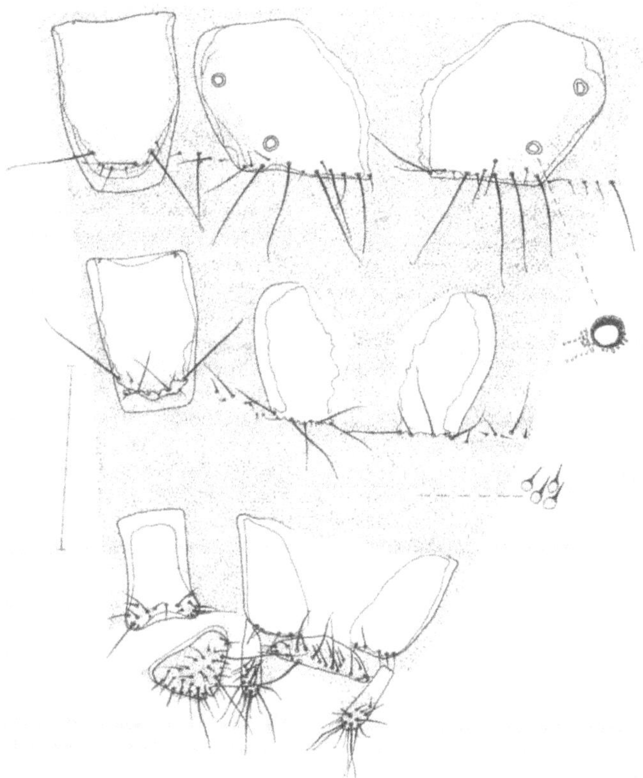

Fig. 685. *Stomorhina lunata* (Fabricius), female. Ovipositor. Stipple indicates extent of microtrichiae. Inset: margin of abdominal spiracle 7 (enlarged). Scale = 0.5 mm. G.pr. 254.

complete row of setae. 1+1 *kepst.* 0+1-2 *acr.* 0+1 *dc.* 0+1 *ia.* 2 *sa.* 1 *prs.* 1-3 *h*, outer one very long. 1 *ph* (the outer), outside a line through *prs.* 3 pairs of marginal scutellar setae. Legs black except for t_2 and t_3 which are brown, this colour also affects the proximal half of the tarsi of the same legs and the knees of the front leg. f_2 without *a* preapical setae. t_1 with 2 *pv.* t_2 with 1 *ad.* t_3 with 2 *av.* Hind coxa bare behind. Costa bare below beyond break in costa just distal to humeral crossvein. Haltere yellow. Abdomen with variable yellow pattern in male in dorsal view (Fig. 674), ventral side of abdomen mostly yellow with black hindmargins to tergites. In female the yellow areas more or less blackened and strongly dusted white, black spots present around setal bases. Lateral margins of ST2 overlapping margins of T2.

♂. Frons extremely narrow, about half as wide as anterior ocellus. t_2 with 0-2 *v* setae. Lower squama yellow with brownish tinge. ST5 with broadly separated lobes, these with inner edges concave and apically rather acute. Cerci cleft in more than apical half, apices diverging, narrow. Surstylus narrow and pointed in lateral view. Bacilliform sclerites fused into a single keeled plate, 2 setae present on each side in the dorsolateral corners. Aedeagus with a "hinge" between basi- and distiphallus, latter almost globular with a strongly sclerotised dorsal shield. Ventral surface of distiphallus with 2 dentate bands. Epiphallus absent. Pregonite low with a few setae posteriorly. Postgonite stout with strong basal seta. Phallapodeme laterally compressed.

♀. Frons 0.267-0.285 times head-width (n=4). Frontal vitta slightly less than 2 times as wide as each fronto-orbital plate. Fronto-orbital plates dusted white or brownish with numerous black spots around setal bases and with 3 glossy black spots along eye margin. Lower squama white. t_2 with *v* seta. In ovipositor T6 and T7 divided, halves separated by microtrichiose dorsal bands, marginal setae continuing laterally beyond margins of tergites. Margin of spiracle 7 with warty protuberances. T8 without microtrichiae. Epiproct very short and broad, microtrichiose. Cerci microtrichiose in distal half. ST8 with concave distal margin. Hypoproct broadly triangular. All pleural and intersegmental margins microtrichiose. Spermathecae bean-shaped, spermathecal ducts subterminal. [1 ovipositor slide examined.]

Length. 5-9 mm (Zumpt 1958).

Distribution. A single ♂ captured on flowerhead of *Tagetes* by W. Nylander in the Botanical garden in Helsinki, Finland, September 29, 1844 (Nylander, 1848) (specimen still in ZMH). No further records are known from Fennoscandia and Denmark and the species does not breed there. - Widely distributed in southern Palaearctic (also northern parts of France, and occasionally Great Britain but not breeding there) and all of Afrotropical Region; in Oriental Region only northern parts; in the Nearctic common on the island of Bermuda, but not elsewhere (Séguy, 1928a; Hall, 1948; Emden, 1954; Zumpt, 1958; Pont, 1980; Dear, 1981; Schumann, 1986).

Biology. Larvae are predatory on the egg capsules of locusts (e.g. "*Dociostaurus maroccanus*", "*Schistocerca gregaria*", "*Schistocerca peregrina*") and seem to be their chief enemy in certain parts of Africa. Larvae have also been found in termite nests (Séguy, 1928a; Hall, 1948). Immature stages described by Hall (1948).

Nomenclature. Fabricius described *Musca lunata* from an unspecified number of specimens in "Mus. Dom. Lund." captured on "Madera" (Madeira) by "Dom. Rahtke". In the Sehested-Tønder Lund collection (in ZMC) are 2♂ under *Musca lunata* (cf. Zimsen, 1964: 488). Both have a red "type" label, and one of the specimens also has a label reading "M:lunata / Mádera Rathke" in Fabricius' hand. This specimen lacks the head and left mid leg, but the other is in good condition. They both agree with the description and I accept them as syntypes. I have labelled and here designate the specimen with head as lectotype, the one without as paralectotype. Both agree with the current interpretation of *Stomorhina lunata* (Fabricius) (e.g. Zumpt, 1956a, 1958). Meigen described *Idia fasciata* from an unspecified number of specimens of unstated sex collected by Baumhauer "bei Marseille und auf dem Gebirge bei Frejus". Meigen Plate 115, fig. 9 shows a ♂ and the head of a ♀ (Morge, 1976a). In Meigen's collection in MNHN are 1♂1♀ in good condition, both labelled (1) "Meigen" and "2077/40", (2) "Idia / fasciata / ♂ [♀]" in Meigen's hand. Following Pont (1986) I accept these specimens as true Baumhauer specimens and therefore as syntypes. I have labelled and here designate the ♂ as lectotype. The ♀ has been labelled as paralectotype. Lehrer's (1979) neotype designation for *Musca lunata* is in violation of the ICZN, Article 75 (d) (3), and is not accepted. *Stomorhina selgae* Lehrer is based exclusively on Hall's (1948) figures of *Stomorhina lunata* and not on material inspected by himself. According to González Mora (1985), who sank *selgae* as a synonym of *lunata*, Hall's figures were prepared from a specimen from Balearic Is.

1

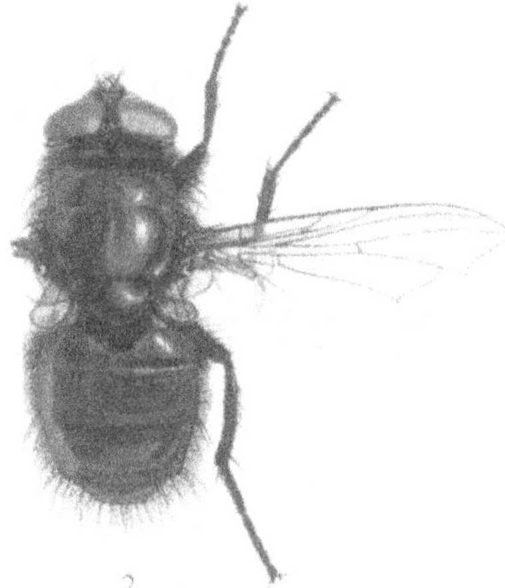

2

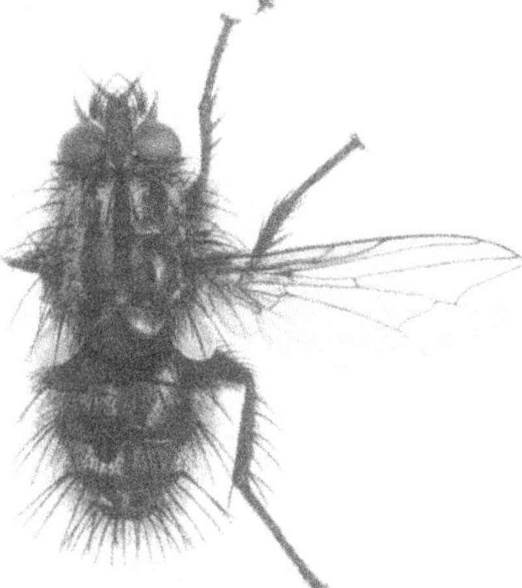

3

4

5

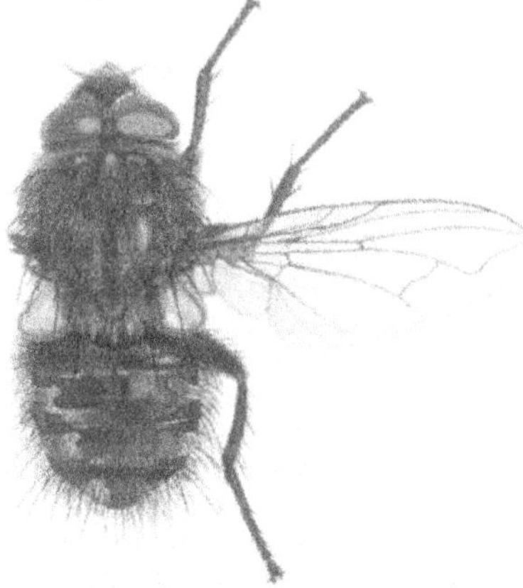

6

Catalogue

←

1: *Calliphora vicina* Robineau-Desvoidy, male; 2: *Protophormia terraenovae* (Robineau-Desvoidy), male; 3: *Eurychaeta palpalis* (Robineau-Desvoidy), male; 4: *Lucilia caesar* (Linnaeus), female; 5: *Eggisops pecchiolii* Rondani, female; 6: *Pollenia rudis* (Fabricius), male. – Steven J. Falk del.

		1. *Bellardia viarum* (R.-D.)	2. *B. vulgaris* (R.-D.)	3. *B. bayeri* (Jac.)	4. *B. pandia* (Walk.)	5. *B. pubicornis* (Zett.)	6. *B. stricta* (Vill.)	7. *B. vespillo* (Fabr.)	8. *Calliphora vicina* R.-D.	9. *C. genarum* (Zett.)	10. *C. loewi* End.	11. *C. stelviana* (B. & B.)	12. *C. subalpina* (Ringd.)	13. *C. uralensis* Vill.	14. *C. vomitoria* (L.)	15. *Cynomya mortuorum* (L.)	16. *Onesia floralis* R.-D.	17. *Phormia regina* (Meig.)	18. *Protocalliphora azurea* (Fall.)	19. *P. chrysorrhoea* (Meig.)	20. *P. nuortevai* Grun.	21. *P. peusi* Gr. & Pov.	22. *P. proxima* Grun.	23. *Protophormia atriceps* (Zett.)	24. *P. terraenovae* (R.-D.)	25. *Trypocalliphora braueri* (Hend.)	26. *Eurychaeta palpalis* (R.-D.)	27. *Lucilia ampullacea* Vill.	28. *L. bufonivora* Mon.	29. *L. caesar* (L.)	30. *L. magnicornis* (Siebke)	31. *L. illustris* (Meig.)	32. *L. regalis* (Meig.)	33. *L. richardsi* Coll.
	Germany	●	●	●	●		●	●	●	●	●		●	●	●	●	●	●	●	●		●			●	●	●	●	●	●		●	●	●
	G. Britain	●	●		●	●			●		●	●	●	●	●	●		●	●						●		●	●	●	●		●		●
DENMARK	SJ	●	●		●				●		●				●	●	●											●		●		●		
	EJ	●	●		●		●		●		●		●		●	●	●								●			●	●	●		●	●	
	WJ	●	●		●		●		●						●	●	●								●			●		●		●		
	NWJ	●	●				●		●		●				●	●	●								●					●		●		
	NEJ	●	●		●		●		●				●		●	●	●								●			●		●		●	●	
	F	●	●		●		●		●				●		●	●	●		●						●			●		●		●	●	
	LFM	●			●				●						●	●	●											●		●		●		
	SZ	●	●		●			●	●						●	●	●										●		●	●		●	●	
	NWZ	●	●		●		●	●	●						●	●	●								●		●	●		●		●	●	
	NEZ	●	●	●	●		●	●	●		●		●		●	●	●		●						●		●	●	●	●		●	●	
	B	●	●		●		●		●						●	●	●								●		●	●		●		●	●	
SWEDEN	Sk.	●	●		●		●	●	●				●		●	●	●	●	●				●		●			●	●	●		●	●	
	Bl.	●	●				●		●							●			●								●	●		●				
	Hall.	●	●						●						●	●	●								●					●		●		
	Sm.	●	●		●		●		●		●		●		●	●	●		●	●			●			●		●	●	●		●		●
	Öl.	●	●		●		●		●						●	●	●								●		●	●	●	●		●	●	●
	Gtl.	●	●		●				●						●	●			●					●	●		●	●		●		●		●
	G. Sand.						●									●																		
	Ög.								●				●		●	●	●		●				●							●				
	Vg.								●						●		●		●											●		●		
	Boh.	●							●						●	●	●								●					●		●		
	Dlsl.								●						●	●									●					●				
	Nrk.		●		●																													
	Sdm.	●	●						●				●				●		●											●		●		
	Upl.	●	●		●				●		●		●		●	●	●		●		●				●		●			●		●		
	Vstm.	●	●													●									●					●				
	Vrm.	●	●						●		●		●		●	●	●			●					●					●		●		
	Dlr.	●	●		●		●		●	●	●		●		●	●	●		●	●	●	●			●		●		●	●	●	●		●
	Gstr.		●						●								●								●									
	Hls.																																	
	Med.		●								●																							
	Hrj.					●	●				●	●	●	●					●		●				●						●	●		
	Jmt.	●	●			●	●		●	●	●	●	●	●	●	●			●	●	●			●	●		●				●	●		
	Ång.	●	●										●														●					●		
	Vb.	●								●	●								●						●							●		
	Nb.	●					●		●		●	●		●	●					●			●		●							●		
	Ås. Lpm.					●	●							●		●																		
	Ly. Lpm.					●				●	●	●	●	●		●			●	●	●			●	●						●			
	P. Lpm.						●																											
	Lu. Lpm.					●	●		●	●	●	●		●		●									●						●	●		
	T. Lpm.					●	●		●	●	●	●	●	●	●	●					●				●						●	●		

		NORWAY																			FINLAND																				USSR		
		Ø + AK	HE (s+n)	O (s+n)	B (ø+v)	VE	TE (y+i)	AA (y+i)	VA (y+i)	R (y+i)	HO (y+i)	SF (y+i)	MR (y+i)	ST (y+i)	NT (y+i)	Ns (y+i)	Nn (ø+v)	TR (y+i)	F (v+i)	F (n+ø)	Al	Ab	N	Ka	St	Ta	Sa	Oa	Tb	Sb	Kb	Om	Ok	Ob S	Ob N	Ks	LkW	LkE	Le	Li	Vib	Kr	Lr
Bellardia viarum (R.-D.)	1	●	◖	●	◖	●	◗	◖	●	●	●		●		◗	●	◖	●			●		●					●			●												
B. vulgaris (R.-D.)	2	●	◖	●	●	●	◗	●	●	●	●	●	◗	◗	◗	●	◖	◖			●	●	●			●	●	●		●	●	●									●	●	
B. bayeri (Jac.)	3								◖																																●		
B. pandia (Walk.)	4						◗														●	●	●		●	●															●	●	
B. pubicornis (Zett.)	5			◗			◗					◗	◖	◗				◖	◖																				●				●
B. stricta (Vill.)	6		◗	◗			◗												◖		●	●	●		●	●			●	●	●	●	●	●		●					●	●	
B. vespillo (Fabr.)	7																																										
Calliphora vicina R.-D.	8	●	●	◗	●	●	◗	●	●	●	●	◗	●	◗	◗	●	◗	●	◖	●	●	●	●	●	●	●	●			●		●	●		●					●	●	●	
C. genarum (Zett.)	9			◗	◗						◗			◗	◗	◗		●	◖	●																				●			
C. loewi End.	10	●	◗	◗	◗		◗	●	◖	◖	●	◗	◗	◗	◗	●		●		◗	●					●				●	●	●	●			●				●		●	
C. stelviana (B. & B.)	11			◗			◗	◗	◗	◖	●	◗		◗	◗	◗	◗	◗	◖																					●			
C. subalpina (Ringd.)	12	●	◗	●	◖		●	●	●	●	●	◗	●		◗	◗	●	●				●	●		●	●			●	●	●	●	●							●			
C. uralensis Vill.	13	◗	◗	◗	◗		◗			◖	◗	◗	◖	◗		●	●	●	●	●	●	●	●		●	●				●		●	●		●					●	●	●	
C. vomitoria (L.)	14	●	◗	●	●	●	●	●	●	●	●	◗	●	◗		●		◖			●	●	●	●	●	●	●			●		●	●		●					●	●		
Cynomya mortuorum (L.)	15	●	●	●	●	●	◗	●	●	●	●	◗	◗	◗	◗	●	●	●	●	●	●	●	●		●	●				●	●	●	●		●					●			
Onesia floralis R.-D.	16	◖																																									
Phormia regina (Meig.)	17																																										
Protocalliphora azurea (Fall.)	18	◗	●	◗	●	●	◗			◖		◖				◖						●				●	●																
P. chrysorrhoea (Meig.)	19		◗		●				◖				◗	◗					◗															●					●	●			
P. nuortevai Grun.	20			●			◗		◗		◖			◗		◗		◗	●	●																			●	●			
P. peusi Gr. & Pov.	21			◗																																							
P. proxima Grun.	22		◗																			●									●											●	
Protophormia atriceps (Zett.)	23			◗			◗					◗																												●			
P. terraenovae (R.-D.)	24	●	●	●	●		◗		●	●	●	◖	●	◗	●	●	◗	●	●	●	●	●	●		●	●	●			●	●	●	●		●					●			
Trypocalliphora braueri (Hend.)	25	◖	◖		◗		◗		◖			◗															●								●	●		●				●	●
Eurychaeta palpalis (R.-D.)	26	●				●	◗									◗					●	●	●	●		●	●															●	
Lucilia ampullacea Vill.	27																																									●	
L. bufonivora Mon.	28			●			◗															●	●			●	●	●				●									●	●	
L. caesar (L.)	29	●	●	●	◖	●	●	◖	●	●	●	◗	●		◗						●	●	●	●	●	●	●	●		●	●	●									●	●	
L. magnicornis (Siebke)	30			◗			●				◗	◗		◗	◗			◗		◗																			●	●			●
L. illustris (Meig.)	31	●	◖	●	●	●	◗	◖	●	●	●	●	●	●	●	●	●	●	●	●	●	●	●	●	●	●	●	●	●	●	●	●	●	●	●	●	●			●	●	●	●
L. regalis (Meig.)	32	◖																																									
L. richardsi Coll.	33	●				●	◗	◖	●												●	●	●	●		●	●			●			●										

		34 *L. sericata* (Meig.)	35 *L. silvarum* (Meig.)	36 *Angioneura acerba* (Meig.)	37 *A. cyrtoneurina* (Zett.)	38 *Eggisops pecchiolii* Rond.	39 *Melanomya nana* (Meig.)	40 *Melinda viridicyanea* (R.-D.)	41 *M. gentilis* R.-D.	42 *Morinia melanoptera* (Fall.)	43 *Pollenia amentaria* (Scop.)	44 *P. angustigena* Wainwr.	45 *P. griseotomentosa* (Jac.)	46 *P. hungarica* Rogn.	47 *P. labialis* R.-D.	48 *P. pallida* Rod.	49 *P. pediculata* Macq.	50 *P. rudis* (Fabr.)	51 *P. vagabunda* (Meig.)	52 *Stomorhina lunata* (Fabr.)
	Germany	●	●	●	●	●	●	●	●	●	●	●	●	●	●		●	●	●	
	G. Britain	●	●	●	●	●	●	●	●		●	●	●		●	●	●	●	●	●
DENMARK	SJ	●	●				●	●	●		●		●		●		●	●		
	EJ	●	●				●	●	●	●	●	●	●		●	●	●	●		
	WJ	●	●									●			●			●		
	NWJ	●	●								●				●			●		
	NEJ	●						●			●	●	●		●		●	●		
	F	●	●			●	●	●	●	●	●	●	●		●		●	●		
	LFM	●	●			●	●	●	●		●	●	●		●		●	●		
	SZ	●	●				●	●	●		●	●	●		●	●	●			
	NWZ	●	●		●			●	●		●	●	●		●	●	●	●	●	
	NEZ	●	●	●	●			●	●	●	●	●	●		●	●	●	●	●	
	B	●	●					●	●		●				●		●	●		
SWEDEN	Sk.	●	●	●			●	●	●	●	●	●	●	●	●	●	●	●	●	
	Bl.	●					●			●	●	●		●	●		●	●	●	
	Hall.	●	●					●			●	●			●		●	●	●	
	Sm.	●	●				●	●	●	●	●	●	●	●	●		●	●	●	
	Öl.	●	●	●				●	●			●	●		●		●	●		
	Gtl.	●	●	●				●			●	●	●		●		●		●	
	G. Sand.																			
	Ög.	●	●	●			●			●	●						●		●	
	Vg.	●					●			●	●				●		●	●	●	
	Boh.	●							●		●	●		●						
	Dlsl.						●				●				●		●	●		
	Nrk.									●	●							●		
	Sdm.		●	●			●	●	●	●	●				●			●	●	
	Upl.	●	●		●		●	●	●	●	●	●	●		●		●	●	●	
	Vstm.		●										●				●	●		
	Vrm.	●	●				●		●		●	●			●		●	●	●	
	Dlr.		●	●			●		●	●	●		●		●		●	●	●	
	Gstr.						●				●		●				●	●		
	Hls.																			
	Med.																			
	Hrj.						●													
	Jmt.						●	●		●			●		●		●	●	●	
	Ång.		●				●			●			●							
	Vb.						●											●		
	Nb.		●				●													
	Ås. Lpm.																			
	Ly. Lpm.									●			●							
	P. Lpm.																			
	Lu. Lpm.																			
	T. Lpm.																			

		NORWAY																			FINLAND																				USSR		
		Ø + AK	HE (s+n)	O (s+n)	B (ø+v)	VE	TE (y+i)	AA (y+i)	VA (y+i)	R (y+i)	HO (y+i)	SF (y+i)	MR (y+i)	ST (y+i)	NT (y+i)	Ns (y+i)	Nn (ø+v)	TR (y+i)	F (v+i)	F (n+ø)	Al	Ab	N	Ka	St	Ta	Sa	Oa	Tb	Sb	Kb	Om	Ok	Ob S	Ob N	Ks	LkW	LkE	Le	Li	Vib	Kr	Lr
L. sericata (Meig.)	34	●			◖	●				◖											●	●	●		●	●	●			●											●		
L. silvarum (Meig.)	35	●	◖			●			◖	◖											●	●	●	●	●	●	●	●		●	●	●	●	●							●	●	
Angioneura acerba (Meig.)	36	◗																					●			●															●	●	
A. cyrtoneurina (Zett.)	37																																										
Eggisops pecchiolii Rond.	38																																										
Melanomya nana (Meig.)	39	◗	◗	●	◖	●	◖			◖	◗	◗										●	●	●	●	●	●	●		●	●	●				●					●	●	
Melinda viridicyanea (R.-D.)	40	●		●	◖	●	◗	◗	◖	●	●					◗							●																		●		
M. gentilis R.-D.	41	●			◖		◗	◖	◖	●	◖																																
Morinia melanoptera (Fall.)	42			◗																		●	●		●					●	●				●						●	●	
Pollenia amentaria (Scop.)	43	◗	◖	●	◖	●			◖												●	●	●			●				●											●	●	
P. angustigena Wainwr.	44	●							◖												●	●																					
P. griseotomentosa (Jac.)	45	●								●	◖	●			◗						●	●	●			●			●	●				●								●	
P. hungarica Rogn.	46	◗																					●			●																	
P. labialis R.-D.	47	●			◗	●	◗	◖	●	●	●		◖	◗							●	●	●		●	●	●		●	●	●			●							●	●	
P. pallida Rod.	48																																										
P. pediculata Macq.	49	●	●	◖	◗	●	◗	◖	●	●	●	●	●			◖					●	●	●			●	●	●													●	●	
P. rudis (Fabr.)	50	●	●	●	●	●	◗	●	●	●	●	●	●	◗	●	●	◗	◖			●	●	●		●	●	●	●		●											●	●	
P. vagabunda (Meig.)	51	●	◖	◖		●		◖	●	●	◗		◗								●	●	●	●		●															●		
Stomorhina lunata (Fabr.)	52																						●																				

Literature

Aldrich, J. M., 1930. New two-winged flies of the the family Calliphoridae from China. - Proc. U. S. natn. Mus. 78 (1): 1-5.

Andersen, S., 1982. Revision of the European species of *Siphona* Meigen (Diptera: Tachinidae). - Ent. scand. 13: 149-172.

- 1983. Phylogeny and classification of Old World genera of Siphonini (Diptera: Tachinidae). - Ent. scand. 14: 1-15.

Andersson, H., 1985. Flugor som lever på fåglar och i fågelbon. Flies (Diptera Brachycera) living on birds or in birds' nests. - Anser 24: 115-124.

Ardö, P., 1957. Studies in the the marine shore dune ecosystem with special reference to the dipterous fauna. - Opusc. ent. Suppl. 14: 255 pp.

Aubertin, D., 1933. Revision of the genus *Lucilia* R.-D. (Diptera, Calliphoridae). - J. Linn. Soc. Lond. (Zool.) 38: 389-436.

Baranov, N., 1939. Sechs neue Raupenfliegen aus der Sammlung Takanos. - Ent. NachrBl., Troppau 12 (1938): 110-112.

- 1943. Calliphoridae Nezavisne Drzave Hrvatske. <Calliphoridae of the independent state of Croatia>. - Vet. Arh. 13: 45-87.

Bezzi, M., 1906. Noch einige neue Namen für Dipterengattungen. - Z. syst. Hymenopt. Dipterol. 6: 49-55.

Bezzi, M. & Stein, P., 1907. Cyclorrapha Aschiza. Cyclorrapha Schizophora: Schizometopa. In: Becker, T., Bezzi, M., Kertész, K. & Stein, P., Katalog der paläarktischen Dipteren, 3: 1-828. Budapest.

Bidenkap, O., 1892. Undersøgelser over Diptera Brachycera i Jarlsberg og Laurvigs amt sommeren 1891. - Ent. Tidskr. 13: 225-246.

- 1898. Diptera Brachycera fra Jarlsberg. - Tromsø Mus. Aarsh. 19: 147-150.

- 1901. Foreløbig oversigt over de i det arktiske Norge hidtil bemærkede Diptera Brachycera. - Tromsø Mus. Aarsh. 23: 13-112.

Bigot, J. M. F., 1860. Diptères de Madagascar. (Suite et fin). Troisième partie. - Annls Soc. ent. Fr. (3) 7 (1859): 533-558.

- 1877. Diptères nouveaux ou peu connus. 8e partie X. Genre *Somomya* (Rondani). *Calliphora*, *Melinda*, *Mufetia*, *Lucilia*, *Chrysomyia* (alias *Microchrysa* Rond.) Robineau-Desvoidy. - Annls Soc. ent. Fr. (5) 7: 243-259.

- 1888a. Diptères nouveaux ou peu connus. MUSCIDI (J.B.) - Bull. Soc. zool. Fr. 12 (5-6) [1887]: 581-617.

- 1888b. [Note in séance of 9 November 1887]. - Annls Soc. ent. Fr. (6) 7 [1887]: clxxx-clxxxii.

Bishopp, F. C., 1916. Fly traps and their operation. - U. S. Dep. Agric. Farmer's Bull. 734: 1-16.

Bonsdorff, E. J., 1866. Finlands tvåvingade insekter (Diptera), förtecknade och i korthet beskrifne. - Bidrag till kännedom af Finlands natur och folk. 7: vi + 306.

Bourke, D. O'D., 1975. Blowflies (*Calliphora erythrocephala* Meig.) (Dipt., Calliphoridae) overwintering as larvae. - Entomol. mon. Mag. 110: 128.

Boyes, J. W. & Shewell, G. E., 1975. Cytotaxonomy of Calliphoridae (Diptera). - Genetica 45: 435-488.

Brander, T., 1964. Lounais-hämeen kärpäsistä, Brachycera (Diptera). - Lounais-Hämeen Luonto 16: 51-69.

Brauer, F., 1893. Vorarbeiten zu einer Monographie der Muscaria schizometopa (exclusive Anthomyidae) von Prof. Dr. Fr. Brauer und Julius Edl. v. Bergenstamm. - Verh. zool.-bot. Ges. Wien 43: 447-525.

- 1899. Beitrag zur Kenntnis der Muscaria Schizometopa. I. Bemerkungen zu den Originalexemplaren der von Bigot, Macquart und Robineau-Desvoidy beschriebenen Muscaria Schizometopa aus der Sammlung des Herrn. G.H. Verrall. - Sber. Akad. Wiss. Wien 108 (Abt.I): 495-529.

Brauer, F. & Bergenstamm, J. E. von, 1889. Die Zweiflügler des Kaiserlichen Museums zu Wien. IV. Vorarbeiten zu einer Monographie der Muscaria Schizometopa (exclusive Anthomyidae). Pars I. - Denkschr. Akad. Wiss., Wien 56: 69-180.

- 1891. Die Zweiflügler des Kaiserlichen Museums zu Wien. V. Vorarbeiten zu einer Monographie der Muscaria Schizometopa (exclusive Anthomyidae). Pars II. - Denkschr. Akad. Wiss., Wien 58: 305-446 [1-142].

- 1893. Die Zweiflügler des Kaiserlichen Museums zu Wien. VI. Vorarbeiten zu einer Monographie der Muscaria Schizometopa (exclusive Anthomyidae). Pars III. - Denkschr. Akad. Wiss., Wien 60: 89-240.

Brinkmann, A., 1976. Blowfly myiasis of sheep in Norway. - Norw. J. Zool. 24: 325-330.

Brumpt, E., 1933. Utilisation des larves de certaines mouches pour le traitement de l'ostéomyélites et de diverses affections chirurgicales chroniques. - Ann. Parasit. hum. comp. 11: 403-420.

- 1934. Recherches expérimentales sur la biologie de la *Lucilia bufonivora*. - Ann. Parasit. hum. comp. 12: 81-97.

Čepelák, J. & Čepelák, S., 1978. *Cynomya gregorpovolnyi*, sp.n. - neue Art der Familie Calliphoridae (Diptera) aus der Tschechoslowakei. - Annls zool. bot., Bratislava 125: 1-4.

Čepelák, J. & Rozkošný, R., 1968. Zur Bionomie der Art *Angioneura cyrtoneurina* Zetterstedt, 1859 (Rhinophorinae, Diptera). - Acta zootechnica, Nitra 17: 181-191.

Chen, Zhi-zi, Fan, Zi-de & Cui, Chang-yuan, 1982. Two

new species of the calliphorid genus *Bellardia* R.-D. (Diptera) from Heilongjiang, China. - Contr. Shanghai Inst. Entomol. 1982-83: 231-233. [In Chinese, with English summary.]
Chernov, Yu. I., 1959. Synanthropic Diptera of the Yugorsky peninsula and of the Vaigach island. - Ént. Obozr. 38: 579-582. [In Russsian; English translation in Ent. Rev., Wash. 38: 521-523, 1960].
- 1961. Kompleks sinantropnýkh dvukrýlýkh v arkticheskikh tundrakh Yakutii. - Nauchnýe dokladý výsshei shkolý- Biologicheskie nauki. 3: 35-38. [In Russian].
- 1965. A complex of synanthropic Diptera in the tundra zone of the USSR. - Ént. Obozr. 44: 74-83. [In Russian; English translation in Ent. Rev., Wash. 44: 39-44, 1965].
Collett, R., 1921. Norges Fugle. Ved Ørjan Olsen under medvirkning av A. Landmark. Bind 1. 596 pp. Kristiania [= Oslo].
Collin, J. E., 1925. Diptera (Cyclorrhapha) from Spitsbergen. Results of the Oxford University Expedition to Spitsbergen 1924. - Ann. Mag. nat. Hist. (9) 16: 332-337.
- 1926. Supplementary notes on the genus *Lucilia* Desv. (Diptera). - Trans. ent. Soc., Lond. 74: 258-260.
- 1931. The Oxford University Expedition to Greenland, 1928. - Diptera (Orthorrhapha Brachycera and Cyclorhhapha) from Greenland. - Ann. Mag. nat. Hist. (10) 7: 67-91.
Coquillett, D. W., 1910. The type-species of the North American genera of Diptera. - Proc. U. S. natn Mus. 37: 499-647.
Cragg, J. B., 1950. Studies on *Lucilia* species under Danish conditions. - Ann. appl. Biol. 37: 66-79.
Crosskey, R. W., 1965. A systematic revision of the Ameniinae (Diptera: Calliphoridae). - Bull. Br. Mus. nat. Hist. (Ent.) 16: 33-140.
- 1977. A review of the Rhinophoridae (Diptera) and a revision of the Afrotropical species. - Bull. Br. Mus. nat. Hist. (Ent.) 36: 1-66.
Crosskey, R. W. (ed.), 1980. Catalogue of the Diptera of the Afrotropical Region. 1437 pp. British Museum (Natural History), London.
Danks, H. V., 1981. Arctic arthropods. A review of systematics and ecology with particular reference to the North American fauna. 608 pp. Ottawa.
Davies, L., 1954. Some Diptera collected at Holandsfjord, Norway. - Norsk ent. Tidsskr. 9: 71-72.
De Geer, C., 1776. Mémoires pour servir à l'histoire des Insectes. 6. viii + 523 pp., 30 pls., Stockholm.
Dear, J. P., 1979. A revision of the Toxotarsinae (Diptera: Calliphoridae). - Papeis avuls. Zool. S. Paulo 32: 145-182.
- 1981. Blowfly recording scheme - an interrim report. - Entomol. mon. Mag. 117: 75-76.
- 1985. A revision of the New World Chrysomyini (Diptera: Calliphoridae). - Rev. Bras. Zool. 3: 109-169.
- 1986. Calliphoridae (Insecta: Diptera). - Fauna of New Zealand 8: 1-86.
Delfinado, M. D. & Hardy, D. E. (eds.), 1977. A catalog of the Diptera of the Oriental Region. Volume III. Suborder Cyclorrhapha (excluding Division Aschiza). x + 854 pp. Honolulu.
Disney, R. H. L., 1973. Diptera and Lepidoptera reared from dead shrews in Yorkshire. - Naturalist, Hull 927: 136.
Dodge, H. R., 1960. An effective, economical flytrap. - J. econ. Ent. 53: 1131-1132.
Downes, W. L., 1965. Tribe Melanomyini. Pp. 932-933 in: Stone, A., Sabrosky, C. W., Wirth, W. W., Foote, R. H. & Coulson, J. R., A catalog of the Diptera of America north of Mexico. - Agric.Handb. 276: iv + 1696 pp., Washington D.C.
- 1986. The Nearctic *Melanomya* and relatives (Diptera: Calliphoridae), a problem in calyptrate classification. - Bull. N. Y. State Mus. 460: v + 1-35.
Draber-Mońko, A., 1971. Niektóre Calyptrata (Diptera) Bieszczadów. - Fragm. faun. Warsz. 17: 483-543.
- 1978. Scathophagidae, Muscinae, Gasterophilidae, Hippoboscidae., Calliphoridae, Sarcophagidae, Rhinophoridae, Oestridae, Hypodermatidae i Tachinidae (Diptera) Pienin. - Fragm. faun. Warsz. 22: 51-229.
Duffield, J. E., 1937. Notes on some animal communities of Norwegian Lappland. An account of the dung and carrion communities and of those insects found in human dwellings. - J. Anim. Ecol. 6: 160-168.
Dufour, L., 1845. Histoire des métamorphoses de la *Lucilia dispar*. - Annls Soc. ent. Fr. (2) 3: 205-214.
Eichler, W., 1980. Grundzüge der veterinärmedizinischen Entomologie. 184 pp. Jena.
Emden, F. I. van, 1954: Diptera Cyclorrhapha Calyptrata (I) Section (a) Tachinidae and Calliphoridae. - Handbk Ident. Br. Insects 10 (4) (a): 1-133.
Enderlein, G., 1903. II. Die Landarthropoden der antarktischen Inseln St. Paul und Neu-Amsterdam. Pp. 249-270 in: Wiss. Ergebnisse der deutschen Tiefsee-Expedition, Valdivia, 3: 197-270.
- 1933. Neue paläarktischen Calliphoriden, darunter Schneckenparasiten (Dipt.). - Mitt. dt. ent. Ges. 4: 120-128.
- 1934. Dipterologica. II. - Sber. Ges. naturf. Freunde Berl. 1934: 133-134, 181-190.
- 1936. 22. Ordnung: Zweiflügler, Diptera. In: Brohmer, P., Ehrmann, P. & Ulmer, G. (eds.), Die Tierwelt Mitteleuropas. Band VI, Insekten. 3. Teil. Abt. XVI: 1-259. Leipzig.
Engel, E. O., 1920. Dipteren, die nicht Pupiparen sind als Vogelparasiten. - Z. Wiss. InsektBiol. 15: 249-258.
Erichsson, W. F., 1851. Hymenopteren, Dipteren und Neuropteren. Pp.60-69 in: Ménétriés, E., Insecten. Pp. 45-76 in: Brandt, F., Erichssson, W. F., Fischer, S., Grube, E., Ménétriés, E. & Middendorff, A. T. (eds.), Dr. A. T. von Middendorff's Reise in den äussersten Norden und Osten Sibiriens ... Band II. Zoologie. Theil 1. Wirbellose Thiere: Annulaten ... 516 pp. St. Petersburg [= Leningrad].
Erzinçlioğlu, Y. Z., 1984. A new parasite record for *Protocalliphora azurea* (Fall.) (Dipt., Calliphoridae). - Entomol. mon. Mag. 120: 172.
- 1985. Immature stages of British *Calliphora* and *Cynomya*, with re-evaluation of the taxonomic characters of larval Calliphoridae (Diptera). - J. nat. Hist. 19: 69-96.

Eschscholz, J. F., 1823. Entomographien. - Naturw. Abhandl. aus Dorpat., 1: 57-186, 2 pls.

Fabricius, J. C., 1775. Systema entomologiae, sistens insectorvm classes, ordines, genera, species, adiectis synonymis, locis, descriptionibvs, observationibvs. 32 832 pp. Flensbvrgi et Lipsiae [= Flensburg and Leipzig].

- 1787. Mantissa insectorvm sistens species nvper detectas, adiectis synonymis, observationibvs, descriptionibvs, emendationibvs, 2. 382 pp. Hafniae [= Copenhagen].

- 1794. Entomologia systematica emendata et aucta secundum classes, ordines, genera, species adjectis synonymis, locis, observationibus, descriptionibus. Vol. 4, 472 pp. Hafniae [= Copenhagen].

- 1805. Systema antliatorum secundum ordines, genera, species adiectis synonymis, locis, observationibus, descriptionibus. xiv + 15-372 + 30 pp. Brunsvigae [= Braunschweig].

Fallén, C. F., 1817. Beskrifning öfver de i Sverige funna fluge arter, som kunna föras till slägtet *Musca*. Första afdelningen. - K. svenska VetenskAkad. Handl. (3) 1816 (2): 226-254.

Fan, Zi-de, 1957. On chinese muscoid flies of the genus *Calliphora* R.-D. (s.lat.). - Acta ent. sin 7: 321-348. [In Chinese, with English summary, key and description of new species.]

- 1965. Key to species of the common synanthropic flies of China. xv + 1-330 pp. Beijing. [In Chinese].

Ferrar, P., 1978. Macrolarviparous reproduction in *Euphumosia* (Diptera: Calliphoridae). - J. Aust. ent. Soc. 17: 13-17.

- 1987a. A guide to the breeding habits and immature stages of Diptera Cyclorrhapha. (Part 1: text). - Entomonograph 8 (1): (4 +) 9-478.

- 1987b. A guide to the breeding habits and immature stages of Diptera Cyclorrhapha. (Part 2: figures). - Entomonograph 8 (2): (4 +) 479-907.

Frey, R., 1909. Finlands *Sarcophaga*- och *Onesia*-arter, bestämda af J. Villeneuve. - Meddn Soc. Fauna Flora fenn. 35: 90-92.

- 1925. [*Protocalliphora azurea* på sädesärla]. - Notul. ent. 5: 62.

Fuller, M. E., 1933. The life history of *Onesia accepta* Malloch (Diptera, Calliphoridae). - Parasitology 25: 342-352, pl.22.

Girschner, E., 1893. Beitrag zur Systematik der Musciden. - Berl. ent. Zeitschr. 38: 297-312.

Gmelin, J. F., 1790. Caroli a Linné, Systema Naturae per regna tria naturae, secundum classes, ordines, genera, species, cum caracteribus, differentiis, synonymis, locis. Editio decima tertia, aucta, reformata. 1 (5): 2225-3020. Lipsiae [= Leipzig].

González Mora, D., 1985. *Stomorhina selgae* Lehrer, 1979, una nueva sinonimia para *Stomorhina lunata* (Fabricius, 1805) (Diptera: Calliphoridae). In: Actas do II Congresso Ibérica de Entomologia - Lisboa Junho de 1985. - Bolm. Soc. port. Ent. 1; Suplemento No. 1: 341-349.

Graham-Smith, G. S., 1938. The generative organs of the blowfly, *Calliphora erythrocephala* L., with special reference to their musculature and movements. - Parasitology 30: 441-476.

Greenberg, B. 1971. Flies and disease. I. Ecology, classification and biotic associations. 856 pp. Princeton, New Jersey.

- 1973. Flies and disease. II. Biology and disease transmission. 447 pp. Princeton, New Jersey.

- 1984. Two cases of human myiasis caused by *Phaenicia sericata* (Diptera: Calliphoridae) in Chicago area hospitals. - J. med. Ent. 21: 615.

Gregor, F. 1987. Calliphoridae. - Acta faun. ent. Mus. natns Pragae 18: 291-293.

Gregor, F. & Povolný, D., 1959. Kritischer Beitrag zur Kenntnis der Tribus Phormiini (Diptera, Calliphoridae). - Cas. csl. Spol. ent. 56: 26-51.

Griffiths, G. C. D., 1972. The phylogenetic classification of Diptera Cyclorrhapha with special reference to the structure of the male postabdomen. - Series ent. 8: 1-340.

- 1982. On the systematic position of *Mystacinobia* (Diptera: Calliphoridae). - Mem. ent. Soc. Wash. 10: 70-77.

Grunin, K. Ya., 1964-1969. 64b. Hypodermatinae. In: Lindner, E. (ed.), Fliegen palaearkt. Reg. 8: 1-160.

- 1966. New and little known Calliphoridae (Diptera), mainly bloodsucking or subcutaneous parasites of birds. - Ént. Obozr. 45: 897-903. [In Russian; English translation in Ent. Rev., Wash. 45: 503-506].

- 1970a. 108. Sem. Calliphoridae - Kalliforidý. Pp. 607-624 in: Bei-Bienko, G. Ya. (ed.), Opredelitel' nasekomýkh evropeiskoi chasti SSSR, V, Dvukrýlýe, Blokhi. Akad. Nauk SSSR, Zool. Inst., Leningr. [In Russian].

- 1970b. Flies of the family Calliphoridae (Diptera) new to the USSR. - Ént. Obozr. 49: 471-483. [In Russian; English translation in Ent. Rev., Wash. 49: 282-289].

- 1971. *Phormiata* Grunin gen.n. - A sixth genus of the tribe Phormiini (Diptera, Calliphoridae). - Ént. Obozr. 50: 444-445. [In Russian; English translation in Ent. Rev., Wash. 50: 253].

- 1972. Beschreibung einer ornithoparasitischen Fliege, *Protocalliphora nuortevai* Grunin, sp.n. (Dipt., Calliphoridae) aus Nord-Finland. - Suomen hyönt. Aikak. 38: 156-158.

Grunin, K. Ya. & Nuorteva, P., 1969. The occurrence of the ornithoparasitic *Protocalliphora* species (Dipt., Calliphoridae) in Finland. - Suomen hyönt. Aikak. 35: 57-58.

Grunin, K. Ya., Nuorteva, P. & Rajala, P., 1969. *Trypocalliphora lindneri* Peus (Dipt., Calliphoridae) as a subcutaneous parasite of the Wheatear in northern Finland. - Suomen hyönt. Aikak. 35: 56-57.

Guimarães, J. H., 1971. 104 Family Tachinidae. In: Papavero, N. (ed.), A Catalog of the Diptera of the Americas south of the United States. 333 pp. Museu de Zoologia, Universidade de São Paulo.

- 1977. A systematic revision of the Mesembrinellidae, stat.nov. (Diptera, Cyclorrhapha). - Archos Zool. S. Paulo 29: 1-109.

Haarløv, N., 1961. Om fluelarver. - Naturens Verden [1961]: 112-117.

Hackman, W., 1963. Studies on the dipterous fauna in burrows of voles (*Microtus*, *Clethrionomys*) in Finland. - Acta zool. fenn. 102: 1-64.
- 1980. A check list of the Finnish Diptera II. Cyclorrhapha. - Notul. ent. 60: 117-162.
Hackman, W. & Väisänen, R., 1985. The evolution and phylogenetic significance of the costal chaetotaxy in the Diptera. - Ann. zool. fenn. 22: 169-203.
Hakanen, R., Grunin, K. Ya. & Nuorteva, P., 1974. Larvae of *Trypocalliphora lindneri* Peus (Dipt., Calliphoridae) as subcutaneous pathogens on nestlings of the Meadow Pipit and Common Redpoll in the subarctic. - Suomen hyönt. Aikak. 40: 15-18.
Haliday, A. H., 1833. Catalogue of the Diptera occurring about Holywood in Downshire. - Ent. Mag. 1: 147-180.
Hall, D. G., 1948. The blowflies of North America. Thomas Say Foundation. 6 + 477 pp., 51 pls., Lafayette, Indiana.
- 1965. Family Calliphoridae. Pp. 922-932 in: Stone, A., Sabrosky, C. W., Wirth, W. W., Foote, R. H. & Coulson, J. R., A Catalog of the Diptera of America north of Mexico. - Agric. Handb. 276: iv + 1696 pp., Washington D.C.
Hall, R. D., Anderson, P. C & Clark, D. P., 1986. A case of human myiasis caused by *Phormia regina* (Diptera: Calliphoridae) in Missouri, USA. - J. med. Entomol. 23: 578-579.
Hall, R. D. & Townsend, L. H., 1977. The blow flies of Virginia (Diptera: Calliphoridae). The Insects of Virginia; No. 11. - Virginia Polytechnic Institute and State University. Research Division Bulletin 123: viii + 1-48.
Hanski, I., 1976. Breeding experiments with carrion flies (Diptera) in natural conditions. - Suomen hyönt. Aikak. 42: 113-121.
Hanski, I. & Kuusela, S., 1977. An experiment on competition and diversity in the carrion fly community. - Suomen hyönt. Aikak. 43: 108-115.
- 1980. The structure of carrion fly communities: differences in breeding seasons. - Ann. zool. fenn. 17: 185-190.
Hanski, I. & Nuorteva, P., 1975. Trap survey of flies and their diel periodicity in the subarctic Kevo Nature Reserve, northern Finland. - Suomen hyönt. Aikak. 41: 56-64.
Harris, M., 1780. An Exposition of English insects. i-viii + 9-166. London.
Harris, T. W., 1869. Descriptions of insects selected from the manuscripts of Dr. Harris. Pp. 325-336 in: Scudder, S. H. (ed.), Entomological corresponence of T. W. Harris. - Occ. Pap. Boston Soc. nat. Hist. 1: 1-375, 4 pls.
Heath, A. C. G., 1986. Fly-strike in New Zealand. Pp. 15-18 in: Dear, J. P., Calliphoridae (Insecta: Diptera). - Fauna of New Zealand 8: 1-86.
Hedström, L., 1964. Lounais-hämeen kärpäsiä, Brachycera (Diptera). - Lounais-Hämeen Luonto 15: 77-89.
Hedström, L. & Nuorteva, P., 1971. Zonal distribution of flies on the hill Ailigas in subarctic northern Finland. - Suomen hyönt. Aikak. 37: 121-125.
Hendel, F., 1901. Beitrag zur Kenntnis der Calliphorinen (Dipt.). - Wien ent. Ztg. 20: 28-33.
- 1911. Über von Professor J. M. Aldrich erhaltene und einige andere amerikanische Dipteren. - Wien ent. Ztg. 30: 19-46.
Hennig, W., 1939. Über Namen und Artenzahl der deutschen "Vogelblutfliegen". - Arb. physiol. angew. Ent. Berl. 6: 359-364.
- 1958. Die Familien der Diptera Schizophora und ihre phylogenetische Verwandtschaftsbeziehungen. - Beitr. Ent. 8: 505-688.
- 1973. 31. Diptera (Zweiflügler). - Handb. Zool. 4(2)2/31: 1-337 (Lieferung 20).
- 1976. 63a. Anthomyiidae. In: Lindner, E. (ed.), Fliegen palaearkt. Reg. 7(1)/1: I-LXXVIII.
Henriksen, K. L., 1939. A revised index of the insects of Grønland. - Meddr Grønl. 119 (10): 1-101.
Herting, B., 1957. Das weibliche Postabdomen der calyptraten Fliegen (Diptera) und sein Merkmalswert für die Systematik der Gruppe. - Z. Morph. Ökol. Tiere 45: 429-461.
- 1961. 64e. Rhinophorinae. In: Lindner, E. (ed.), Fliegen palaearkt. Reg. 9: 1-36.
- 1974. Revision der von Robineau-Desvoidy beschriebenen europäischen Tachiniden und Rhinophoriden. - Stuttg. Beitr. Naturk. (A) 264: 1-46.
- 1976. Revision der von Macquart beschriebenen paläarktischen Tachiniden und Rhinophorinen (Diptera). - Stuttg. Beitr. Naturk. (A) 289: 1-10.
Hicks, E. A., 1959. Check-list and bibliography on the occurrence of insects in birds' nests. 681 pp. State College Press, Iowa.
- 1962. Check-list and bibliography on the occurrence of insects in birds' nests. Supplement I. - Iowa St. J. Sci. 36: 233-348.
- 1971. Check-list and bibliography on the occurrence of insects in birds' nests. Supplement II. - Iowa St. J. Sci. 46: 122-338.
Holloway, B. A., 1976. A new bat-fly family from New Zealand (Diptera: Mystacinobiidae). - New Zealand J. Zool. 3: 279-301.
Hori, K., 1961. Comparative anatomy of the internal organs of the calyptrate muscoid flies II. Female internal sexual organs of the adult flies. - Sci. Rep. Kanazawa Univ. 7: 61-101.
Hori, K. & Iwasa, M., 1988. Biology of two species of the *Protocalliphora* (Diptera: Calliphoridae) in Tokachi, Hokkaido, Japan: relationship between the occurrence and three species of host birds. - Appl. Ent. Zool. 23: 298-303.
Horsfield, D., 1984. *Delia caledonica* Assis Fonseca (Dipt., Anthomyiidae) and other Diptera from the Fannich Hills, West Ross. - Entomol. mon. Mag. 120: 162.
Hough, G. de N., 1898a. Two new American species of *Cynomyia*, - a study in chaetotaxy. - Ent. News 9: 105-111.
- 1898b. A third American species of *Cynomyia*. - Ent. News 9: 165-166.
- 1899a. Synopsis of the Calliphorinae of the United States. - Zool. Bull. 2: 283-290.

- 1899b. Some North American genera of the dipterous group Calliphorinae Girschner. - Ent. News 10: 62-66.
Ibrahim, S. H., 1984. A study on a dipterous parasite of honeybees. - Z. ang. Ent. 97: 124-126.
Irwin, A. G., 1976. *Calliphora uralensis* Villeneuve (Diptera, Calliphoridae) new to Ireland. - Entomol. mon. Mag. 111: 62.
Ishijima, H., 1967. Revision of the third stage larvae of synanthropic flies of Japan (Diptera: Anthomyiidae, Muscidae, Calliphoridae and Sarcophagidae). - Jap. J. sanit. Zool. 18: 47-100.
Itämies, J. & Koskela, P., 1980. *Cynomya mortuorum* myiasis in *Lepus timidus*. - Suomen hyönt. Aikak. 46: 24.
Itämies, J. & Merilä, E., 1984. *Lucilia*-kärpästoukkia elävällä kurjella (Abstract: A *Lucilia*-myiasis in a living crane (*Grus grus*)). - Luonnon Tutkija 88: 138.
Itämies, J. & Ojanen, M., 1975. *Trypocalliphora lindneri* Peus (Dipt. Calliphoridae) infesting a great tit nestling (*Parus major* L.). - Suomen hyönt. Aikak. 41: 42.
Iwasa, M. & Hori, K., 1988. A subcutaneous bird-parasitic blowfly of the genus *Trypocalliphora* Peus newly recorded from Japan (Diptera: Calliphoridae). - Jap. J. sanit. Zool. 39: 267-270.
Jacentkovský, D., 1936a. Les Tachinaires (Tachinidae Dipt.) des forêts dans le terrain marécageux de Lednice. Résumé. - Bull. int. Acad. tcheque Sci. 37: 8-10.
- 1936b. Beitrag zur Kenntnis der Raupenfliegen (Tachinariae, Diptera) Bulgariens. - Izv. tsarsk. přírodonauch. Inst. Sof. <Mitt. naturw. Inst. Sofia> 9: 109-134.
- 1937a. Kuklice (Tachinidae) luzních lesu území lednického. - Rozpr. české Akad. Věd. Umění (2) 46 (8): 1-20.
- 1937b. Příspěvek k studiu biologické obrany proti lesním škudcum <Contribution à l'étude de la défense biologique contre les ravageurs des forêts.>. - Sborník vysoké školy zemědělské v Brně, ČSR., (Fakulta lesnická) <Bull. inst. nat. agron. Brno. Fac. silvicult.> (D), 24: 1-54.
- 1939. Über einige interessante Sarcophagiden (Tachinariae, Diptera) aus Bulgarien. - Izv. tsarsk. přírodonauch. Inst. Sof. <Mitt. naturw. Inst. Sofia.> 12: 188-192.
- 1941a. *Pollenia mayeri* n.sp. (Calliphoridae, Tachinoidea) nová kuklice z Moravy <*Pollenia mayeri* n.sp. (Calliphoridae, Tachinoidea) eine neue Raupenfliege aus Mähren>. - Entomologické listy <Folia entomologica> 4: 14-16.
- 1941b. Kuklice (Tachinoidea, Diptera) Moravy a Slezska. - Práce moravské prírodovědecké společnosti <Acta societatis scientiarum naturalium moravicae> 13 (4), fasc 129: 1-64.
- 1941c. Moravské kuklice z rodu *Pollenie* [sic] R.D. (Calliphoridae, Diptera) <Mährische Arten aus der Raupenfliegen Gattung *Pollenia* R.D. (Calliphoridae, Diptera)>. - Sborník klubu přírodovědeckého v Brně 23 (1940): 28-32.
- 1942. Beitrag zur Kenntnis der Gattung *Pollenia* R.D. (Calliphoridae, Tachinoidae, Diptera). - Zap. Nauch-no-issled. Ob'ed. russk. svob. Univ. Prag <Abhandl. russ. Forschungsges. Prag> 11 (16) no.84: 193-225 [1-33].
- 1944a. Další příspěvek k zvířeně kuklic (Tachinoidea, Dipt.) Moravy <Weiterer Beitrag zur Fauna der Raupenfliegen (Tachinoidea, Dipt.) Mährens>. - Entomologické listy <Folia entomologica> 7: 45-49.
- 1944b. Beitrag zur Fauna der Raupenfliegen der Mark Brandenburg (Diptera: Tachinoidea). - Arb. morph. taxon. Ent. Berl. 11: 116-122.
Jaennicke, F., 1867. Neue exotische Dipteren. - Abh. senckenb. naturforsch. Ges. 6: 311-408.
James, M. T., 1955. The blowflies of California (Diptera: Calliphoridae). - Bull. Calif. Ins. Survey 4(1): 1-34.
- 1966. The blow flies (Diptera, Calliphoridae) collected in the Philippine Islands by the Noona Dan expedition. - Ent. Meddr 34: 463-488.
- 1970. 102 Family Calliphoridae. In: Papavero, N. (ed.), A Catalog of the Diptera of the Americas south of the United States. 28 pp. Museu de Zoologia, Universidade de São Paulo.
- 1971. New species and records of Australasian Calliphorinae, with special reference to the fauna of New Guinea (Diptera: Calliphoridae). - Pac. Ins. 13: 1-12.
- 1977. Family Calliphoridae. Pp. 526-556 in: Delfinado, M. D. & Hardy, D. E. (eds.), A catalog of the Diptera of the Oriental Region. Volume III Suborder Cyclorrhapha (excluding Division Aschiza). x + 854 pp. Honolulu.
James, M. T. & Kurahashi, H., 1976. Blow flies of the genus *Onesia* from the Bismarck Islands (Insecta, Diptera, Calliphoridae). - Steenstrupia 4: 19-22.
Kamal, A. S., 1958. Comparative study of thirteen species of sarcosaprophagous Calliphoridae and Sarcophagidae (Diptera). I. Bionomics. - Ann. ent. Soc. Am. 51: 261-271.
Kano, R., 1962. Notes on flies of medical importance in Japan. Part XVI. Three new species of the genus *Melinda* (Calliphoridae, Diptera) from Japan. - Jap. J. sanit. Zool. 13: 1-6.
Kano, R., Okazaki, T., Utsumi, K. & Hagi, K., 1959. Notes on flies of medical importance in Japan. [Part XV] Finding of *Calliphora subalpina* (Ringdahl, 1931) in the mountains of Honshu, Japan. - Bull. Tokyo Med. Dent. Univ. 6: 413-415.
Kano, R. & Sato, K., 1952. Notes on the flies of medical importance in Japan. (Part VI) Larvae of Luciliini in Japan. - Jap. J. Exp. Med. 22: 33-42.
Kano, R. & Shinonaga, S., 1968. Calliphoridae (Insecta: Diptera). - Fauna japon. 181 pp, Pls. I-XXIII, Figs. 1-45. Tokyo.
Keilin, D., 1915. Recherches sur les larves de Diptères cyclorhaphes. Cycle évolutif de "*Pollenia rudis* Fabr.", parasite d'"*Allolobophora chlorotica*". - Biologie comparée des larves des Diptères. - Bull. sci. Fr. Belg. 49: 15-198, Pls. I-XVI.
- 1919. On the life history and larval anatomy of *Melinda cognata* Meigen (Diptera Calliphorinae) parasite in the snail *Helicella (Heliomanes) virgata* DaCosta, with an account of the other Diptera living upon mol-

luscs. - Parasitology 11: 430-455, Pls. XXII-XXV.
- 1921. Supplementary account of the dipterous larvae feeding upon molluscs. - Parasitology 13: 180-183.
Kerney, M. P. & Cameron, R. A. D., 1979. A field guide to the land snails of Britain and North-west Europe. 288 pp. London.
Khitsova, L. N., 1979: A new species of the genus *Bellardia* (Diptera, Calliphoridae) from the Caucasus. - Zool. Zh. 58: 1245-1246. [In Russian, with English summary].
- 1983. A new species of the genus *Morinia* (Diptera, Rhiniphoridae [sic] from the Caucasus. - Zool. Zh. 62: 1588-1590. [In Russian, with English summary].
King, J. H. L. & Kurahashi, H., 1977. A new *Phumosia* species from East Malaysia, Borneo (Diptera: Calliphoridae). - Sarawak Museum Journal 25: 221-225.
Koskela, P., Itämies, J. & Pasanen, S., 1974. *Lucilia bufonivora* Moniez (Dipt., Calliphoridae), a lethal parasite in *Rana temporaria* L. (Anura). - Ann. Zool. fenn. 11: 105-106.
Kovalev, V. G. & Verves, Yu. G., 1987. Dvukrýlýe (Diptera, Calliphoridae) iz ghezd vorob'inýkh ptits (Passeriformes) v Kandalakshskom Zapovednike. - The Problems of molecular and general biology, Kiev University Press 6: 59-61. [In Russian.]
Kramer, H., 1910. Zur näheren Kenntnis der Dipterengattung *Lucilia* R.D. - Ent. Vereinsbl. Nr.6, Beil. Ent. Rdsch. 27: 34-35.
- 1911. Die Tachiniden der Oberlausitz. - Abh. naturforsch. Ges. Görlitz 27: 117-166, 3 pls.
- 1917. Die Musciden der Oberlausitz. - Abh. naturforsch. Ges. Görlitz 28: 257-352.
- 1928. Zur Gattung *Calliphora* R.D. - Konowia 7: 62-64.
Krivosheina, N. P., 1961. Diptera larvae feeding on earthworms. - Zool. Zh. 40: 715-718. [In Russian, with English summary].
Kurahashi, H., 1964. Studies on the calypterate muscoid flies from Japan. III. Revision of the genus *Polleniopsis* (Diptera, Calliphoridae). - Kontyû 32 (4): 484-489.
- 1970. Tribe Calliphorini from Australian and Oriental regions, I. *Melinda*-group (Diptera: Calliphoridae). - Pac. Ins. 12: 519-542.
- 1971. The tribe Calliphorini from Australian and Oriental Regions II. *Calliphora*-group (Diptera: Calliphoridae). - Pac. Ins. 13: 141-204.
- 1972a. A new species of the genus *Tricycleopsis* from Japan (Diptera: Calliphoridae). - Kontyû 40: 24-26.
- 1972b. Tribe Calliphorini from Australian and Oriental Regions, IV. *Onesia*-group: genus *Polleniopsis* (Diptera: Calliphoridae). - Pac. Ins. 14: 709-724.
- 1977. The tribe Luciliini from Australian and Oriental Regions. I. Genus *Hypopygiopsis* Townsend (Diptera; Calliphoridae). - Kontyû 45: 553-562.
- 1978. Tribe Calliphorini from Australian and Oriental regions. V. *Onesia*-group: genus *Tainanina* (Diptera: Calliphoridae). - Pac. Ins. 18: 1-8.
Kurahashi, H. & Beaver, R. A., 1979. *Nepenthomyia malayana* gen.n, sp.n. A new calliphorid fly bred from pitchers of *Nepenthes ampullaria* in West Malaysia (Diptera, Calliphoridae). - Annls Soc. ent. Fr. (N.S.) 15: 25-30.
Kuusela, S., 1983. Community structure of carrion flies along an island gradient. - Holarct. Ecol. 6: 372-380.
- 1984. Suitability of carrion flies for field experiments on reproductive behaviour. - Suomen hyönt. Aikak. 50: 1-6.
Kuusela, S. & Hanski, I., 1982. The structure of carrion fly communities: the size and the type of carrion. - Holarct. Ecol. 5: 337-348.
Laitinen, E., Nuorteva, P. & Renkonen, O. V., 1970. A case of wound myiasis in man caused by *Lucilia illustris* (Meig.) (Dipt., Calliphoridae). - Suomen hyönt. Aikak. 36: 123.
Lehrer, A. Z., 1963. Études sur les Diptères Calliphorides. I. - La classification des Polleniinae paléarctiques et leur dispersion en Roumanie. - Bull. Annls Soc. r. ent. Belg. 99: 285-310.
- 1967. Études sur les Diptères calliphorides. II. Revision de la nomenclature de la tribu Polleniini. - Bull. Annls Soc. r. ent. Belg. 103: 255-259.
- 1970. Considérations phylogénétiques et taxonomiques sur la famille Calliphoridae (Diptera). - Annotnes zool.-bot., Bratisl. 61: 1-51.
- 1972. Diptera Familia Calliphoridae. - Fauna Repub. soc. România 11 (12): 1-245.
- 1979. *Stomorhina lunata* (Fabricius) et *S. selgae* nom. nov. (Dipt., Calliphoridae). - Eos, Madrid 53 (1977): 81-90.
Lewis, C. T. & Pollock, J. N., 1975. Engagement of the phallosome in blowflies. - J. Ent. (A) 49: 137-147.
Lindner, E., 1957. Vogelblutfliegen (*Protocalliphora*, Calliphorinae, Diptera). Ein Beitrag zur Kenntnis ihrer Systematik und Biologie. - Vogelwarte 19: 84-90.
Lindroth, C. H., 1957. The faunal connections between Europe and North America. 344 pp. Stockholm, New York.
Linnaeus, C., 1758. Systema Naturae per regna tria naturae, secundum classes, ordines, genera, species, cum caracteribus, differentiis, synonymis, locis. 10th edition, 1. 824 pp. Holmiae [= Stockholm].
- 1761. Fauna Svecica sistens animalia sveciae regni. 2nd edition [revised]. 578 (+ 48) pp., 2 plates. Stockholmiae.
Lioy, P., 1864. I ditteri distribuiti secondo un nuovo metodo di classificazione naturale (Part). Atti Ist. veneto Sci. (3) 9: 879-910.
Lobanov, A. M., 1971. On the biology and morphology of *Melinda agilis* Mg. (Diptera, Calliphoridae). - Ént. Obozr. 50: 513-517. [In Russian; English translation in Ent. Rev., Wash. 50: 290-292].
- 1976. Opredelitel' samok sem. Calliphoridae (Diptera) faunÿ SSSR po yaitsekladam. - Opredelitel' po faune SSSR, Izdavamye Zool. Inst. Akad. Nauk SSSR, Leningr. 109: 1-48.
Lopes, H. de Souza, 1941. Sobre o aparelho genital feminino dos "Sarcophagidae" e sua importância na classificaçao (Diptera). - Rev. Brasil. Biol. 1: 215-221.
- 1955. Considerações sobre os gêneros de Sarcophagidae (Diptera) propostos por Robineau-Desvoi-

dy em 1830 e 1863. - Arquivos do Museu nacional 42: 265-272.
- 1969. 103 Family Sarcophagidae. In: Papavero, N. (ed.), A Catalog of the Diptera of the Americas south of the United States. 88 pp. Museu de Zoologia, Universidade de São Paulo.
- 1981. On *Chrysogramma* Rohdendorf and *Sarcotachina* Portschinsky (Diptera, Sarcophagidae). - Rev. Brasil. Biol. 41: 205-209.
- 1982. The importance of the mandible and clypeal arch of the first instar larvae in the classification of the Sarcophagidae. - Revta bras. Ent. 26: 293-326.
Lopes, H. de Souza & Albuquerque, D. de O., 1982. Notes on Neotropical Calliphoridae (Diptera). - Rev. Brasil. Biol. 42: 63-69.
Løken, A., 1973. Studies on scandinavian bumble bees (Hymenoptera, Apidae). - Norsk ent. Tidsskr. 20: 1-218.
Lundbeck, W., 1927. Diptera Danica. Part VII. Platypezidae, Tachinidae. 560 + 11 pp. København.
Lyneborg, L., 1965. Hansted-Reservatets Entomologi. 9. Diptera, Brachycera & Cyclorrhapha - Fluer. - Ent. Meddr 30: 201-262.
- 1968. XLIIa. Diptera Brachycera et Cyclorrhapha. In: Spärck, R. & Tuxen, S. L. (eds.), The zoology of the Faroes, 2 (2), 79 pp.
MacLeod, J., 1937. The species of Diptera concerned in cutaneous myiasis of sheep in Britain. - Proc. R. ent. Soc. London (A) 12: 127-133.
- 1943a. A survey of British sheep blowflies. - Bull. ent. Res. 34: 65-88.
- 1943b. A survey of British sheep blowflies. II.Relation of strike to host and edaphic factors. - Bull. ent. Res. 34: 95-111.
Macquart, J., 1834. Insectes Diptères du Nord de la France. Athéricères: Créophiles, Oestrides, Myopaires, Conopsaires, Scénopiniens, Céphalopsides. - Mém. Soc. Sci. Agric. Lille 1833: 137-368.
- 1835. Histoire Naturelle des Insectes. Diptères. 2: 1-703 (or 710). Paris.
- 1843. Diptères exotiques nouveaux ou peu connus. - Mém. Soc. Sci. Agric. Lille 1842: 162-460 [5-304].
- 1851. Diptères exotiques nouveaux ou peu connus. Suite du 4e supplément. - Mém. Soc. Sci. Agric. Lille 1850: 134-294 [161-336].
- 1854. Nouvelles observations sur les Diptères d'Europe de la tribu des Tachinaires (Suite) (1). - Annls Soc. ent. Fr. (3) 2: 733-754.
- 1855. Nouvelles observations sur les Diptères d'Europe de la tribu des Tachinaires. - Annls Soc. ent. Fr. (3) 3: 21-47.
Malloch, J. R., 1919. The Diptera collected by the Canadian Arctic Expedition, 1913-1918 (excluding the Tipulidae and Culicidae). Pp. 34-90, pls. 7-10 in: Anderson, R. M. (ed.), Report of the Canadian Arctic Expedition 1913-18. Vol. 3: Insects, Pt. C: Diptera, 90 pp., 2 figs., 10 pls. Ottawa.
- 1926. Exotic Muscaridae (Diptera). - XVIII. - Ann. Mag. nat. Hist. (9) 17: 489-510.
- 1927. Notes on Australian Diptera. No.xi. - Proc. Linn. Soc. New South Wales 52: 299-335.
McAlpine, J. F., 1965. Insects and related terrestrial invertebrates of Ellef Ringnes Island. - Arctic 18: 73-103.
- 1981. Morphology and terminology - adults. Pp. 9-63 in: McAlpine, J. F., Peterson, B. V., Shewell, G. E., Teskey, H. J., Vockeroth, J. R. & Wood, D. M. (eds.), Manual of Nearctic Diptera, 1. Res. Branch Agric. Canada Monogr. 27: vi + 1-674. Ottawa.
McAlpine, J. F., Peterson, B. V., Shewell, G. E., Teskey, H. J., Vockeroth, J. R. & Wood, D. M., 1981. Introduction. Pp. 1-7 in: McAlpine, J. F., Peterson, B. V., Shewell, G. E., Teskey, H. J., Vockeroth, J. R. & Wood, D. M. (eds.), Manual of Nearctic Diptera, 1. Res. Branch Agric. Canada Monogr. 27: vi + 1-674. Ottawa.
Mehl, R., 1970. Om innsamling av insekter og midd på fugler og pattedyr. - Fauna, Oslo 23: 237-252.
Meigen, J. W., 1824. Systematische Beschreibung der bekannten europäischen zweiflügeligen Insekten. 4: xii + 1-428. Hamm.
- 1826. Systematische Beschreibung der bekannten europäischen zweiflügeligen Insekten. 5: xii + 1-412. Hamm.
- 1830. Systematische Beschreibung der bekannten europäischen zweiflügeligen Insekten. 6: xi + 1-401. Hamm.
- 1838. Systematische Beschreibung der bekanntnen europäischen zweiflügeligen Insekten. 7 (Supplementsband): xii + 1-434. Hamm.
Meinert, F., 1888. En spyflue, *Lucilia nobilis*, snyltende hos mennesket. - Ent. Meddr 1: 119-122.
- 1890. *Philornis molesta*, en på fugl snyltende Tachinarie. - Vidensk. Medd. dansk naturh. Foren. 1889: 304-317.
Mercier, L., 1929. Contribution à la connaissance de l'epèce chez les myodaires supérieurs. - Bull. Biol. Fr. Belg. 63: 399-421.
- 1930. Variation de certaines pièces de l'armature génitale mâle de *Pollenia rudis* F. (Diptère Calliphorinae); importance de cette variation pour la notion d'espèce chez les Myodaires supérieurs. - C. r. h. Acad. Sci., Paris 190: 320-322.
Messersmith, D. H., 1982. A report on a collection of Diptera from Iceland and Greenland. - Fauna norv. Ser. B. 29: 36-39.
Michelsen, V., 1979. Notes on the identity and type-material of the Anthomyiidae, Fanniidae, and Muscidae (Diptera) described by J. C. Fabricius. - Steenstrupia 5: 181-196.
- 1983. The Anthomyiidae (Diptera) described by Fallén, with a review of European *Emmesomyia* Malloch. - Ent. scand. 14: 109-120.
Mihályi, F., 1965. Rearing flies from faeces and meat, infected under natural conditions. - Acta zool. Acad. sci. hung. 11: 153-164.
- 1967. The danger-index of the synanthropic flies. - Acta zool. Acad. sci. hung. 13: 373-377.
- 1976. Contribution to the the knowledge of the genus *Pollenia* R.-D. (Diptera, Calliphoridae). - Acta zool. Acad. sci. hung. 22: 327-333.
- 1977. A new key to Hungarian *Lucilia* species (Dipte-

ra, Calliphoridae). - Ann. hist.-nat. Mus. natns hung. 69: 181-184.
Mik, J., 1884. Fünf neue österreichische Dipteren. - Verh. zool-bot. Ges. Wien 33 (2) (1883): 251-262.
Miller, D., 1932. The bucco-pharyngeal mechanism of a blow-fly larva (*Calliphora quadrimaculata* Swed.). - Parasitology 24: 491-499.
Moniez, R. L., 1876. Un Diptère parasite du crapaud (*Lucilia bufonivora* n.sp.). - Bull. Sci. Dep. Nord. Lille 8 (1877): 25-27.
Morge, G., 1975. Dipteren-Farbtafeln nach den bisher nicht veröffentlichten Original-Handzeichnungen Meigens: "Johann Wilhelm Meigen: Abbildung der europaeischen zweiflügeligen Insecten, nach der Natur". Pars I. Vorwort des Herausgebers - Faksimiles von Original-Legenden und -Index: "Erklaerung der Fig." und "Alphabetisches Verzeichniss der Gattungen und Arten" - Farbtafeln I-LXXX. - Beitr. Ent. 25: 383-500, pls. 1-80.
- 1976a. Dipteren-Farbtafeln nach den bisher nicht veröffentlichten Original-Handzeichnungen Meigens: "Johann Wilhelm Meigen: Abbildung der europaeischen zweiflügeligen Insecten, nach der Natur". Pars II: Farbtafeln LXXXI-CLX. - Beitr. Ent. 26: 441, pls. 81-160.
- 1976b. Dipteren-Farbtafeln nach den bisher nicht veröffentlichten Original-Handzeichnungen Meigens: "Johann Wilhelm Meigen: Abbildung der europaeischen zweiflügeligen Insecten, nach der Natur". Pars III: Farbtafeln CLXI-CCCV. - Beitr. Ent. 26: 543, pls. 161-305.
Mueller, A., 1922. Über den Bau des Penis der Tachinarier und seinen Wert für die Aufstellung des Stammbaumes und die Artdiagnose. - Arch. Naturgesch. 88 A (2): 45-168.
Natvig, L. R., 1950. Helseskadelige og blodsugende insekter og midder. Pp. 166-203 in: Føyn, B., Ruud, G. & Røise, H. (eds.), Norges Dyreliv, Bind IV: Hvirvelløse dyr. Oslo.
- 1959. Helseskadelige og blodsugende insekter og midder. Pp. 168-206 in: Føyn, B., Ruud, G. & Røise, H. (eds.), Norges Dyreliv, Bind IV: Hvirvelløse dyr. Ny utgave. Oslo.
Nielsen, B. O., 1963. Om forekomsten af Diptera på Alm. Stinksvamp (*Phallus impudicus* Pers.) og fund af *Phaonia errans* Mg. og *Helomyza fuscicornis* Zett., nye for den danske fauna. - Flora og Fauna 69: 126-134.
- 1967. *Calliphora subalpina* (Ringdahl) ny for Danmark og andre *Calliphora*-arter (Diptera: Calliphoridae) på stinksvamp (*Phallus impudicus* Pers.). - Flora og Fauna 73: 68-70.
- 1968. Studies on the dipterous fauna of the stinkhorn (*Phallus impudicus* Pers.). - Natura Jutlandica 14: 89-98.
- 1984a. Spyflue-myiasis hos får i Danmark (Diptera: Calliphoridae). - Ent. Meddr 51: 119-124.
- 1984b. Myiasis i danske fårebesætninger. - Dansk Vet-Tidsskr. 67: 718-723.
Nielsen, B. O. & Nielsen, S. A., 1976. Schmeissfliegen (Calliphoridae) und vakuumverpackter Schinken. - Anz. Schädlingskde Pflanzenschutz Umweltschutz 49: 113-115.
Nielsen, J. C., 1917. [Without title, in minutes of meeting]. - Vidensk. Meddr dansk naturh. Foren. 68: xix-xxi.
Nielsen, P., Ringdahl, O. & Tuxen, S. L., 1954. 48a. Diptera 1 (exclusive of Ceratopogonidae and Chironomidae). - Zool. of Iceland 3 (48a): 1-189.
Nielsen, S. A., Nielsen, B. O. & Walhovd, H., 1978. Blowfly myiasis (Diptera: Calliphoridae, Sarcophagidae) in the hedgehog (*Erinaceus europaeus* L.). - Ent. Meddr 46: 92-94.
Norris, K. R., 1965. The bionomics of blow flies. - Ann. Rev. Ent. 10: 47-68.
Nowicki, M., 1868. Der Kopaliner Heerwurm und die aus ihm hervorgehende *Sciara militaris* n.sp. - Verh. naturf. Ver. Brünn 6 (1867): 3-69.
Nuorteva, P., 1958. Some peculiarities of the seasonal occurrence of poliomyelitis in Finland. - Ann. Med. Exper. Fenn. 36: 335-342.
- 1959a. A case of wound myiasis in the nightjar. - Ornis fennica 36: 8-10.
- 1959b. Studies on the significance of flies in the transmission of poliomyelitis I. The occurrence of the *Lucilia* species (Dipt., Calliphoridae) in relation to the occurrence of poliomyelitis in Finland. - Suomen hyönt. Aikak. 25: 1-24.
- 1959c. Studies on the significance of flies in the transmission of poliomyelitis II. The composition of the annual blowfly population as compared with the incidence of poliomyelitits in England during the years 1949-1952. - Suomen hyönt. Aikak. 25: 25-120.
- 1959d. Studies on the significance of flies in the transmission of poliomyelitis III. The composition of the blowfly fauna and the activity of the flies in relation to the weather during the epidemic season of poliomyelitis in South Finland. - Suomen hyönt. Aikak. 25: 121-136.
- 1959e. Studies on the significance of flies in the transmission of poliomyelitis IV. The composition of the blowfly fauna in different parts of Finland during the year 1958. - Suomen hyönt. Aikak. 25: 137-162.
- 1960a. *Protocalliphora chrysorrhoea* (Meig.) kärpäsen toukat törmäpääskyn loisina. - Ornis fennica 37: 122-124.
- 1960b. Förekomsten av asflugan *Phormia terraenovae* R.-D. i Finland. - Notul. ent. 40: 38-45.
- 1960c. Flugornas betydelse för polions epidemiologi. - Notul. ent. 40: 19-27.
- 1960d. Studies on the significance of flies in the transmission of poliomyelitis. VI. On the influence of icosaeonic climatic fluctuations on the incidence of poliomyelitis and on the occurrence of *Lucilia* species in Finland. - Suomen hyönt. Aikak. 26: 273-280.
- 1961. Studies on the significance of flies in the transmission of poliomyelitis VII. Attraction of blowflies to a house by the honeydew of the aphid *Phorodon humuli* (Schrk.). - Suomen hyönt. Aikak. 27: 51-53.
- 1962. The occurrence of pollenine and calliphorine flies (Diptera, Calliphoridae) in Urjala during the year 1961. - Lounais-Hämeen Luonto 13: 3-7.

- 1963. Synanthropy of blowflies (Dipt., Calliphoridae) in Finland. - Suomen hyönt. Aikak. 29: 1-49.
- 1964a. The occurrence of pollenine and calliphorine flies (Diptera, Calliphoridae) in Urjala during the year 1962. - Lounais-Hämeen Luonto 15: 15-19.
- 1964b. Differences in the ecology of *Lucilia caesar* (L.) and *Lucilia illustris* (Meig.) (Diptera, Calliphoridae) in Finland. - Wiadomosci parazytologiczne 10: 583-587.
- 1964c. The zonal distribution of blowflies (Dipt., Calliphoridae) on the arctic hill Ailigas in Finland. - Suomen hyönt. Aikak. 30: 218-226.
- 1966a. The flying activity of *Phormia terraenovae* R.-D. (Dipt., Calliphoridae) in subarctic conditions. - Ann. zool. fenn. 3: 73-81.
- 1966b. Local distribution of blowflies in relation to human settlement in an area around the town of Forssa in South Finland. - Suomen hyönt. Aikak. 32: 128-137.
- 1966c. The occurrence of *Phormia terraenovae* R.-D. (Dipt., Calliphoridae) and other blowflies in the archipelago of the subarctic lake Inarinjärvi. - Suomen hyönt. Aikak. 32: 240-251.
- 1967. 8. Observations on the blowflies (Dipt., Calliphoridae) of Spitsbergen. - Suomen hyönt. Aikak. 33: 62-64.
- 1969. The occurrence of pollenine and calliphorine flies in Urjala during the years 1963-64, with some conclusions from a five-year period of observations. - Lounais-Hämeen Luonto 34: 3-14.
- 1970. Histerid beetles as predators of blowflies (Diptera, Calliphoridae) in Finland. - Ann. zool. fenn. 7: 195-198.
- 1971a. Annoying mass occurence of *Phormia terraenovae* R.-D. (Diptera, Calliphoridae) in the surroundings of a rendering plant in southwestern Finland. - Ann. zool. fenn. 8: 336-339.
- 1971b. Duration of development of *Calliphora alpina* (Zett.) (Dipt., Calliphoridae) in subarctic northern Finland. - Suomen hyönt. Aikak. 37: 209.
- 1972. A three-year survey of the duration of development of *Cynomya mortuorum* (L.) (Dipt., Calliphoridae) in the conditions of a subarctic fell. - Suomen hyönt. Aikak. 38: 65-74.
- 1977. Sarcosaprophagous insects as forensic indicators. Pp. 1072-1095 in: Tedeschi, C. G., Eckert, W. G. & Tedeschi, L. G. (eds.), Forensic medicine: A study in trauma and environmental hazards, vol. 2. Philadelphia, London, Toronto.
- 1987. Empty puparia of *Phormia terraenovae* R.-D. (Diptera, Calliphoridae) as forensic indicators. - Suomen hyönt. Aikak. 53: 53-56.

Nuorteva, P. & Auvinen, E., 1968. A case of intestinal myiasis caused by *Calliphora vicina* (R.-D.) (Dipt. Calliphoridae) in a baby. - Suomen hyönt. Aikak. 34: 244.

Nuorteva, P. & Häsänen, E., 1972. Transfer of mercury from fishes to sarcosaprophagous flies. - Ann. zool. Fenn. 9: 23-27.

Nuorteva, P., Isokoski, M. & Laiho, K., 1967. Studies on the possibilities of using blowflies (Dipt.) as medicolegal indicators in Finland. 1. Report of four indoor cases of the city of Helsinki. - Suomen hyönt. Aikak. 33: 217-225.

Nuorteva, P. & Järvinen, U., 1961a. The insect fauna of the nests of sand martin (*Riparia riparia* L.) in Finland. - Suomen hyönt. Aikak. 27: 197-204.

Nuorteva, P., Kotimaa, T., Pohjolainen, L. & Räsänen, T., 1964. Blowflies (Dipt., Calliphoridae) on the refuse depot of the city of Kuopio in Central Finland. - Suomen hyönt. Aikak. 30: 94-104.

Nuorteva, P. & Laurikainen, E., 1964. Synanthropy of blowflies (Dipt., Calliphoridae) on the island of Gotland, Sweden. - Suomen hyönt. Aikak. 30: 187-190.

Nuorteva, P. & Räsänen, T., 1968. The occurrence of blowflies (Dipt., Calliphoridae) in the archipelago of the lake Kallavesi, Central Finland. - Ann. zool. fenn. 5: 188-193.

Nuorteva, P., Schumann, H., Isokoski, M. & Laiho, K., 1974. Studies on the possibilities of using blowflies (Dipt., Calliphoridae) as medicolegal indicators in Finland. 2. Four cases where species identification was performed from larvae. - Suomen hyönt. Aikak. 40: 70-74.

Nuorteva, P. & Skarén, U., 1960. Studies on the significance of flies in the transmission of poliomyelitis V. Observations on the attraction of blowflies to the carcasses of micro-mammals in the commune of Kuhmo, East Finland. - Suomen hyönt. Aikak. 26: 221-226.

Nuorteva, P. & Vesikari, T., 1966. The synanthropy of blowflies (Diptera, Calliphoridae) on the coast of the Arctic Ocean. - Ann. Med. exp. fenn. 44: 544-548.

Nylander, W., 1848. Strödda anteckningar. - Notiser Sällsk. Fauna Flora fenn. Forhandl. 1: 289-304.

O'Hara, J. E., 1982. Classification, phylogeny and zoogeography of the North American species of *Siphona* Meigen (Diptera: Tachinidae). - Quaest. ent. 18: 261-380.

Ojala, O. & Nuorteva, P., 1966. Isolering av *Salmonella*-bakterier hos flugor i södra Finland. - 10th Nordic Veterinary Congress, Stockholm 15-18.Aug.1966. 2: 622-627.

Omodeo, P., 1963. Distribution of the terricolous oligochaetes on the two shores of the Atlantic. Pp. 127-152 in: Löve, Á. & Löve, D. (eds.), North Atlantic biota and their history. xii + 430 pp. Oxford, London, New York, Paris.

Ozerov, A. L., 1986. Calliphoridae and Carnidae of the Zeja State Natural Reserve. - Trudý zool. Inst. Akad. Nauk, SSSR 146: 79-86. [In Russian].

Økland, K. A., 1981. Inndeling av Norge til bruk i biogeografiske oppgaver - et revidert Strand-system. - Fauna, Oslo 34: 167-178.

Pandellé, L., 1896. Études sur les Muscides de France. IIe partie. - Revue Ent. 15: 1-230.

Panzer, G. W. F., 1804. Faunae insectorum germanicae initiae oder Deutschlands Insecten, Heft. 86, 24 pp, 24 pls., Nürnberg.

Pape, T., 1986. A phylogenetic analysis of the woodlouse-flies (Diptera, Rhinophoridae). - Tijdschrift voor Entomologie 129 (2): 1-34.

- 1987a. An annotated check-list of Finnish flesh-flies (Diptera: Sarcophagidae). - Notul. ent. 67: 43-46.
- 1987b. The Sarcophagidae (Diptera) of Fennoscandia and Denmark. - Fauna ent.scand. 19: 1-203.
- 1988. First record of *Melanomya* Rondani (Diptera, Calliphoridae) from Japan. - Kontyû 56: 641-644.
Patton, W. S., 1935. Studies on the higher Diptera of medical and veterinary importance. A revision of the genera of the family Muscidae Testaceae R-D based on a comparative study of the male and female terminalia. The genera *Adichosia* Surcouf and *Auchmeromyia* Brauer and von Bergenstamm (s.lat.). - Ann. trop. Med. Parasit. 29: 199-230.
- 1936. Studies on the higher Diptera of medical and veterinary importance. A revision of the genera of the family Muscidae Testaceae R-D based on a comparative study of the male and female terminalia. The genus *Cordylobia* Grünberg (sens.lat.). - Ann. trop. Med. Parasit. 30: 57-69.
Persson, P. I., 1983. Flugor på Ölands Stora alvar. - Ent. Tidskr. 104: 151-164.
Peus, F., 1960. Zur Kenntnis der ornithoparasitischen Phormiinen (Diptera, Calliphoridae). - Dt. Ent. Z. (N.F.) 7: 193-235.
Pokorny, E., 1889. Beitrag zur Dipterenfauna Tirols. - Verh. zool.-bot. Ges. Wien 39: 543-574.
Pont, A. C., 1972. 97 Family Muscidae. In: Papavero, N. (ed.), A Catalog of the Diptera of the Americas south of the United States. 111 pp. Museu de Zoologia, Universidade de São Paulo.
- 1974. 96a Family Anthomyiidae. In: Papavero, N. (ed.), A Catalog of the Diptera of the Americas south of the United States. 21 pp. Museu de Zoologia, Universidade de São Paulo.
- 1976. 81. Calliphoridae. Pp. 107-108 in: Kloet, G. S. & Hincks, W. D. (eds.), A check list of British insects. Second edition. Part 5: Diptera and Siphonaptera. - Handb Ident. Br. Insects 11 (5): ix + 139 pp.
- 1980. 90. Family Calliphoridae. Pp. 779-800 in: Crosskey, R. W. (ed.), Catalogue of the Diptera of the Afrotropical Region. 1437 pp. British Museum (Natural History), London.
- 1983. *Mya* Rondani, 1850, and *Somomya* Bertoloni, 1861 (Insecta, Diptera): designation of type-species, and proposed suppression of *Somomya* under the plenary powers. Z.N.(S.) 2127. - Bull. zool. Nom. 40: 106-110.
- 1984. A revision of the Fanniidae and Muscidae (Diptera) described by Fallén. - Ent. scand. 15: 277-297.
- 1986. A revision of the Fanniidae and Muscidae described by J. W. Meigen (Insecta: Diptera). - Ann. naturhist. Mus. Wien (B) 87: 197-253.
Portschinsky, I. A., 1875. Enumération des espèces du genre *Cynomyia* du gouvernement de Mohilew. - Hor. Soc. ent. ross. 11: 37-38.
- 1881. Diptera europaea et asiatica nova aut minus cognita. Pars 1ma. - Hor. Soc. ent. ross. 16: 136-145.
Povolný, D. & Rozsypal, J., 1968. Towards the autecology of *Lucilia sericata* (Meigen, 1826) (Dipt., Call.) and the origin of its synanthropy. - Acta sc. nat. Brno 2(8): 1-32.
Pulliainen, E., 1974. The flying activity of *Phormia terraenovae* R.-D. (Dipt., Calliphoridae) in late winter in northeastern Lappland. - Suomen hyönt. Aikak. 40: 26-28.
Quo, F., 1952. On the Chinese species of the genus *Lucilia* R.-D. (Diptera, Calliphoridae). - Acta ent. sin. 2: 113-118.
Rasmussen, K., 1934. Mindeudgave. Udgivet af Peter Freuchen, Therkel Mathiassen, Kaj Birket-Smith. 1. Bind. 348 pp. København.
Reinhard, H. J., 1929. Notes on the muscoid flies of the genera *Opelousia* and *Opsodexia* with the description of three new species. - Proc. U. S. natn. Mus. 76 (20): 1-9.
Ringdahl, O., 1921. Bidrag till kännedomen om de skånska stranddynernas insektfauna. - Ent. Tidskr. 42: 21-40, 65-92.
- 1931. Einige Mitteilungen über lappländische Dipteren. - Ent. Tidskr. 52: 171-174.
- 1933. Tachiniden und Musciden aus Nordost-Grønland. - Skrifter om Svalbard og Ishavet 53: 15-18.
- 1934. Bidrag till kännedomen om en del av Zetterstedts tachinid-typer (Diptera). - Ent. Tidskr. 55: 266-272.
- 1937. Bidrag till kännedomen om de svenska tachinidernas utbredning. - Ent. Tidskr. 58: 31-38.
- 1939. Diptera Brachycera i Regio alpina. - Ent. Tidskr. 60: 37-50.
- 1942. Neue schwedische Tachiniden-Gattungen und Arten. - Opusc. ent. 7: 62-65.
- 1944a. Revision av Vilhelm Storms Diptersamling I. Familjen Tachinidae. - K. Norske Vidensk. Selsk. Forh. 17 (19): 79-81.
- 1944b. Contributions to the knowledge of the tachinids and muscids of Norway. - Tromsø Mus. Aarsh. 65 (1942) (4): 1-27.
- 1945a. Förteckning över de av Zetterstedt i Insecta Lapponica och Diptera Scandinaviae beskrivna tachiniderna med synonymer jämte anteckningar över en del arter. - Opusc. ent. 10: 26-35.
- 1945b. Översikt över de hittills från Sverige kända arterna av familjen Tachinidae (Diptera). - Ent. Tidskr. 66: 177-210.
- 1947. Förteckning över flugor från Ölands alvar. - Ent. Tidskr. 68: 21-28.
- 1951. Flugor från Lapplands, Jämtlands och Härjedalens fjälltrakter (Diptera Brachycera). - Opusc. ent. 16: 113-186.
- 1952. Catalogus Insectorum Sueciae. - XI. Diptera Cyclorrapha: Muscaria Schizometopa. - Opusc. ent. 17: 129-186.
- 1954. Nya fyndorter för norska Diptera. - Norsk ent. Tidsskr. 9: 46-54.
- 1958. Dipterologiska notiser 16 och 17. 16. Norrländska Diptera brachycera. 17. Värmländska Diptera brachycera. - Opusc. ent. 23: 90-94.
- 1960. Flugfaunan på Kullaberg och Hallands Väderö. - Kullabergs Natur 2: 1-40.
Roberts, M. J., 1971. The structure of the mouthparts of some dipteran larvae in relation to their feeding habits. - Acta zool. 52: 171-188.
Robineau-Desvoidy, A. J. B., 1830. Essai sur les Myo-

daires. - Mém. prés. div. Sav. Acad. Sci. Inst. Fr. (2) 2: 1-813.
- 1849. [Notes on *Lucilia dispar* and *Phormia regina*.] - Annls Soc. ent. Fr. (2) 7: (Bull.) iv-v.
- 1863. Histoire naturelle des Diptères des environs de Paris. Oeuvre posthume du Dr Robineau-Desvoidy publiée par les soins de sa famille, sous la direction de M. H. Monceaux. 2: 1-920. Paris, Leipzig, London.
Rodendorf, B. B., 1924. Calliphorinen-Studien I. (Dipt., Tachinidae). - Ent. Mitt. 13: 281-285.
- 1926. Morphologisches Studium an äusseren Genitalorganen der Calliphorinen (Diptera). - Zool. Zh. 6: 83-128. [In Russian, with German summary].
- 1928. Calliphorinen-Studien II (Dipt.). - Ent. Mitt. 17: 336-338.
- 1931. Calliphorinen-Studien IV (Dipt.). Eine neue Calliphorinengattung aus Ostsibirien. - Zool. Anz. 95: 175-177.
Rodendorf, E. B., 1957. On the parasite flies, injurious to nestling singing birds. - Ént. Obozr. 36: 116-124. [In Russian, with English summary].
Rodhain, J. & Bequaert, J., 1916. Matériaux pour une étude monographique des diptères parasites de l'Afrique. Première partie. Histoire de "Passeromyia heterochaeta" Villen. et de "Stasisia (Cordylobia) rodhaini" Ged. - Bull. Sci. Fr. Belg. 49: 236-289, Plate XIX.
Rognes, K., 1980. The blow-fly genus *Lucilia* Robineau-Desvoidy (Diptera, Calliphoridae) in Norway. - Fauna norv. Ser. B. 27: 39-52.
- 1981. A recent record of *Lucilia bufonivora* Moniez, 1876 from Norway (Dipt., Calliphoridae). - Fauna norv. Ser. B. 28: 119.
- 1982. A small collection of calypterate Diptera (Tachinidae, Sarcophagidae, Calliphoridae, Muscidae) from the Dovre mountains, Southern Norway. - Fauna norv. Ser. B. 29: 110-114.
- 1983a. Some Diptera (Tachinidae, Calliphoridae, Fanniidae, Muscidae) from the mountains of the Finse area, Southern Norway. - Fauna norv. Ser. B. 30: 25-33.
- 1983b. Additions to the Norwegian fauna of calypterate Diptera (Tachinidae, Calliphoridae, Muscidae). - Fauna norv. Ser. B. 30: 88-93.
- 1983c. The identity of *Sarcophaga magnicornis* Siebke, 1863 (Diptera: Calliphoridae). - Ent. scand. 14: 405-406.
- 1985a. Revision of the bird-parasitic blowfly genus *Trypocalliphora* Peus, 1960 (Diptera: Calliphoridae). - Ent. scand. 15: 371-382.
- 1985b. A check-list of Norwegian blowflies (Dipt., Calliphoridae). - Fauna norv. Ser. B. 32: 89-93.
- 1986a. The Sarcophagidae (Diptera) of Norway. - Fauna norv. Ser. B. 33: 1-26.
- 1986b. The identity of *Sarcophaga alpina* Zetterstedt, 1838 and the interpretation of *Acrophaga* Brauer & Bergenstamm, 1891 (Diptera: Calliphoridae). - Ent. scand. 16: 351-354.
- 1986c. The systematic position of the genus *Helicobosca* Bezzi with a discussion of the monophyly of the calyptrate families Calliphoridae, Rhinophoridae, Sarcophagidae and Tachinidae (Diptera). - Ent. scand. 17: 75-92.
- 1986d. The Rhinophoridae or woodlouse-flies (Diptera) of Norway. - Fauna norv. Ser. B. 33: 64-68.
- 1987a. A new species in the *intermedia*-group and a new synonymy in the genus *Pollenia* Robineau-Desvoidy, 1830 (Diptera: Calliphoridae). - Syst. Ent. 12: 381-388.
- 1987b. The taxonomy of the *Pollenia rudis* species-group in the Holarctic Region (Diptera: Calliphoridae). - Syst. Ent. 12: 475-502.
- 1988. The taxonomy and phylogenetic relationship of the *Pollenia semicinerea* species-group (Diptera: Calliphoridae). - Syst. Ent. 13: 315-345.
Rognes, K. & Blackith, R.E., 1990. *Calliphora vicina* Robineau-Desvoidy, 1830 (Insecta, Diptera): proposed conservation of the specific name. - Bull. zool. Nom. 47: 187-190.
Rondani, C., 1856. Dipterologiae Italicae Prodromus, I. Genera Italicae ordinis Dipterorum ordinatim disposita et distincta et in familias et stirpes aggregata. 1-226 + 2 pp. Parmae [= Parma].
- 1861. Dipterologiae Italicae Prodromus, 4. Species italicae ordinis dipterorum in genera characteribus definita, ordinatim collectae, methodo analatica [sic] distinctae, et novis vel minus cognitis descriptis. Pars Tertia. Muscidae Tachininarum complementum. 174 pp. Parmae [= Parma].
- 1862. Dipterologiae Italicae Prodromus, 5. Species italicae ordinis dipterorum in genera characteribus definita, ordinatim collectae, methodo analitica distinctae et novis vel minus cognitis descriptis. Pars Quarta. Muscidae Phasiinae - Dexiinae - Muscinae - Stomoxidinae. 239 pp. Parmae [= Parma].
- 1865. Diptera italica non vel minus cognita descripta vel annotata observationibus nonnullis additis (Fasc. II). - Atti Soc. ital. Sci. nat. 8: 193-231.
Sabrosky, C. W., 1956. The nomenclature of *Protocalliphora* (Diptera: Calliphoridae). - Proc. R. ent. Soc. Lond. (B) 25: 175-179.
- 1967. Corrections to A catalog of the Diptera of America north of Mexico. - Bull. ent. Soc. Am. 13: 115-125.
- 1989. *Protocalliphora* Hough, 1899 (Insecta, Diptera) and its type species *Musca azurea* Fallén, 1817: proposed conservation of usage by designation of a replacement lectotype. - Bull. zool. Nom. 46: 126-129.
Sabrosky, C. W. & Bennett, G. F., 1958. The utilization of morphological, ecological, and life history evidence in the classification of *Protocalliphora* (Diptera: Calliphoridae). - Proc. 10th Int. Congr. Ent. (1956) 1: 163-164.
Sabrosky, C. W., Bennett, G. F. & Whitworth, T. L., 1989. Bird blow flies (*Protocalliphora*) in North America (Diptera: Calliphoridae), with notes on the Palearctic species. 312 pp. Washington, D. C.
Sabrosky, C. W. & Crosskey, R. W., 1970. The type-material of Muscidae, Calliphoridae, and Sarcophagidae described by N. Baranov. - Proc. ent. Soc. Wash. 72: 425-436.
Sailer, R. I. & Lienk, S. E., 1951. Blow flies in Alaska. - Can. Ent. 83: 208-211.

Schiner, J. R., 1862. Fauna Austriaca. Die Fliegen (Diptera). I. Theil. LXXX + 1-674, 2pls. Vienna.
Schmitz, H., 1910. Zur Lebensweise von *Helicobosca muscaria* Mg. - Z. wiss. InsektBiol. 6: 107-109.
- 1917. Biologische Beziehungen zwischen Dipteren und Schnecken. - Biol. Zbl. 37: 24-43.
Schumann, H., 1954. Morphologisch-systematische Studien an Larven von hygienisch wichtigen mitteleuropäischen Dipteren der Familien Calliphoridae - Muscidae. - Wiss. Z. Univ. Greifswald Jahrgang III, 1953/54 Mathematisch-naturwissenschaftliche Reihe Nr. 4/5: 245-274.
- 1963. Beitrag zur Kenntnis der Dipteren im Wohnbereich des Menschen. - Dt. ent. Z. (N.F.) 10: 315-322.
- 1964. Revision der Gattung *Onesia* Robineau-Desvoidy, 1830 (Diptera: Calliphoridae). - Beitr. Ent. 14: 915-938.
- 1965. Die Schmeissfliegengattung *Calliphora*. - Merkbl. angew. Parasitenk 11: 1-14.
- 1971. Die Gattung *Lucilia* (Goldfliegen). - Merkbl. angew. Parasitenk 18: 1-20.
- 1973. Revision der palaearktischen *Melinda*-Arten (Diptera: Calliphoridae). - Dt. ent. Z. (N.F.) 20: 293-314.
- 1974a. Bemerkungen zum Status der Gattungen *Onesia*, *Melinda* und *Bellardia* (Diptera, Calliphoridae). - Mitt. zool. Mus. Berlin 49 (1973): 333-344.
- 1974b. Revision der palaearktischen *Bellardia*-Arten (Diptera, Calliphoridae). - Dt. ent. Z. (N.F.) 21: 231-299.
- 1986. Family Calliphoridae. Pp. 11-58 in: Soós, Á. & Papp, L. (eds.), Catalogue of Palaearctic Diptera, 12, Calliphoridae - Sarcophagidae. 265 pp. Budapest.
Schøyen, W., 1889. Supplement til H. Siebke's Enumeratio Insectorum Norvegicorum, Fasc. IV. (Diptera). - Christiania Vidensk. Selsk. Forh. 1889 (12): 1-15.
Scopoli, I. A., 1763. Entomologia carniolica exhibens insecta carnioliae indigena et distributa in ordines, genera, species, varietates, methodo linnaeana. (36 +) 420 pp. Vindobonae [= Vienna].
Séguy, E., 1925a. Étude sur quelques Muscides exotiques à larves parasites. - Bull. Soc. Path. exot. 18: 732-735.
- 1925b. Sur les *Lucilia* et les *Chrysomyia* de Macquart et de Robineau-Desvoidy. II. - Encycl. ent. (B II), Dipt. 2: 93-94.
- 1928a. Études sur les mouches parasites. Tome I Conopides, Oestrides et Calliphorines de l'Europe occidentale. Recherches sur la morphologie et la distribution géographique des Diptères à larves parasites. - Encycl. ent. (A) 9: 1-251.
- 1928b. Étude sur *Pollenia Hasei*. - Z. Angew. Ent. 14: 369-375.
- 1934. Diptères de Chine de la collection de M. J. Hervé-Bazin. - Encycl. ent. (B II), Dipt. 7: 1-28.
- 1935. Caractères particuliers des Calliphorines. - Encycl. ent. (B II), Dipt. 8: 121-150.
- 1941. Études sur les mouches parasites. Tome II. Calliphorides. Calliphorines (suite), Sarcophagines et Rhinophorines de l'Europe occidentale et méridionale. Recherches sur la morphologie et la distribution géographique des Diptères à larves parasites. - Encycl. ent. (A) 21: 1-436.
Senior-White, R., 1923. The Muscidae Testaceae of the Oriental Region (With descriptions of those found within Indian limits). - Spolia zeylan. 12: 294-314.
Senior-White, R., Aubertin, D. & Smart, J., 1940. Diptera Vol. VI. Family Calliphoridae. - Fauna Br. India: xiii + 288 pp. London.
Shannon, R. C., 1923. Genera of Nearctic Calliphoridae, blowflies, with revision of the Calliphorini. - Insecutor Inscit. menstr. 11 (7-9): 101-118, pls. VI-VIII.
- 1924. Nearctic Calliphoridae, Luciliini (Diptera). - Insecutor Inscit. menstr. 12: 67-81.
- 1926. Synopsis of the American Calliphoridae (Diptera). - Proc. ent. Soc. Wash. 28: 115-133.
Shannon, R. C. & Dobroscky, I. D., 1924. The North American bird parasites of the genus *Protocalliphora* (Calliphoridae, Diptera). - J. Wash. Acad. Sci. 14: 247-253.
Shewell, G. E., 1961. Notes on three European Diptera recently dicovered in Canada. - Can. Ent. 93: 1044-1047.
- 1987. 106 Calliphoridae. Pp. 1133-1145 in: McAlpine, J. F., Peterson, B. V., Shewell, G. E., Teskey, H. J., Vockeroth, J. R. & Wood, D. M. (eds.), Manual of Nearctic Diptera, 2. Res. Branch Agric. Canada Monogr. 28: vi + 675-1332. Ottawa.
Siebke, J. H. S., 1853. Beretning om en i sommeren 1850 foretagen entomologisk reise i en deel af Gudbrandsdalen. - Nyt Mag. Naturv. 7: 253-306.
- 1863. Beretning om en i Sommeren 1861 foretagen entomologisk Reise. - Nyt Mag. Naturv. 12: 105-192.
- 1866a. Entomologisk reise i Romsdals amt i sommeren 1864. - Nyt Mag. Naturv. 14: 375-388.
- 1866b. Entomologiske undersøgelser, foretagne i sommeren 1865. - Nyt Mag. Naturv. 14: 389-420.
- 1870. Beretning om en i sommeren 1869 foretagen entomologisk reise gjennem Ringerike, Hallingdal og Valders. - Nyt Mag. Naturv. 17: 246-314.
- 1873. Bidrag til Norges insektfauna. Beretning om en i Østerdalen foretagen reise i 1870. - Nyt Mag. Naturv. 19: 39-102.
- 1877. Enumeratio insectorum norvegicorum. Fasciculum IV, Catalogum dipterorum continentem. xiv + 255 pp. Christianiae [= Oslo].
Singh, K. I., Kurahashi, H. & Kano, R., 1979. A preliminary key to the common calliphorid flies of peninsular Malaysia (Insecta: Diptera). - Bull. Tokyo med. dent. Univ. 26: 5-24.
Singh, P., 1986. Rearing Calliphoridae. Pp. 19-20 in: Dear, J. P., Calliphoridae (Insecta: Diptera). - Fauna of New Zealand 8: 1-86.
Smith, K. G. V. (ed.), 1973. Insects and other arthropods of medical importance. xiv + 561 pp., 12 plates and 217 figs. British Museum (Natural History), London.
Smith, K. G. V., 1975. First New World occurrence of the Prosthetosomatinae (?Muscidae), an enigmatic termitophilous "subfamily" based on larval forms. - Studia ent. 18: 91-94.

- 1986. A manual of forensic entomology. 205 pp. British Museum (Natural History). London.
Soot-Ryen, T., 1925. g) Makkflueundersøkelsene. - Aarsberetn. vedk. Norges Fiskerier 1925 (1): 140-150.
Soulsby, E. J. L., 1982. Helminths, arthropods and protozoa of domesticated animals. 809 pp. London.
Stackelberg, A. A., 1962. List of Diptera of the Leningrad region. VI. Diptera Calyptrata, Part 1. Trudý zool. Inst. Akad. Nauk, SSSR 31: 318-388. [In Russian].
- 1970. 110. Sem. Rhinophoridae. Pp. 670-673 in: Bei-Bienko, G. Ya. (ed.), Opredelitel' nasekomýkh evropeiskoi chasti SSSR, V, Dvukrýlýe, Blokhi. Akad. Nauk SSSR, Zool. Inst., Leningr. [In Russian.]
Stewart, S. & Foote, R. H., 1974. An unusual infestation by *Phaenicia sericata* (Mg.) (Diptera: Calliphoridae). - Proc. ent. Soc. Wash. 76: 466.
Stone, A., Sabrosky, C. W., Wirth, W. W., Foote, R. H. & Coulson, J. R., 1965. A catalog of the Diptera of America north of Mexico. - Agric. Handb. 176: iv + 1696 pp., Washington D.C.
Storm, V., 1891. Diptera brachycera, quae observavit determinavitque. - K. norske Vidensk. Selsk. Skr. 1891: 283-313.
- 1896. Dipterologiske undersøgelser. - K. norske Vidensk. Selsk. Skr. 1895: 225-241.
- 1907. Supplerende iagttagelser over Insecta Diptera ved Trondheim. - K. norske Vidensk. Selsk. Skr. 1907 (5): 1-11.
Strand, A., 1943. Inndeling av Norge til bruk ved faunistiske oppgaver. - Norsk ent. Tidsskr. 6: 208-224.
Strand, E., 1900. Bidrag til Hallingdals og Lyngørs insektfauna. - Nyt Mag. Naturv. 37: 46-72.
- 1903. Nye lokaliteter for Diptera. - Christiania Vidensk. Selsk. Forh. 1903 (3): 3-11.
- 1906. Nye bidrag til Norges hymenopter- og dipterfauna. III. Nye norske lokaliteter for Diptera. - Nyt Mag. Naturv. 44: 102-104.
- 1914. XVIII. Weiteres über von mir gesammelte Diptera. - Nyt Mag. Naturv. 51: 323-329.
Strobl, G., 1894. Die Dipteren von Steiermark. II. Teil. - Mitt. naturw. Ver. Steierm. 30 (1893): 1-152.
Swederus, N. S., 1787. Fortsättning af Beskrifningen på 50 nya Species af Insecter. - Kungl. Sv. Vetensk.-akad. Handl. 1787: 276-290.
Sýchevskaya, V. I., 1970. Zonal distribution of coprophilous and schizophilous flies (Diptera) in Middle Asia. - Ént. Obozr. 49: 819-831. [In Russian, with English summary; English translation in Ent. Rev., Wash. 49: 498-505].
Tawfik, M. F. S. & El-Husseini, M. M., 1971. Life-history of *Pollenia dasypoda* Portochisky [sic], a parasite of the earth worm, *Allolobophora caliginosa* (Sav.) (Diptera: Calliphoridae). - Bull. Soc. ent. Egypte 55: 275-287.
Teskey, H. J., 1981a. Morphology and terminology - larvae. Pp. 65-88 in: McAlpine, J. F., Peterson, B. V., Shewell, G. E., Teskey, H. J., Vockeroth, J. R. & Wood, D. M. (eds.), Manual of Nearctic Diptera, 1. Res. Branch Agric. Canada Monogr. 27: vi + 1-674. Ottawa.
- 1981b. Key to families - larvae. Pp. 125-147 in: McAlpine, J. F., Peterson, B. V., Shewell, G. E., Teskey, H. J., Vockeroth, J. R. & Wood, D. M. (eds.), Manual of Nearctic Diptera, 1. Res. Branch Agric. Canada Monogr. 27: vi + 1-674. Ottawa.
Theowald, B., 1957. Notes on Calliphoridae (Diptera) of New Guinea I. - Nova Guinea, n.ser. 8: 157-161.
Thomas, H. T., 1951. Some species of the blow-fly genera *Chrysomyia* R.-D., *Lucilia* R.-D., *Hemipyrellia* Tnsd. and *Calliphora* R.-D. from south-eastern Szechuan, China. - Proc. zool. Soc., Lond. 121: 147-200.
Thompson, W. R., 1921a. Recherches sur les Diptères parasites. I. - Les larves des Sarcophagidae. - Bull. biol. Fr. Belg. 54: 313-463, pls. VI-XVIII.
- 1921b. Contributions à la connaissance des formes larvaires des Sarcophagides (Dipt.) I. *Engyops Pecchiolii* Rond. - Bull. Soc. ent. Fr. 1921: 27-31.
Thomson, A. J. & Davies, D. M., 1973a. The biology of *Pollenia rudis*, the cluster fly (Diptera, Calliphoridae). I. Host location by first-instar larvae. - Can. Ent. 105: 335-341.
- 1973b. The biology of *Pollenia rudis*, the cluster fly (Diptera, Calliphoridae). II. Larval feeding behaviour and host specificity. - Can. Ent. 105: 985-990.
- 1974. The biology of *Pollenia rudis*, the cluster fly (Diptera, Calliphoridae). III. The effect of soil conditions on the host-parasite relationship. - Can. Ent. 106: 107-110.
Thomson, C. G., 1869. Diptera. Species novas descripsit C. G. Thomson. Pp. 443-614, plate 9, in Kongliga svenska fregatten Eugenies Resa omkring jorden under befäl af C. A. Virgin, Åren 1851-1853. 2 (zoologi), 1, Insecta. 617 pp. Stockholm, "1868".
Tiensuu, L., 1938. [Note at meeting of 6 May 1938]. - Suomen hyönt. Aikak. 4: 195 (Finnish), 197 (German).
- 1941. Brachycera (Muscidae, Tachinidae). Pp. 32-43 in: Enumeratio Inectorum Fenniae. VI. Diptera. Helsingin Hyönteisvaihtoyhdistys <Helsingfors Entomologiska Bytesförening>, Helsinki.
- 1964. Huomion arvoisia kärpäslöytöjä Lounais-hämeestä. - Lounais-Hämeen Luonto 15: 76-77.
Tikjøb, G. & Haarløv, N., 1985. Myiasis. - Ugeskr. Læger 147: 1200-1201.
Torgerson, R. L. & James, M. T., 1967. Revision of the world species of *Euphumosia* (Diptera: Calliphoridae). - Pac. Ins. 9: 205-242.
Townsend, C. H. T., 1908. The taxonomy of the muscoidean flies, including descriptions of new genera and species. - Smithson. misc. Collns 51 (2): 1-138.
- 1915a. Proposal of new muscoid genera for old species. - Proc. biol. Soc. Wash. 28: 19-23.
- 1915b. Synonymical notes on Muscoidea. - Insecutor Inscit. menstr. 3: 115-122.
- 1916a. New genera and species of Australian Muscoidea. - Can. Ent. 48: 151-160.
- 1916b. Designations of muscoid genotypes, with new genera and species. - Insecutor Inscit. menstr. 4: 4-12.
- 1917. Indian flies of the subfamily Rhiniinae. - Rec. Indian Mus. 13: 185-202.
- 1919. New genera and species of muscoid flies. - Proc. U. S. natn. Mus. 56: 541-592.

- 1926. New holarctic Muscoidea (Diptera). - Insecutor Inscit. menstr. 14: 24-41.
- 1931. Notes on Old-World Oestromuscoid types. Part I. - Ann. Mag. nat. Hist. (10) 8: 369-391.
- 1934. Muscoid notes and descriptions. - Revta Ent. 4: 110-112.
- 1935a. New muscoid genera, mainly from the Neotropical Region. - Revta Ent. 5: 68-74.
- 1935b. Manual of Myiology in twelve parts. Part 2, Muscoid classification and habits. 296 pp. São Paulo.
- 1937. Manual of Myiology in twelve parts. Part 5, Muscoid generic diagnoses and data, Glossinini to Agriini. 234 pp. Sao Paulo.

Tschorsnig, H.-P., 1985a. Die Struktur des männlichen Postabdomens der Rhinophoridae (Diptera). - Stuttg. Beitr. Naturk. (A) 375: 1-18.
- 1985b. Taxonomie forstlich wichtiger Parasiten: Untersuchungen zur Struktur des männlichen Postabdomens der Raupenfliegen (Diptera, Tachinidae). - Stuttg. Beitr. Naturk. (A) 383: 1-137.

Tumrasvin, W., Kurahashi, H. & Kano, R., 1979. Studies on medically important flies of Thailand VII. Report on 42 species of calliphorid flies, including the taxonomic keys (Diptera: Calliphoridae). - Bull. Tokyo med. dent. Univ. 26: 243-272.

Van, Chi-sheng & Ma, Shao-yuan, 1978. A new species of the genus *Calliphora* R.-D. from Chinghai province, China (Diptera: Calliphoridae). - Acta ent. sinica 21: 194-196. [In Chinese, with English summary.]

Verves, Yu. G., 1980. The composition of the subfamily Paramacronychiinae (Diptera, Sarcophagidae) of the Old World. - Zool. Zh. 59: 1476-1482. [In Russian, with English summary].
- 1982a. 64h. Sarcophaginae. In: Lindner, E. (ed.), Fliegen palaearkt. Reg. 11: 235-296. Stuttgart.
- 1982b. On the fauna of Sarcophagidae (Diptera) of the Mongolian People's Republic. IV. New data on sarcophagid flies from Mongolia and South Siberia. - Nasekomye mongolii 8: 545-568. [In Russian.]
- 1986. Family Sarcophagidae. Pp. 58-193 in: Soós, Á. & Papp, L. (eds.), Catalogue of Palaearctic Diptera, 12, Calliphoridae - Sarcophagidae. 265 pp. Budapest.

Villeneuve, J., 1907. Observations et notes synonymiques concernant quelques Tachinaires types de Pandellé (Dipt.). - Annls Soc. ent. Fr. 76: 379-398.
- 1910. Diptères nouveaux du Nord de l'Afrique. - Dt. ent. Z. 1910: 150-152.
- 1914. Descriptions de nouveaux Calliphorinae africains (Dipt.). - Bull. Soc. ent. Fr. 1914: 305-308.
- 1919. Descriptions de Tachinaires nouveaux [Dipt.]. - Bull. Soc. ent. Fr. 1919: 305-308.
- 1920a. Sur deux espèces appartenant à la tribu des Muscidae Calliphorinae (Dipt.). - Bull. Soc. ent. Fr. 1920: 295-296.
- 1920b. Diptères inédits. - Annls Soc. r. ent. Belg. 60: 199-205.
- 1922. Descriptions de Tachinides nouveaux (Dipt. Musc.). - Bull. Mus. natn. Hist. nat. Paris 28: 514-516.
- 1924. Contribution à la classification des "Tachinidae" paléarctiques. - Annls Sci. nat. (Zool.) (10) 7: 5-39.
- 1926. Espèces nouvelles du genre *Onesia* R.-D. (Dipt.). - Konowia 5: 130-133.
- 1927. Sur *Onesia genarum* (Zett.). - Bull. Annls Soc. r. ent. Belg. 66 (1926): 357.
- 1928. Quelques mots sur les Calliphorinae palearctiques. - Bull. Annls Soc. r. ent. Belg. 68: 147-151.
- 1930. Propos diptérologiques (suite). - Bull. Annls Soc. r. ent. Belg. 70: 41-45.

Villers, C. de, 1789. Caroli Linnaei entomologia, faunae suecicae descriptionibus aucta. 4: 556 + i-ccxiii pp., 1 pl. Lugduni [= Lyon].

Wainwright, C. J., 1928. The British Tachinidae (Diptera). - Trans. ent. Soc. Lond. 76: 139-254.
- 1932. The British Tachinidae (Diptera). First supplement. - Trans. ent. Soc. Lond. 80: 405-424.
- 1940. The British Tachinidae (Diptera): Second supplement. - Trans. R. ent. Soc., Lond. 90: 411-448.

Walker, F., 1849. List of the specimens of Dipterous insects in the collection of the British Museum, 4. Pp. 689-1172. London.
- 1852. Diptera [part]. Vol. 1, pp. 157-252. In: Saunders, W. W. (ed.), Insecta Saundersiana. London, "1856".
- 1853. Diptera [part]. Vol. 1, pp. 253-414. In: Saunders, W. W. (ed.), Insecta Saundersiana. London, "1856".

Wallis, D. I., 1962. The sense organs of the ovipositor of the blowfly, *Phormia regina* Meigen. - J. Ins. Physiol. 8: 453-467.

Wang, Chang-jing, Lu, Yin-lian, Chen, Zhi-zi & Fan, Zi-de, 1981. Four new Calypterate flies from Shanxi, China (Diptera: Muscidae, Calliphoridae, Sarcophagidae). - Contr. Shanghai Inst. Ent. 2: 253-258. [In Chinese, with English summary.]

Wiedemann, C. R. W., 1819. Beschreibung neuer Zweiflügler aus Ostindien und Afrika. - Zool. Mag., Kiel. 1 (3): 1-39.
- 1820. Munus rectoris in Academia Christiano-Albertina iterum aditurus nova dipterorum genera. viii + 23 pp. Kiliae [= Kiel].
- 1830. Aussereuropäische zweiflügelige Insekten. 2: xii + 1-684. Hamm.

Wulp, F. M. van der, 1896. Fam. Muscidae. Pp. 273-344 in: Godman, F. D. & Salvin, O. (eds.), Biologia Centrali-Americana. Zoologia. Insecta. Diptera. Vol.2, 489 pp., 11 figs., 13pls. London.

Xue, Wanqi, 1983. A new species of *Melinda* R.-D. (Diptera: Calliphoridae). - Entomotaxonomia 5: 109-112. [In Chinese, with English summary.]

Yahnke, W. & George, J. A., 1972. Rearing and immature stages of the cluster fly (*Pollenia rudis*) (Diptera: Calliphoridae) in Ontario. - Can. Ent. 104: 567-576.

Zetterstedt, J. W., 1833. Resa genom Umeå Lappmarker i Vesterbottnes Län, förrättad år 1832. xviii + 398 pp., 2 maps. Örebro [Facsimile edition 1980, Umeå].
- 1838. Sectio Tertia. Diptera. Pp. 477-868 in: Insecta Lapponica. vi + 1140pp. Lipsiae [= Leipzig], "1840".
- 1844. Diptera scandinaviae disposita et descripta. 3: 895-1280. Ex officina lundbergiana, Lundae [= Lund].
- 1845. Diptera scandinaviae disposita et descripta. 4: 1281-1738. Ex officina lundbergiana, Lundae [= Lund].
- 1859. Diptera Scandinaviae disposita et descripta.

[Part] Tomus tridecimus seu Supplementum Quartum, continens addenda, corrigenda & emendanda tomis duodecim prioribus, una cum conspectu omnium generum. 13: i-xvi, 4943-6190. Ex officina lundbergiana, Lundae [= Lund].
Zimsen, E., 1964. The type material of I. C. Fabricius. 655 pp. Copenhagen.
Zinov'ev, A. G., 1981. On the classification of Palaearctic flies of the genus *Phaonia* R.-D. (Diptera, Muscidae). - Ént. Obozr. 60: 686-698. [In Russian, with English summary; English translation in Ent. Rev., Wash. 61: 148-158, 1982.]
Zumpt, F., 1952. Remarks on the classification of the Ethiopian Sarcophaginae with descriptions of new genera and species. - Proc. R. ent. Soc. Lond. (B) 21: 1-18.
- 1953. A preliminary contribution to the taxonomy of the genera *Hemigymnochaeta* and *Tricyclea* (Diptera: Calliphoridae). - Trans. R. ent. Soc. Lond. 104: 481-520.
- 1956a. 64i. Calliphorinae. In: Lindner, E. (ed.), Fliegen palaearkt. Reg. 11: 1-140.
- 1956b. Calliphoridae (Diptera Cyclorrhapha). Part I: Calliphorini and Chrysomyiini. - Explor. Parc. natn. Albert Miss. G. F. de Witte 87: 1-200.
- 1957. Some remarks on the classification of the Oestridae s.lat. (Diptera). - J. ent. Soc. S. Afr. 20: 154-161.
- 1958. Calliphoridae (Diptera Cyclorrhapha). Part II: Rhiniini. - Explor. Parc. natn Albert Miss. G. F. de Witte 92: 1-207.
- 1965. Myiasis in man and animals in the Old World. A textbook for physicians, veterinarians and zoologists. iv + 267 pp., London.

Taxonomic index

The index contains the names of taxonomic and nomenclatural significance listed in the main text. No references are given to the Figures, Plate-figures and Catalogue. Synonyms (including emendations, misidentifications or nomina nuda) are in italics. Suprageneric names are in uppercase. Author, original generic assignment and erroneous use of a name are indicated for identical names given separate entries to distinguish different (nominal) taxa. Numbers in boldface, given only for valid names, refer to the main treatment of the taxon concerned.

Abago 32
abdominalis 188
abina 75, 80
Abonesia 60
accepta 36
accincta 106
acerba 188, **189**, 191
Acronesia 60, 61, 68, 75
Acrophaga 59-61, 71, 75
Acrophagella 147
aculeata 37, 40
affinis 88
agilis Harris 40
agilis Meigen 40, 44
agilis: Schumann (pars) 59
alaskensis Shannon (Francilia) 147, 164
alaskensis Shannon (Steringomyia) 75, 84
albiceps Meigen 158, 164
albiceps Robineau-Desvoidy 102, 103
aldrichia 60
Aldrichiella Hendel 60
Aldrichiella Rodendorf 60
Aldrichiella Vaughan 60
Aldrichina 60
alpicola 194
alpina: authors 75
alpina Zetterstedt 61, 68, 71
Ambodicria 34, 54
AMENIINAE 14, 15
amentaria 7, 212-215, **218**, 221
americana 97
amplectens 44
ampullacea 21, 28, 31, 147, 148, 150, **152**
anana 75
Angioneura 7, 14, 23, 26, 27, 185, **188**, 189, 191, 194, 197
Angioneurilla 188
ANGIONEURINI 15, 185
angustifrons 158
angustigena 213, 214, 216, **221**, 226, 235, 238
anthracina: authors 203
anthracina Meigen 200, 203
Anthracomya 209, 210
Apaulina 110
appendicifera 147
argenticincta 209
aristatus 128, 129
atramentaria 14, 21, 103, 212, 215, 221
atriceps 128, **129**, 133
Auchmeromyia 15
AUCHMEROMYIINAE 15
aucta 63
augur 60
aurisquama 106
Australocalliphora 60, 61
austriaca 97, 98
Avihospita 110, 137
avium 110
azurea Fallén 63, 105, 106, 110, 112, **113**, 116, 119, 128, 140,
azurea Robineau-Desvoidy (Lucilia) 158
azurea Robineau-Desvoidy (Phaenicia) 168
barberi 178
bayeri 35, 36, 39, **44**, 47
Bellardia 7, 8, 22, 23, 29-32, **34**-37, 40, 44, 47, 50, 54, 56, 59, 61, 98, 102, 185, 203, 214, 215
Bengalia 15
BENGALIINAE 15
biseta 47, 50
bisetosa 40, 44
bisulca 213
boganidae 129
Booponus 15
Boreellus 128, 129
borotinensis 56, 59
braueri 10, 110, 113, **137**
brevistylata 34
brunibarbis 88
brunnicosa 181
Bufolucilia 147
bufonivora 10, 21, 22, 28, 147, 148, 152, **156**, 158, 181
Bureschiella 212
Buresiella 212
cadaverina Linnaeus 158
cadaverina Robineau-Desvoidy 92, 93, 97
caerulea De Geer 88
caerulea Malloch 129
caerulea Robineau-Desvoidy 113, 129
caerulea Séguy 147
caesar 28, 147, 148, 152, **158**, 163, 164, 168

Caesariceps 147
Callidesia 15
Calliphora 7, 10, 21-23, 27, 28, 32, 34, 37, 40, 44, **59**, 60, 61, 63, 68, 71, 75, 80, 84, 88, 91, 93, 97, 98, 102, 113, 137, 168, 171, 175, 205
CALLIPHORIDAE 7-9, 11, 12, 14, 15, 21-23, 27, 28, 113, 141, 185
CALLIPHORINAE 7, 12, 14, 15, 22, 31, **32**, 116, 185
CALLIPHORINI 15
CALLIPHOROIDEA 11
calliphoroides 60
Calobataemyia 188, 189, 209
Calobatemyia 209, 210
Calobatémyie 209
capensis 178
Carcinomyia 92
carinata 230
carnaria Linnaeus 88
carnaria Scopoli 88
carnivora 63, 68
Catapicephala 15
Cephenemyia 11
Cephysa 212
Chaetophaenicia 147
Chaetopollenia Enderlein 212, 218, 221
Chaetopollenia Jacentkovsky 212
chinghaiensis 75
chloris 178
Chloroprocta 103
chrysocephala 93
Chrysomya 103-105, 178
CHRYSOMYINAE 7, 12, 14, 21, 22, 31, **103**, 137
CHRYSOMYINI 14, 103
Chrysopyrellia 103
chrysorrhoea 110-113, **116**, 119, 127
cinerea 241
claripennis 98, 102
clausa 34
Cochliomyia 103
coerulea: authors 200
coerulea Meigen 199, 203, 205
coerulea Wiedemann 203
coerulescens 233, 234
coerulescens: authors 234
cognata: authors 40, 200, 205
cognata Meigen 203, 205
collini 68
compacta 137
concinna 147, 178
Cordylobia 15
corusca 133
crassipalpis 12
cuprea 106
cuprina 150
curvipes 51
Cyanus 32
Cynomya 10, 22, 23, 27, 28, 32, 33, 75, **92**, 93, 97
Cynomyia 75, 92
Cynomyomima 32
Cynomyopsis 92
Cynophaga 92
cyrtoneurina 188, 189, **191**, 193
Dasylucilia 147
dasyphthalma 59
dasypoda 212-214
Dasypollenia Jacentkovsky 212
Dasypollenia Lehrer 213
depressa Gmelin 228
depressa Meigen 228
depressa Swederus 228
depressa Villers 228
Dexia 51, 189, 191, 197, 199
Dexopollenia 206, 215
DIPTERA 3, 8, 9, 14, 23, 61, 68, 71, 113, 119, 136, 167, 181, 221
discolor 241
dispar 113
Disticheria 209
distinguenda 142
dolens 51
Dyscritomyia 147
echinosa 75, 84
Eggisops 7, 8, 14, 22, 23, 185, 187, 188, **193**, 194, 197
Elephantolaemus 15
elongata 158
Engyops 193, 194
Engyzops 193
enigmatica 191
equestris 168
erectiseta 68, 71
erythrocephala De Geer 63
erythrocephala Fabricius 63
erythrocephala Meigen 63, 68
erythrocephala Villers 63
Eucalliphora 60
Euidiella 241
Euphormia 105
Euphumosia 15
Eupollenia 212
Eurychaeta 8, 10, 22, 23, **141**, 142, 197
excarinata 228
familiaris Harris 235
familiaris Panzer 235
fasciata 241, 245
Feideria 34, 36
fimbriata 188
flavipennis Kramer 147, 152
flavipennis Macquart 152, 178
floralis 35, 97, **98**, 102, 103
Francilia 147, 164
franzi 80, 84
fulgida 178, 181
fulvibarbis 88
fulvicornis Robineau-Desvoidy 230
fulvicornis Zetterstedt 230
fulvifacies 106
fumicosta 147
fuscipalpis 147, 164, 167
fuscipennis 210
GASTEROPHILINAE 12
Gasterophilus 12
genarum 21, 60-62, **68**, 71

genei 209, 210
geneji 209, 210
gentilis 184, 199, 200, **203**, 205
germanorum 71
germanorum: Kramer 205
giraulti 178
Glutoxys 185
grahami 60
gregorpovolnyi 93, 97
griseotomentosa 7, 212-216, **224**-226
groenlandica 113, 133, 136
halterata 209
hasei 238
Helicobosca 8, 141, 142
HELICOBOSCINAE 7, 12, 14, 22, 23, 28, 30, **140**
Helicophagella 28
Hemigymnochaeta 15
Hemilucilia 103
Hemipyrellia 147, 150
hendeli 97
hirta 92, 93, 97
hirudo 137
hokkaidensis 97
hungarica 213, 216, **226**, 235, 238
HYPODERMATINAE 11, 12
Hypopygiopsis 147, 150
Idia Hübner 241
Idia Wiedemann 241, 245
Idiopsis 209
illustris 28, 148, 152, 164, **168**
infernalis 147
insidiosa 63
intermedia: authors 7, 21, 215, 228
intermedia Macquart 228
intestinalis 12
Jacentkovskyiomyia 212
jeddensis 158
johanseni 128, 129
kowarzi 97
krameri Mueller 37
krameri Séguy 152
labialis 7, 212-216, **228**
lagyra 178
laoshanensis 152
laticornis 68
latifrons Hough 60
latifrons Schiner 178
latigena 34
lepida 158
LEPIDOPTERA 141, 241
lindneri 137
littoralis 63
loewi 7, 10, 21, 63, **71**, 84, 205
longelobata 37
longilobata 171, 175
lucens 106
luciensis 230
Lucilia 8, 10, 21-23, 27, 28, 31, 106, 113, 133, **147**, 148, 150, 152, 156, 158, 164, 168, 171, 175, 178, 181
Luciliella 147
LUCILIINAE 7, 12, 15, 31, **147**
LUCILIINI 15, 147
lucilioides 181
lunata **241**, 245
Macronesia 97
macronyx 194
Macrophallus 97
maculata 241
magnicornis 21, 147, 148, 151, **164**, 168
Mallochomyia 128, 129
Maravigna 34
Mariomyia 212
Marsilia 97
mayeri 212, 225
Medoria 188, 189, 191
Melanodexia 206, 209
Melanomya 7, 14, 27, 185, 187, 188, 194, **197**
Melanomyia 197
MELANOMYINAE 7, 12, 14, 15, 21-23, 26, 28, 29, 31, 32, **184**, 185, 206
melanoptera 188, 209, **210**, 211
melanorhina 241
melanura 28
Melinda 7, 8, 14, 23, 28, 32, 40, 44, 97, 102, 103, 184, 185, **199**, 200, 203, 205
Melindopsis 34, 36
menechma 34, 36
MESEMBRINELLINAE 15
Miaspia 34
micans 218, 221
Microchaetina 141
Micronitellia Enderlein 212
Micronitellia Lehrer 213
MILTOGRAMMATINAE 12
Mimodexia 15
MIMODEXIINAE 16
minima 197, 199
minimus 88, 91
minutissima 54
modesta 178
mohileviana 75
mollis 106
monspeliaca 63
moravica 218, 221
Morinia 7, 14, 27, 31, 188, 194, 197, 199, 206, **209**, 210
morticia 71
mortuorum 28, 75, 92, **93**, 97
Musca 7, 8, 34, 37, 39, 40, 44, 47, 50, 51, 56, 59, 60, 63, 68, 88, 91-93, 98, 102, 105, 106, 110, 113, 116, 129, 133, 147, 158, 164, 168, 171, 175, 178, 180, 181, 183, 197, 199, 200, 203, 205, 209, 210, 212, 213, 215, 218, 221, 228, 230, 233, 235, 238, 241, 245
musca 63
muscaria 141, 142
muscidea 212
muscoidea 241
Mya 61
myoidea 241
MYSTACINOBIINAE 15
nana Meigen 197, 199
nana Robineau-Desvoidy 63
Nasonia 113, 127

Neocalliphora 59
Neocordylobia 15
Neomyia 116
Nepenthomyia 206
Nesodexia 206
nidicola 137
nigerrima 209
nigra 209, 210
nigrapex 142
nigrina Bigot 106
nigrina Fallén 218
nigrina Meigen 218, 221
nigripalpis Robineau-Desvoidy 133
nigripalpis Townsend 181
Nitellia 59, 212, 215
nitens Villers 218
nitens Zetterstedt 218, 221
nobilis 178, 180, 181
noelleri 203
norwegica 60, 75
nuortevai 112, **119**
Nyctia 209
obscoena 88
obscura Bigot 233
obscura Fabricius 233, 235
obscura Townsend 188
obsoleta 34
oceaniae 60
Oedemagena 11
OESTRIDAE 11, 12
OESTROIDEA 11
Oestrus 12
Onesia 7, 8, 22, 23, 29, 32, 34-37, 40, 44, 47, 54, 56, 59, 61, 71, **97**, 98, 102, 103, 181, 185, 200, 203, 225
onesioidea 60
Onesioides 32, 97, 98
Onesiomima 32
Opelousia 188
Opsodexia 184, 185, 188
Opsodexiopsis 188
Orizia 212
Orneocalliphora 110
osetica 44, 47
ovis 12
Pachychoeromyia 15
pallida 212-216, **230**
palpalis 10, 141, **142**, 147
pandia 34, 37, 40, **47**, 59, 102
papuensis 150
Papuocalliphora 60, 61
Paracalliphora 60, 61
Paradichosia 185, 200
Paralucilia 103
PARAMACRONYCHIINAE 12
Parapollenia 212
Parasarcophaga 12
parva 197
parvula 168
Pavlovskiomyia 15
pecchiolii 193, **194**
pediculata 213, 214, 216, **233**, 234
Pellonesia 97
Pericallimyia 32
petiolata 194
peusi 112, **122**
Phaenicia 147, 168, 178
philadelphica 106
Phormia 14, 22, 23, 27, 103, **105**, 106, 113, 128, 129, 133
Phormiata 103, 137
PHORMIINI 14, 103
Phumonesia 147
Phumosia 15
pilosiventris 147
pionia 147
polita 34, 54
Pollenia 7-9, 14, 21, 23, 26, 27, 29-31, 36, 59, 103, 150, 206, 209, **212**-215, 218, 221, 224, 226, 228, 230, 233-235, 238
Polleniella 7, 212, 224-226
POLLENIINAE 7, 12, 14, 15, 21, 22, 31, 185, **206**
POLLENIINI 15
Polleniomyia 212
Polleniomyma 212
Polleniopsis 32, 34, 36
Pollenomyia 212, 215
popoffana 68
PROSTHETOSOMATINAE 15
Protocalliphora 10, 22, 23, 26, 28-30, 63, 103-**106**, 110, 113, 116, 119, 122, 127, 137, 140
Protophormia 10, 21-23, 27, 28, 103, 105, 110, 113, 116, **128**, 129, 132, 133, 137
proxima Grunin 22, 111-113, 116, **127**
proxima Walker 106
pruinosa: Grunin 44, 47
pruinosa Meigen 178, 181
pseuderythrocephala 84
pseudobisulca 218, 221
Pseudonesia 34
Pseudopollenia 212
Pseudopsodexia 185
pseudorudis 233, 234
pseudovomitoria 88
pterygoides 97
Ptilonesia 32
puberula 34, 50, 51
pubescens 178
pubicornis 7, 31, 34, 36, **50**, 51
pumicata 54
pusilla Gmelin 37
pusilla Meigen 37, 39, 40,
pusilla: Lehrer 44
pusilla: Schumann 44, 47
pygmaea 189, 191
quadrimaculata 59
regalis 21, 148, 151, **171**, 175, 177
regina 22, 103, 105, **106**
retrocurva 98, 102
RHINIINAE 7, 12, 14, 22, 31, 103, **241**
RHINIINI 14
RHINOPHORIDAE 14, 185, 206
RHINOPHORINAE 7, 11, 12, 14, 16, 22, 29, 31

richardsi 21, 28, 148, 151, **175**, 177, 178
rohdendorfi 71
Rohdendorfiomyia 212
rostrata 241
Roubaudiella 147
rubrifrons 88
rudis 21, 212-216, 221, 224, 226, 228, 230, 233, **235**, 238
rufifacies 63, 68
rufigena 106
rufipalpis 106
rupicola 106
Sachtlebeniola 212
Sarcophaga 29, 61, 68, 71, 84, 128, 129, 141, 142, 147, 164, 167, 168
SARCOPHAGIDAE 8, 11, 12, 28, 29, 141, 185, 209
SARCOPHAGINAE 12
Sarcotachina 12
sayi 178
SCHIZOPHORA 12
schumanni 47
scutellata 63
Seguyiomyia 213
selgae 241, 245
semicinerea 21, 215
septentrionis 200
sepulchralis error 35
sepulchralis Linnaeus 98
sepulcralis Meigen 47, 50, 98, 102, 103
sericata 21, 22, 27, 28, 147, 148, 150, 151, 168, 171, 175, **178**, 180, 181
siciliensis 44, 47
sicillensis 44
silvarum 21, 147, 148, 151, 156, 164, **181**, 183
simulatrix 168
sinensis 215
Sinolucilia 147
skufyini 209
Somomya 61, 158
Somomyia 106
sordida 113, 116, 119
soudeki 218, 221
spitzbergensis 63
splendida 158, 164
squalens 106
stelviana 59-62, 71, **75**, 80, 84, 97
Steringomyia 59, 60, 80
stigmaticalis 106
Stobbeola 60, 75
Stomorhina 26, 29, **241**, 245
Stomorhyna 241
stricta 36, **54**
stylifera 59
subalpina 7, 60, 61, 71, 75, **80**, 84
subapennina 98
subcylindrica 12
surcoufi 113, 116, 119
syrphoidea 241
Tachina 34, 50, 51, 188, 191, 193
TACHINIDAE 11, 12, 40, 141, 218
Tainanina 32
tarandi 11
tegularia 178
tenuiforceps 21, 215
terraenovae Macquart (Calliphora) 88
terraenovae Macquart (Lucilia) 133
terraenovae Robineau-Desvoidy 21, 105, 110, 113, 116, 128, 132, **133**
thalassina 106
Thelesina 97
Theria 141, 142, 147
thuscia 63
townsendi 44, 47
toxopeusi 60
TOXOTARSINAE 15, 103
Triceratopyga 60
Trichopollenia 212
Tricyclea 15
TRICYCLEINAE 15
Tricycleopsis 32, 185
trompe 11
Trypocalliphora 8, 10, 22, 28, 29, 31, 103-105, 110, 113, **137**
turanica 63, 68, 88
unxia 47
uralensis 21, 63, 68, **84**, 88
vagabunda 212-215, **238**
varia: authors 224
varia Gmelin 235
varia Meigen 213, 235, 238
velox 209, 210
vera 212, 228
vernalis 34
vespillo: authors 59, 212, 215, 218, 221
vespillo Fabricius 7, 8, 37, 40, 44, **56**, 59, 212, 215
vetusta 188, 189
viarum 7, **37**, 39, 40, 44
vicina 21, 27, 28, 62, **63**, 68, 84, 88, 91, 113
villeneuvi 203, 205
violacea 113, 116
viridicaerulea 158
viridicyanea 7, 40, 200, 203
Viridinsula 147
vittata 106
vomitoria 21, 28, 59, 62, 68, 84, **88**, 91
vomitorina 68, 84
vulgaris 7, 34, 37, **40**, 44, 47, 56, 59
Wilhelmina 206
Xanthotryxus 206, 215
Xenocalliphora 32
Xerophilophaga 37, 54, 199, 203
xinganensis 54
zetterstedti 37, 40
Zumptiomyia 213

Biological index

The index contains names of hosts, parasites, victims of myiasis, specific substrates for larval development and related biological information.

Allolobophora caliginosa 35
Allolobophora chlorotica 35
Alopex lagopus 133
amphibians 28, 157, 180, 183
Anthus pratensis 140
ants 15
arctic fox 133
arctic hare 133
birds 28, 29, 63, 110, 113, 119, 127, 137, 140, 177
Bufo bufo 157, 180
bull-frog 183
Calcarius lappopicus 119
Canis familiaris 133
Caprimulgus europaeus 28, 177
Carduelis flammea 140
cattle 133
Cepaea hortensis 142
Chondrostegia maghrebica 238
Corvus cornix 127
Corvus corone 127
crane 28
crustaceans 29
decaying wood 211
Dicrostonyx groenlandicus 133
Dicrostonyx torquatus 133
Discus rotundatus 203
Dociostaurus maroccanus 245
Eisenia foetida 35, 47
Eisenia rosea 226, 233, 238
Erinacaeus europaeus 28, 63, 152, 164, 168
Erithacus rubecula 113
Ficedula hypoleuca 113, 140
garbage 27, 28, 180
Grus grus 28
hare 28, 97, 133
hedgehog 28, 63, 152, 164, 168
Helicella candidula 203
Helicella ericetorum 203
Helicella virgata 203
Helicidae 29, 141, 203
Hirundo rustica 63, 140
Homo sapiens 15, 27, 28, 60, 63, 71, 84, 106, 133, 136, 150, 164, 168, 180
  baby 28, 63
  cadavers 136
  external auditory meatus 28, 164, 180
  ulcus cruris 180
  woman 28, 168
  honeybees 214
lambs 133
Lemmus lemmus 97, 133
Lemmus obensis 133
Lemmus sibiricus 133
Lepidoptera 214, 238
Lepus arcticus 133
Lepus timidus 28, 97
locusts 22, 245
Luscinia luscinia 140
mammals 15, 28, 97, 133, 177
Microtus oeconomus 75, 97
Miliaria calandra 140
Motacilla alba 119, 140
mouse-nest 113
mushrooms 15
myiasis 7, 8, 15, 22, 28, 60, 63, 91, 93, 97, 106, 133, 150, 152, 164, 168, 177, 180
Nasonia vitripennis 113, 127
nightjar 28, 177
Odobaenus rosmarus 133
Oenanthe oenanthe 140
Oxyloma pfeifferi 193
Parus ater 113
Parus caeruleus 113
Parus cinotus 140
Parus major 63, 113, 140
Patula rotundata 203
Phallus impudicus 7, 68, 71, 84, 91, 97, 152, 164, 183
Phoenicurus phoenicurus 140
Phylloscopus sibilatrix 140
Phylloscopus trochilus 113, 119, 140
Pica pica 127
poliomyelitis 27, 150, 180
Rana catesbiana 183
Rana temporaria 157
refuse 28, 88, 136, 175
reindeer 88, 133
Riparia riparia 116
rodents 71
Schistocerca gregaria 245
Schistocerca peregrina 245
Sesamia nonagrioides 238
sheep 8, 28, 68, 91, 133, 164, 168, 180
shrew carcasses 84
snails 22, 28, 98, 141, 185, 200
Sorex araneus 75, 97
stinkhorn 7, 68, 71, 84, 91, 97, 152, 164, 183
Succinea (Oxyloma) elegans 193
termite nests 245
termites 15
toad 157, 180
Turdus iliacus 113, 119
Turdus torquatus 119
walrus 133
Zea mais 238

www.ingramcontent.com/pod-product-compliance
Ingram Content Group UK Ltd.
Pitfield, Milton Keynes, MK11 3LW, UK
UKHW021409091026
15624UKWH00009B/75

9 789004 093041